Emerging and Eco-Friendly Approaches for Waste Management

Ram Naresh Bharagava • Pankaj Chowdhary
Editors

Emerging and Eco-Friendly Approaches for Waste Management

Editors
Ram Naresh Bharagava
Department of Environmental Microbiology
Babasaheb Bhimrao Ambedkar University
(A Central University)
Lucknow, Uttar Pradesh, India

Pankaj Chowdhary
Department of Environmental Microbiology
Babasaheb Bhimrao Ambedkar University
(A Central University)
Lucknow, Uttar Pradesh, India

ISBN 978-981-10-8668-7   ISBN 978-981-10-8669-4   (eBook)
https://doi.org/10.1007/978-981-10-8669-4

Library of Congress Control Number: 2018942202

© Springer Nature Singapore Pte Ltd. 2019
This work is subject to copyright. All rights are reserved by the Publisher, whether the whole or part of the material is concerned, specifically the rights of translation, reprinting, reuse of illustrations, recitation, broadcasting, reproduction on microfilms or in any other physical way, and transmission or information storage and retrieval, electronic adaptation, computer software, or by similar or dissimilar methodology now known or hereafter developed.
The use of general descriptive names, registered names, trademarks, service marks, etc. in this publication does not imply, even in the absence of a specific statement, that such names are exempt from the relevant protective laws and regulations and therefore free for general use.
The publisher, the authors and the editors are safe to assume that the advice and information in this book are believed to be true and accurate at the date of publication. Neither the publisher nor the authors or the editors give a warranty, express or implied, with respect to the material contained herein or for any errors or omissions that may have been made. The publisher remains neutral with regard to jurisdictional claims in published maps and institutional affiliations.

Printed on acid-free paper

This Springer imprint is published by the registered company Springer Nature Singapore Pte Ltd.
The registered company address is: 152 Beach Road, #21-01/04 Gateway East, Singapore 189721, Singapore

Preface

Rapid industrialization and urbanization is a serious concern for healthy and green environment. The contamination of environments (soil/water/air) with various toxic pollutants released from natural as well as anthropogenic activities and their adverse effects on living organisms needs focused research to mitigate and solve these problems. This book provides a detailed knowledge on the different types of emerging toxic environmental pollutants discharged from various natural and anthropogenic sources, their toxicological effects in environments, humans, animals, and plants, as well as their biodegradation and bioremediation approaches by various emerging and eco-friendly approaches such as anammox technology, advance oxidation processes, membrane bioreactors, microbial degradation (bacteria, fungi, algae, etc.), and phytoremediation. Hence, this book provides a detailed knowledge on various types of emerging and eco-friendly approaches for the degradation and detoxification of pollutants, which is urgently required for the safety of environment, human, animal, and plants health.

Chapter 1 conveys an impression on the usage, advantages, and shortcomings of current conventional, emerging treatment technologies available for the management of waste products generated by industrial activities. It is the overall introduction about different wastewater producing industries and toxic organic and inorganic pollutants generated and their adequate treatment methodologies before its final disposal into the environment. Chapter 2 highlights the emerging contaminants constituting a group of natural and synthetic chemicals and microorganisms, which have proved to cause serious effects in laboratory organisms and therefore a threat to human, animal, and aquatic organisms. These are assumed to act as endocrine disruptors, are carcinogenic and teratogenic, and interfere with sex and reproduction of organisms. Chapter 3 provides an overview of the application of bioremediation of contaminated sites owing to municipal solid waste. The application of bioremediation technologies and well-organized mechanisms for environmental safety measures of these methods were discussed. Because some pollutants can seriously affect the environment, this chapter furthermore suggests strategies for better remediation of contaminated sites. Chapter 4 describes the anammox start-up and enrichment process in a submerge bioreactor and discusses about the performance of anammox bacteria for treating industrial wastewaters with relevant updated information. In this chapter, several reports showed that anaerobic membrane bioreactors (AnMBRs) were operated for a period of 100 to 500 days under completely anaerobic conditions and fed with synthetic media. This work demonstrated that anammox cultures can be enriched from sludges conveniently sourced from anaerobic digesters in a reasonable time frame using the AnMBR, thus enabling the next stage of utilizing the biomass for treatment of industrial wastewaters.

Chapter 5 explores the various positive and negative impacts of distillery industries (DIs) wastes. DI is one of the most alarming dangers that faces living being including human and animals. The wastewater generated from distilleries creates a hazardous problem and causes serious environmental issues. Moreover, distillery wastewater (DWW) or spent wash has not only negative impact on environment but also positive aspects such as ferti-irrigation, energy generation, concentration by evaporation, production of many value-added products, etc. Chapter 6 describes about polycyclic aromatic hydrocarbons (PAHs), which are a class of diverse organic compounds with two or more intertwined benzene rings in a linear, angular, or bunch arrangement. Expulsion of PAHs is crucial as these are persevering toxins with ubiquitous event and adverse natural impacts. There are several remedial techniques, which are productive and financially savvy in elimination or removal of PAHs from the contaminated environment. However, biological approaches appear to be the most efficient and cost-effective environmental-friendly method to decontaminate PAHs from source. Chapter 7 mainly highlights that industrial wastewaters such as distillery, pulp and paper, and tannery have endocrine-disrupting chemicals (EDCs), which cause serious effects in living organisms. EDC is a chemical agent, which interferes the synthesis, secretion, transport, binding, or elimination of natural hormones in human and animal body that play a key role for the maintenance of various physiological and cellular functions such as homeostasis, reproduction, development, and behavioral activities. These also play significant role in metabolism, sexual development, hormones production and their utilization in growth, stress response, gender behavior, and reproduction even in development of living organisms. Therefore, there is an urgent need of awareness and critical research on the endocrine-disrupting chemicals present in industrial wastewaters.

Chapter 8 highlights arsenic (As) toxicity and several conventional and emerging remediation methods. Arsenic is necessary for living beings. However, it is also an emerging issue by virtue of the toxicity it causes in living beings including human and animals. Basically, the groundwater contaminated by arsenic, coming from sources including arsenic-affected aquifers, has severely threatened humanity around the world. Arsenic poisoning is worse in Bangladesh and Uttar Pradesh where As(III) is found in higher concentration in groundwater. The dissolution process caused by oxidation and reduction reactions leads to natural occurrence of arsenic in groundwater. Microalgae possess a unique metabolic process making them highly efficient for the removal of nutrients as well as organic pollutants from industrial wastewater such as alcohol distillery wastewater. Microalgae generate dissolved oxygen increasing the efficiency of alcohol distillery wastewater treatment. Microalgal cells are represented to take up and store large amounts of N and phosphorus as described in Chap. 9. Chapter 10 mainly highlights the endophytes and their significant role in pollutants' mitigation or removal. These exist in tissues of host plant and have traditionally been studied for their plant growth-promoting properties, biocontrol activities, and production of bioactive compounds. Their bioremediation potential is new and has tremendous opportunity for research and development. Endophytes are therefore interesting microorganisms in our effort to discover new tools for the bioremediation of pollutants. In this chapter, the nature of endophytes, their tolerance to pollutants, and their application and mechanisms in removing pollutants such as toxic metals and triphenylmethane dyes have been discussed in details. Chapter 11 mainly focuses on the textile wastewater toxicity profile and various remediation strategies. This wastewater severely affects photosynthetic function of plant as well as aquatic flora and fauna by eutrophication. So, the textile wastewater must be treated before their discharge. In this chapter, different treatment methods to treat the textile wastewater have been discussed such as physical methods (adsorption, ion exchange, and membrane filtration), chemical methods (chemical precipitation, coagulation and flocculation, chemical oxidation), and biological methods (aerobic and anaerobic) discussed in detail.

Chapter 12 focuses on environmental exposure to toxic chemicals such as pesticides and fungicides, which are a significant health risk for humans and other animals. Worldwide pesticides, fungicide, insecticides, herbicides, molluscicides, nematicides, and rodenticides are used as plant growth regulators. The pesticides, which are presently used, include a large composition of chemical compounds, which have great differences in their mode of action, absorbance by the body, metabolism, removal from the body, and the adverse effect on humans, animal, and other living entities. Chapter 13 provides the overall information on sources, characteristics, toxicity, and various physicochemical and biological treatment technologies for the management of pulp and paper mill wastewaters. Pulp and paper mill wastewater at high concentration reduces the soil texture; inhibits seed germination and growth and depletion of vegetation while in aquatic system; blocks the photosynthesis and decreases the dissolved oxygen (DO) level, which affects both flora and fauna; and causes toxicity to aquatic ecosystem. The advantages and disadvantages of several treatment methods (physicochemical and biological) have been also discussed in this chapter to upgrade the knowledge on various treatment methods. Chapter 14 mainly focused on the phytoremediation, using wild-type or transgenic plants and their attendant rhizospheric and endophytic bacteria to remove metal pollutants from contaminated sites. This environment friendly, cost-effective, and plant-based technology is expected to have significant economic, aesthetic, and technical advantages over traditional engineering approaches. Chapter 15 highlights on peroxidases, which obtained from several plant sources such as horseradish, turnip, soybean seed coat, pointed gourd, white radish, and some microbial sources have successfully been immobilized on/in various types of organic, inorganic, and nano supports by using different methods and employed for the treatment of industrial wastewaters having phenolic pollutants in batch as well as in continuous reactors. Chapter 16 mainly focuses on the nanotechnology and profitable production of nanoparticles. It has the prospective to progress the environment, both through direct utilization of those materials to distinguish, avoid, and eliminate contamination, likewise indirectly by applying the nanotechnology to intention cleaner manufacturing processes and generate environmentally accountable products. Chapter 17 describes the novel biphasic treatment approaches to remove pollutants from industrial wastewater. Different biphasic treatment systems including liquid-liquid two-phase partitioning and solid-liquid partitioning systems have proved successful for the cleaning of effluents containing textile dyes, heavy metals, organic contaminants, pharmaceutical ingredients, and many other xenobiotic compounds. Chapter 18 mainly focuses on phycoremediation technology and their several remediation as well as beneficial aspects for environmental sustainability. It is well known that algae are widely distributed on the earth and are adapted to variety of habitats. This unique feature of their fast adaptation allows the algae to develop wide range of adoptable toward many environmental conditions, suited for wastewater remediation and production of biofuel and other value-added products, including food, feed, fertilizer, pharmaceuticals, and, of late, biofuel.

Furthermore, this book also boosts up students, scientists, and researchers working in microbiology, biotechnology, and environmental sciences with the fundamental and advance knowledge about the environmental challenges. In addition, readers can also get valuable information/awareness related to various environmental problems and their solutions.

Lucknow, Uttar Pradesh, India   Ram Naresh Bharagava
                                                                        Pankaj Chowdhary

Contents

1 **Conventional Methods for the Removal of Industrial Pollutants, Their Merits and Demerits**... 1
Sandhya Mishra, Pankaj Chowdhary, and Ram Naresh Bharagava

2 **Toxicological Aspects of Emerging Contaminants**............................. 33
Miraji Hossein

3 **An Overview of the Potential of Bioremediation for Contaminated Soil from Municipal Solid Waste Site**.................... 59
Abhishek Kumar Awasthi, Jinhui Li, Akhilesh Kumar Pandey, and Jamaluddin Khan

4 **Anammox Cultivation in a Submerged Membrane Bioreactor**.......... 69
M. Golam Mostafa

5 **Toxicity, Beneficial Aspects and Treatment of Alcohol Industry Wastewater**... 83
Pankaj Chowdhary and Ram Naresh Bharagava

6 **Bioremediation Approaches for Degradation and Detoxification of Polycyclic Aromatic Hydrocarbons**................... 99
Pavan Kumar Agrawal, Rahul Shrivastava, and Jyoti Verma

7 **Endocrine-Disrupting Pollutants in Industrial Wastewater and Their Degradation and Detoxification Approaches**..................... 121
Izharul Haq and Abhay Raj

8 **Arsenic Toxicity and Its Remediation Strategies for Fighting the Environmental Threat**.. 143
Vishvas Hare, Pankaj Chowdhary, Bhanu Kumar, D. C. Sharma, and Vinay Singh Baghel

9 **Microalgal Treatment of Alcohol Distillery Wastewater**..................... 171
Alexei Solovchenko

10	Endophytes: Emerging Tools for the Bioremediation of Pollutants 189
	Carrie Siew Fang Sim, Si Hui Chen, and Adeline Su Yien Ting
11	Textile Wastewater Dyes: Toxicity Profile and Treatment Approaches 219
	Sujata Mani, Pankaj Chowdhary, and Ram Naresh Bharagava
12	Pesticide Contamination: Environmental Problems and Remediation Strategies 245
	Siddharth Boudh and Jay Shankar Singh
13	Recent Advances in Physico-chemical and Biological Techniques for the Management of Pulp and Paper Mill Waste 271
	Surabhi Zainith, Pankaj Chowdhary, and Ram Naresh Bharagava
14	Role of *Rhizobacteria* in Phytoremediation of Metal-Impacted Sites.. 299
	Reda A. I. Abou-Shanab, Mostafa M. El-Sheekh, and Michael J. Sadowsky
15	Remediation of Phenolic Compounds from Polluted Water by Immobilized Peroxidases 329
	Qayyum Husain
16	Nanoparticles: An Emerging Weapon for Mitigation/Removal of Various Environmental Pollutants for Environmental Safety 359
	Gaurav Hitkari, Sandhya Singh, and Gajanan Pandey
17	Biphasic Treatment System for the Removal of Toxic and Hazardous Pollutants from Industrial Wastewaters 397
	Ali Hussain, Sumaira Aslam, Arshad Javid, Muhammad Rashid, Irshad Hussain, and Javed Iqbal Qazi
18	Phycotechnological Approaches Toward Wastewater Management 423
	Atul Kumar Upadhyay, Ranjan Singh, and D. P. Singh

Contributors

Reda A. I. Abou-Shanab Department of Environmental Biotechnology, Genetic Engineering and Biotechnology Research Institute, City of Scientific Research and Technological Applications, New Borg El Arab City, Alexandria, Egypt

Department of Soil, Water and Climate, Biotechnology Institute, University of Minnesota, St. Paul, MN, USA

Pavan Kumar Agrawal Department of Biotechnology, G.B. Pant Engineering College, Ghurdauri, Pauri, Garhwal, Uttarakhand, India

Shruti Agrawal Department of Microbiology, Sai institute of Paramedical and Allied Science, Dehradun, India

Sumaira Aslam Microbiology and Biotechnology Laboratory, Department of Zoology, Government College Women University, Faisalabad, Pakistan

Abhishek Kumar Awasthi Mycological Research Laboratory, Department of Biological Sciences, Rani Durgavati University, Jabalpur, Madhya Pradesh, India

School of Environment, Tsinghua University, Beijing, People's Republic of China

Vinay Singh Baghel Department of Environmental Microbiology (DEM), Babasaheb Bhimrao Ambedkar University (A Central University), Lucknow, Uttar Pradesh, India

Ram Naresh Bharagava Department of Environmental Microbiology, Babasaheb Bhimrao Ambedkar University (A Central University), Lucknow, Uttar Pradesh, India

Siddharth Boudh Department of Environmental Microbiology (DEM), Babasaheb Bhimrao Ambedkar University (A Central University), Lucknow, Uttar Pradesh, India

Si Hui Chen School of Science, Monash University Malaysia, Bandar Sunway, Selangor Darul Ehsan, Malaysia

Pankaj Chowdhary Department of Environmental Microbiology, Babasaheb Bhimrao Ambedkar University (A Central University), Lucknow, Uttar Pradesh, India

Mostafa M. El-Sheekh Botany Department, Faculty of Science, Tanta University, Tanta, Egypt

Izharul Haq Environmental Microbiology Laboratory, Environmental Toxicology Group, CSIR-Indian Institute of Toxicology Research (CSIR-IITR), Lucknow, Uttar Pradesh, India

Vishvas Hare Department of Environmental Microbiology (DEM), Babasaheb Bhimrao Ambedkar University (A Central University), Lucknow, Uttar Pradesh, India

Gaurav Hitkari Department of Applied Chemistry, Babasaheb Bhimrao Ambedkar University (A Central University), Lucknow, Uttar Pradesh, India

Miraji Hossein Chemistry Department, School of Physical Sciences, College of Natural Sciences, University of Dodoma, Dodoma, Tanzania

Qayyum Husain Department of Biochemistry, Faculty of Life Sciences, Aligarh Muslim University, Aligarh, India

Ali Hussain Applied and Environmental Microbiology Laboratory, Department of Wildlife and Ecology, University of Veterinary and Animal Sciences, Lahore, Pakistan

Irshad Hussain General Chemistry Laboratory, Faculty of Fisheries and Wildlife, University of Veterinary and Animal Sciences, Lahore, Pakistan

Arshad Javid Applied and Environmental Microbiology Laboratory, Department of Wildlife and Ecology, University of Veterinary and Animal Sciences, Lahore, Pakistan

Jamaluddin Khan Mycological Research Laboratory, Department of Biological Sciences, Rani Durgavati University, Jabalpur, Madhya Pradesh, India

Bhanu Kumar Pharmacognosy and Ethnopharmacology Division, CSIR-National Botanical Research Institute, Lucknow, Uttar Pradesh, India

Jinhui Li School of Environment, Tsinghua University, Beijing, People's Republic of China

Sujata Mani Laboratory for Bioremediation and Metagenomics Research (LBMR), Department of Environmental Microbiology (DEM), Babasaheb Bhimrao Ambedkar University (A Central University), Lucknow, Uttar Pradesh, India

Sandhya Mishra Laboratory for Bioremediation and Metagenomics Research (LBMR), Department of Environmental Microbiology, Babasaheb Bhimrao Ambedkar University (A Central University), Lucknow, Uttar Pradesh, India

Contributors

M. Golam Mostafa Institute of Environmental Science, University of Rajshahi, Rajshahi, Bangladesh

Akhilesh Kumar Pandey Mycological Research Laboratory, Department of Biological Sciences, Rani Durgavati University, Jabalpur, Madhya Pradesh, India

Madhya Pradesh Private Universities Regulatory Commission, Bhopal, Madhya Pradesh, India

Gajanan Pandey Department of Applied Chemistry, Babasaheb Bhimrao Ambedkar University (A Central University), Lucknow, Uttar Pradesh, India

Javed Iqbal Qazi Microbial Biotechnology Laboratory, Department of Zoology, University of the Punjab, Lahore, Pakistan

Abhay Raj Environmental Microbiology Laboratory, Environmental Toxicology Group, CSIR-Indian Institute of Toxicology Research (CSIR-IITR), Lucknow, Uttar Pradesh, India

Muhammad Rashid General Chemistry Laboratory, Faculty of Fisheries and Wildlife, University of Veterinary and Animal Sciences, Lahore, Pakistan

Michael J. Sadowsky Department of Soil, Water and Climate, Biotechnology Institute, University of Minnesota, St. Paul, MN, USA

D. C. Sharma Department of Microbiology, Dr Shakuntala Mishra National Rehabilitation University, Lucknow, Uttar Pradesh, India

Carrie Siew Fang Sim School of Science, Monash University Malaysia, Bandar Sunway, Selangor Darul Ehsan, Malaysia

D. P. Singh Department of Environmental Science, Babasaheb Bhimrao Ambedkar Central University, Lucknow, India

Jay Shankar Singh Department of Environmental Microbiology (DEM), Babasaheb Bhimrao Ambedkar University (A Central University), Lucknow, Uttar Pradesh, India

Ranjan Singh Department of Environmental Science, Babasaheb Bhimrao Ambedkar Central University, Lucknow, India

Sandhya Singh Department of Applied Chemistry, Babasaheb Bhimrao Ambedkar University (A Central University), Lucknow, Uttar Pradesh, India

Alexei Solovchenko Biological Faculty of Lomonosov Moscow State University, Moscow, Russia

Rahul Shrivastava Department of Biotechnology & Bioinformatics, Jaypee University of Information Technology, Waknaghat, Solan, Himachal Pradesh, India

Adeline Su Yien Ting School of Science, Monash University Malaysia, Bandar Sunway, Selangor Darul Ehsan, Malaysia

Atul Kumar Upadhyay Department of Environmental Science, Babasaheb Bhimrao Ambedkar Central University, Lucknow, India

Jyoti Verma Department of Biotechnology, G.B. Pant Engineering College, Ghurdauri, Pauri, Garhwal, Uttarakhand, India

Surabhi Zainith Laboratory for Bioremediation and Metagenomics Research (LBMR), Department of Environmental Microbiology (DEM), Babasaheb Bhimrao Ambedkar University (A Central University), Lucknow, India

About the Editors

Ram Naresh Bharagava was born in 1977 and completed school education from government schools at Lakhimpur Kheri, Uttar Pradesh (UP), India. He received his B.Sc. (1998) in Zoology, Botany, and Chemistry from University of Lucknow, Lucknow, UP, India, and M.Sc. (2004) in Molecular Biology and Biotechnology from Govind Ballabh Pant University of Agriculture and Technology (GBPUAT), Pantnagar, Uttarakhand, India. He received his Ph.D. (2010) in Microbiology jointly from Environmental Microbiology Division, Indian Institute of Toxicology Research (IITR) Council of Scientific and Industrial Research (CSIR), Lucknow, and Pt. Ravishankar Shukla University, Raipur, Chhattisgarh, India. He was Junior Research Fellow (JRF) during his Ph.D. and qualified twice (2002 and 2003) the CSIR-National Eligibility Test (NET) and Graduate Aptitude Test in Engineering (GATE) in 2003. His major research work during Ph.D. focused on the bacterial degradation of recalcitrant melanoidin from distillery wastewater. He has authored one book entitled *Bacterial Metabolism of Melanoidins* and edited two books entitled *Bioremediation of Industrial Pollutants* and *Environmental Pollutants and Their Bioremediation Approaches*. He has authored and coauthored many research/review papers and one book review in national and international journal of high impact factor. He has also written many chapters for national and international edited books. He has published many scientific articles and popular science articles in newspapers and national and international magazines, respectively. He has been presenting many papers relevant to his research areas in national and international conferences. He has also been serving as a potential reviewer for various national and international journals in his respective areas of research. He was awarded a postdoctoral appointment at CSIR-IITR, Lucknow, but left the position after a while and subsequently joined (2011) Babasaheb Bhimrao Ambedkar (Central) University, Lucknow, UP, India, where he is currently Assistant Professor of Microbiology, and is actively engaged in teaching at postgraduate and doctoral level and research on the various Government of India (GOI) sponsored projects in the area of environmental toxicology and bioremediation at Laboratory for Bioremediation and Metagenomics Research (LBMR) under Department of Environmental Microbiology (DEM). His research has been supported by University Grants Commission (UGC) and Department of Science and Technology (DST), India. He has been the advisor to 40 postgraduate students and the mentor to 1 project fellow and 6 doctoral students. His major thrust areas of research are the biodegradation and bioremediation of environmental pollutants in industrial wastewaters, metagenomics, and wastewater microbiology. He is a member of the Academy of Environmental Biology (AEB), Association of Microbiologists of India (AMI), Biotech Research Society (BRSI), and Indian Science Congress Association (ISCA), India. In spare time, he enjoys traveling to peaceful environments and spends maximum time with his family. He lives in south Lucknow with his wife (Ranjana) and three kids (Shweta, Abhay, and Shivani). He can be reached at bharagavarnbbau11@gmail. com, ramnaresh_dem@bbau.ac.in.

Pankaj Chowdhary is a coeditor and a senior Ph.D. research scholar in the Department of Environmental Microbiology at Babasaheb Bhimrao Ambedkar University (a central university), Lucknow, Uttar Pradesh, India. He received his B.Sc. (2009) in Zoology, Botany, and Chemistry from St. Andrews College, Gorakhpur, Uttar Pradesh (UP), India, and postgraduation degree (2011) in Biotechnology from Deen Dayal Upadhyaya Gorakhpur University Uttar Pradesh (UP), India. His Ph.D. topic is "Study on the Bacterial Degradation and Detoxification of Distillery Wastewater Pollutants for Environmental Safety." The main focus of his research work is the role of ligninolytic enzyme producing bacterial strains in the degradation of coloring compounds from distillery wastewater. He has published original research and review articles in peer-reviewed journals of high impact. He has also published many national and international book chapters and magazine articles on the biodegradation and bioremediation of industrial pollutants. He is a life member of the Association of Microbiologists of India (AMI) and Indian Science Congress Association (ISCA) Kolkata, India.

Chapter 1
Conventional Methods for the Removal of Industrial Pollutants, Their Merits and Demerits

Sandhya Mishra, Pankaj Chowdhary, and Ram Naresh Bharagava

Abstract Release of unprocessed and incompletely treated industrial wastes signifies a severe environmental threat regarding anthropological and ecological health concerns. An appropriate treatment approach that may be fruitful for waste management in a cost-effective and eco-friendly manner is becoming more important and a critical issue all over the world. There are various conventional, emerging, advanced, and biological treatment methodologies available that are suitable for wastewater remediation. In recent years, researchers have made notable achievements by means of conventional, biotechnological applications for the treatment and degradation of hazardous pollutants and noxious organic and inorganic combinations. Moreover, promising outcomes were achieved by these conventional treatment technologies, but they also have some drawbacks. Therefore, this chapter delivers an impression of the usage, advantages, and shortcomings of current technologies available for the effective management of waste products generated by industrial activities and their economically feasible alternatives.

Keywords Environmental pollutants · Industrial wastewater · Conventional treatment technologies

1 Introduction

Industrial and economic developments involve environment by discharging huge amount of wastewater containing high concentrations of noxious and lethal pollutants with low degradability that rigorously affect the climate and living standards of modern life. Speedy developmental progression of various industrial activities (Fig. 1.1), including electroplating, iron and steel, mines and quarries, pharmacies,

S. Mishra · P. Chowdhary · R. N. Bharagava (✉)
Laboratory for Bioremediation and Metagenomics Research (LBMR), Department of Environmental Microbiology, Babasaheb Bhimrao Ambedkar University (A Central University), Lucknow, Uttar Pradesh, India

© Springer Nature Singapore Pte Ltd. 2019
R. N. Bharagava, P. Chowdhary (eds.), *Emerging and Eco-Friendly Approaches for Waste Management*, https://doi.org/10.1007/978-981-10-8669-4_1

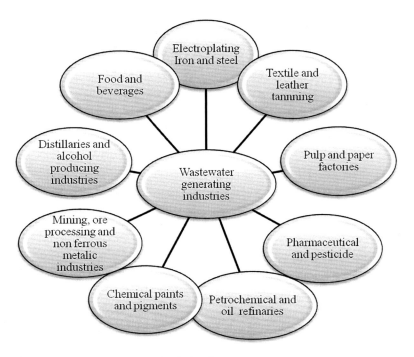

Fig. 1.1 Different wastewater-generating industries

distilleries, textiles and leather processing industries, paper, food and agricultural activities producing polluted, contaminated and unclean water that is particularly used by poor people in their daily routine (Marshall et al. 2007; Mishra and Bharagava 2016).

Wastewater discharged from such industries loaded with numerous complex and toxic organics, metals, dyes, pathogenic organisms, recalcitrant chlorine and phenolic products and many perilous inorganic pollutants that are continuously used enters into the environment, with a deleterious effect on the natural resources and causes severe health hazards (Zainith et al. 2016; Yadav et al. 2017). Massive generation and release of raw industrial wastewater (IWW) are constantly polluting environmental resources and becoming a big problem for maintaining a safe and sustainable environment (Gricic et al. 2009). Since, ancient times, there have been many different conventional treatment methods such as adsorption, ion exchange, precipitation, biosorption, coagulation, electro-dialysis (ED) and many more that are applicable for the treatment and management of waste produced by industrial activities (Narmadha and Kavitha 2012; Shivajirao 2012). However, these approaches are less suitable and less efficient regarding their higher cost and energy requirement. Thus, new, more sophisticated and advanced technologies are being developed and bring additional superior analytical functioning of wastewater characterization for improved performance over conventional approaches. Currently, several green technologies known as bioremediation and phytoremediation are also popular and are extensively used for remediation of IWW because of their low cost and safety with regard to environmental protection and are liable to be more proficient in the dumping of hazardous waste (Rajasulochana and Preethy 2016; Bharagava et al. 2017). Therefore, this chapter is a meaningful approach that provides brief information about wastewater discharging industries, their pollutants/waste products that are typically found in IWW, with a review and selection of proper conventional/emerging, operative and significantly suitable treatment processes for adequate management of IWW before its final disposal into the environment.

2 Various Types of Wastes and Their Contamination in the Environment

Various types of waste are generated by many anthropogenic activities and exposures. Basically, there are two major sources that are primarily responsible for the contamination of the environment: municipal and industrial waste. Instead, on the basis of origin and pollutant types, waste can be categorised involving other important sources, as shown in Table 1.1.

2.1 Municipal Waste

Municipal waste is a prominent source of various pollutants including suspended particulate matter and dissolved solids (DS). It serves as a rich source of carbon and nitrogenous nutrients persistent in organic compounds, toxic metals and salts. It also contains grit debris, solid particles, disease-causing microbial pathogenic organisms and various hazardous chemicals at higher concentrations than their

Table 1.1 Classification of different types of waste on the basis of their origin and source

Waste source	Generated waste products
Domestic and household waste	Garbage, paper and plastic waste, food waste, tin and glass pieces
Commercial and industrial waste	Toxic metals, effluents from textiles, tanneries, distilleries, pulp industries, acids, ash, and hazardous chemical waste
Municipal waste	Household discharge, public toilets and sewage sludge waste
Agricultural waste	Spoiled food waste, pesticides and insecticidal waste, fertilizer waste and crop residues
Biomedical waste	Hospital, clinical, chemical drugs and surgical and laboratory waste
Radioactive and nuclear waste	Coal ash, radioactive elements such as uranium and radium, hazardous gases and fuel products
Electric waste	Cell phones, televisions, keyboards, laptops, scanners, copiers and other electronic devices

recommended permissible limits. When municipal wastewater is discharged directly into bodies of water and their soil surroundings, it can cause severe aesthetic problems related to odour and discolouration. The presence of disease-causing pathogens (bacteria and virus) such as *E. coli, Staphylococcus* sp., *Bacillus* sp., *Klebsiella* sp., *Vibrio* sp., can make water unfit for utilization.

2.2 Industrial Waste

Industrial activities are the foremost and cohort sources of wastewater generation and contaminate aquatic systems without a proper waste disposal system. There are millions of factories, mills and industries that produce a significant amount of toxic and hazardous waste that is dangerous for the safety and health of our environment (Dsikowitzky and Schwarzbauer 2013). Industrial waste can be divided into two main categories, biodegradable and non-biodegradable, comprising numerous chemical, solid, liquid, toxic, and inflammable, corrosive, infectious, hazardous metals, pesticides and organic and hydrocarbon waste that can de degraded or are non-degradable depending on their complex nature.

- *Biodegradable waste:* This kind of waste is produced by textile mills, distilleries, leather, food and paper processing industries and can disintegrate into non-toxic material with the help of microorganisms. Biodegradable waste is similar to household waste and can be easily handled with treatment technologies.
- *Non-biodegradable hazardous waste:* Non-biodegradable waste is completely toxic, causes environmental pollution and life-threatening illness to living organisms, as it cannot decompose into non-poisonous materials. It is produced from plastic, glass industries, fertilizers, iron and steel industries, radioactive activities, chemical and drug industries. This kind of hazardous waste needs proper disposal treatment technologies because of their injurious and dangerous effects.

2.2.1 Tannery

The tannery industry, designated as a red category industry, is extremely contaminating and the biggest producer of processed leather materials. It is a significant contributor to the socio-economic equity of a country because of its great potential for export and employment (Nandy et al. 1999; Yadav et al. 2016a; Chowdhary et al. 2017b). It is the oldest technology that is used for the conversion of raw skin and hides into valuable items, but, the processing of raw skin/hides is a highly water-consuming process that utilises large volumes of fresh water (~35 to 40 L) and discharges much of this water as wastewater in huge quantities, which is a great challenge for the environment (Chowdhury et al. 2013; Kumari et al. 2016). Tannery wastewater contains substantial aggregates of sulphates, chlorides, suspended constituents, toxic metals (mainly chromium), coloured complexes and other pollutants (Bharagava and Mishra 2017; Chowdhary et al. 2017b; Yadav et al. 2017). However, tannery industries are the most significant cause of chromium pollution, as chrome tanning is used for the production of cheap and high-quality leather (Alfredo et al. 2007). Tannery effluent is extremely toxic for the ecology of marine and human life, causing severe carcinogenic, mutagenic and toxicological irreversible changes to the vital body structure (Mishra and Bharagava 2016; Yadav et al. 2016b; Thacker et al. 2007). High exposure to chromium is noticeably toxic and produces adverse effects on plants and other living organisms (Yadav et al. 2017; Kumari et al. 2014).

2.2.2 Distillery

Alcohol-producing industries are increasing greatly because of their extensive use in chemicals, pharmaceuticals, cosmetics, food and the perfumery industry etc. These are high pollution-causing industries, producing ~8 to 15 L wastewater/L of total liquor production (Saha et al. 2005). According to the Central Pollution Control Board (CPCB) (2011), in India, there are ~300 molasses-based distilleries, which are mainly responsible for producing ~40 billion litres of wastewater annually. Generally, molasses is used as a raw material for alcohol production owing to its cheap cost in distilleries that generated waste known as molasses-spent wash (MSW). MSW is creating serious environmental problems and contains enormous amounts of melanoidins, a high organic load and low degradability (Pant and Adholeya 2007; Bharagava and Chandra 2010a; Chowdhary et al. 2017c). Melanoidins are complex polymeric compounds that give a dark brown colour to wastewater, having concentrations of mineral salts, phenolics and organic compounds, high chemical oxygen demand (COD; 80,000–160,000 mg/L) and low pH (3.7–4.5) and DS (Basu et al. 2015; Chen et al. 2003; Mohana et al. 2009). Melanoidins are recalcitrant compounds that notably influence the production and quality of this industry (Inamdar 1991). IWW remarkably leads to a high level of soil pollution, eutrophication, preventing photosynthesis and acidification of improperly treated wastes (Mohammad et al. 2006; Chowdhary et al. 2017a). It also causes manganese deficiency when disposed of in soil, resulting in soil infertility, reducing soil alkalinity, inhibiting seed germination and damaging agricultural crops (Agarwal and Pandey 1994; Bharagava and Chandra 2010b). The huge quantity of IWW generated from distilleries, characterized by the presence of toxic and persistent pollutants that may become dangerous for marine and soil environments owing to the existence of putrescible organics, sulphur and phenolic compounds, shows lethal effects on microbial communities, fishes and other planktons (Pant and Adholeya 2007; Mahimraja and Bolan 2004).

2.2.3 Pulp and Paper

Pulp and paper mills are high chemical processing industries that are rich in recalcitrant organic pollutants (ROPs) and pose a major threat to the environment. It is a high water- and energy-consuming industry that requires quite a large amount of water: ~4000 to 12,000 gallons per tonne of pulp (Ince et al. 2011). Of the total amount of fresh water used, around 75% emerges as wastewater (Yadav and Garg 2011). Paper-making industries consist of two main production steps; namely, pulping and bleaching. Pulping is the major and primary source of waste pollution generation (Billings and Dehaas 1971). This industry contains various types of toxic and hazardous chemicals and pollutants high concentrations that are released every year into the atmosphere without proper disposal (Cheremisinoff and Rosenfeld 1998).Wastewater discharged from the paper and pulp industries poses a serious concern regarding their toxicity and pollutant profile, containing high concentrations of sodium, calcium, carbonates, sulphide, salts and acids such as Na(OH), Na_2CO_3, $Ca(OH)_2$, Na_2S, HCl, HNO_3 sulphur compounds, various gases such as NOx are emitted into the air, chlorinated and organic compounds, nutrient and metals (EPA 2002; Sumathi and Hung 2006; Singh et al. 2016). The wastewater is highly rich in organic content and chlorinated hydrocarbons, absorbable organic halides (AOX) that adversely affect the environment, humans and animals. Various pollutants from pulp and paper mills accumulate in the food chain and are difficult to degrade because of their chemical complexity and high reactivity. Such pollutants are very harmful and can cause severe physiological, neural and carcinogenic health problems and impairments (Ince et al. 2011; Savant et al. 2006).

2.2.4 Textile Industry

Dyes and pigments are the pollutants most frequently released from textile industries, which is the greatest consumer of higher quality fresh water and has created huge pollution problems. The textile, clothing and dying industry uses approximately 700,000 tonnes of various kinds of dyes and more than 8000 chemicals in the manufacturing of textile products (Mani and Bharagava 2016). Wastewater emanating from textiles and fabric-processing industries can provoke serious environmental effects, comprising toxic reactive dyes and their dark coloration, heavy metals, chlorine residues, persistent organic pollutants (POPs), dioxins and pesticides etc. (Zollinger 1987). The textile yard wastewaters are very dark in colour and have a high biochemical oxygen demand (BOD) and solids, mainly total dissolved solids (TDS) owing to the excessive use of pigmented raw material and dyes. The impacts of the direct discharge of such a type of textile effluent on environmental aspects notably responsible for the increased pollution and deterioration of water and soil parameters would be often dangerous for plants, animals and mankind (Chavan 2001). A huge quantity of water, ~30 to 50 L/kg, is required for the colouring and tinting of clothes. According to the World Health Organisation (WHO), 17–20% of IWW comes from textile industries that contain hazardous toxic waste, organic dyes and inorganic materials and compounds that cause significant contamination and pollution (HSRC 2005). Mostly synthetic dyes are used, which include a maximum of 60–70% of Azo dyes for their regular use in dyestuffs (ETAD 2003). These are highly toxic and difficult to decolourize relative to their complex and recalcitrant nature. Synthetic dyes and their intermediates including the aromatic compounds naphthalene, benzidine, etc. are potential carcinogenic and mutagenic agents that can deleteriously affect the food chain and show harmful effects on photosynthetic organisms and crops (Anjaneyulu et al. 2005).

2.2.5 Food and Beverages

Food and beverage units are intensive water users in the industrial sector and hold prominent positions in economic development. This industry waste consists primarily of remainders of organic substances of used materials and requires huge volumes of water; thus, wastewater discharged from these factories is highly variable from industry to industry and also within the industry sector according to its products and production procedures (Jayathilakan et al. 2012). The food industry comprises various sectors, including meat processing, dairy, brewery and wine production, marine industry, grain processing, fruit and vegetable industry (Helkar et al. 2016). Wastewater emanating from such industries bears a high level of pollution, alkaline to acidic in nature, containing a high proportion of persistent pollutants, solids, hazardous materials, extracted residues, carbon-dioxide and sweeteners such as sugar or syrup and flavours dissolved into the water (Bohdziewicz and Sroka 2005; Pathak et al. 2015). This is why BOD and COD concentrations are much higher in wastewater and the direct discharge of this into environment causes serious pollution problems (Russ and Pittroff 2004). The meat industry causes degradation of the atmosphere to a large extent because of the extensive chemicals and water usage associated with a significant effluent management issue that can negatively damage the environment (Water saving fact sheet 2014).

3 Types and Characterization of Pollutants Generated from Various Industrial Processes

3.1 Organic Pollutants

3.1.1 Phenolics

Phenolic compounds and their waste products are produced from various pharmaceuticals, tanneries, distilleries, textiles, and wood preserving industries, and the unnecessary usage of pesticides in farming sectors. Catechol is a very well-known phenolic compound that is extensively used in the photography and drug industries. It is highly toxic, and ranked as a priority pollutant causing endocrine disruption, reducing reproductive hormonal activity and blocking neurons signals (Cohen et al. 2009). Chlorinated and methylated substituted catechol is usually discharged from pulp, tanneries and oils refineries, which are extremely detrimental and have biocidal effects on human beings and plant biota (Schweigert et al. 2001).

3.1.2 Dyes

Dyes are broadly used in various industrial activities such as in printing, tanning of leather, manufacturing of pigments and paints, textiles, dyeing and cosmetics, and in photographic film as a colouring agent (Ahmad 2009; Nelson and Hites 1980). Excess quantities of synthetic dyes: acidic, basic triphenylmethanes, anthraquinone, nitro, methane, quinolone and azo dyes etc. are used in textiles, among which ~15% of the dyes remain in the wastewater in their residual form and escape into the environment (Mani and Bharagava 2016; Jadhave and Govindwar 2006). Dyes can easily be soluble in water in a tiny quantity that affects the water's quality and contaminates water bodies. Synthetic dyes and their intermediates create severe threats to the environment and affecting and disturbing physiological activities causing carcinogenic and mutagenic changes in humans and animals (Mittal et al. 2010; Senthilkumar et al. 2006). Dyes are also retarding plant growth by reducing and affecting the activity of rhizospheric micro-organisms and soil productiveness (Parshetti et al. 2011; Sujata and Bharagava 2016).

3.1.3 Volatile Organic Compounds

Volatile organic compounds (VOCs) such as gasoline, benzene, formaldehyde, toluene and xylene are organic substances discharged into the atmosphere from various regular and anthropogenic activities. These are highly reactive at high pressure and normal temperature. Metal mining and finishing, petrochemicals and automobiles, and certain chemical production industries are the reason for the generation of such compounds (Ramirez et al. 2011). VOCs are highly toxic at low concentrations of parts per billion (ppb) and affect human health severely. They cause respiratory disorders, birth defects, irritation and inflammation, reduced reproductive potency and neurotoxic and genotoxic effects (WHO 2000). According to a study by the United States Environmental Protection Agency (USEPA), ambient VOCs are the reason for 35–55% of the cancer risk associated with outdoor air (USEPA 1990; Liu et al. 2008). VOCs, when mixed with nitrogen oxide react to form ground level ozone or smog, which is greatly responsible for critical climatic conditions.

3.2 Inorganic Pollutants

3.2.1 Suspended and Dissolved Solids

Various impurities exist in the form of suspended and DS in wastewater produced from industries and households. Suspended solids (SS) are the remainder particles of pollutants that could not pass through a standard filter, whereas DS are the remains that pass through a standard filter. SS include colloidal organic particles, clay, slit and microscopic organisms, which give rise to turbidity of the water (Chowdhury et al. 2013). Salinity is a factor that is responsible for DS. Occurrences of high suspended and DS in industrial and municipal waste are predominantly detrimental owing to their toxic and negative impacts on the oxygen content of water (Jeyasingh and Philip 2005). Excess quantity of DS causes salinity problems of soil and water, results in osmotic stress and disturbs osmotic regulatory functions in plants and animals (Thacker et al. 2007).

3.2.2 Nitrogenous Compounds

Nitrogen is primarily found in the forms of organic nitrogen, ammonium, nitrate and nitrite in the environment. Predominant applications of urea-, ammonia- and nitrate-rich artificial fertilizers in agricultural practices are conspicuous causes of high nitrogen concentrations in pollution in the atmosphere (Nagajyoti et al. 2010). The excess load of nitrogenous compounds in water bodies is mainly responsible for eutrophication, resulting in unnecessary plant growth that consequently decreases oxygen levels and ultimately causes death of aquatic life (Dey and Islam 2015). The occurrence of higher concentrations of nitrogenous compounds in wastewater is detrimental regarding their particulate and soluble nature, adversely affecting public health and the environment (Hurse and Connor 1999). Methaemoglobinaemia is the most dangerous health problem caused by the superfluous amount of nitrate content in the water and the presence of ammonia is tremendously poisonous to fishes (Jenkins et al. 2003).

3.2.3 Sulphides

Wastewater generated from tanneries, paper industries, oil refineries, distilleries and rock residues contain large amounts of sulphate (Mahendra and Gregory 2000). Tannery effluent contains sulphate in a very high concentration (~4000 mg/L) owing to the widespread use of salts of sodium sulphide and many other auxiliary reagents and substances used in the processing of raw hides/skins (unhairing and liming) (Chowdhury et al. 2015). The CPCB (2013) recommended that the permissible limit for sulphate is >250 mg/L to discharge into the public streams. However, nowadays the direct discharge of sulphate-rich waste into the environment is undesirable and leads to an environmental threat. Sulphate present into the wastewater is converted into H_2S under anaerobic conditions and highly toxic, malodorous and causes corrosion problems (Rao et al. 2003). H_2S also has inhibitory effects on aquatic organisms and excessive sulphate may cause purgative effects (Purushotham et al. 2011).

3.2.4 Chlorides

Chlorides are naturally present in water in an associated arrangement of NaCl. Chlorides coupled with sodium are responsible for the saline flavour of water when its concentration is >250 mg/L. Chloride and its salts are commonly used for domestic, industrial purposes, which could possibly be a reason for concern regarding chloride contamination, as chloride is added into the groundwater by such municipal, drainage and manufacturing undertakings (Venkatesan and Swaminathan 2009). Chloride-containing compounds and substances are often employed in tanning and textiles for bleaching purposes and are tremendously toxic for workers and consumers.

3.2.5 Heavy Metals

Metals exist in the earth's crust and persist for many years in the environment. Various toxic metals such as, Cr, Cd, Hg, Pb, Ni and Zn are reported in high concentrations in IWW because of their extensive use at various manufacturing stages (Mishra and Bharagava 2016; Hare et al. 2017). Because of the persistent nature, metals cannot be degraded and accumulate in soil and water environments because of their excess concentration and cause toxic, genotoxic, mutagenic and carcinogenic effects. Thus, a metal-contaminated environment represents a serious matter for the whole world (Yadav et al. 2017). Various industries including leather tanning, electroplating, textiles and dyes metallurgical, chemical, refractory, wood furnishing, and mining are the major sources of metal pollution in the environment (Nagajyoti et al. 2010). Metals also affect fetuses and neonates, human beings, mainly industrial workers who face the maximal risk of metal exposure. Regular contact with metal ions disrupts the nervous system, creating behavioural and physiological problems, malfunctioning of the brain and damage to the red blood cells (Mishra and Bharagava 2016).

4 Conventional Treatment Technologies

4.1 Physical Treatment Technologies

4.1.1 Adsorption

Adsorption is an ancient conventional physical treatment process that is efficiently used for the treatment and recycling of the chlorinated, metallic and organic impurities of IWW (Worch 2012). It involves a phase transfer process that generally removes contaminated substances from the fluid phase (gaseous or liquid) onto a solid surface and becomes bound through physical or chemical connections (Kurniawan and Babel 2003). It is a physico-sorption or chemisorption process that occurs because of Van Der Waals, covalent and electrostatic forces. Various low-cost and modified adsorbents are best suited for IWW treatment, including cyclodextrin-containing polymers, that act as an efficient adsorbent technique for organic pollutant removal owing to the simple designing, cheap cost and specific affinity (Gillion 2007; Romo et al. 2008). Cellulose acetate with triolin (CA-triolin) is frequently relevant for the degradation of POPs. Activated carbon is a significant absorbent material that is simply appropriate for micro-pollutants, pharmaceuticals, personal care products, leachates, and a broad range of noxious organic pollutants (Moreno-Castilla 2004; Rashed 2013).

4.1.2 Electro-Dialysis

Electro-dialysis (ED) is a membrane separation technique used for the separation of anions and cations with the help of two charged membranes of anode and cathode. It is a very appropriate and suitable process for the management of inorganic and noxious pollutants (Pedersen 2003). The membranes are usually ion-exchange resins that selectively transfer ions. They are created from a polymer of materials, such as styrene or polyethylene, incorporating fixed and mobile charged groups (Oztekin and Atlin 2016; Chang et al. 2009). During this process, positively charged ions move to a negatively charged cathode plate when electric current passes over an aqueous metal solution (Chen 2004). ED also requires a higher frequency of pre-treatment before the process. The working efficiency of ED depends on various factors, such as quality of membrane, current density, pH, ED cell structure, water ion concentration and flow rate (Mohammadi et al. 2004). ED deals with the detrimental impurities that are present in the form of metals, solids and other forms of waste in the industrial effluent of water. Furthermore, it is used for the recovery of some useful metals such as chromium and copper. A noticeable drawback of this technique is the corrosion and fouling of membrane affected by solid particles or colloids and biomass that reduce the ion transportation (Lee et al. 2002).

4.1.3 Ion Exchange

Ion exchange is a different, more attractive process for the treatment of terminal and undesirable anions and cations from IWW. Ion exchange resin comprises a mineral and carbon-based network structure with attached functional groups, but synthetic organic resins are commonly used matrices for ion exchange (Barakat 2011). Resins that exchange positive ions are called cationic and those that exchange negative ions are anionic. These resins are made by the polymerisation of carbon-based material into a porous three-dimensional structure. Cationic resins containing acidic functional groups, such as sulphonic groups, are exchanged for hydrogen or sodium, whereas anions are exchanges for hydroxyl ions and they have basic functional groups, such as amines. This treatment method is highly selective for the detoxification of certain heavy metals ions Cr^{+6}, Pb^{+2}, Zn^{+2}, Cu^{+2} and Cd^{+2} (Bose et al. 2002). It is extensively used in the plating industry, where chromium recovery and water reuse have resulted in considerable savings.

4.1.4 Sedimentation

Sedimentation is a widely used physical, operational treatment unit that involves gravitational settling to remove large and heavy particles suspended in wastewater (Omelia 1998). It proceeds in a settling tank known as a clarifier. It primarily works on the principle of gravitation force and the density gradient difference of elements. It is commonly used for the differentiation of bulky solid pollutants, and grit from the effluent in relation to their increasing or decreasing densities and settles them down in the primary settling tank. Three main types of sedimentation designs are used for the treatment of IWW: solid contact, horizontal flow and inclined surface (USEPA 2004).

4.2 Chemical Treatment Technologies

4.2.1 Coagulation/Flocculation

The coagulation procedure is applied for the clarification of IWW containing colloidal and SS. IWW containing emulsified oils and paperboard waste simply clarifies through coagulation by using alum as a coagulating agent. There are several varieties of compounds that work as coagulating agents, such as aluminium sulphate (AlSO4), ferrous sulphate (FeSO4), silica, ferric chloride ($FeCl_3$) and pyroelectrolyte, that have been reported for the deduction and elimination of lethal and poisonous contaminants and solid residues from IWW (Lofrano et al. 2013). Coagulating agents are pH-specific and at low pH values, they can frequently break down an emulsion.

4.2.2 Precipitation and Filtration

Precipitation and filtration are mainly employed for the differentiation of suspended and DS, sulphates, carbonates, chloride impurities and metallic content from IWW. Metals are generally precipitated as hydroxide through lime (NaOH). Chemical precipitation commonly involves metal salts such as aluminium sulphate and ferric chloride as precipitating agents that form complexes with contaminated compounds (Mirbagherp and Hosseini 2004). Chemical precipitation is a cheap and reliable method but largely depends on the pH parameter of the wastewater (Nassef 2012). Phosphorus is excessively removed by this well-established technology. Further, the precipitates are eliminated from the wastewater by using a separate filtration or clarification process.

4.2.3 Oxidation

The oxidation method related to Fenton's reagent is gaining importance owing to the effective degradation of phenolic, chlorinated and organic pollutants that predominantly exist in industrial effluents (Beekeepers 2000). Fenton's oxidation process is a simple and attractive method that increases the biodegradability of organic halides, polyphenols, total organic carbon and intractable pollutants because of its ability to work at low pH values. Fenton's process is significantly employed for the potential decolourisation of textile dyes, the colour removal of lignocellulosic and for refining wastewater (Covinch et al. 2014; Bianco et al. 2011). The main advantage of this process is its easy handling, as there is no need for any other complex tool for oxidation and its reagents are safe for our environment, influencing the high removal rate of toxic chromophoric compounds (Gogate and Pandit 2004; Krishnan et al. 2016).

4.3 Biological Treatment Technologies

Biological treatment approaches in particular, use microorganisms (mainly bacteria and fungus) for the management of contaminated polluted organic waste products into flocculent/settable materials and gases that can easily be removed. Biological treatment processes are more suitable and safe rather than physical and chemical processes.

4.3.1 Activated Sludge Process

The activated sludge process (ASP) is an aerobic process that is extensively used for the treatment of enormous amounts of wastewater primarily for solid and organic contaminates. In developing countries, ASP is the most commonly employed method at sewage treatment plants (STPs)/common and combined effluent treatment

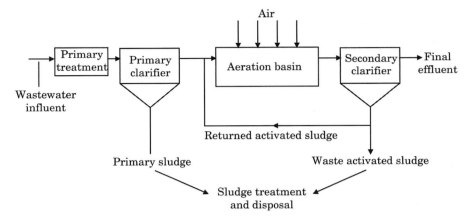

Fig. 1.2 Schematic representation of an activated sludge treatment process

plants and effluent treatment plants of industry. It involves a suspension of bacterial, fungal or algal biomass that is principally liable for the degradation and decomposition of organic substances (Rajasulochana and Preethy 2016). Microorganisms are generally grown in an aeration tank, having dissolved oxygen and waste organic material on which they can feed and causing floc formation, by which bulky solid particles settle at the bottom of the tank and clear liquid wastewater is obtained at the top. It is a very good practice, but sludge bulking and foam formation are two main drawbacks of this efficient technology (Fig. 1.2).

4.3.2 Anaerobic Process

Anaerobic waste treatment processes have been employed on a large scale for many years to recover non-conventional energy profits of the municipal, industrial and agricultural sectors. Anaerobic digesters have been principally recommended for the degradation of organic compounds and substances in a more effective and economically feasible manner for the treatment of solid and sludge waste from STPs and urban waste, and effluents that are discharged from various dangerous industrial activities (Fig. 1.3).

4.3.2.1 Fixed Film Reactors

Fixed film reactors are appropriate and useful treatment anaerobic reactors that are applicable for virtually all types of industrial and commercial waste comprising a wide range of contaminates. It involves the use of microorganisms that are fixed to an inert medium, which can be any of the media known as aerobic trickling filters. The effluent passes the bed with a downstream flow. In all configurations, a generous

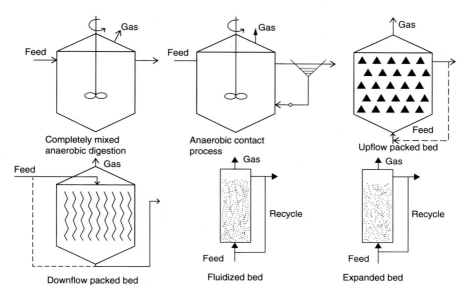

Fig. 1.3 Different types of anaerobic bioreactors/digesters for wastewater treatment

amount of the biomass is present as suspended flocs become entrapped in the voids of the inert media. In general, this is operated without recycling (Young and Dahab 1982). Nowadays, four main types of fixed film reactors, namely trickling filter, rotating biological contractor, integrated fixed film activated sludge and moving bed biofilm reactor, are regularly employed in wastewater management plants.

4.3.2.2 Anaerobic Baffled Reactors

Anaerobic baffled reactors (ABRs) are the most appropriate and powerful anaerobic digesters for low-strength wastewater containing particulate organic substances with horizontal and vertical baffles (Abbasi et al. 2016; Ayaz et al. 2012). It works as a two-segment device that separates the acidogenesis and methanogenesis processes without any control problems and reduces heavy sediment solids (Weiland and Rozzi 1991). ABR techniques are good methods for the treatment of cassava wastewater, which shows 92% removal of organic material at a higher COD level of 2000–5000 mg/L in just 3–5 days.

4.3.2.3 Modified Hybridized Anaerobic Inclining Baffled Reactor

Modified hybridized anaerobic inclining baffled (MHAI-B) reactor is an advanced innovative treatment device that includes both fixed and baffled reactor features. The most significant benefit of this bioreactor is its multi-stage anaerobic ability and low cost, permitting suspended growth of different bacterial groups to develop in such additional conditions (Hassan et al. 2013). It does not require any special sludge separator because of reduced sludge expansion. MHAI-B bioreactors similarly act as a Upflow Anaerobic Sludge Blanket (USAB) and as a trickling submerged fixed film reactor (Hassan et al. 2013).

4.3.2.4 Anaerobic Contact Process

The anaerobic contact process is an activated sludge digester procedure. It is also called anaerobic digester, as the sludge is reutilized by the clarifier of the reactor. This process offers a great solution for large-scale municipal and industrial sludge generation and treatment applications. The key function of this method is the washout of active anaerobic bacterial biomass coming out from the machine being controlled by a sludge separation and recycling system. This is commonly used for concentrated industrial effluents such as distillery and tannery wastewater.

4.4 Emerging Treatment Technologies

Nowadays, the cumulative applications of chemicals and harmful substances in the developing sector and agricultural practices result in the generation of a vast variety of noxious pollutants and contaminants that are difficult to treat with conventional treatment activities; thus, conventional methods have gradually become unsuitable for the effective treatment of the wastewater/effluent. Therefore, the introduction of newer and emerging technologies including advanced oxidation process (AOPs), sonication, photocatalysis redox mediators, enzymatic treatments and membrane filtration. These emerging technologies have the potential to provide better alternatives to the conventional treatment of IWW and public health.

4.4.1 Advanced Oxidation Process

Advanced oxidation processes are chemical treatment processes that generate hydroxyl radicals in adequate amounts for the oxidation of chemical compounds. AOPs use vigorous oxidizing agents (ozone/H_2O_2) or catalysts such as Fe^{+2}/Fe^{+3}, Mn, TiO_2NiSO, CCl, $CuSO_4$ and high-energy radiation that significantly enhance the degradation (Dixit et al. 2015; Naumczyk and Rusiniak 2005; Srinivasam et al. 2012; Gogate and Pandit 2004). AOPs have been growing as emerging treatment technologies in the waste management industry, exploiting the high energy of hydroxyl radicals through various oxidation processes, including Fenton oxidation, photo-oxidation, photo-Fenton oxidation, ozonation, photocatalysis and electrochemical treatment processes that are applied to treat IWW (Rameshraja and Suresh

2011; Lofrano et al. 2013). These processes are mainly applicable in the treatment of organic molecules, aromatic and halogenated compounds, polyphenols, and resins subsequently resulting from the putrefaction of biological and toxic compounds, as attacked by the hydroxyl radical (Alnaizy and Akgerman 2000; Matta et al. 2007; Eskelinen et al. 2010). AOPs can destroy organic pollutants, aromatic rings, VOCs, pesticides, nitrites, sulphides and other inorganic pollutants at a continuous speed of 106–109 M^{-1} s^{-1} (Glaze and Kang 1989). AOPs are extremely appropriate for the reduction and treatment of organic pollution load and toxicity from IWW to such an extent that the treated IWW may be reintroduced and reused during the process. Sometimes AOPs reduce the concentration of the pollutant from a few hundred parts per million to less than 5 ppb (Santos et al. 2009).

4.4.2 Photo-Catalysis

Photo-catalysis is a very innovative and efficient treatment approach for the reduction of extremely harmful pollutants present in IWW. It is commonly used for the removal of organic and metallic waste products (Ananpattarachai and Kumket 2015). Presently, photo-catalytic oxidation appears to be one step ahead in relation to development than any other oxidation procedure. Many organic contaminants absorb UV energy and slightly decompose owing to direct photolysis or become excited and more reactive with chemical oxidants. The photo-catalytic process is a novel and promising technique for the efficient destruction of toxic contaminants for a sustainable environment (Skubal et al. 2002).

4.4.3 Membrane Filtration

Membrane filtration is an advanced physical separation technique that is used for the removal of microbials, solids, undesirable particles and natural organic material from wastewaters. Membranes are thin-layered and semipermeable and act as a physical barrier that can separate the feed stream into retentate and permeate fractions (Van der Bruggen et al. 2003). There are several types of membrane filtration process, such as reverse osmosis (RO), ultrafiltration (UF), nanofiltration (NF), microfiltration, which are dependent upon the type of membrane material, pore size and charge used for liquid/solid and liquid/liquid treatment. These UF and NF membranes offer great advantages for the treatment of surface and industrial wastewater because of their high flux, pore size and good chemical resistance compared with other membrane practices (Zhou and Smith 2002). However, inorganic contaminants are largely treated by RO depending on ionic diffusion. Membranes possess high permeability and selectivity and are composed of organic polymers such as polysulphones, polyethylene, cellulose esters and acetates, polyimide/polyamide, polypropylene and polyetherketones etc. (Zhou and Smith 2002; Van der Bruggen et al. 2003). Membrane processes are effectively useful in various net operations of tanning processes for colour removal of aqueous solutions. The possible use of NF membranes was reported by Koyuncu (2002) for the removal of >99% toxic and reactive dyes from textiles industries with increasing salt concentration. Frank et al. (2002) developed a two-step NF process that was capable of reducing 99.8% of colour content and recovering up to 90% of waste.

4.4.4 Enzymatic Treatments

Enzymes are biotic reagents that are known as catalytic agents for the transformation of the substances into products, providing favourable conditions that reduce the activation energy of the reaction. Biological processes in addition to enzymatic digestion are much better, safer and more effective in comparison with chemical and physical treatments for wastewater remediation (Rao et al. 2014). Enzymes are highly specific and carry a low risk of forming secondary pollutants, and are simply applicable for enormous concentrations of recalcitrant toxic pollutants with an extensive assortment of substrates, pH values, temperatures, and low biomass generation (Feng et al. 2013; Nicell 2003). From the environmental perspective, enzymes are extremely efficient and more acceptable because of their versatility, selectivity and biodegradability for the degradation of target pollutants (Nelson and Cox 2004; Adam et al. 1999). Researchers have recently shown that the degradative enzymes of plant and microbial origins, including oxidoreductases, ligninases, laccases, peroxidases, tyrosinases, cellulases, lipases and oxygenases, are efficiently applicable and useful for IWW treatments. These enzymes transform and reduce phenols, chlorinated, non-chlorinated, toxic, even recalcitrant and estrogenic compounds (Guenther et al. 1998; Toumela and Hatakka 2011; Chandra and Chowdhary 2015). Oxidoreductase enzymes have the capability to detoxify phenolic, anilinic, toxic xenobiotic compounds by polymerization, copolymerization and binding with humic or other substrates (Park et al. 2006). *Arthromyces ramosus* peroxidase and soybean peroxidase have also been reported to be potential enzymes because of its low cost and prospect of commercial availability for treatment applications (Biswas 2004; Feng et al. 2013). Cytochrome P450 enzymes have shown promising biotechnological and environmental applications in catalysing several chemical reactions: dealkylation, dehalogenation, reduction and transformation/ removal of contaminants (Barnhardt 2006). Tannins and caffeine present in the manufacturing waste of coffee and tanneries easily convert into less toxic forms with the application of fungal and bacterial enzymes (Murthy and Naidu 2012). Organophosphorus hydrolase is applicable for the treatment of organophosphates (Kapoor and Rajagopal 2011). Laccase, manganese peroxidase, and lignin peroxidase, released from fungal mycelium, are used for the detoxification of phenolic substances (Chandra and Chowdhary 2015). Extracellular hydrolytic enzymes have miscellaneous potential applications in various commercial areas (Porro et al. 2003).

4.5 Remediation of Industrial Pollutants with Green Technologies

4.5.1 Bioremediation

Bioremediation is a transforming, fortunate, cost-effective and environmentally friendly opted approach for waste management for the decontamination of polluted soil and wastewater containing noxious and poisonous pollutants. This is a natural practice that utilises the metabolic potential of living agents (mainly microorganisms) for the reduction and degradation of hazardous pollutants by mineralizing or transforming them into less harmful substances under ex situ or in situ conditions (Okonko and Shittu 2007). Bioremediation is a vast term that is usually related to various other eco-friendly technologies such as phytoremediation, bioventing, landfarming, composting, bioaugmentation, rhizofiltration, biostimulation and bioleaching. Microorganisms (bacteria and fungi) able to grow and survive in polluted environments can be remarkably affected via various abiotic and biotic factors in a direct or indirect manner, providing the primary information about bioremediation and helping to process this effectively. According to Diez (2010), bioremediation shared three main phases: attenuation; biostimulation; bioaugmentation. In the attenuation phase, contaminants and waste products are first reduced by indigenous natural microbes devoid of any anthropogenic activities. Second, the biostimulation process employed with nutrients and oxygen that improves the effectiveness of enhancing the biodegradation process. In the course of the bioaugmentation phase, extra microorganisms are added to the systems that are more effective than the native flora for the potential degradation of target pollutants. Bacteria such as *Pseudomonas*, *Bacillus*, *Alcaligenes*, *Micrococcus*, *Sphingomonas*, and *Mycobacterium* are reported for their potential degradation and removal of pesticides, hydrocarbon, polyaromatic compounds and detoxification of toxic metals Cr, Pd, Cu, Ni, As (Megharaj et al. 2011).

4.5.2 Phytoremediation

Phytoremediation may also be denoted as a plant-mediated remediation (Chaney et al. 1997) that uses the whole plant and their specific parts as an alternative technology to the remediation and removal of pollutants. It is basically a plant-based green technology that can accumulate large concentrations of noxious chemicals. Phytoremediation technology is a useful tool for environmental clean-up through the five different physiological plant mechanisms: phytoextraction, rhizofiltration, phytostabilization, phytodegradation, and phytovolatilization. It is a moderately innovative and advantageous technology that understands plant-microbe association and uses such healthy microbial relationships in the rhizospheric region of plants, for the removal/degradation or detoxification of hazardous pollutants from the environment. Various hyper-accumulator plants such as *Thlapsi* sp., *Arabidopsis* sp., and

Sedum alfredii sp. have great potential to accumulate high concentrations of poisonous metals/components and translocate them into specific harvestable parts for their maximum reduction (Kabeer et al. 2014). The use of hyper-accumulator plants offers a cost-effective and frequently abundant way because of their solar-driven mechanism (Rai 2012). Phytoremediation notably presents substantial assistance for a pollution-free, green environment.

4.5.3 Engineered Wetlands Systems

The constructed wetlands (CWs) are environment-friendly, man-engineered systems that are planned to remove highly toxic pollutants from IWW in an eco-friendly manner. CWs are used for the treatment of IWW to recover the quality of wastewater by using natural biogeochemical processing of soil microorganisms. In CWs, plant-associated micro-organisms play a crucial role in the detoxification and degradation of poisonous pollutants into less poisonous substances (Kabra et al. 2012; Oliveira 2012). It is an aesthetically pleasant and cost-effective approach to remediating a diverse array of waste products from domestic wastewater and IWW (Vyzmazal 2011). Wetlands are rich in a rhizospheric microbial community that plays a major role in the removal of both nutrients and non-essential contaminates from IWW (Bai et al. 2014). Earlier studies have reported that CWs are effective in the abatement of heavy metal pollution. Aquatic macrophytes such as *Typha, Phragmitis, Eichornia, Azolla* and many others are considered for potent wetlands as they have the potential for heavy metal detoxification or removal (Rai 2008).

5 Merits and Demerits of Conventional and Emerging Treatment Technology

Every treatment technology has having some limitation regarding its usage for wastewater management. Hence, the merits and demerits of conventional and emerging treatment methodologies are presented in Table 1.2.

6 Policies and Legal Frame Work Criteria

- Strict legal policies and standard guidelines should be developed by environmental protection agencies such as USEPA and the WHO for the protection and control of variable point sources for their direct discharge into water bodies.
- Standards and recommended permissible values of chemicals and contaminants play a fundamental role in the characterization of wastewater, for example, the National Pollutant Discharge Elimination System (NPDES) permit programme

1 Conventional Methods for the Removal of Industrial Pollutants, Their Merits…

Table 1.2 Merits and demerits of various wastewater treatment technologies

Wastewater treatment technology			Merits	Demerits
Physical treatment process	Adsorption		Low-cost	Low selectivity
			Easy operating conditions	Production of waste products
			Wide pH range	
			High metal binding capacities	
	Electro dialysis		High separation selectivity	High operational
				Membrane fouling
			Useful for recovery and rejection of undesirable pollutants from water	High energy consumption
	Ion exchange		Regeneration with low loss of adsorbents	Not applicable for concentrated metal solution
				Easily fouled by organics and other solids
	Sedimentation		Effectively used for the destruction of toxic waste and non-biodegradable effluents	Only settable solids, such as sands, silts and larger microbes settle efficiently
Chemical treatment process	Coagulation/flocculation		Short detention time	High cost of chemicals for pH adjustment
			Low capital cost	
			Good removal efficiencies	Dewatering and sludge handling problems
	Precipitation/filtration		Low capital cost	High sludge generation
			Easy handling	Extra operation cost for sludge disposal
			Simple operation	
	Oxidation		Low temperature requirement	High cost
			No additional chemical requirement	
			End products are non-hazardous	

(continued)

Table 1.2 (continued)

Wastewater treatment technology			Merits	Demerits
Biological treatment processes	Aerobic	Activated sludge process	Highly efficient treatment method	High cost
			Requires little land area	Requires sludge disposal area (sludge is usually land-spread)
			Applicable to small communities for local-scale treatment and to big cities for regional-scale treatment	Requires technically skilled manpower for operation and maintenance
		Trickling filter	Minimal land requirements; can be used for household-scale treatment	Requires mechanical devices
			Relatively low cost and easy to operate	
		Stabilization lagoons	Low capital cost	Requires a large area of land
			Low operation and maintenance costs	May produce undesirable odours
			Low technical manpower requirement	
	Anaerobic	Anaerobic contact process	Biological mediated process	Energy required
			Recovery of non-conventional energy	Sludge disposal is minimal
			Methane recovery	
			Capable of being simulated for treating wastes emanating from municipal, agricultural, and industrial activities	
		Anaerobic baffled reactor	High reduction of biochemical oxygen demand	Long start-up phase
			Low sludge production	Low reduction of pathogens and nutrients
			No electrical energy is required	
			Simple to operate	Effluent and sludge require further treatment

1 Conventional Methods for the Removal of Industrial Pollutants, Their Merits...

Emerging treatment process	Advanced oxidation process	High reactivity	Interferes with the chemical oxygen demand
		Low oxidation selectivity	Reduces the reaction kinetics
	Photo-catalysis	The photo-catalytic process can be utilized as the final step of purification of pre-treated wastewaters, because it is the most effective solution with a small amount of pollutants	Long duration time
			Limited applications
	Membrane filtration	Useful for water recycling	Membrane fouling
			Inefficient in reducing the dissolved solid content
	Enzymatic treatments	Ecologically sustainable technique owing to its biological origin	Prohibitively high costs
		Effective at a low pollutant level	Enzyme inactivation
Green eco-friendly treatment processes	Bioremediation	Useful for the transformation and reduction of toxic and hazardous compounds into less toxic products	Limited application to only biodegradable compounds
		Eco-friendly and environmentally safe	Slow processing rate
	Phytoremediation	Eco-friendly traditional method	Slow processing of environmental clean-up
		Low cost	Affected by climatic change
		Requires no specialized equipment	
		Useful for the reclamation of contaminated environment	
	Constructed wetlands	Cost-effective technology	Installation cost is quite high
		Applicable with huge volumes of wastewater	Requires expertise
			Management becomes difficult during monsoon

Source: Chowdhary et al. (2017a, b, c), Barakat (2011), and Anjaneyulu et al. (2005)

in 1972 to control by regulating point sources that discharge pollutant into waters.
- The 74th Constitution Amendment Act 1992 provides a framework and devolves upon the urban local bodies for providing water supply and sanitation facilities in urban areas.
- Water Prevention and Control of Pollution Rules, 1975.
- National Environment Policy, 2006.
- Hazardous Waste (Management and Handling) Rules, 1989.

7 Conclusion

Increased industrialisation, fast growth and development have generated an enormous volume of wastewater containing high concentrations of numerous toxic compounds, directly discharged into fresh aquatic bodies (rivers, ponds) creating severe problems for the environment. The last few decades' emphasis on wastewater management and treatment methodologies is also responsible for the production of poisonous and toxic remains. Therefore, it is necessary to ensure the safety, suitability and working efficiency of an appropriate treatment technology that has been used for wastewater to remediate the complex toxic compounds into less toxic or simpler compounds to improve water quality and reduce pollutants. The vital aim of these treatment technologies is the protection of the environment, public health and socio-economic concerns in an eco-friendly manner, to conserve natural water resources used for the supply of drinking water.

Acknowledgment The authors are extremely grateful to the University Grant Commission (UGC), Government of India, New Delhi for providing fellowship.

References

Abbasi HN, Lu X, Xu F (2016) Seasonal performance and characteristics of ABR for low strength wastewater. Appl Ecol Environ Res 15(1):263–273

Adam W, Lazarus M, Saha-Mollera C et al (1999) Biotransformation with peroxidase. Adv Biochem Eng Biotechnol 63:73–107

Agarwal CS, Pandey GS (1994) Soil pollution by spent wash discharge: depletion of manganese (II) and impairment of its oxidation. J Environ Biol 15:49–53

Ahmad R (2009) Studies on adsorption of crystal violet dye from aqueous solution onto coniferous pinus bark powder (CPCB). J Hazard Mater 171:763–773

Alfredo C, Leondina DP, Enrico D (2007) Ind Eng Chem Res 46:6825

Alnaizy R, Akgerman U (2000) Advanced oxidation of phenolic compounds. Adv Environ Res 4:233–244

Ananpattarachai J, Kumket P (2015) Chromium (VI) removal using nano-TiO2/chitosan film in photocatalytic system. Int J Environ Waste Manag 16(1):55

Anjaneyulu Y, SreedharaChryz N, Samuel Suman Raj D (2005) Decolourization of industrial effluents-available methods and emerging technologies – a review. Rev Environ Sci Bio/Technol. https://doi.org/10.1007/s11157-005-1246-z

Ayaz SC, Akca L, Aktas O, Findik N, Ozturk I (2012) Pilot scale anaerobic treatment of domestic wastewater in upflow anaerobic sludge bed and anaerobic baffled reactors at ambient temperatures. Desalination Wastewater 46(1–3):60–67

Bai Y, Liang J, Liu R, Hu C, Qu J (2014) Metagenomic analysis reveals microbial diversity and function in the rhizosphere soil of a constructed wetland. Environ Technol 35(20):2521–2527

Barakat MA (2011) New trends in removing heavy metals from industrial wastewater. Arab J Chem 4:361–377

Barnhardt R (2006) Cytochromes P450 as versatile biocatalyst. J Biotechnol:124–128

Basu S, Mukherjee S, Kushik A, Batra VS (2015) Integrated treatment of molasses distillery wastewater using microfiltration. J Environ Manag 158:55–60

Beekeepers Y (2000) Arising from reactive dyes in textile industry colour Fenton process remedy with ITU institute of science, M.Sc. Istanbul

Bharagava RN, Chandra R (2010a) Biodegradation of the major color containing compounds in distillery wastewater by an aerobic bacterial culture and characterization of their metabolites. Biodegradation 21:703–711

Bharagava RN, Chandra R (2010b) Effect of bacteria treated and untreated post-methanated distillery effluent (PMDE) on seed germination, seedling growth and amylase activity in *Phaseolus mungo* L. J Hazard Mater 180:730–734

Bharagava RN, Mishra S (2017) Hexavalent chromium reduction potential of *Cellulosimicrobium sp.* isolated from common effluent treatment industries. Ecotoxicol Environ Saf 147:102–109

Bharagava RN, Chowdhary P, Saxena G (2017) Bioremediation an eco-sustainable green technology, its applications and limitations. In: Bharagava RN (ed) Environmental pollutants and their bioremediation approaches. CRC Press, Taylor & Francis Group, Boca Raton, pp 1–22

Bianco B, Michelis DI, Veglio F (2011) Fenton treatment of complex industrial wastewater: optimization of process condition by surface response method. J Hazard Mater 186:1733–1738

Billings RM, Dehaas GG (1971) Pollution control in the pulp and paper industry. In: Lund HF (ed) Industrial pollution control handbook. McGraw-Hill, New York, pp 18–28

Biswas MM (2004) Removal of reactive azo dyes from water by Feo reduction followed by peroxidase catalysed polymerization. University of Windsor, Windsor

Bohdziewicz J, Sroka E (2005) Treatment of wastewater from the meat industry applying integrated membrane systems. Process Biochem 40:1339–1346

Bose P, Bose MA, Kumar S (2002) Critical evaluation of treatment strategies involving adsorption and chelation for wastewater containing copper, zinc, and cyanide. Adv Environ Res 7:179–195

Chandra R, Chowdhary P (2015) Properties of bacterial laccases and their application in bioremediation of industrial wastes. Environ Sci: Process Impacts 17:326–342

Chang DI, Chook H, Jung et al (2009) Foulant identification and fouling control with iron oxide adsorption in electro dialysis for the desalination of secondary effluent. Desalination 236(1–3):152–159

Chaney RL, Malik M, Li YM, Brown SL, Brewer EP, Scott Angle J, Baker AJM (1997) Phytoremediation of soil metals. Curr Opin Biotechnol 8(3):279–284

Chavan RB (2001) Indian textile industry- environmental issue. Indian J Fibre Text Res 26:11–21

Chen GH (2004) Electrochemicals technologies in wastewater treatment. Sep Purif Technol 38(1):11–41

Chen W, Westerhoff P, Leenheer JA, Booksh K (2003) Fluorescence excitation-emission matrix regional integration to quantify spectra for dissolved organic matter. Environ Sci Technol 37:5701–5710

Cheremisinoff NP, Rosenfeld PE (1998) The best practices in the wood and paper industries. Elsevier, Burlington

Chowdhary P, Yadav A, Kaithwas G, Bharagava RN (2017a) Distillery wastewater: a major source of environmental pollution and its biological treatment for environmental safety. In: Singh R, Kumar S (eds) Green technologies and environmental sustainability. Springer, Cham, pp 409–435

Chowdhary P, More N, Raj A, Bharagava RN (2017b) Characterization and identification of bacterial pathogens from treated tannery wastewater. Microbiol Res Int 5(3):30–36

Chowdhary P, Raj A, Bharagava RN (2017c) Environmental pollution and health hazards from distillery wastewater and treatment approaches to combat the environmental threats: a review. Chemosphere 194:229–246

Chowdhury M, Mostafa MG, Biswas TK et al (2013) Treatment of leather industries effluents by filtration and coagulation process. Water Res Ind 3:11–22

Chowdhury M, Mostafa MG, Biswas TK et al (2015) Characterization of the effluents from leather processing industries. Environ Process 2:173–187

Cohen S, Belinky P, Hadar Y, Dosoretz C (2009) Characterization of catechol derivative removal by lignin peroxidase in aqueous mixture. Bioresour Technol 100(7):2247–2253

Covinch LG, Bengoechea DI, Fenoglio RJ, Area MC (2014) Advanced oxidation process for wastewater in pulp and paper industry: a review. Am J Environ Eng 4(3):56–70

CPCB Central Pollution Control Board (2011) Central zonal office Bhopal, A report on assessment of grain based fermentation technology. Waste treatment options disposal of treated effluents

CPCB Central Pollution Control Board (2013) Pollution assessment: river Ganga. Status of grossly polluting industries (GPI)

Dey S, Islam A (2015) A review on textile wastewater characterization in Bangladesh. Resour Environ 3(1):15–44

Diez MC (2010) Biological aspects involved in the degradation of organic pollutants. J Soil Sci Plant Nutr 10(3):244–267

Dixit S, Yadav A, Dwivedi PD, Das M (2015) Toxic hazard of leather industry and technologies to combat threat: a review. J Clean Prod 87:39–49

Dsikowitzky L, Schwarzbauer (2013) Organic contaminants from industrial wastewaters: identification, toxicity and fate in the environment. In: Lichtfouse E, Schwarzbauer J, Robert D (e) (eds) Pollutant diseases, remediation and recycling. Springer, Cham, pp 45–101. https://doi.org/10.1007/978-3-319-02387-8Environmental chemistry for a sustainable world

EPA (2002) Office of compliance sector notebook project, profile of pulp and paper industry.2nd edn. U.S. Environmental Protection Agency, Washington, DC

Eskelinen K, Sarkka H, Agustiono T et al (2010) Removal of recalcitrant contaminants from bleaching effluents in pulp and paper mills using ultrasonic irradiation and Fenton like oxidation, electrochemical treatment and/or chemical precipitation: a comprehensive study. Desalination 255(1–3):179–187

ETAD (2003) ETAD information on the 19th amendment of the restriction on the marking and use of certain azo colourant. ETAD- Ecology and toxicology association of dyes and organic pigment manufactures

Feng W, Taylor KE, Biswas N, Bewtra JK (2013) Soybean peroxidase trapped in product precipitate during phenol polymerization retains activity and may be recycled. Chem Technol Biotechnol 88(8):1429–1435

Frank MJW, Westerink JB, Schokker A (2002) Recycling of industrial waste water by using a two-step nanofiltration process for the removal of colour. Desalination 145(1-3):69–74

Gillion RJ (2007) Pesticide in US streams and groundwater. Environ Sci Technol 41(10):3408–3414

Glaze WH, Kang JW (1989) Advanced oxidation process. Description of a kinetic model for the oxidation of hazardous materials in aqueous media with ozone and hydrogen peroxide in a semi batch reactor. Ind Eng Chem Res 28:1573–1580

Gogate PR, Pandit AB (2004) A review of imperative technologies for wastewater treatment I: oxidation technologies at ambient conditions. Adv Environ Res 8(3–4):501–551

Gricic I, Vujevic D, Sepcic J, Koprivanac (2009) Minimization of organic content in simulated industrial a wastewater by Fenton type processes: a case study. J Hazard Mater 170:954–961

Guenther T, Sack U, Hofrichter M, Laetz M (1998) Oxidation PAH and PAG derivatives by fungal and plant oxidoreductases. J Basic Microbiol 38:113–122

Hare V, Chowdhary P, Baghel VS (2017) Influence of bacterial strains on *Oryza sativa* grown under arsenic tainted soil: accumulation and detoxification response. Plant Physiol Biochem 119:93–102

Hassan SR, Zwain HM, Dahlan I (2013) Development of anaerobic reactor for industrial wastewater treatment: an overview, present stage and future prospect. J Adv Sci Res 4(1):07–12

Hassan SR, Zaman NQ, Dahlan I (2015) Effect of organic loading rate on anaerobic digestion: case study on recycled papermill effluent using modified anaerobic baffled (MHAB) reactor. KSCE J Civ Eng 19(5):1271–1276

Helkar PB, Sahoo AK, Patil NJ (2016) Review: food industry by products functional food ingredients department of technology. Int J Waste Resour 6:3

HSRC (2005) Hazardous Substances Research Centre/south and south west outreach program. Environmental hazards of the textile industry. Environmental update #24, Business week

Hurse JT, Connor AM (1999) Nitrogen removal from wastewater treatment lagoons. Water Sci Technol 39(6):191–198

Inamdar S (1991) The distillery industry growth prospect for the 90s. Chem Eng World XXVI:43

Ince BK, Zeynep C, Ince O (2011) Pollution prevention in the pulp and paper industries, environmental management in practice. Broniewicz E (ed), InTech, China

Jadhave JP, Govindwar SP (2006) Biotransformation of malachite green by Saccharomyces cerevisiae. Yeast 23:315–323

Jayathilakan K, Sultan K, Radhakrishnan K, Bawa AS (2012) Utilization of by products and waste material from meat, poultry and fish processing industries: a review. J Food Sci Technol 49(3):278–293

Jenkins D, Richard M, Daigger G (2003) Manual of the causes and control of activated sludge bulking and foaming.2nd edn. Lewis publishers, Boca Roton

Jeyasingh J, Philip L (2005) Bioremediation of chromium contaminated soil: optimization of operating parameters under laboratory conditions. J Hazard Mater 118:113–120

Kabra AN, Khandare RV, Govindwar SP (2012) Development of a bioreactor for remediation of textile effluent dye mixture: a plant-bacterial synergistic strategy. Water Res 47:1036–1048

Kabeer R, Varghese R, Kannan VM, Thomas JR, Poulose SV (2014) Rhizosphere bacterial diversity and heavy metal accumulation in Nymphaea pubescens in aid of phytoremediation potential. J BioSci Biotech 3(1):89–95

Kapoor M, Rajagopal R (2011) Enzymatic bioremediation of organophosphorus insecticides by recombinant organophosphorus hydrolase. Int Biodeter Biodegr 65:896–901

Koyuncu I (2002) Reactive dye removal in dye/salt mixture by nanofiltration membranes containing vinyl sulphone dyes effects of food concentration and cross flow velocity. Desalination 143:243–253

Krishnan S, Rawindran H, Sinnathambi CM, Jim JW (2016) Comparison of various advanced oxidation processes used in remediation of industrial wastewater laden with recalcitrant pollutants. 29th Symposium of Malaysian chemical engineers IOP conference. Series: Material Science And Engineering 206

Kumari V, Sharad K, Izharul H, Yadav A, Singh VK, Ali Z, Raj A (2014) Effect of tannery effluent toxicity on seed germination á-amylase activity and early seeding growth of mung bean (Vigna Radiata) seeds. Int J Lat Res Sci Technol 3(4):165–170

Kumari V, Yadav A, Haq I, Kumar S, Bharagava RN, Singh SK, Raj A (2016) Genotoxicity evaluation of tannery effluent treated with newly isolated hexavalent chromium reducing Bacillus cereus. J Environ Manag 183:204–211

Kurniawan TA, Babel S (2003) A research study on Cr(VI) removal from contaminated wastewater using low cost adsorbents and commercial activated carbon. In: Second international conference on energy and technology towards clean environment (RCETE) VI. 2. Phuket, 2:1110–1117

Lee HJ, Moon SH, Tsai SP (2002) Effects of pulsed electric field on membrane fouling in electro dialysis of NaCl solution containing humate. Sep Purif Technol 27(2):89–95

Liu PWG, Yao YC, Tsai JH et al (2008) Sci Total Environ 398:154

Lofrano G, Meric S, Zengin GE, Orhon D (2013) Chemical and biological treatment technologies for leather tannery chemicals and wastewater: a review. Sci Total Environ 461–462

Mahendra J, Gregory JZ (2000) Anaerobes, industrial uses. In: Flickinger MC, Drew SW (eds) Encyclopedia of bioprocess technology: fermentation, biocatalysis and bioseparation1st edn. Wiley, New York, p 166

Mahimraja S, Bolan NS (2004) Problems and prospects of agricultural use of distillery spent wash in India. In: Super soil, 3rd Australian New Zealand soils conference, 5–9 December 2004. University of Sydney, Australia

Megharaj M, Ramakrishnan B, Venkateswarlu K, Sethunathan N, Naidu R (2011) Bioremediation approaches for organic pollutants: a critical perspective. Environ Int 37(8):1362–1375

Mani S, Bharagava RN (2016) Exposure to crystal violet, its toxic genotoxic and carcinogenic effects on environment and its degradation and detoxification for environmental safety. Rev Contam Toxicol 273:71–104

Marshall FM, Holden J, Ghose et al (2007) Contaminated irrigation water and food safety for the urban and peri urban poor: appropriate measures for monitoring and control from the field research in India and Zambia, Inception report DFID Enkar r8160. SPRU, University of Sussex, Brighton

Matta R, Hanna K, Chiron S (2007) Fenton like oxidation of 2,4,6-trinitrotoluene using different iron minerals. Sci Total Environ 385:242–251

Mirbagherp SA, Hosseini SN (2004) Pilot plant investigation on petrochemical wastewater treatment for the removal of copper and chromium with the objective of reuse. Desalination 171:85–93

Mishra S, Bharagava RN (2016) Toxic and genotoxic effects of hexavalent chromium in environment and its bioremediation strategies. J Environ Sci Health Part C 34(1):1–34

Mittal A, Mittal J, Malviya A et al (2010) Adsorption of hazardous crystal violet from wastewater by waste materials. J Colloid Interface Sci 343:463–473

Mohammad P, Azarmidokht H, Fatollah M, Mahboubeh B (2006) Application of response surface methodology for optimization of important parameters in decolorizing treated distillery wastewater using *Aspergillus fumigatus* U_B60

Mohammadi T, Razmi A, Sadrzadeh M (2004) Effect of operating parameters on Pb^{+2} separation from wastewater using electrodialysis. Desalination 167:379–385

Mohana S, Acharya BK, Madamwar D (2009) Distillery spent wash: treatment technologies and potential applications. J Hazard Mater 163(1):12–25

Moreno-Castilla C (2004) Adsorption of organic molecules from aqueous solutions on carbon materials. Carbon 42:83–94

Murthy PS, Naidu MM (2012) Sustainable management of coffee industry by-products and value addition – a review. Resour Conserv Recycl 66:45–58

Nagajyoti PC, Lee KD, Sreekanth TVM (2010) Heavy metals, occurrence and toxicity for plants. Environ Chem Lett 8:199–216

Nandy T, Kaul SN, Shastry S, Manivel U, Deshpande CV (1999) Wastewater management in cluster of tanneries in Tamil Nadu through implementation of common effluent treatment plants. J Sci Ind Res 58:475–516

Narmadha D, Kavitha SVM (2012) Treatment of domestic wastewater using natural flocculants. Int J Life Sci Biotechnol Pharma Res 1(3):206

Nassef E (2012) Removal of phosphates from industrial wastewater by chemical precipitation. Eng Sci Technol Int J 2(3):409–413

Naumczyk J, Rusiniak M (2005) Physiochemical and chemical purification of tannery wastewaters. Polish J Environ Stud 14(6):789–797

Nelson C, Cox M (2004) Principles of biochemistry.4th edn. W. H. Freeman, New York, pp 47–50

Nelson CR, Hites RA (1980) Aromatic amines in and near the Buffalo river. Environ Sci Technol 14:1147–1149

Nicell JA (2003) Enzymatic treatment of waters and wastes. In: Tarr MA (ed) Chemical degradation methods for wastes and pollutants: environmental and industrial applications. CRC Press, Boca Raton, pp 384–428

Schweigert N, Zehnder AJB, Eggen RIL (2001) Chemical properties of catechols and their molecular modes of toxic action in cells, from microorganisms to mammals. Minireview. Environ Microbiol 3(2):81–91

Okonko IO, Shittu OB (2007) Bioremediation of wastewater and municipal water treatment using latex exudate from *Calotropis procera* (Sodium apple). Elec J Env Agricult Food Chem 6(3):1890–1904

Oliveira H (2012) Chromium as an environmental pollutant: insights on induced plant toxicity. J Bot 2012:1–8

Omelia C (1998) Coagulation and sedimentation in lakes, reservoirs and water treatment plants. Water Sci Technol 37(2):129

Oztekin E, Atlin S (2016) Wastewater treatment by electro dialysis system and fouling problems. Online J Sci Technol 6(1):91–99

Pant D, Adholeya A (2007) Biological approaches for treatment of distillery wastewater: a review. Bioresour Technol 98:2321–2334

Park JW, Park BK, Kim JE (2006) Remediation of soil contaminated with 2,4-dicholrophenol by treatment of minced shepherd's purse roots. Arch Environ Contam Toxicol 50(2):191–195

Parshetti GK, Parshetti SG, Telke AA et al (2011) Biodegradation of crystal violet by *Agrobacterium radiobacter*. J Environ Sci 23:1384–1393

Pathak N, Hagare P, Guo W, Ngo HH (2015) Australian food processing industry and environmental aspect– a review. International conference on biological civil and engineering (BCEE-2015) Feb, 3-4. Bali, Indonesia

Pathak N, Hagare P, Guo W, Ngo HH (2015) Australian food processing industry and environmental aspect- a review. international conference on biological civil and engineering (BCEE-2015) Feb, 3-4. Bali, Indonesia

Pedersen AJ (2003) Characterization and electrolytic treatment of wood combustion fly ash for the removal of cadmium. Biomass Bioenergy 25(4):447–458

Porro SC, Martin S, Mellado et al (2003) Diversity of moderately halophilic bacteria producing extracellular hydrolytic enzymes. J Appl Microbiol 94(2):295–300

Purushotham D, Narsing Rao A, Ravi Prakash M, Shakeel A, Ashok Babu G (2011) Environmental impact on groundwater of Maheshwaram Watershed, Ranga Reddy district, Andhra Pradesh. J Geolo Soc India 77(6):539–548

Rai PK (2008) Heavy-metal pollution in aquatic ecosystem and its phytoremediation using wetland plants: an eco-sustainable approach. Int J Phyto 10(2):133–160

Rai PK (2012) An eco-sustainable green approach for heavy metals management: two case studies of developing industrial region. Environ Monit Assess 184(1):421–448

Rajasulochana P, Preethy V (2016) Comparison on efficiency of various techniques in treatment of waste and sewage water- a comprehensive review. Resour Efficient Technol 2:175–184

Rameshraja D, Suresh S (2011) Treatment of tannery wastewater by various oxidation and combined processes. Int J Environ Res 5(2):349–360

Ramirez N, Marce RM, Borrull F (2011) Determination of volatile organic compounds in industrial wastewater plant air emission by multi-sorbent adsorption and thermal desorption-gas chromatography-mass spectrometry. Int J Environ Anal Chem 91(10):911–928

Rao AG, Prashad K, Naidu V, Rao C, Sharma PN (2003) Removal of sulphide in integrated anaerobic-aerobic wastewater treatment system. Clean Tech Environ Policy 6:66–71

Rao MA, Scelza R, Acevedo F, Diez MC, Gianfreda L (2014) Enzymes as useful tools for environmental purposes. Chemosphere 107:145–162

Rashed MN (2013) Adsorption technique for the removal of organic pollutants from water and wastewater. Organic pollutants- monitoring, risk and treatment. InTech, New York, pp 167–194

Romo-Hualde A, Penas FJ, Isasi JR et al (2008) Extraction of phenols from aqueous solutions by beta-cyclodextrin polymers. Comparison of sorptive capacities with other sorbents. React Funct Polym 68(1):406–413

Rufus L Chaney, Minnie Malik, Yin M Li, Sally L Brown, Eric P Brewer, J Scott Angle, Alan JM Baker, (1997) Phytoremediation of soil metals. Current Opinion in Biotechnology 8(3):279-284

Russ W, Pittroff RM (2004) Utilizing waste products from the food production and processing industries. Crit Rev Food Sci Nutr 44(2):57–62

Saha NK, Balakrishnan M, Batra VS (2005) Improving industrial water use: case study for an Indian distillery. Resour Conserv Recycl 43:163–174

Santos WDL, Ramosa T, Pozyak I, Chairez I, Cordova R (2009) Remediation of lignin and its derivatives from pulp and paper industry wastewater by the combination of chemical precipitation and ozonation. J Hazard Mater 169:428–434

Savant DV, Abdul-Rahman R, Ranade DR (2006) Anaerobic degradation of adsorbable organic halides (AOX) from pulp and paper industry wastewater. Bioresour Technol 97:1092–1104

Senthilkumar S, Kalaamani P, Subburaam CV (2006) Liquid phase adsorption of crystal violet onto activated carbon derives from male flowers of coconut tree. J Hazard Mater 136:800–808

Shivajirao AP (2012) Treatment of distillery wastewater using membrane technologies. Int J Adv Res Stud 1(3):275–283

Singh C, Chowdhary P, Singh JS, Chandra R (2016) Pulp and paper mill wastewater and coliform as health hazards: a review. Microbiol Res Int 4(3):28–39

Skubal LR, Meshkov NK, Rajh T, Thurnauer (2002) Cadmium removal from water using thiolactic acid-modified titanium dioxide nanoparticles. Photochem Photobiol A Chem 148:–393, 397

Srinivasam SV, Mary GPS, Kalyanaraman C, Sureshkumar PS, Balakmeswari KS (2012) Combined advanced oxidation and biological treatment of tannery effluent. Clean Techn Environ Policy 14(2):251–256

Sujata M, Bharagava RN (2016) Microbial degradation and decolourization of dyes from textile industry wastewater. In: Bharagava RN, Saxena G (eds) Bioremediation of industrial pollutants. Write and Print Publication, Delhi, pp 307–331

Sumathi S Hung YT (2006) Treatment of pulp and paper industry.2nd edn. Washington, DC

Thacker U, Parikh R, Shouche Y, Madamwar D (2007) Reduction of chromate by cell-free extract of *Brucella* sp. isolated from Cr(VI) contaminated sites. Bioresour Technol 98:1541–1547

Toumela M, Hatakka A (2011) Oxidative fungal enzymes for bioremediation. In: Agathos A, Moo-Young M (eds) Comprehensive biotechnology: environmental biotechnology and safety, 2 6:133–196, Amsterdam: Elsevier

USEPA United States Environmental Protection Agency (1990) Office of air quality planning cancer risk from outdoor exposure to air toxics EPA-450/1-90-004a, Washington, DC

USEPA United States Environmental Protection Agency (2004) Primer for municipal wastewater treatment systems. Document no. EPA832-R-04-001, Washington, DC

Van der Bruggen B, Vandecasteele C, Gastel TV et al (2003) A review of pressure- driven membrane processes in wastewater treatment and drinking water production. Environ Prog 22(1):46–56

Venkatesan G, Swaminathan G (2009) Review of chloride and sulphate attenuation in ground water nearby solid-waste landfill sites. J Environ Landsc Manag 17(1):1–7

Vyzmazal J (2011) Plants used in constructed wetlands with horizontal subsurface flow: a review. Hydrobiologia 674:133–156

Water Saving Factsheet (2014) Meat and meat product manufacturing. www.aigroup.com.au

Weiland P, Rozzi A (1991) The start-up operation and monitoring of high-rate anaerobic treatment systems: discussion report. Water Sci Technol 24(8):257–277

WHO World Health Organization (2000) Air quality guidelines for Europe.2nd edn. WHO Regional Publications. European series no. 29, Copenhagen

Worch E (2012) Adsorption technology in water treatment fundamental processing and modelling. Walter de Gruyter& Co. KG, Berlin, pp 1–345

Yadav BR, Garg A (2011) Treatment of pulp and paper mill effluent using physicochemical processes. IPPTA J 23:155–160

Yadav A, Mishra S, Kaithwas G, Raj A, Bharagava RN (2016a) Organic pollutants and pathogenic bacteria in tannery wastewater and their removal strategies. In: Singh JS, Singh DP (eds) Microbes and environmental management. Studium Press (India) Pvt. Ltd, New Delhi, pp 101–127

Yadav A, Raj A, Bharagava RN (2016b) Detection and characterization of a multidrug and multimetal resistant Enterobacterium *Pantoea sp.* from tannery wastewater after secondary treatment process. Int J Plant Environ 1(2):37–41

Yadav A, Chowdhary P, Kaithwas G, Bharagava RN (2017) Toxic metals in environment, threats on ecosystem and bioremediation approaches. In: Das S, Dash HR (eds) Handbook of metal-microbe interactions and bioremediation. CRC Press, Taylor & Francis Group, Boca Raton, p 813

Young JC, Dahab MF (1982) Effect of media design on the performance of fixed bed anaerobic reactors. Water Sci Technol 15:369–368

Zainith S, Sandhya S, Saxena G, Bharagava RN (2016) Microbes an ecofriendly tool for the treatment of industrial wastewater. In: Singh JS, Singh DP (eds) Microbes and environmental management. Studium Press (India) Pvt. Ltd, New Delhi, pp 78–103

Zollinger H (1987). VCH publishers, New York, 92–100

Zhou H, Smith DW (2002) Advanced technologies in water and wastewater treatment. J Environ Eng Sci 1(4):247–264

Chapter 2
Toxicological Aspects of Emerging Contaminants

Miraji Hossein

Abstract The emerging contaminants field suffers from insufficient global information, especially in Africa, and even less from Asia. Emerging contaminants constitute a group of natural, synthesised chemicals and microorganisms that have been proved to cause serious effects on the laboratory organisms and therefore a threat to human and aquatic species. They are presumed to be endocrine disruptors, carcinogenic, teratogenic, and interfere with the sexual and reproductive behaviour of some small aquatics. This being the case, the discovery of emerging contaminants stands to be the finest toxicological investigation owing to improved science and technology. Nevertheless, in future, the world will come up with new environmental discoveries because of the continuous production of materials, which by themselves may be harmful, or their life cycle may be a threat.

1 Introduction

Emerging contaminants (ECs) are the potentially global, dynamic type of contaminants, representing unregulated chemical and microbial contaminants, which are anticipated to occur in the public water system, but lack public monitoring regulations. The circumstantial risks are uneven, vague and the advancement of science and technology extends the production of new contaminants that are expected to create more ecological risks. The limited traditional ecotoxicology approaches to environmental contaminants were not able to depict the occurrence of ECs until the discovery of advanced technologies, such as gas chromatography. These contaminants are not termed as emerging because they are new in terms of discovery, or the use of new advanced detection and treatment methods, but because they had been previously unrecognised, and there is a lack of standards and guidelines for their environmental monitoring. Recently, they have gained scientific attention because of their effects on health (Nosek et al. 2014).

M. Hossein (✉)
Chemistry Department, School of Physical Sciences, College of Natural Sciences, University of Dodoma, Dodoma, Tanzania

© Springer Nature Singapore Pte Ltd. 2019
R. N. Bharagava, P. Chowdhary (eds.), *Emerging and Eco-Friendly Approaches for Waste Management*, https://doi.org/10.1007/978-981-10-8669-4_2

Table 2.1 Emerging contaminants

Paracetamol in µg/L	0.211 (USA), 10 (USA), 10.19 (Spain), (Richardson 2007; Kolpin et al. 2002; Rivera-Utrilla et al. 2013)
Metronidazole in µg/L	0.176 (Spain), 0.9 (Switzerland) (Urtiaga et al. 2013; Qi et al. 2015)
Amoxicillin in µg/L	0.12 (Spain), 2.69–31.71 (Tanzania), (Rivera-Utrilla et al. 2013; KASEVA et al. 2008)
Artemether/ lumefantrine	No immediate report as this medication is mainly used in Tanzania
Estrone in µg/L	0.0004 (USA), 0.0001–0.00157 (France), 0–0.67 (Ecuador), 0.0001–0.017 (USA) (Lee et al. 2014; Wu et al. 2010; Voloshenko-Rossin et al. 2015; Richardson 2007)
Estriol in µg/L	0.0004 (USA), 0.0049–0.0121 (France), 0.005 (USA), 0.005 (USA), (Lee et al. 2014; Wu et al. 2010; Kolpin et al. 2002; Bradley and Journey 2014)
Sunscreen µg/L	19, 125 ng/L of benzophenone-3 in Swiss, 10 ng/L of sunscreen in Greece (Richardson and Ternes 2011), between 5 and 5154 ng/L from Spain (Urtiaga et al. 2013), 0.03 µg/L of triclosan from Zambia (Sorensen et al. 2014), 0.43 µg/L of triclosan from Poland (Nosek et al. 2014)
Artificial sweeteners	47–1640 ng/L, 610–3200 ng/L in Germany and 2800–6800 ng/L in Swiss (Tran et al. 2014). About 12.9 µg/L of sucralose was reported from EU countries (Comero et al. 2013)
Perfluorinated surfactants	0.08–0.87 ng/g (ww) was reported in French, 0.04–0.08 in Asia and Brazil, 2.08 ng/g in Korea and 0.11–0.13 in China and Japan (Munschy et al. 2013). Fawell and Ong 2012 reported a concentration range of 0.12–0.92 ng/L in China (Fawell and Ong 2012), 1290–2440 µg/L from human blood in USA, 950–1930 µg/L in Belgium, 7–40 ng/L in German and 31 µg/L in Norwegian serum (Sturm and Ahrens 2010)
Microcystin/L	50–75 µg microcystin/L in USA intoxicate aquatic organisms

The public occurrence of ECs has being ongoing since the creation of the universe because of inorganic ECs such as cobalt, germanium and manganese. Nonetheless, reports on the scientific discovery date back to 1735. Again, organic ECs such as industrial solvents, pharmaceuticals, pesticides, water disinfection by-products and perfluorinated surfactants were discovered after World War II. The first comprehensive research and reports of the environmental occurrence of ECs led by the US government under the environmental protection agency (US EPA) commenced in the 1970s (Snyder 2014). Significant amounts of ECs were reported worldwide, as summarised here under (Table 2.1).

The discovery of cobalt in 1735 for glasses and ceramics, germanium in 1886 for transistors, ziram in the 1940s as a pesticide, the first use of contraceptive pills in 1960 as an endocrine-disrupting hormone and the discovery of nanomaterials around the 1980s mark the potential history of the rise of ECs. However, in 1998, the first contaminant candidate list (CCL 1) was published by the US EPA with 10 microbes and 50 chemical contaminants. The next 2005 CCL 2 contained the same contaminants as CCL 1, but with integrated stakeholders' recommendations. The CCL 3 of 2008 included 104 chemicals and 12 microbial contaminants (EPA 2009), whereas the latest CCL 3 of November 2016 contains about 94 chemicals and 12 microbial contaminants (EPA 2015).

The EC field suffers from insufficient global information, especially in Africa, and even less on Asia, the EU and later the US (Miraji et al. 2016). As for the USA, the EPA preferably uses the term "candidate" to express the essence of periodic (5 years) review of ECs based on their public health concern and requirements for regulatory decisions (US EPA 2016a). Apart from the scarcity of information some effects of ECs worth mentioning include hormonal interference in fishes, genotoxicity, carcinogenicity in laboratory animals, endocrine disruption and immune toxicity (Mortensen et al. 2014).

2 Sources of Emerging Contaminants

Subsequently, all sources of ECs, including military sites, mine tailings, agricultural fields, industrial units, waste treatment plants, pharmaceutical, clinical, construction and demolition waste, electronic waste and municipal wastewaters prevail in our environment; therefore, the public addressing of ECs is inevitable. Unforeseen sources of ECs are include landfills, accidental and intended spillage, improper disposal of cosmetics and waste drugs and swimming pools. To increase general ecological health and safety, and preventing future emergencies, the analytical world needs to develop analytical and clinical techniques for the identification of ECs in all matrices. Furthermore, by identifying the environmental occurrence, establishment, understanding of current and presumed future sources, transportation pathway and ecological risks upon exposure to ECs are needed. Later, we need to address eco-friendly techniques for the remediation of ECs in all environmental matrices.

In the traditional way, the natural and previously mentioned sources of ECs have lasted for decades to the point that they are common sources. In the modern way of classifying sources of ECs, the previously mentioned sources are termed as primary sources. Continuous release of ECs into the aquatic systems saturates absorbing sediments in the streams and receivers. Yet, the reversed process leads to the release of ECs from saturated sediments, thus becoming a secondary source, also referred as a sink (USDHHS 2015).

3 General Characteristics of Emerging Contaminants

Emerging contaminants (ECs) constitute a classical group of contaminants that have divergent chemistries, but share occurrences and lack primary regulations. The chemistry of ECs is characterised by slightly higher polarity, acidity, and alkalinity than natural environmental chemicals, thus making them unique. They are characterised by low levels, persistence, multiple sources, disease outbreaks, lack of standard biological tests, and entering the aquatic environment from the point and non-point sources. The organic nature of most ECs makes them hydrophobic; hence, they accumulate in the lipid-rich tissues (Ross and Ellis 2004). Their transport behavior and fate are not clear and policies to address the challenges are yet to be fixed. The previous and later arguments make ECs versatile to transportation through the ecological and food chains. In the presence of unpredicted human and natural changes in addition to yet to be discovered technologies such as zero emissions, environmental discoveries are expected. In such an unpredictable future, the next era of environmental pollutants resulting from the reactions among the existing and yet to be identified materials is to come soon or later. Conversely, once released into the environment, they undergo bioaccumulation, bioconcentration, persistence and wide range transport in the aquatic environment (Clarke and Cummins 2014). Despite the fact that there are no fully established national and international monitoring regulations, their suspected availability brings the potential for unclear health and ecological risks (Petrisor 2004; Richardson 2007; Lee et al. 2014; Qi et al. 2015).

4 Mobility of Emerging Contaminants

A typical non-polar compound such as some of the organic-type ECs is less reactive in aqueous media because of their equal sharing of electrons between atoms and its symmetrical arrangement of the polar bonds, making them hydrophobic (Stasinakis et al. 2013). Unless there are changes in the thermodynamic stability of a stable species, they will persist in the environment. Persistence is considered when the half-life of a chemical is more than 2 days in the air, 182 days in the water, 365 days in the sediments and 182 days in the soil (Government of Canada 1995). The persistence is associated with long periods of environmental existence, long-range transported bioaccumulation and thus adverse effects on the ecosystem (Bergman 2005).

The solubility facet of ECs, especially organic ones, is merely negligible; therefore, it plays a peculiar role in the persistence of ECs. Once in the aquatic system, most ECs remain suspended while moving with flowing water or remain still in the stagnant water. Surface adsorption on the rough sands, sediments, gravels, underlying bedrocks and settling at the bottom of stagnant water bodies are common scenarios. Surface adsorption on the aquatic and irrigated crops and absorption through direct feeding or through the skin of aquatic organisms are expected too. Human consumption of contaminated water through bathing, washing, and other anthropogenic needs, eating contaminated crops, fishes and animals results in accumulation in fatty tissues. Primary and secondary periodic releases, evaporations, sewage sludge and leaching re-concentrate the same contaminants as they circulate in the aquatic organisms, water, terrestrial organisms and soils, and then back to water (Clarke and Cummins 2014).

5 Ecological Risks of Emerging Contaminants

Simple scientific facts on the essence of ECs in the aquatic system may involve antimalarial resistance to malaria parasites. Their DNA may have been altered by the presence of the type of drugs residual in the reproduction site. Likewise, continuous release of endocrine-disrupting hormones may alter the sexual behavior of aquatic organisms as the results interfere with breeding. Algal blooming resulted from nutrient pollution in the stagnant water releases algal toxins as an excretory waste, and when algae die their cells break down to release the same toxicant. The existing studies proved health effects in laboratory animals, meaning that short- or long-term exposure is a threat to humans too (Commettee Report 1999; Richardson et al. 2006; Guidotti 2009; Hansen 2007; Gothwal and Shashidhar 2014).

Conventional water treatment methods were considered effective to the point of releasing its treated water into rivers and public consumption before the discovery of ECs. In fact, there was no blame as the monitoring regulations accounted for only known contaminants. In such a situation, collected sludge was an effective source of farming nutrients. Unfortunately, the agricultural application of biosolids from wastewater treatment sludge causes the accumulation of ECs in the soil and then subsequently translocation into the food chain (Clarke and Cummins 2014). This state of affairs is similar to public treated drinking water, which was chlorinated, without knowing the presence of natural organic acids such as humic acids could result in the formation of harmful disinfestation by-products.

Once toxicants are in the human body, they undergo biotransformation via enzymatic oxidation, reduction and hydrolysis, followed by synthesis. In these processes, the body merely intends to detoxify, contrary to some resulting products being even more toxic than the original one. Consequently, they attack metabolic enzymes, cause cell membrane damage and uncoupling of oxidative phosphorylation (Bradberry et al. 2000). The extent of toxicity depends on the acute and chronic effects. In either case, the effect may be widespread or locally affect specific organs such as the central nervous system, circulatory system, liver, kidney, lungs, and skin. The biochemical attack varies among the toxicants, for example, the DNA-attacking agents such as benzopyrene and protein-attacking agents such as diclofenac can lead to mutation and/or activate tumor suppressors, which activate tumor followed by cancer. Apart from the above-mentioned effects, the failure of ATP synthesis and functioning is caused by cocaine slow-down or interfere with energy production in the body (Hussain 2013). Some toxicants may inactivate enzymes, attacking the immune system and leading to cellular changes.

6 Classification of Emerging Contaminants

Naturally occurring and synthetic ECs were all designed for a good ecological balance. The presence of guiding and monitoring standards is basically for environmental protection. Products such as prescribed drugs, biogenic hormones, nanomaterials, personal care products and artificial sweeteners are formulated for human, animal and plant consumption (Schultz et al. 2014). Unfortunately, the aquatic systems may contain chemical products such as perfluorinated surfactants, benzotriazoles, perchlorate, sunscreen/UV filters, flame reductants, algal toxins and emerging microbes that have accidentally arrived there.

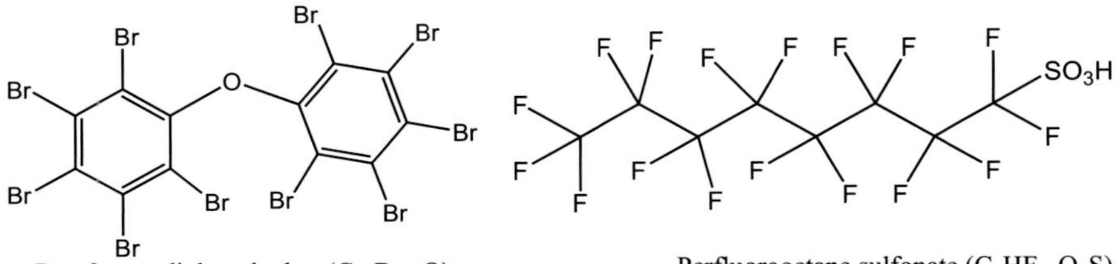

Decabromodiphenyl ether ($C_{12}Br_{10}O$) Perfluorooctane sulfonate ($C_8HF_{17}O_3S$)

Fig. 2.1 Decabromodiphenyl ether and perfluorooctane sulphonate

Once in the environment, the classification of ECs brings challenges, merely because of the various classification approaches. For example, Madhumitha et al. (2013) classified ECS with regard to suspected health effects. A more comprehensive and supported approach is grouping them based on the sources, effects, uses and their chemistry (Richardson and Ternes 2011; Thomaidis 2012; Fawell and Ong 2012; Yang et al. 2014). Similarly applied to the US EPA, ECs are mainly classified into microbial contaminants and chemical contaminants. The latter can be subclassified into inorganic- and organic-based contaminants. The organic forms of ECs are also classified into pharmaceuticals, industrial solvents, food stuffs such as artificial sweeteners etc.

In this context, the following groups of ECs are briefly covered: polybrominated diphenyl ethers (PBDEs), nitrosodimethylamine (NDMA), artificial sweeteners, personal care products, algal toxins, microbes, inorganic elements, illicit drugs, endocrine-disrupting hormones and pharmaceuticals.

6.1 Polybrominated Diphenyl Ethers

Polybrominated diphenyl ethers (PBDEs) are commercially synthesised by bromination of diphenyl ethers resulting in mixtures of brominated diphenyl ethers, with similar chemical structures to polychlorinated biphenyls (Fig. 2.1) (Siddiqi and Clinic 2003). The environmental toxicological effects of PBDEs include strong binding to the soil, dust and sediments; yet, they also reside in the aquatic environment (Bennett et al. 2015). They enter living organisms through dermal contact, inhalation, and ingestion and basically through the food chain gets, and are then deposited in the fatty tissues. They are likely carcinogenic to humans, as they have shown positive results in rats and mice.

Other potential adverse effects of PBDEs include endocrine/thyroid disruption, teratogenicity, weight-gain, decrease in glucose oxidation, interference with gene expression in a metabolic pathway, and other prenatally associated effects (Miraji et al. 2016; Chevrier et al. 2010; Vuong et al. 2016; US EPA 2014b). The use of PBDEs was banned and withdrawn from the market in Europe in 2003, followed by the USA in 2004 (USDHHS 2015).

6.2 Nitrosodimethylamine

Nitrosodimethylamine (NDMA) was at first synthesised and patented by (Richard and George 1957) through the reaction of dimethylamine and aqueous sodium nitrate in the presence of hydrochloric or sulfuric acid. NDMA is also produced as a by-product in wastewater chlorination and chlorination. NDMA is also industrially produced unintentionally in tanneries, pesticide, and rubber production through the reaction of alkylamines with nitrite salts and nitrous and nitrogen oxide (US EPA 2014a).

Apparently, NDMA is produced for research purposes; however, its natural production during industrial processes, metabolism and during water treatment contributes to environmental contamination. It is rarely found in the soil as it soon leaches into groundwater. Thus, its miscibility in water and nearly equal densities with water makes water the perfect residing medium. When exposed in the sunlight and/or biological processes, NDMA degrades with an estimated half-life of 16 min. Although NDMA is not persistent, it has been widely found in water sources (Richardson et al. 2006; Pal et al. 2014; Koumaki et al. 2015). It is through the food chain, direct food stuffs, drinking water, and whiskey, cosmetics such as shampoos and workplace exposure enable NDMA to get into the human body. Through studies in mice and rats, in which NDMA led to tumors in the liver, kidney, and lungs, NDMA is categorically considered carcinogenic, mutagenic and clastogenic to humans. That being the case, 0.7 ng/L concentration has been set as a cancer risk level for direct human consumption (Liteplo et al. 2002; US EPA 2014a; Ooka et al. 2016).

6.3 Artificial Sweeteners

Sucralose is a non-calorific artificial sweetener with a sweetness about 600 times that of sucrose, aspartame and saccharin (Fig. 2.2). It was discovered in 1976, synthesised in the laboratory by the conversion of sucrose, a refined form of sucrose, to sucrose-6 acetate, followed by selective chlorination and then deacetylation (Luo et al. 2008; Yu and Cn 2011; Wikipedia 2016a). The presence of three chlorine atoms, which replaced hydroxyl ions, make it indigestible in humans, mice, dogs and rats and is hence non-calorific.

Artificial sweeteners such as sucralose are widely consumed by diabetics. As they affect glucose regulation, they are presumed and associated with the development of metabolic imbalances that lead to obesity, cardiovascular disease, lymphomas, bladder and brain cancer, leukaemia and fatigue syndrome (Suez et al. 2014; Boullata and Mccauley 2008). They are also associated with premature birth and/or health risks to the mother or child, though these are yet to be confirmed (French Agency for Food 2015). Artificial sweeteners have proved to incur weight gain in cows and pigs. Some studies show no chronic or acute effects when some organisms

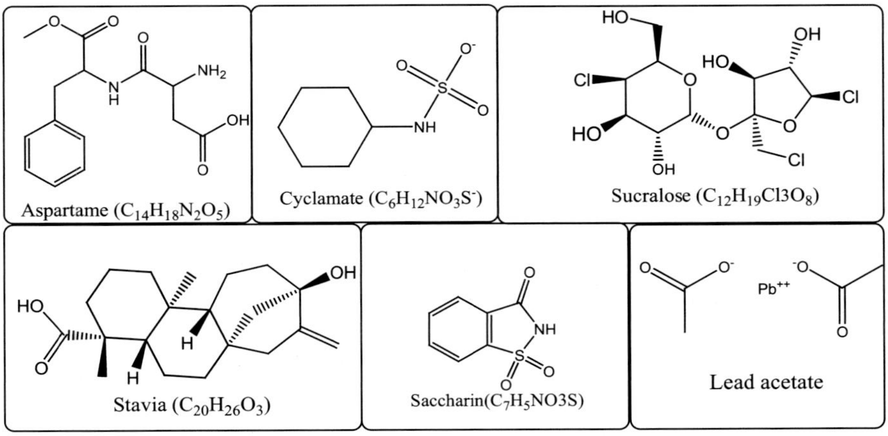

Fig. 2.2 Examples of artificial sweeteners

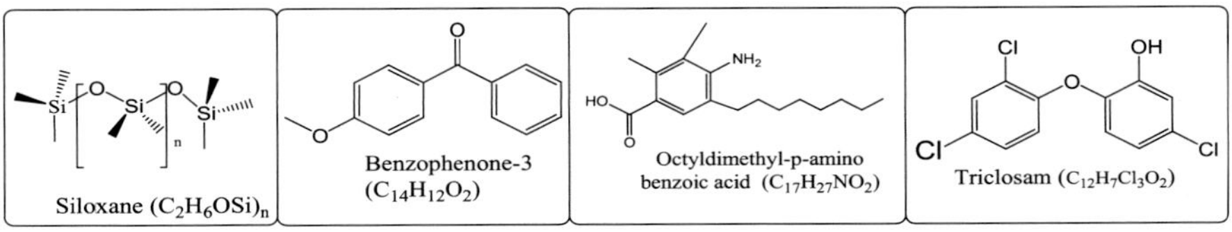

Fig. 2.3 Examples of commercial additive chemicals in domestic and cosmetic products

were subjected to various doses of sucralose, just minor behavior changes in some aquatics (Bergheim et al. 2015; Perkola 2014). Conversely, these speculations have presented challenges in monitoring because of the need for long-term monitoring and metabolic and environmental variables (Shearer and Swithers 2016). Some literature has reported sucralose to be safe for consumption and that it lacks fatal effects (Brahmini et al. 2012). On the other hand, the presence of artificial sweeteners in the aquatic environment is a clear justification of fecal contamination, as sucralose is not metabolised in the body unless there is a direct disposal, which is less common.

6.4 Personal Care Products

Triclosan is a commercial additive chemical in domestic and cosmetic products such as soaps, detergents, toothpaste, hand washes, deodorants and toys purposely as an antifungal and antibacterial (Fig. 2.3). Its synthesis is a three-step process involving dehydration of 1-(2-hydroxyethyl)pyrrolidin-2-one with either zinc or calcium oxide to 1-vinylpyrrolidin-2-one, followed by reaction with 5-chloro-2-(2,4-dichlorophenoxy)phenyl acrylate in n-heptane to form triclosan (Wikipedia 2016b). Triclosan is a lipophilic white solid with a half-life of about 11 h from urine and 21 h from plasma, mostly detected in the urine rather than feces (Arbuckle et al. 2015). Triclosan enters the environment from a wide range of sources, including laundry and domestic waste waters, municipal waste, and commercial and out-of-use products (Chen et al. 2015). It is contained in sunscreens, where it protects human from exposure to harmful UV light up to 20% (Peng et al. 2016).

Toxicological investigation shows that triclosan interferes with oestrogen responses, thus it is speculated that it causes breast cancer. Experimental tests in rats indicated suppressed thyroid; therefore, being an endocrine disruptor it can induce female infertility in both test animals and humans (Yuan et al. 2015). Further investigations in the aquatic environment revealed that a concentration range of 0.26–0.54 mg/L is lethal to fishes, 0.13–0.39 mg/L is acute to crustaceans and 1.4–10 µg/L inhibits the growth of algae. Including its lipid solubility, triclosan undergoes bioaccumulation, and hence bioconcentration (Norwegian Scientific Committee for Food and Safety 2002). Following aerobic degradation, triclosan undergoes biotransformation into several products, including methyl-triclosan (Chen et al. 2015). No bacteria can biodegrade methyl-triclosan, making it more persistent than the parent form (Lozano et al. 2012). Finally, when triclosan merely resides in the sediments and soils in the aquatic systems owing to lipophilic and density factors, the quantity prevailing in the water is only in transit.

6.5 Algal Toxins

Algal toxins are varieties of toxicants either released from live cells or dead and ruptured cells, mostly occurring in the stagnant or slowly moving waters (US EPA 2016b). The blue–green algae (cyanobacteria) are the most infectious group of algae residing in the freshwaters (US EPA 2015). Other harmful species include *Alexandrucitronella,* which produces paralytic toxins to shellfish, *Pseudo-nit zs chia*, producing domoic acid, which is poisonous, *Dinophysis*, producing diarrhetic poisoning, *Heterosigma, Ciguatera, Lingulodinium polyedrum* and *Gonyaulax spinifera* (CIMT 2006). Harmful algal bloom releases algal toxins such as saxitoxins and anatoxins that are quietly found bioaccumulated in shellfish, and are responsible for acute neurotoxicity and possibly carcinogenicity (Zervou et al. 2016). The common exposure routes to humans include skin contact through swimming, inhalation, and ingestion of contaminated seafood and water. Fish toxicity, increased water viscosity, oxygen depletion, blocking sunlight, irritation of gills, food poisoning, bioaccumulation and public health concerns from consuming contaminated waters have occurred (UK Marine SACs Project 2001; Zhang and Zhang 2015).

6.6 Microbes

There are about 72 serotypes of enteroviruses isolated from the human body; yet, the only enteroviruses affecting humans that are homologous to the rhinovirus belongs to polioviruses, coxsackieviruses and enterocytopathic categories (Royston and Tapparel 2016). They enter the human body through the alimentary canal, via the fecal–oral route and contaminated food stuffs, and then reproduce in the gastrointestinal tract, followed by spreading and affecting body organs such as the nervous system, the heart and the skin. The incubation period ranges from 2 to 14 days. Epidemiological investigations revealed that enteroviruses have been responsible for type 1 diabetes as they were found in the pancreas of a diabetic person. These viruses are prospective suspects for damaging insulin-producing cells (Kondrashova 2014). Its world discovery is increasing tremendously; the same virus was detected in children with acute respiratory infections in the Philippines (Puppe et al. 1999; Imamura and Oshitani 2015). The environmental occurrence of enteroviruses had been reported (Kuroda et al. 2015; Han et al. 2015; Richardson and Ternes 2011), which can be quantified by using the EPA method 1615 (Fout et al. 2016).

6.7 Inorganic Elements

Some elements, including cobalt, germanium, strontium, tellurium and vanadium have been classified and categorised among ECs (EPA 2009). Ge and Te are diagonally related, V and Co are transition elements and Sr is an alkali earth metal. Co, Ge and V are under the same period 4, whereas Sr and Te are under period 5. Apart from industrial uses such as steel additive, pigment for ceramics and glasses, they play a dietary role in humans in very small amounts (Royal Society of Chemistry 2016; Yadav et al. 2017; Hare et al. 2017).

Inorganic elements can enter the environment through natural processes such as rock degradation, mining, wearing of metals, consumer products and domestic and municipal waste. Formerly, cobalt was used in drugs and as a germicide. Vanadium is widely used as a catalyst; therefore, chemical laboratory waste is among the sources. Electronic waste is a major source of germanium waste in the environment. Wash-outs from garages, car washes, and welding areas can be sources too. Vanadium is a water-soluble element that forms vanadium complexes. Vanadium oxide and vanadium chloride irritate the eyes and skin, are harmful on inhalation and contact with skin, and cause burns (Chemical Book 2016b). Germanium is flammable, irritating, and carries other properties of heavy metals (Chemical Book 2016a). These elements have dietary values; however, continuous exposure in various doses for a very long time may lead to bioaccumulation, which later results in unexpected chronic diseases.

6.8 Illicit Drugs

Illicit drugs are the classes of drugs that have been internationally banned from public use except for medical and research use. They include but are not limited to cocaine, heroin, marijuana, non-benzoylecgonine, methamphetamine and 3–4-methylendioxyamphetamine. The active component of marijuana is the cannabinoids, which, when abused, become illegal, but again they have medicinal value including decreasing pain, inflammation, controlling epileptic seizures and treating addiction and spasticity. The worse side of marijuana includes aggressive behavior, drugged drivers, impaired mental health, lung diseases, bladder cancer, delinquent behavior and low IQ in a baby when smoked by a pregnant woman. Collectively, wastewaters, urine, feces, runoffs from marijuana gardens and sewage networks are the primary sources of illicit drugs in the environment (Sarkar et al. 2009).

The pharmacological effects exerted by drugs of abuse represents contagious outcomes to humans and ecology as whole; therefore, further intensive research is needed. Drugs of abuse have been detected in the surface and wastewaters, which reflects its uses. Cultivation of coca and marijuana requires and destroys fertile lands, causes deforestation; destruction of protected lands, land cleaning, i.e. soil erosion. Utilisation of fertilisers and pesticides also causes a change in biodiversity. Nutrient runoffs cause algal bloom, which has devastating effects, as previously discussed (US Department of Justice 2014). Zebra mussels exposed to cocaine experienced oxidative stress, DNA damage and an inactive diffusive response (Pharand et al. 2015).

6.9 Endocrine-Disrupting Hormones

Estradiol and progesterone are sex hormones, but operate in the opposite sexes (Fig. 2.4). Its four-ring structure resembles steroids such as cholesterol, estradiol and testosterone (male sex hormone) (Hu et al. 2016). Steroids are responsible for membrane fluidity, carbohydrate regulation, inflammatory responses, bone metabolism, stress responses, cardiovascular fitness, behavior, mineral balancing and steroid receptors (Salimetrics Europe 2014). Not only in females, estradiol is also found in very small amounts in fishes, vertebrates, crustaceans, men and other animals.

Estradiol is a commercial hormonal replacement therapy, used in the treatment of infertility in women, stimulating female puberty and the treatment of prostate cancer. It comes in the form of oral, transdermal, topical, injectable and vaginal insertions. Its direct consumption together with its active metabolites may lead to a wide range of effects, including ovarian cancer, breast cancer, mood disturbances, increased blood pressure and many others (Linda 2013; Seeger and Mueck 2010). The most common routes of hormones in the environment include body excretions, dumping and accidental releases. As conventional water treatment methods were

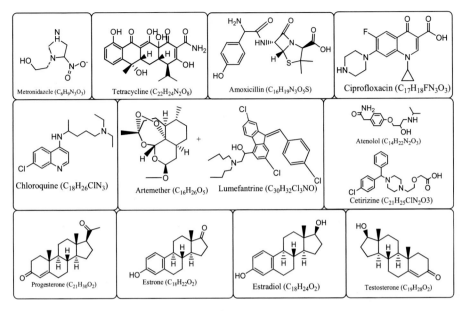

Fig. 2.4 Examples of endocrine-disrupting hormones

not established for removal of hormones, therefore these hormones may occur in the drinking water. **Through food chain, they inter human body and then bioaccumulate in the liver. In addition to the food chains, they enter human bodies and then bioaccumulate in the liver** (Narender and Cindy 2009). Early exposure to estradiol led to masculinisation in females and exposure to testosterone led to the expression of male genitalia in female offspring (Grober et al. 1998). The overall outcomes of hormonal releases in the environment are endocrine disruption, genotoxicity, metabolism alteration, cancerous cases, diabetes, birth defects, obesity, cardiopulmonary diseases, affected brain development and abnormal embryogenesis, which interferes with reproduction (Filby et al. 2007; Bellanger et al. 2015; Sanches et al. 2016; Jobling and Owen 2012).

6.10 Pharmaceuticals

Pharmaceuticals are materials or substances, which, when taken into the body by injection, inhalation, skin absorption or oral intake, change the physiology of the body. Their classification is too broad, ranging from their origin, mode of administration, type of diseases, therapeutic effects etc. Nearly all of them are organic and containing one or more cyclic rings. Not all consumed drugs are absorbed in the body; 30–90% of active drugs or metabolites are excreted (Heath et al. 2016; Jones et al. 2003).

Either in dissolved or suspended form, drugs are transported over a long distance and can be deposited in the lipid/fatty tissues, thus moving from one trophic level to another. Their environmental occurrence as an antibacterial, antidepressant, antihistamine, antiepileptic, hormone, lipid regulator, anti-malarial and/or analgesia in the aquatic environment present a threat to both aquatic and human lives. The receiving organism experiences physiological changes despite unnoticeable concentrations, as drugs are very active, even at very low concentrations (Donk et al. 2016). Some of the realistic effects include increased resistance of bacteria to antibiotics (Jones et al. 2003). Antibiotics can further affect cyanobacteria in the water, and thus have an effect on the food chains of all aquatic species. Research into the response to cancer drugs of zebra fish indicated histopathological changes, impaired kidney, liver and DNA, and interfered with plant reproduction capacity (Heath et al. 2016). Following exposure to complex mixtures of pharmaceuticals in the environment, worse chronic diseases can be promoted in human organs (National Toxics Network 2015).

7 Quantification Approaches

Emerging contaminants exist in the environment in very small quantities, which require effective methods for their isolation, identification and quantification. Solid phase extraction is a recent popular method widely used for the separation, extraction and pre-concentration of ECs. The adsorbent materials used bind reversibly with organic contaminants, which are then washed with polar solvent followed by analysis (Nosek et al. 2014; Matamoros et al. 2016; Hanigan et al. 2016). Ultrasound-assisted extraction is reported to be an effective technique for the extraction of pharmaceuticals and illicit drugs (Gago-ferrero and Thomaidis 2016). The analysis process is achieved by using high-resolution mass spectrometry (Zendonga et al. 2015), ultra-performance liquid chromatography-tandem mass spectrometry (Yanlong et al. 2016), inductively coupled plasma-mass spectrometry for the elemental determination of trace elements (EPA 1994) and gas chromatography mass spectrometry.

8 Environmental Remediation of Emerging Contaminants

8.1 Classical Water Treatments

The first recorded history of water treatment through filtration dates back to the seventeenth century when Sir Francis Bacon attempted to desalinate seawater, although it was unsuccessful. Since then, water treatment has employed physical, chemical, biological, instrumental or natural processes to remove natural and induced contaminants to meet specific requirements. The conventional and advanced water remediation approaches are intended to purify water for safe human consumption, ecological purposes and special uses such as medical or regulatory requirements. Traditional and household methods of water treatment include

Table 2.2 Classical water treatment methods

	Technique	Advantages	Disadvantages
1	Boiling	100% pathogen reduction	Does not remove toxic metals, tannins, suspended particles and organic waste
2	Distillation	Removes all contaminants	Energy intensive
3	Solar disinfection	Reduces some pathogens	Requires effective pre-filtering
			Does not affect chemical contaminants
4	Sedimentation	Reduces suspended solids	Does not remove pathogens, very fine particles, and most chemicals
5	Chlorination	Pathogen reduction	Affected by time, pH, disinfectant concentration, temperature and turbidity
			Not 100% effective
6	Activated carbons	Removes most contaminants	Energy intensive for preparation of activated carbon

boiling, activated carbon filters, cloth filtration, solar disinfection and chlorination. Essential techniques for public water treatment shown in Table 2.2 are filtration, sedimentation, flocculation and chlorination, and sometimes activated carbons. Traditional methods simply move contaminants elsewhere and obviously create significant risks in the excavation, handling and transport of hazardous material. The isolated containment areas require monitoring and maintenance, which involves unnecessary costs.

Classical water treatment methods include but are not limited to activated sludge methods for the treatment of pharmaceuticals (Pei et al. 2015), carbon xerogels reported by (Álvarez et al. 2015) in the removal of ECs such as caffeine and diclofenac from aqueous solution, with a very promising output. Membrane bioreactors, ozonisation, photocatalysis (UV disinfection), artificial recharging, constructed wetlands, phytoremediation by using duckweeds (Allam et al. 2015), modified mesoporous silica (Ortiz-Martínez et al. 2015) and treatment with clay minerals (Styszko et al. 2015) have been reported to be effective approaches against ECs. Bioremediation, reverse osmosis and ultrafiltration membranes are useful too in decreasing the amount of ECs in the water (Table 2.3).

Public water treatment is a large scale project that requires intensive investments, commercial scale waste removal techniques, fast running procedures, state-of-art experienced workers and moreover retains value for money. Consequently, water containment, contaminants screening, pH adjustments, temporary storage and pre-chlorination to reduce fouling organisms are preliminary activities. Flocculation with alum (Al_2SO_4) or iron III salt ($FeCl_3$) coagulates suspended particles that speed sedimentation. Accumulation of sediments at the bottom forms sludge, which is eventually removed periodically, either treated or without treatments. At a particular point, dissolved air flotation is used to remove fine particles that were not removed by flocculants. It is achieved by applying air at the bottom of the tank, creating bubbles, which trap fine particles, making them floating masses. Either rapid or slow sand filtration or wetlands are used to remove remaining suspended particles.

2 Toxicological Aspects of Emerging Contaminants

Table 2.3 Final CCl$_4$ for selected chemical contaminants

Substance name	Discovery	Selected uses
1,1-Dichloroethane	1794	Industrial solvent
1,1,1,2-Tetrachloroethane	1986	Industrial solvent
1,2,3-Trichloropropane	1980s	Industrial solvent
1,4-Dioxane	1980s	Used as a solvent
1-Butanol	1950s	Solvent
Methanol	1661	Used as an industrial solvent
Nitrobenzene	1847	Used in the production of aniline, solvent
Nitroglycerin	1891	For explosives and rocket propellants
Hydrazine	1875	Ingredient in rocket propellants, plastics
Aniline	1826	Used as an industrial chemical and as a solvent
Ethylene glycol	1856	Used as antifreeze, in textiles
Chloromethane	1835	Used as a foaming agent
Perfluorooctanoic acid	1940s	Resistant to water, grease, stains, fire-fighting
Nonylphenol	1940	In detergents, cleaners, degreasers, paints
Formaldehyde	1859	Ozonation disinfection by-product
Urethane	1930s	Paint and coating ingredient
Estradiol (17-beta estradiol)	1951	Found in some pharmaceuticals
Mestranol	1956	Veterinary and human pharmaceuticals
Quinoline	1834	Production of pharmaceutical (anti-malarial)
2-Methoxyethanol	N. A	Used in cosmetics, perfumes, fragrances
Acephate	1970s	Insecticide
Dicrotophos	1956	Insecticide
Triethylamine	1980	Stabiliser in herbicides and pesticides
Ethylene oxide	1859	Fungicidal and insecticidal fumigant
α-Hexachlorocyclohexane	N.A	Insecticide
Tebufenozide	N.A	Insecticide
Thiodicarb	<1985	Insecticide
Oxyfluorfen	1979	Herbicide
Permethrin	1973	Insecticide
Profenofos	<1975	Insecticide and acaricide
Dimethipin	1982	Herbicide and plant growth regulator
Diuron	1974	Herbicide
Ziram	1956	Fungicide
Tebuconazole	1986	Fungicide
Thiophanate-methyl	2014	Fungicide
Chlorate	NA	Chlorate compounds are used in agriculture
Cobalt	1735	Part of the vitamin B12 molecule
Germanium	1886	Used in diodes, as a dietary supplement
Manganese	1774	Essential nutrient found in vitamins
Vanadium	1801	Essential nutrient
Molybdenum	1778	Essential dietary nutrient

(continued)

Table 2.3 (continued)

Substance name	Discovery	Selected uses
Tellurium	1782	Used as sodium tellurite in bacteriology
Cyanotoxins	1878	Cyanotoxins
N-nitrosodiethylamine	2008	Disinfection by-product
N-nitrosodimethylamine	1978	Disinfection by-product

NA not available

At this point, conventional water treatment is mostly finalised by water disinfection. Yet, ECs that exist in very small concentrations are still there. Therefore, advanced remediation techniques are employed to remove ECs.

8.2 Modern Water Treatments

8.2.1 Bioremediation

Bioremediation is a cost-effective technique that uses the metabolic action of microorganisms to remove contaminants from environmental matrices. The method plays an effective role by removing hazardous organic and inorganic contaminants in the aquatic, sludge, soil, sediments and wastewater via eco-friendly, publically accepted and useful in onsite application. Microorganisms cannot remove inorganic contaminants from the environment; nevertheless, their oxidation reduction ability mobilises inorganic contaminants to less harmful waste. Bioremediation is therefore attained as the result of microorganisms using contaminants as their sources of food and energy. It requires less equipment, a small amount of labour and energy, and is thus very useful. Bioremediation has proved to be useful in the cleaning of blood, body fluids and communicable diseases at crime scenes (Bharagava et al. 2017).

8.2.1.1 Chemistry of Bioremediation

Biodegradation is a redox reaction occurring in organic matter facilitated by microbes. It is either a metabolic approach where bacteria consume released energy for its growth, or co-metabolism where the energy released during biodegradation is not consumed by bacteria for their growth. At lower carbon and contaminant concentrations oligotrophic bacteria dominate whereas eutrophic bacteria dominate at higher concentrations.

Bacteria speed up chemical reactions that transfer electrons from electron donors such as organic matter to electron acceptors. The energy released in redox reactions are consumed by bacteria for their growth. Some bacteria such as hydrogenotrophic and methanogens use inorganic contaminants through anaerobic respiration and fermentation to generate energy. Other examples are nitrifiers and sulphur-oxidising bacteria.

Simplified bioremediation equation:

$$\text{Organic contaminant} + O_2^- \rightarrow H_2O + CO_2 + \text{Cell material} + \text{Energy}$$

Suitable conditions are not limited to:

- The presence of suitable microbes
- The availability of C, H, O_2, N and P as nutrients
- A temperature range between 15 and 45 °C
- A pH range between 5.5 and 8.5
- A moisture level between 40 and 80%,
- An oxygen concentration greater than 2 mg/L

Under aerobic conditions, bacteria use oxygen to oxidise organic matters by removing electrons and converting them to carbon dioxide and water. In the absence of oxygen bacteria, nitrate, manganese (IV), iron (III), sulphate and carbonate are used as electron acceptors. Once electron acceptors are completely consumed, the bacteria switch to fermentation where organic wastes act as electron donor and acceptor (EPA 2013).

Bioremediation involves a combination of processes such as biostimulation for environmental modification to stimulate the bacteria responsible for bioremediation. Stimulation is achieved by adding nutrients and electron acceptors such as phosphorus, nitrogen, oxygen or carbon, in the form of molasses. When dealing with halogenated contaminants in anaerobic conditions, stimulation is achieved through the addition of electron donors such as organic substrates straw, sawdust, or corn cobs. Injection wells are used to accomplish the process. Bioaugumentation is employed to enhance biostimulation, which involves the introduction of specially prepared bacterial culture to increase the density of the bacteria, achieving a specific goal such as degrading complex organic compounds and increasing the overall removal rate. It is then followed by the bioaccumulation of live cells, biosorption of dead biomasses, phytoremediation by plants and then rhizoremediation through plant–microbe interactions. Other useful examples of bioremediation technologies are venting, bioleaching, land farming, bioreactors, composting and rhizofiltration.

Under suitable aerobic conditions such as balanced pH, moisture and temperature, microbes secrete enzymes that break down contaminants to a consumable level. While taking food and energy, they release water, carbon dioxide and nonharmful by-products such as amino acids. The performance of this technique is limited to the presence of microorganisms capable of degrading pollutants, the accessibility of contaminants to the microorganisms, the type of soil, temperature, pH, and the presence of oxygen or other electron acceptors and nutrients. Not all bacteria are capable of biodegradation, however. Some potential reportedly effective bacteria include *Rhodococcus*, *Pseudomonas*, *Sphingomonas*, *Alcaligenes* and *Mycobacterium*. Fungi such as the white rot fungus *Phanerochaete chrysosporium* have the ability to degrade persisting environmental pollutants, including polychlorinated biphenyls in river sediments, dechlorination of the solvent trichloroethylene and chloroform. *Methylotroph* is an aerobic bacterium that grows by using methane for carbon and energy. Sometimes an injection of air under pressure below the water table is performed to increase groundwater oxygen. This process is referred to as biosparging. Biopile treatment is a full-scale technology in which excavated soils are mixed with soil amendments, placed in a treatment area, and bioremediation takes place using forced aeration. Moisture, heat, nutrients, oxygen and pH are controlled to enhance biodegradation. Preliminarily, contaminated soils require excavation before being placed in the bioreactor (Ruiz-Aguilar et al. 2002). Bioreactors are special industrially manufactured controlled vessels in which microorganisms carry out biochemical reactions.

8.2.1.2 Evidence of Biodegradation

The ratio of biological oxygen demand and chemical oxygen demand are the key parameters for the establishment of biodegradation. Biological oxygen demand is a measure of oxygen required by microbes to break down a given organic contaminant in a sample of water. Chemical oxygen demand is an indirect measure of the amount of organic contaminant present in a sample of water. The method has proved useful upon remediation of ECs (Gothwal and Shashidhar 2014).

Membrane bioreactors constitute a wastewater remediation technique involving a combination of bioreactors and membranes in the remediation of water (Nguyen et al. 2013).

8.2.2 Membrane Filtration

Membrane filtration is a physical process that replaces traditional processes such as sedimentation, flocculation and adsorption through sand filters and active carbon filters, ion exchangers, extraction and distillation. The membrane is made up of semi-permeable materials that selectively allow the passage of water while particulates and microbes are retained with high productivity if suitable conditions prevail. It makes use of relatively low energy and no need for the addition of chemicals. Implementation of membrane filters can adapt either plate or tubular membrane systems.

Its mechanism of action can involve the maintenance of concentration gradients, the application of high pressure or the application of the electric potential. Either way, membrane filtration is divided into microfiltration (0.03–10 μm) and ultrafiltration (0.002–0.1 μm) for the removal of large particles, nanofiltration (0.001 μm) and reverse osmosis or hyperfiltration used for the removal of salts from water. the former methods principally rely on the pore sizes with less pressure while the latter depend on the diffusion and high pressure. Membrane filters present a reasonable opportunity for cost-effective and environmentally friendly approaches; yet, reversible or irreversible fouling caused by water quality, process design and control, membrane type and materials interferes. Eventually, membrane flushing, chemical cleaning or membrane replacement affects the outcome of the method.

The invention of membrane filtration technology around the 1960s changed the course of water treatment. The membranes are prepared from either polymeric organic material such as polypropylene, polyvinylchloride, polycarbonate, polyester, polysulphone, polytetrafluoroethylene, cellulose acetate of polysulphone, or from inorganic materials such as metals or commonly ceramics.

8.2.2.1 Operationalisation of Membranes

Various forms of energy are required for the proper functioning of membrane filters. These include:

- **Temperature gradient membranes** such as membrane distillation, which allows only the passage of the vapour phase while blocking the liquid phase.
- **Pressure-operated membranes** such as microfiltration, ultrafiltration, nanofiltration and reverse osmosis.
- **Electric potential-driven membranes** such as membrane electrolysis, electrodialysis, electrodeionisation, electrofiltration and fuel cells.
- **Concentration-based membranes** such as gas separation, forward osmosis, dialysis, pervaporation and artificial lungs.

Water purification and wastewater treatment plants widely use nanofiltration and reverse osmosis, whereas micro- and ultrafiltration membranes are commonly used in the food and beverage industries.

The membranes are porous sheets capable of selectively reclaiming portable water from microbes, organic materials that could react with disinfectants to form water disinfection by-products with rational outputs. Membrane fouling, production of polluted water via backwashing and regular replacements of the membrane are among hands-on challenges. However, a high-performance, space saving, simple operation and automatic disinfection create more opportunities.

Nanofiltration and reverse osmosis proved to remove sufficient amounts of ECs in the water (Snyder et al. 2007). In this case, there are possibilities of other remediation techniques to remove ECs from water, but it depends on what the analyst requires.

8.2.3 Ozonation

Ozonation is a chemical process of water treatment in an eco-friendly advanced technology. The mechanism of ozone treatment involves ozone decomposition to release hydroxyl radicals that react with organic particulate contaminants, which are then converted to small biodegradable molecules. In this process, hydrogen peroxide is useful in speeding up the process by adding extra hydroxyl radicals.

The ozonation process is an in-situ chemical oxidation process also known as advanced oxidation or UV disinfection. The production of radicals is the result of ozone and hydrogen peroxide decomposition, oxygen, an ultraviolet energy source or inorganic catalysts such as titanium oxide. Moreover, sulphate radical-based oxidation may appear during the treatment and therefore contribute to the reduction of pollutants. The method has been proven to remediate contaminants, including volatile organic compounds, pesticides and aromatics with high efficiency. This technique has been in existence since 1987; however, its commercial application is yet to be fully implemented because of the high running costs, despite its efficacy. Ozonation remediation, either in sludge or in water, has been proven to treat methyl tert-butyl ether, tetrachloroethene, NDMA, 1,4 dioxins and chlorinated organic compounds. Ozonation is less effective for the removal of pathogens because its half-life is short (Mckie et al. 2016).

8.2.3.1 Chemistry of Ozonation

Once ozone is generated in situ by using ozone generators, it is then released into the water to oxidise double bonds, amino groups and aromatic systems. Performance of this method depends on the amount of ozone, contact time and susceptibility of the contaminants. Ozone is very unstable, such that after its generation, it soon decomposes back to oxygen (Janna 2011). The mode of action includes:

- Breakage of carbon–nitrogen bonds leading to depolymerisation
- Reactions with radical by-products of ozone decomposition
- Direct oxidation/destruction of the cell wall with leakage of cellular constituents outside of the cell
- Damage to the constituents of the nucleic acids (purines and pyrimidines)

Disadvantages of Ozonation:

- Ozonation is more complex than chlorination as it requires a special steel vessel
- Ozone is very reactive and corrosive; hence the need for corrosion-resistant containers
- Effective after secondary treatment,
- Higher treatment costs than chlorine
- Ozone is highly irritating

Advantages of Ozonation:

- No harmful residual such as disinfection by-products formed after chlorination
- Ozone is more effective at killing pathogens compared with chlorine
- Short-term requirements
- Ozonation increases the amount of dissolved oxygen
- Ozone acts like a microflocculant
- Ozone is generated in situ
- There is no re-growth of microorganisms etc.

Snyder (2014) and Jobling and Owen (2012) reported effective ozonation treatment of water, despite the higher running costs. Ozonation is advantageous owing to its oxidative and disinfectant abilities (Snyder et al. 2003).

References

Allam A, Tawfik A, Negm A et al (2015) Treatment of drainage water containing pharmaceuticals using duckweed (Lemna Gibba). Energy Procedia 74:973–980. Available at https://doi.org/10.1016/j.egypro.2015.07.734

Álvarez S et al (2015) Chemical engineering research and design synthesis of carbon xerogels and their application in adsorption studies of caffeine and diclofenac as emerging contaminants. Chem Eng Res Des 95:229–238. Available at https://doi.org/10.1016/j.cherd.2014.11.001

Arbuckle TE et al (2015) Exposure to free and conjugated forms of bisphenol a and Triclosan among pregnant women in the MIREC cohort. Environ Health Perspect 123(4):277–284

Bellanger M et al (2015) Costs of exposure to endocrine-disrupting Chemicals in the European Union. J Clin Endocronol Metab 100(April):1256–1266

Bennett DH et al (2015) Polybrominated diphenyl ether (PBDE) concentrations and resulting exposure in homes in California: relationships among passive air, surface wipe and dust concentrations, and temporal variability. Indoor Air 25(2):220–229

Bergheim M et al (2015) Antibiotics and sweeteners in the aquatic environment: biodegradability, formation of phototransformation products, and in vitro toxicity. Environ Sci Pollut Res 22(22):18017–18030

Bharagava RN, Chowdhary P, Saxena G (2017) Bioremediation an eco-sustainable green technology, its applications and limitations. In: Bharagava RN (ed) Environmental pollutants and their bioremediation approaches. CRC Press, Taylor & Francis Group, Boca Raton, pp 1–22

Boullata J, Mccauley LA (2008) The potential toxicity of artificial sweeteners. Contin Educ 56(6):251–259

Bradberry SM et al (2000) Mechanisms of toxicity, clinical features, and management of acute chlorophenoxy herbicide poisoning: a review. J Toxicol Clin Toxicol 38(2):111–122. Available at http://www.ncbi.nlm.nih.gov/pubmed/10778907

Bradley PM, Journey CA (2014) Assessment of endocrine-disrupting chemicals attenuation in a coastal plain stream prior to wastewater treatment plant closure. J Am Water Resour Assoc 50(2):388–400. https://doi.org/10.1111/jawr.12165

Brahmini M et al (2012) Myths and facts about aspartame and sucralose: a critical review. IJRAP 3(3):373–375

Chemical Book (2016a) Germanium CAS#_ 7440–56-4. Public Domain. Available at http://www.chemicalbook.com/ProductChemicalPropertiesCB7733835_EN.htm#MSDSA

Chemical Book (2016b) VANADIUM (IV) OXIDE. Public Domain. Available at http://www.chemicalbook.com/ProductChemicalPropertiesCB7691007_EN.htm#MSDSA

Chen X et al (2015) Science of the total environment identification of triclosan-O-sulfate and other transformation products of triclosan formed by activated sludge. Sci Total Environ 505:39–46. Available at https://doi.org/10.1016/j.scitotenv.2014.09.077

Chevrier J et al (2010) Polybrominated diphenyl ether (PBDE) flame retardants and thyroid hormone during pregnancy. Environ Health Perspect 118(10):1444–1449

CIMT (2006) Harmful Algal Blooms (HABs). CeNCOOS, p 2

Clarke RM, Cummins E (2014) Evaluation of "classic" and emerging contaminants resulting from the application of biosolids to agricultural lands: a review. Hum Ecol Risk Assess Int J 21(2):492–513. Available at http://www.tandfonline.com/doi/abs/10.1080/10807039.2014.930295

Comero S et al (2013) EU-wide monitoring survey on emerging polar organic contaminants in wastewater treatment plant effluents. Water Res 47:6475–6487. https://doi.org/10.1016/j.watres.2013.08.024

Committee Report (1999) Emerging pathogens: viruses, protozoa and algal toxins. AWWA 91(9):110–121

EPA (1994) Determination of trace elements in waters and wastes by inductively coupled plasma – mass spectrometry. Available at https://www.epa.gov/sites/production/files/2015-08/documents/method_200-8_rev_5-4_1994.pdf

EPA (2009) Fact sheet: final third drinking water Contaminant Candidate List 3 (CCL 3)

EPA (2013) Introduction to in situ sioremediation of groundwater. Available at https://www.clu-in.org/download/remed/introductiontoinsitubioremediationofgroundwater_dec2013.pdf

EPA (2015) Fact sheet: drinking water contaminant candidate list 4 – Draft, Available at: http://www2.epa.gov/ccl

Fawell J, Ong CN (2012) Emerging contaminants and the implications for drinking water emerging contaminants and the implications for drinking water. Water Resour Dev 28(2):247–263. https://doi.org/10.1080/07900627.2012.672394

Filby AL et al (2007) Health impacts of estrogens in the environment, considering complex mixture effects. Environ Health Perspect 115(12):1704–1710

Fout GS et al (2016) EPA method 1615. Measurement of enterovirus and norovirus occurrence in water by culture and RT-qPCR. Part III Virus Detection by RT-qPCR. J Vis Exp 107(January):1–13. Available at http://www.jove.com/video/52646

French Agency for Food (2015) 2015 report on the safety of artificial sweeteners from the French Agency for Food, Environment and Occupational., (January), p 1. Available at http://www.medscape.com/viewarticle/839455

Gago-Ferrero P, Thomaidis NS (2016) Simultaneous determination of 148 pharmaceuticals and illicit drugs in sewage sludge based on ultrasound-assisted extraction and liquid chromatography – tandem mass spectrometry. Anal Bioanal Chem 407(15):4287–4297

Gothwal R, Shashidhar T (2014) Antibiotic pollution in the environment: a review. CLEAN Soil Air Water 42(9999):1–11

Government of Canada (1995) Toxic substances management policy Reprint of. Government of Canada, Ottawa. Available at http://publications.gc.ca/collections/Collection/En40-499-1-1995E.pdf

Green N, Bergman A (2005) Chemical reactivity as a tool for estimating persistence. Environ Sci Technol, 39(23), 23480A–23486A. Available at http://pubs.acs.org/doi/pdf/10.1021/es053408a

Grober MS et al (1998) The effects of estradiol on gonadotropin-releasing hormone neurons in the developing mouse brain. Gen Comp Endocrinol 112:356–363

Guidotti TL (2009) Emerging contaminants in drinking water: what to do? Arch Environ Occup Health 64(2):1–3

Han N, Gin KY, Hao H (2015) Science of the total environment fecal pollution source tracking toolbox for identification, evaluation and characterization of fecal contamination in receiving urban surface waters and groundwater. Sci Total Environ 538:38–57. Available at https://doi.org/10.1016/j.scitotenv.2015.07.155

Hanigan D et al (2016) Sorption and desorption of organic matter on solid-phase extraction media to isolate and identify N -nitrosodimethylamine precursors. J Sep Sci 9(14):2796–2805

Hansen P (2007) Risk assessment of emerging contaminants in aquatic systems. Trends Anal Chem 26(11):5

Hare V, Chowdhary P, Baghel VS (2017) Influence of bacterial strains on *Oryza sativa* grown under arsenic tainted soil: accumulation and detoxification response. Plant Physiol Biochem 119:93–102

Heath E et al (2016) Fate and effects of the residues of anticancer drugs in the environment. Environ Res Lett 23(15):14687–14691

Hu D et al (2016) Actions of estrogenic endocrine disrupting chemicals on human prostate stem/progenitor cells and prostate carcinogenesis. Open Biotechnol J 10(77):76–97

Hussain S (2013) Mechanisms of toxicity. University of California. Available at http://nature.berkeley.edu/~dnomura/pdf/Lecture6Mechanisms3.pdf. Accessed 13 Aug 2016

Imamura T, Oshitani H (2015) Global reemergence of enterovirus D68 as an important pathogen for acute respiratory infections. Rev Med Virol 25:102–114

Janna H (2011) Occurrence and removal of emerging contaminants in wastewaters. Brunel

Jobling S, Owen R (2012) 13 Ethinyl oestradiol in the aquatic environment

Jones OAH, Voulvoulis N, Lester JN (2003) Potential impact of pharmaceuticals on environmental health. Bull World Health Organ 81(10):768–769

Kaseva ME, Mwegoha WJS, Kihampa C, Matiko S (2008) Performance of a waste stabilization pond system treating domestic and hospital wastewater and its implications to the aquatic environment-a case study in Dar es Salaam, Tanzania. J Build Land Dev 15(1–2):14

Kolpin D, Furlong E, Zaugg S (2002) Pharmaceuticals, hormones and other organic wastewater contaminants in U. S. Streams, 1999–2000: a National Reconnaissance. US Geological Survey, pp 1999–2000

Kondrashova A, Hyöty H (2014) Role of viruses and other microbes in the pathogenesis of type 1 diabetes. Int Rev Immunol 33(4):284–295

Koumaki E et al (2015) Chemosphere degradation of emerging contaminants from water under natural sunlight: the effect of season, pH, humic acids and nitrate and identification of photodegradation by-products. Chemosphere 138:675–681. Available at https://doi.org/10.1016/j.chemosphere.2015.07.033

Kuroda K et al (2015) Science of the total environment pepper mild mottle virus as an indicator and a tracer of fecal pollution in water environments: comparative evaluation with wastewater-tracer pharmaceuticals in Hanoi, Vietnam. Sci Total Environ 506–507:287–298. Available at https://doi.org/10.1016/j.scitotenv.2014.11.021

Lee KE, Barber LB, Schoenfuss HL (2014) Spatial and temporal patterns of endocrine active chemicals in small streams. J Am Water Resour Assoc 50(2), 19

Linda G (2013) Steroids and hormonal science estradiol synthesis and metabolism and risk of ovarian cancer in older women taking prescribed or plant-derived estrogen supplementation. J Steroids Hormon Sci S12(3):2157–7536

Liteplo RG, Meek ME, Windle W (2002) Concise international chemical assessment document 38 N-nitrosodimethylamine first. Available at http://www.who.int/ipcs/publications/cicad/en/cicad38.pdf

Lozano N et al. (2012) Fate of triclosan and methyltriclosan in soil from biosolids application. Environ Pollut 160:103–108. Available at https://doi.org/10.1016/j.envpol.2011.09.020

Luo Y, Xu L, Sun X (2008) Synthesis of strong sweetener sucralose. Mod Appl Sci 2(3):13–15

Madhumitha R, Eden S, Catharine Mitchell, BW (2013) Contaminants of emerging concern in water, Arizona

Matamoros V, Rodríguez Y, Albaig J (2016) A comparative assessment of intensive and extensive wastewater treatment technologies for removing emerging contaminants in small communities. Water Res 88:777–785

McKie MJ, Andrews SA, Andrews RC (2016) Science of the total environment conventional drinking water treatment and direct biofiltration for the removal of pharmaceuticals and artificial sweeteners: a pilot-scale approach. Sci Total Environ 544:10–17. Available at https://doi.org/10.1016/j.scitotenv.2015.11.145

Miraji H et al 2016 Research trends in emerging contaminants on the aquatic environments of Tanzania. Scientifica 2016:7. Available at https://doi.org/10.1155/2016/3769690

Mortensen A et al (2014) Levels and risk assessment of chemical contaminants in byproducts for animal feed in Denmark levels and risk assessment of chemical contaminants in byproducts for animal feed in Denmark. J Environ Sci Health B 49:797–810

Munschy C et al (2013) Levels and trends of the emerging contaminants HBCDs (hexabromocyclododecanes) and PFCs (perfluorinated compounds) marine shellfish along French coasts. Chemosphere 91(2):233–240

Narender K, Cindy L (2009) Water quality guidelines for pharmaceutically-active-compounds (PhACs): 17α-ethinylestradiol (EE2), Provience of British Columbia

National Toxics Network (2015) Pharmaceutical pollution in the Environment: issues for Australia, New Zealand and Pacific Island countries, Australia

Nguyen LN et al (2013) Removal of emerging trace organic contaminants by MBR-based hybrid treatment processes. Int Biodeterior Biodegrad 85:474–482

Norwegian Scientific Committee for Food and Safety (2002) Risk assessment on the use of triclosan in cosmetics. In Risk assessment on the use of triclosan in cosmetics. pp 4–6

Nosek K, Styszko K, Golas J (2014) Combined method of solid-phase extraction and GC-MS for determination of acidic, neutral, and basic emerging contaminants in wastewater (Poland). Int J Environ Anal Chem 94(10):961–974

Ooka M et al (2016) Cytotoxic and genotoxic profiles of benzo[a]pyrene and N-nitrosodimethylamine demonstrated using DNA repair deficient DT40 cells with metabolic activation. Chemosphere 144:1901–1907

Ortiz-Martínez K et al (2015) Transition metal modified mesoporous silica adsorbents with zero microporosity for the adsorption of contaminants of emerging concern (CECs) from aqueous solutions. Chem Eng J 264:152–164

Pal A et al (2014) Emerging contaminants of public health significance as water quality indicator compounds in the urban water cycle untreated water sewer system. Environ Int 71:46–62. Available at https://doi.org/10.1016/j.envint.2014.05.025

Pei J et al (2015) Bioresource technology effect of ultrasonic and ozone pre-treatments on pharmaceutical waste activated sludge's solubilisation, reduction, anaerobic biodegradability and acute biological toxicity. Bioresour Technol 192:418–423. Available at https://doi.org/10.1016/j.biortech.2015.05.079

Peng X et al (2016) Persistence, temporal and spatial profiles of ultraviolet absorbents and phenolic personal care products in riverine and estuarine sediment of the Pearl River catchment, China. J Hazard Mater. Available at https://doi.org/10.1016/j.jhazmat.2016.05.020

Perkola N (2014) Fate of artificial sweeteners and perfluoroalkyl acids in aquatic environment. University of Helsinki

Petrisor IG (2004) Emerging contaminants – the growing problem. Environ Forensic 5:183–184

Pharand P et al (2015) Effects of various illicit drugs on immune capacity of blue mussel (Mytilus edulis). J Xenobiotics 5(5770):1–3

Puppe W et al (1999) Rapid identification of nine microorganisms causing acute respiratory tract infections by single-tube multiplex reverse transcription-PCR: feasibility study. J Clin Microbiol 37(1):1–7

Qi W et al (2015) Elimination of polar micropollutants and anthropogenic markers by wastewater treatment in Beijing, China. Chemosphere 119:1054–1061. Available at https://doi.org/10.1016/j.chemosphere.2014.09.027

Richard D, George E (1957) Synthesis of nitrosodimethylamine. p 5. Available at https://www.google.com/patents/US3136821

Richardson SD (2007) Water analysis: emerging contaminants and current issues. Anal Chem 79(12):4295–4324

Richardson SD, Ternes TA (2011) Water analysis: emerging contaminants and current issues. Anal Chem 83:4614–4648

Richardson SD, Exposure N, Agency USEP (2006) Environmental mass spectrometry: emerging contaminants and current issues. Anal Chem 78(12):4021–4046

Rivera-utrilla J et al. (2013) Pharmaceuticals as emerging contaminants and their removal from water. A review. Chemosphere, in press(in press). https://doi.org/10.1016/j.chemosphere.2013.07.059

Ross PS, Ellis GM (2004) PBDEs, PBBs, and PCNs in three communities of free-ranging killer whales (Orcinus orca) from the northeastern Pacific Ocean. Environ Sci Technol 38(16):4293–4299

Royal Society of Chemistry (2016) Vanadium – element information, properties and uses. Public Domain. Available at http://www.rsc.org/periodic-table/element/23/vanadium. Accessed 25 Aug 2016

Royston L, Tapparel C (2016) Rhinoviruses and respiratory Enteroviruses: not as simple as ABC. Viruses 8(16):23

Ruiz-Aguilar GML et al (2002) Degradation by white-rot fungi of high concentrations of PCB extracted from a contaminated soil. Adv Environ Res 6(4):559–568

Salimetrics Europe (2014) High sensitivity salivary 12B–estradiol enzyme immunoassay kit

Sanches S et al (2016) Comparison of UV photolysis, nanofiltration, and their combination to remove hormones from a drinking water source and reduce endocrine disrupting activity. Environ Sci Pollut Res 23(11):11279–11288

Sarkar PK et al (2009) Toxicity and recovery studies of two ayurvedic preparations of iron. Indian J Exp Biol 47(12):987–992

Schultz AG et al (2014) Aquatic toxicity of manufactured nanomaterials: challenges and recommendations for future toxicity testing. Environ Chem 11(3):207–226. Available at http://www.publish.csiro.au/?paper=EN13221

Seeger H, Mueck AO (2010) Estradiol metabolites and their possible role in gynaecological cancer. J Reproduktionsmed Endokrinol 7(1):62–66

Shearer J, Swithers SE (2016) Artificial sweeteners and metabolic dysregulation: lessons learned from agriculture and the laboratory. Rev Endocr Metab Disord. Available at https://doi.org/10.1007/s11154-016-9372-1

Siddiqi MA, Clinic M (2003) Polybrominated diphenyl ethers (PBDEs): new pollutants – old diseases. Clin Med Res 1(4):281–290. Available at http://www.ncbi.nlm.nih.gov/pmc/articles/PMC1069057/pdf/ClinMedRes0104-0281.pdf

Snyder SA (2014) Emerging chemical contaminants: looking for greater harmony. Am Water Works Assoc 108(8):14

Snyder SA et al (2003) Pharmaceuticals, personal care products and endocrine disruptors in Water : implications for the water industry. Environ Eng Sci 20(5):21

Snyder SA et al (2007) Role of membranes and activated carbon in the removal of endocrine disruptors and pharmaceuticals. Desalination 202:156–181

Sorensen JPR et al (2014) Emerging contaminants in urban groundwater sources in Africa. Water Res 72:1–13. https://doi.org/10.1016/j.watres.2014.08.002. Elseviere Ltd

Stasinakis AS et al (2013) Science of the total environment contribution of primary and secondary treatment on the removal of benzothiazoles, benzotriazoles, endocrine disruptors, pharmaceuticals and perfluorinated compounds in a sewage treatment plant. Sci Total Environ 463–464:1067–1075. Available at https://doi.org/10.1016/j.scitotenv.2013.06.087

Sturm R, Ahrens L (2010) Trends of polyfluoroalkyl compounds in marine biota and in humans. Environ Chem 7:457–484. https://doi.org/10.1071/EN10072

Styszko K et al (2015) Preliminary selection of clay minerals for the removal of pharmaceuticals, bisphenol A and triclosan in acidic and neutral aqueous solutions. C R Chim 18(10):1134–1142. Available at https://doi.org/10.1016/j.crci.2015.05.015

Suez J et al (2014) Artificial sweeteners induce glucose intolerance by altering the gut microbiota. Nature 514(7521):181–186. Available at https://doi.org/10.1038/nature13793

Thomaidis N(2012) Emerging contaminants: a tutorial mini-review. Global NEST …, 14(1), pp 72–79. Available at http://journal.gnest.org/sites/default/files/Journal Papers/72-79_823_Thomaidis_14–1.pdf. Accessed 13 Nov 2014

Tran NH et al (2014) Suitability of artificial sweeteners as indicators of raw wastewater contamination in surface water and groundwater. Water Res 48:443–456. https://doi.org/10.1016/j.watres.2013.09.053. Elsevier Ltd

UK Marine SACs Project (2001) Toxic substance profile_ Algal toxins and algae-related fish kills. Public Domain. Available at http://www.ukmarinesac.org.uk/activities/water-quality/wq8_51.htm. Accessed 22 Aug 2016

Urtiaga AM et al (2013) Removal of pharmaceuticals from a WWTP secondary effluent by ultrafiltration/reverse osmosis followed by electrochemical oxidation of the RO concentrate. Desalination 331:26–34. https://doi.org/10.1016/j.desal.2013.10.010. Elsevier B.V

US Department of Justice (2014) The dangers and consequences of marijuana abuse. (May), p 45. Available at www.DEA.gov

US EPA (2014a) Technical fact sheet – N-nitroso-dimethylamine. Available at https://www.epa.gov/sites/production/files/2014-03/documents/ffrrofactsheet_contaminant_ndma_january2014_final.pdf

US EPA (2014b) Technical fact sheet – polybrominated diphenyl ethers (PBDEs) and polybrominated biphenyls (PBBs) technical fact sheet – PBDEs and PBBs at a glance

US EPA (2015) Algal toxin risk assessment and management strategic plan for drinking water, Kansas. Available at www.waterone.org

US EPA (2016a) Basic information on the CCL and regulatory determination drinking water contaminant candidate list (CCL) and regulatory determination _ US EPA. EPA

US EPA (2016b) The effects human health nutrient pollution. Public Domain. Available at: https://www.epa.gov/nutrientpollution/effects. Accessed 22 August 2016

USDHHS (2015) Draft: toxicological profile for polybrominated diphenyl ethers (PBDEs), Atlanta. Available at http://www.atsdr.cdc.gov/toxprofiles/tp207.pdf

Voloshenko-Rossin A, Gasser G, Cohen K, Gun J, Cumbal-Flores L, Parra-Morales W, Sarabia F, Ojeda F, Lev O (2015) Emerging pollutants in the Esmeraldas watershed in Ecuador: discharge and attenuation of emerging organic pollutants along the San Pedro–Guayllabamba–Esmeraldas rivers. Environ Sci Process Impacts Roy Soc Chem 17(1):41–53 https://doi.org/10.1039/C4EM00394B.

van Donk E et al (2016) Pharmaceuticals may disrupt natural chemical information flows and species interactions in aquatic systems: ideas and perspectives on a hidden global change. Rev Environ Contam Toxicol 235:15

Vuong AM et al (2016) Prenatal polybrominated diphenyl ether exposure and body mass index in children up to 8 years of age. Environ Health Perspect, (February). Available at http://ehp.niehs.nih.gov/wp-content/uploads/advpub/2016/6/EHP139.acco.pdf

Wikipedia (2016a) Sucralose

Wikipedia (2016b) Triclosan

Wu J, Zhang L, Yang Z (2010) A review on the analysis of emerging contaminants in aquatic environment. Crit Rev Anal Chem 40:234–245. https://doi.org/10.1080/10408347.2010.515467

Yadav A, Chowdhary P, Kaithwas G, Bharagava RN (2017) Toxic metals in environment, threats on ecosystem and bioremediation approaches. In: Das S, Dash HR (eds) Handbook of metal-microbe interactions and bioremediation. CRC Press, Taylor & Francis Group, Boca Raton, p 813

Yang G, Fan M, Zhang G (2014) Emerging contaminants in surface waters in China – a short review. Environ Res Lett 74018:13

Yan-long W et al (2016) Determination of eight typical lipophilic algae toxins in particles suspended in seawater by ultra performance liquid chromatography – tandem mass spectrometry. Chin J Anal Chem 44(3):335–341. Available at https://doi.org/10.1016/S1872-2040(16)60911-8

Yu Y, Cn N (2011) Method of sucralose synthesis yield. Patent 2(12):4

Yuan M et al (2015) Preimplantation exposure to bisphenol a and Triclosan may lead to implantation failure in humans. Biomed Res Int 2015:9

Zendonga Z, McCarronb P, Christine Herrenknecht MS, Amzila Z, Coled RB, Hess P (2015) High resolution mass spectrometry for quantitative analysis and untargeted screening of algal toxins in mussels and passive samplers. J Chromatogr A 1416:10–21

Zervou S et al (2016) New SPE-LC-MS/MS method for simultaneous determination of multi-class cyanobacterial and algal toxins. J Hazard Mater. Available at: https://doi.org/10.1016/j.jhazmat.2016.07.020

Zhang C, Zhang J (2015) Environmental analytical chemistry current techniques for detecting and monitoring algal toxins and causative harmful algal blooms. J Environ Anal Chem 2(1):1–12

Chapter 3
An Overview of the Potential of Bioremediation for Contaminated Soil from Municipal Solid Waste Site

Abhishek Kumar Awasthi, Jinhui Li, Akhilesh Kumar Pandey, and Jamaluddin Khan

Abstract The soil contamination due to open disposal of municipal solid waste has become a serious issue particularly in the developing countries. Several studies have revealed variable impacts of pollutant toxicity on the environment and exposed inhabitants. This chapter provides an overview of the application of bioremediation of sites contaminated owing to municipal solid waste. The application of bioremediation technologies and well-organized mechanisms for environmental safety measures of these methods were discussed. Because some pollutants can be seriously affect to the environment, the chapter furthermore suggests strategies for better remediation of site. In addition, more detail studies exploring the linkage between the fates, and environmentally important factors are necessary to better understand the parameters on using bioremediation technologies.

Keywords Municipal solid waste · Soil contamination · Heavy metals · Pollutants · Bioremediation

---

A. K. Awasthi (✉)
Mycological Research Laboratory, Department of Biological Sciences,
Rani Durgavati University, Jabalpur, Madhya Pradesh, India

School of Environment, Tsinghua University, Beijing, People's Republic of China

J. Li
School of Environment, Tsinghua University, Beijing, People's Republic of China

A. K. Pandey
Mycological Research Laboratory, Department of Biological Sciences,
Rani Durgavati University, Jabalpur, Madhya Pradesh, India

Madhya Pradesh Private Universities Regulatory Commission,
Bhopal, Madhya Pradesh, India

J. Khan
Mycological Research Laboratory, Department of Biological Sciences,
Rani Durgavati University, Jabalpur, Madhya Pradesh, India

© Springer Nature Singapore Pte Ltd. 2019
R. N. Bharagava, P. Chowdhary (eds.), *Emerging and Eco-Friendly Approaches for Waste Management*, https://doi.org/10.1007/978-981-10-8669-4_3

1 Introduction

Municipal solid wastes (MSW) mainly include domestic waste generated from community or local municipality. In most of countries, the MSW are produced from mainly three sources: (a) Waste from households and public areas, including waste collected from residential buildings, litter bins, streets, marine areas, and country parks- known as domestic solid waste (b) Waste from shops, restaurants, hotels, offices, and markets in private housing estates known as commercial solid waste, and (c) Waste from industries, excluding hazardous waste- known as industrial solid waste (Chen et al. 2016; Stenuit et al. 2008).

In this context, it has been estimated that approximately 1.3 billion tons of MSW are generated every year worldwide, which is growing quickly as result of rapid growth and development, urbanization, resource consumption, and "Use & Throw" lifestyles become more common. The total volume of MSW production worldwide is estimated to be double in 2025, mainly in developing countries (Hoornweg and Bhada 2012).

Open dumping is a common practice among in several developing countries. In detail, an open dumping is a process where solid wastes are being disposed-off in a way which does not take care the surrounding environment, and is resulting exposed to the human health risk. MSW contain a different type of pollutants such as, heavy metals, and organic pollutants (Gautam et al. 2012). Also, the degradation of the MSW releases gases for example; volatile organic compounds (VOCs) and benzene, toluene, ethyl-benzene, and xylene isomers by oxidation of CO_2, CH_4, and their derivatives. The contaminants turn out in the leachate form and holding a significant level of pollutants is a common incidence in most of the MSW site of developing countries (Gautam et al. 2012; Mani and Kumar 2014).

Therefore, with rapid rising of population and decrease natural resources, it is very important and required to adopt safe disposal methodologies and develop appropriate remediation technologies for soil (Stenuit et al. 2008; Rayu et al. 2012; Bharagava et al. 2017). Soil remediation is the process of returning their functional that existed before to contamination. Different techniques exist for remediation of soil contaminated. This could be through physical, chemical and biological approaches. Beside, several unsuccessful remediation technologies have been reported owing to the use of inappropriate technologies (Zabbey et al. 2017). Therefore, it is important to explore an approach that would be applicable as well as sustainable for the environments.

2 Environmental Pollution and Health Risk from Municipal Solid Waste

Improper management of MSW is one of the sources for environmental pollution in towns, cities and municipalities. Most of the cities do not enforced MSW regulations properly in particularly in developing countries. The improper handling of

3 An Overview of the Potential of Bioremediation for Contaminated Soil...

Table 3.1 The Public health risk via different environmental media affected due to improper management of municipal solid waste (Lee and Lee 1994; UNEP 2007; Selin 2013)

Routes of exposure	These pollutants can be found in different environmental media such as, air, soil and water, and possibly will find their way to reach human body via, ingestion, inhalation, absorption
Public health effects	Skin problem: – fungal infection, allergic dermatitis, and skin cancer
	Respiratory problem – bacterial upper respiratory tract infections (such as: pharyngitis, laryngitis and rhinitis), chronic bronchitis and asthma
	Intestinal problems – bacterial enteritis, helminthiasis, amoebiasis, liver cancer, kidney and renal failure
	Skeletal muscular systems problem – back pain
	Central nervous system problem – impairment of neurological development, peripheral nerve damage and headaches
	Eye problem – allergic conjunctivitis, bacterial eye infections
	Others problems – septic wounds and congenital abnormalities, cardiovascular diseases and lung cancer etc.

MSW can lead several public health risk (Table 3.1) to nearby residents 'owing to its appearances as infectious, or toxic nature. Additional environmental impacts are damage of the environmental system by pollution of air, water, and soil. Such improper management of MSW poses a high risk to human health (Cointreau 2006; Lee and Lee 1994).

3 Bioremediation with Historical Insight

Historically, biological methods are being extensively applied across the world. The rapid damage of biologically rich natural ecosystems has accounted to a regular loss of information almost native biodiversity. The traditional societies along by ethnobiologists have ample of information of biodiversity and their usage although the regular socio-economic transformation in their life routine because of rapid globalization, and their immense knowledge should be connected in the developing area of bioremediation. When the whole world is discussion on several areas about the benefits and risks of scientific development in terms of biotechnological advances, at that time the ethno-sciences are debating the option of involving scientific investigation to human priorities (particularly to help traditional societies those are historically excluded), the vital requirements of environmental safety and the more cost effective and environmental friendly application of biological approaches in bioremediation courses (Kavamura and Esposito 2010). The studied suggests that bioremediation has great potential developments in modern era-although these areas are still on the way of constructing a comprehensive theoretical knowledge and integrated methodology. In terms of qualitative perspective, but the development is still required in methodological point of view, quantitative methods and taxonomic accuracy. Also, the bioremediation presently meets to several challenges, and few of them vital matters consist of the founding of well-organized discussions among various areas that edge with biotechnology, and genomics; qualitative advances in research techniques in relative to the findings and the procedures applied; and also the progress of monitoring plans established thorough research interested in the sustainable consumption of natural resources (Chakraborty et al. 2012).

4 Exiting Methods for Soil Remediation

4.1 Physical Remediation Method

The physical method mainly involves soil replacement and thermal treatment, the process is costly, and only appropriate for small polluted sites. This indicates it could be inappropriate for bigger-scale pollution (Dua et al. 2002; Atagana et al. 2003).

4.2 Chemical Remediation Method

Chemical method includes washing of polluted soil by consuming clean water and chemicals that can leach the contaminants from the soil (Fulekar et al. 2012). This method could be attained by chemical leaching, electrokinetic remediation, chemical fixation, vitrify technology, photo-degradation and chemical immobilization among others. The chemical approach is also expensive and has the potential to cause secondary pollution (Iqbal and Ahemad 2015) while the method is comparatively faster to clean-up of chemicals.

4.3 Bioremediation

Bioremediation is an approach that involves biodegradation process of contaminates by using the available nutrients and oxygen essential for microbes. Un-doubtfully, the bioremediation approaches are both resource conservative and economical feasible methods. According to the United States Environmental Protection Agency, the bioremediation is an approach that applies indigenous microorganisms to transform the hazardous substances into lesser toxic substances. The examples for bioremediation technologies are such as: phytoremediation, bioreactor, bioleaching, biostimulation, rhizofiltration, and composting, bioaugmentation (Rayu et al. 2012; Bharagava et al. 2017).

Numerous microbes (fungi, bacteria, algae, yeast, etc.) have ability to remove different heavy metals from soil. The remediation of the contaminants by microbes generally involved via bioaccumulation, biosorption and biodegradation process (Carro et al. 2013; Ghosh and Das 2014). The functional groups are present on the cell wall for example carboxyl, phosphate amine, sulfhydryl and hydroxyl, groups are mainly responsible for binding the contaminants on the microbial cell surface (Ghosh et al. 2015, 2016). For example, fungi have greater ability to resistance to heavy metals and also have higher surface area as well as higher biomass yield then other microbes. These fungi are extremely capable for biodegradation of different dyes owing to the occurrence of several oxido-reductive enzymes such as, peroxidases, lignin peroxidases manganese peroxidases and laccases (Ghosh et al. 2015; Chandra and Chowdhary 2015; Karigar and Rao 2011).

5 Factors Affecting Bioremediation

There are several factors such as, the nutrients supplementation, microbial diversity, pH, and temperature etc., can affect the bioremediation and the bioavailability of contaminates, thus triggering comprehensive changes in the toxic nature of contaminates towards microbes (Chakraborty et al. 2012; Kavamura and Esposito 2010). In this chapter we take up some of the factors briefly discussed as following.

5.1 The Nutrients Supplementation

Supplementation of proper nutrients is one of the important factors for bioremediation, if the nutrients are not enough for proper cellular growth and metabolism of the microbes in the polluted sites. As, in the polluted sites, the organic carbons content is high, and these possibly will be depleted in the course of microbial metabolism (Jing et al. 2017). Various nutrient sources for example potassium, nitrogen, and phosphate to the polluted site can stimulate the microbial growth which increase the bioremediation. Generally, the need of carbon-nitrogen and carbon-phosphorous ratio is 10:1, and 30:1, respectively, for bioremediation (Kensa 2011).

5.2 Temperature

Temperature has a vital role in the bioremediation method of contaminates. The solubility of pollutants such as, PAHs and heavy metals rises with the increase of temperature, which increases the bioavailability of pollutants (Zhang et al. 2006; Liu et al. 2017). Also, the microbial actions rise simultaneously with the rise of temperature in the suitable range, as it can increase the metabolism and the enzymatic activity of microorganisms, which will speed up the bioremediation procedure of contaminates. For instance, the amount of collective O_2 at 43 °C, which is a key for microbial action during the composting, is unusually higher than that at 22–36 °C (Liang et al. 2003). Additionally, temperature can directly affect mechanism of the adsorption and desorption of contaminates on particles or microorganisms. The adsorption ability and strength will rise with the growth of temperature level (Liang et al. 2003). The improved adsorption of contaminates may confine the adsorption of pollutants, as the adsorption area on the microbes is comparatively constant. Remarkably, the co-existing contaminates may help metals adsorption as the contaminants can reallocate among strongly and weakly bound fractions.

5.3 Microbial Diversity

The different species of microbes are being able to affect bioremediation activity. The existence of contaminates such as, heavy metals can also effect microbial diversity. And, the microbial groups are required to adapt the condition to the hazardous environment. The microbial strains isolated and recovered from these polluted sites commonly show great potential against the contamination of heavy metals and PAHs (Paul et al. 2005). The genomic technology advances the approach of remediation potentials of contaminates by microbes (Chen et al. 2009).

The microbial diversity of the MSW site for instance: *Alternaria alternata, Acremonium butyri, Aspergillus clavatus, Aspergillus flavus, Aspergillus candidus, Aspergillus luchuensis, Aspergillus fumigatus, Aspergillus nidulans, Aspergillus niger, Aspergillus terreus, Chaetomium* sp., *Chrysosporium* sp., *Cladosporium* sp., *Curvularia lunata, Drechslera* sp., *Fusarium oxysporum, Fusarium roseum, Gliocladium* sp., *Humicola* sp., *Mucor* sp., *Myrothecium* sp., *Paecilomyces* sp., *Penicillium digitatum, Rhizopus* sp., *Sclerotium rolfsii, Trichoderma viride*, etc. have great potential to degradation of MSW (Gautam et al. 2012; Jing et al. 2017; Cui et al. 2017).

5.4 pH

It is usually known that pH is a one of the main factor in bioremediation efficiency of contaminates. Microbes are affected by pH, as the optimum pH for diverse species is changing (Meier et al. 2012). As a result, contaminates such as, heavy metals and PAHs can toughly effect bacteria diversity, their enzyme activity and morphological structure as a result of changing pH, oxygen availability and also other environmental features, in the meantime equally effect the bioremediation of contaminates (Brito et al. 2015; Guo et al. 2010). In addition, pH has influences on the redox potential and solubility of metals. Difference valence states and different forms of metals cause diverse toxic impacts on microbes, which effect heavy metal remediation efficiency at the end. The in situ microbes" are inhibited under alkaline or acidic conditions, and cannot transform heavy metals, on the other hand they are extra tolerance ability to adverse conditions and still have the potential to survive with contaminates (such as, heavy metals) in sub-optimal situations. Thus, amending the pH at contaminated sites could be a best effort (Bamforth and Singleton 2005).

6 Integrating via Microbial Application to Improve Metal Uptake by Plants-Microbes Interaction

Microbes based phytoremediation is a very important bioremediation approach (Becerra-Castro et al. 2011; Yadav et al. 2017). For example, we draw an outline for integrated microbial radiation as shown in (Fig. 3.1) In detail, the bacterial/fungal consortia in the soil to rehabilitate environments polluted with hazardous chemicals, because they cooperatively form the microbial inocula, which have advantageous features, for example heavy metal tolerance ability, solubilization of mineral phosphate, ability of nitrogen fixation, and the ability for bio-mineralization (Iqbal and Ahemad 2015; Passatore et al. 2014). For example, Zimmer et al. (2009) studied the properties of applied ectomycorrhizal bacteria such as, *Micrococcus luteus* and *Sphingomonas* sp. as well as ectomycorrhizal fungi, such as, *Hebeloma crustuliniforme* on the development and metal accumulation of polluted soil. On the other hand, those bacteria, isolated from fungal sporocarps, they ability to improved plant growth because of the mixed inoculation of bacterial and fungal.

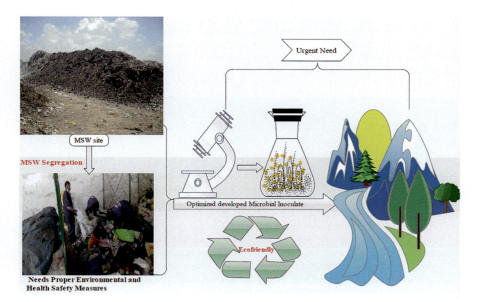

Fig. 3.1 Proposed Systematic brief outline for bioremediation of soil from MSW site

The number of bioremediation studies is subjected to by examination of specific phenomena (Ma et al. 2016). Comparatively some studies report interactions between diverse bioremediation approaches can be integrated in order to make expectations of field level performance. In the context of bioremediation, the process engineering includes the combination of site historic information, geologic description, hydrologic information, chemical as well as microbiological features, both laboratory and field statistics, and potential remedial process in order to estimates and make strategy base conclusions. Execution of integration is very crucial and perhaps the most challenging stage in order to application.

7 Concluding Remarks

There are a number of remediation methods existing, but for the optimal selection, it is very important, earliest, to do an advance examination of the conditions of the contaminated soil and the pollutants. For example, several sites can be contaminated with numerous pollutants, it is important to integrate techniques for remediation systems in order to improve the remedial action. Microbe-assisted integrated remediation can be best option for that specific contaminated site having comparatively low pollution that are adaptable to the option, for instance-biomineralization, phytodegradation, mycoremediation, cyanoremediation, phytostabilization, and hyperaccumulation, for an ecofriendly and sustainable approach. Therefore, there is essential to pursue, integrated advance biotechnological study in field of bioremediation.

Acknowledgments The first (corresponding) author is grateful to Dr. P.S. Bundela Regional Officer, Madhya Pradesh Pollution Control Board, (M.P.) India for valuable advice. The authors gratefully acknowledge the esteemed reviewers/editors for their critical assessment and valuable suggestions.

References

Atagana HI, Haynes RJ, Wallis FM (2003) Optimization of soil physical and chemical conditions for the bioremediation of creosote-contaminated soil. Biodegradation 14:297–307

Bamforth SM, Singleton I (2005) Bioremediation of polycyclic aromatic hydrocarbons: current knowledge and future directions. J Chem Technol Biotechnol 80(7):723–736

Becerra-Castro C, Kidd PS, Prieto-Fernández Á, Nele W, Acea MJ, Jaco V (2011) Endophytic and rhizoplane bacteria associated with *Cytisus striatus* growing on hexachlorocy-clohexane-contaminated soil: isolation and characterisation. Plant Soil 340:413–433

Bharagava RN, Chowdhary P, Saxena G (2017) Bioremediation an eco-sustainable green technology, its applications and limitations. In: Bharagava RN (ed) Environmental pollutants and their bioremediation approaches. CRC, Taylor & Francis Group, USA, pp 1–22

Brito EM, De la Cruz BM, Caretta CA, Goni-Urriza M, Andrade LH, Cuevas-Rodríguez G, Malm O, Torres JP, Simon M, Guyoneaud R (2015) Impact of hydrocarbons, PCBs and heavy metals on bacterial communities in Lerma River, Salamanca, Mexico: investigation of hydrocarbon degradation potential. Sci Total Environ 521:1–10

Carro L, Barriada JL, Herrero R, Sastre de Vicente ME (2013) Surface modifications of Sargassum muticum algal biomass for mercury removal: a physicochemical study in batch and continuous flow conditions. Chem Eng J 229:378–387

Chakraborty R, Wu CH, Hazen TC (2012) Systems biology approach to bioremediation. Curr Opin Biotechnol 23:1–8

Chandra R, Chowdhary P (2015) Properties of bacterial laccases and their application in bioremediation of industrial wastes. Environ Sci: Processes Impacts 17:326–342

Chen B, Liu X, Liu W, Wen J (2009) Application of clone library analysis and real-time PCR for comparison of microbial communities in a low-grade copper sulfide ore bioheap leachate. J Ind Microbiol Biotechnol 36:1409–1416

Chen XW, Wong James TF, Wai Ng CW, Wong MH (2016) Feasibility of biochar application on a landfill final cover – a review on balancing ecology and shallow slope stability. Environ Sci Pollut Res 23(8):7111–7125

Cointreau S (2006) Occupational and Environmental Health Issues of Solid Waste Management Urban Papers. The World Bank Group, Washington, DC. UP-2, JULY 2006. http://www.worldbank.org/urban/

Cui Z, Zhang X, Yang H, Sun L (2017) Bioremediation of heavy metal pollution utilizing composite microbial agent of Mucor circinelloides, Actinomucor sp. and Mortierella sp. J J Environ Chemical Eng 5:3616–3621

Dua M, Sethunathan N, Johri AK (2002) Biotechnology bioremediation success and limitations. Appl Microbiol Biotechnol 59(2–3):143–152

Fulekar MH, Sharma J, Tendulkar A (2012) Bioremediation of heavy metals using biostimulation in laboratory bioreactor. Environ Monit Assess 184(12):7299–7307

Gautam SP, Bundela PS, Pandey AK, Jamaluddin Awasthi MK, Sarsaiya S (2012) Diversity of cellulolytic microbes and the biodegradation of municipal solid waste by a potential strain. Int J Microbiol 2012:325907., 12 pages. https://doi.org/10.1155/2012/325907

Ghosh A, Das P (2014) Optimization of copper adsorption by soil of polluted wasteland using response surface methodology. Indian Chem Eng 56:29–42

Ghosh A, Ghosh Dastidar M, Sreekrishnan TR (2015) Recent advances in bioremediation of heavy metals and metal complex dyes: review. J Environ Eng C4015003:1–14

Ghosh A, Dastidar MG, Sreekrishnan TR (2016) Response surface optimization of bioremediation of acid black 52 (Cr complex dye) using Aspergillus tamarii. Environ Technol 38:1–12

Guo H, Luo S, Chen L, Xiao X, Xi Q, Wei W, Zeng G, Liu C, Wan Y, Chen J, He Y (2010) Bioremediation of heavy metals by growing hyperaccumulator endophytic bacterium *Bacillus* sp. L14. Bioresour Technol 101(22):8599–8605

Hoornweg D, Bhada-Tata P (2012) What a waste: a global review of solid waste management. In: Urban development series. Urban Development and Local Government Unit. Sustainable Development Network. The World Bank. Washington, DC, 20433. USA. March, 2012, No.5. Website: www.worldbank.org/urban; https://siteresources.worldbank.org/INTURBANDEVELOPMENT/Resources/336387-1334852610766/What_a_Waste2012_Final.pdf

Iqbal J, Ahemad M (2015) Recent advances in bacteria-assisted phytoremediation of heavy metals from contaminated soil. In: Chandra R (ed) Advances in biodegradation and bioremediation of industrial waste. CRC, Boca Raton, pp 401–423

Jing Q, Zhang M, Liu X, Li Y, Wang Z, Wen J (2017) Bench-scale microbial remediation of the model acid mine drainage: effects of nutrients and microbes on the source bioremediation. Int Biodeterior Biodegrad 00(00):1–5. https://doi.org/10.1016/j.ibiod.2017.01.009

Karigar CS, Rao SS (2011) Role of microbial enzymes in the bioremediation of pollutants: a review. Enzyme Res. Article ID 805187. https://doi.org/10.4061/2011/805187

Kavamura VN, Esposito E (2010) Biotechnological strategies applied to the decontamination of soils polluted with heavy metals. Biotechnol Adv 28:61–69

Kensa MV (2011) Bioremediation: an overview. J Ind Pollut Control 27(2):161–168

Lee GF, Lee AJ (1994) Impact of municipal and industrial non-hazardous waste landfills on public health and the environment: an overview'. Prepared for California EPA Comparative Risk Project, Sacramento, May (1994). http://www.gfredlee.com/Landfills/cal_risk.pdf

Liang C, Das K, McClendon R (2003) The influence of temperature and moisture contents regimes on the aerobic microbial activity of a biosolids composting blend. Bioresour Technol 86:131–137

Liu SH, Zeng GM, Niu QY, Liu Y, Zhou L, Jiang LH, Tan XF, Xu P, Zhang C, Cheng M (2017) Bioremediation mechanisms of combined pollution of PAHs and heavy metals by bacteria and fungi: a mini review. Bioresource Technology 224:25–33

Ma XK, Ding N, Peterson EC, Daugulis AJ (2016) Heavy metals species affect fungal-bacterial synergism during the bioremediation of fluoranthene. Appl Microbiol Biotechnol 100:7741–7750

Mani D, Kumar C (2014) Biotechnological advances in bioremediation of heavy metals contaminated ecosystems: an overview with special reference to phytoremediation. Int J Environ Sci Technol 11:843–872

Meier J, Piva A, Fortin D (2012) Enrichment of sulfate-reducing bacteria and resulting mineral formation in media mimicking pore water metal ion concentrations and pH conditions of acidic pit lakes. FEMS Microbiol Ecol 79:69–84

Passatore L, Rossetti S, Juwarkar AA, Massacci A (2014) Phytoremediation and bioremediation of polychlorinated biphenyls (PCBs): state of knowledge and research perspectives. J Hazard Mater 278:189–202

Paul D, Pandey G, Pandey J, Jain RK (2005) Accessing microbial diversity for bioremediation and environmental restoration. Trends Biotechnol 23:135–142

Rayu S, Karpouzas DG, Singh BK (2012) Emerging technologies in bioremediation: constraints and opportunities. Biodegradation 23:917–926

Selin E (2013) Solid waste management and health effects – a qualitative study on awareness of risks and environmentally significant behavior in Mutomo, Kenya. http://www.diva-portal.org/smash/get/diva2:607360/FULLTEXT02

Stenuit B, Eyers L, Schuler L, Agathos SN, George I (2008) Emerging high-throughput approaches to analyse bioremediation of sites contaminated with hazardous and/or recalcitrant wastes. Biotechnol Adv 26:561–575

UNEP (2007) Environmental pollution and impacts on public health: implications of the Dandora Municipal Dumping Site in Nairobi

Yadav A, Chowdhary P, Kaithwas G, Bharagava RN (2017) Toxic metals in environment, threats on ecosystem and bioremediation approaches. In: Das S, Dash HR (eds) Handbook of metal-microbe interactions and bioremediation. CRC, Taylor & Francis Group, USA, p 813. http://www.unep.org/urban_environment/pdfs/dandorawastedump-reportsummary.pdf 2012-11-20

Zabbey N, Sam K, Onyebuchi AT (2017) Remediation of contaminated lands in the Niger Delta, Nigeria: prospects and challenges. Sci Total Environ 586:952–965

Zhang XX, Cheng SP, Zhu CJ, Sun SL (2006) Microbial PAH-degradation in soil: degradation pathways and contributing factors. Pedosphere 16:555–565

Zimmer D, Baum C, Leinweber P, Hrynkiewicz K, Meissner R (2009) Associated bacteria increase the phytoextraction of cadmium and zinc from a metal-contaminated soil by mycorrhizal willows. Int J Phytoremediation 11:200–213

Chapter 4
Anammox Cultivation in a Submerged Membrane Bioreactor

M. Golam Mostafa

Abstract A submerged anaerobic membrane bioreactor (AnMBR) was seeded with digester sludge and operated aiming at demonstrating a practical rapid anammox start-up and enrichment process. The study has aimed at the anammox start-up and enrichment process in a submerged bioreactor and discussed about the performance of anammox bacteria for treating industrial wastewater with relevant references. Several reports showed that anaerobic membrane bioreactors (AnMBRs) were operated for a period of 100–500 days under a completely anaerobic environment and fed with synthetic media. The maximum ammonium nitrogen removal rate was achieved to be 3.12 kg-N/m^3/day under anaerobic nitrogen loading rate of 4.1 kg-N/m^3/day in a granular anammox reactor. Mixed liquor suspended solids (MLSS) were 1.5 g/L after 90 days, and regular membrane backwashing maintained transmembrane pressure (TMP) below 30 kPa. This work demonstrated that anammox cultures can be enriched from sludges conveniently sourced from anaerobic digesters in a reasonable time frame using the AnMBR, thus enabling the next stage of utilising the biomass for treatment of industrial wastewaters.

Keywords Anammox · Anaerobic · Nitrogen · Removal · AnMBR · Wastewater

1 Introduction

Anammox, an abbreviation for anaerobic ammonia oxidation, produces dinitrogen (N$_2$) gas from ammonia-rich water by means of nitrite as the electron acceptor in anaerobic conditions (Mulder et al. 1995; Van de Graaf et al. 1996; Strous et al. 1977). The anammox process is considered to be a promising method of removing nitrogen from ammonia-rich wastewater with a low carbon to nitrogen (C/N) ratio (Van Loosdrecht and Jetten 1998; Ciudad et al. 2005).

M. G. Mostafa (✉)
Institute of Environmental Science, University of Rajshahi, Rajshahi, Bangladesh
e-mail: mgmostafa@ru.ac.bd

© Springer Nature Singapore Pte Ltd. 2019
R. N. Bharagava, P. Chowdhary (eds.), *Emerging and Eco-Friendly Approaches for Waste Management*, https://doi.org/10.1007/978-981-10-8669-4_4

In this process, the nitrification reaction is carried out in two steps. Firstly, ammonia is converted to nitrite by ammonia-oxidising bacteria (AOB). Secondly, the denitrification of nitrite is carried out by anammox bacteria where ammonium acts as the electron donor to yield nitrogen gas. This process of partial nitrification (Abeliovich 1992) and anammox reaction (Strous et al. 1998) proceed by the following two reactions:

$$NH_4^+ + 0.75O_2 \rightarrow 0.5NH_4^+ + 0.5NO_2^- + H^+ + 0.5H_2O \tag{4.1}$$

$$NH_4^+ + 1.32NO_2^- + 0.066HCO_3^- + 0.13H^+ \rightarrow 1.02N_2 + 0.26NO_3^- \\ + 0.066CH_2O_{0.5}N_{0.15} + 2.03H_2O \tag{4.2}$$

The autotrophic process has shown promise in being part of improving the sustainability of wastewater treatment in that it reduces the need for carbon addition, oxygen consumption and the emission of nitrous oxide during oxidation of ammonia (Jetten et al. 1997). In practical terms, these features have interested the wastewater treatment industry because they translate to lower energy consumption, lower sludge production and environmental friendly gas generation. It offers several advantages over the well-established industrial equivalent, nitrification-denitrification, including higher ammonium loading rate, lower operational cost and less space requirement (Schmidt et al. 2003; Chowdhary et al. 2017). Several studies have been conducted in the last two decades on various ammonium-rich wastewaters (van Dongen et al. 2001; Fux et al. 2002, 2004; van der Star et al. 2007; Reginatto et al. 2005; Ahn and Kim 2004; Hwang et al. 2005; Toh and Ashbolt 2002; Chen et al. 2007; Mostafa et al. 2011).

Anammox also has great potential in reducing CO_2 emissions by up to 90% due to no organic carbon addition is needed and is considered to be a "green" process and considering the need to reduce our emissions, with several full-scale anammox reactors being set up (Van der Star et al. 2007).

Much like any microbial process, anammox reactors need specific conditions to create the environment needed for the bacterial growth and its activities during the operation period. Researchers are therefore targeting rapid anammox culture expansion and enrichment processes as well as high nitrogen removal rates. The anammox growth depends on the seed sludge characteristic, anammox species, feeding media, temperature, pH and other physiochemical environments. Several reports showed that full-scale operation for anammox start-up process varies from 60 to 400 days (Van der Star et al. 2007; Samik et al. 2010; Lopez et al. 2008; Huosheng et al. 2012). A membrane bioreactor (MBR) is a commercial technology for aerobic water treatment (Fenu et al. 2010; Yeon et al. 2005; Chang et al. 2001) and could be useful in addressing the issue preventing wider industrial uptake of anammox of slow cultivation. In the MBR, the membranes are highly effective at retaining biomass, particularly since they do not require the solids to possess the settling property as would be needed for gravity-based separators (e.g. clarifiers). A MBR configuration for culturing anammox bacteria showed that a solids retention time (SRT) as low as 12 days can be achieved, i.e. doubling times are less than 10 days, compared to up to 30 days for conventional anammox reactors (Van der Star et al. 2008). But this work featured seeding of the MBR with a confirmed anammox culture. However, accessing anammox cultures is not always convenient, so the concept of using the MBR to rapidly grow anammox from more convenient sources is highly desirable. Wang et al. (2009) successfully utilised an anaerobic MBR (AnMBR) to cultivate anammox starting from more readily available aerobic and nitrifying activated sludge. While AnMBR as a useful tool for cultivating anammox seeded from a variety of convenient sources has progressed, membrane fouling has not been a major focus, yet features such as flux and cleaning frequency are tied closely to the economics of MBRs. Several strategies have been used to minimise membrane fouling, including hydraulic backwashing, interval operation, subcritical flux operation, periodic physical or chemical cleaning, etc. (Chang et al. 2002; Jeison and Lier 2006a; Liao et al. 2006; Mostafa 2012). This study aims to demonstrate the rapid cultivation of anammox in the AnMBR from naturally occurring sources using a state-of-the-art hollow fibre MBR membrane and then explore fouling via TMP rise at different fluxes. The flux limit of the anammox cultivating an MBR using the most common in situ cleaning technique, regular hydraulic backwashing, can then be compared to the state-of-the-art AnMBR with a view to propose further membrane fouling studies for anammox cultivation.

2 Anammox Bioreactors

A lab-scale AnMBR with a definite working volume can be operated in a particular period (100–500 days). This study considered an anaerobic anammox submerged bioreactor operating for about 150 days, and the results are discussed in the literature reports to understand the potentiality of anammox bacteria in removing nitrogen from industrial wastewater. A glass bioreactor equipped with ports for gas inlet, gas outlet, feed inlet (influent), and effluent outlet is set up to form the AnMBR as shown in Fig. 4.1. The reactor is equipped with water jacket for temperature control. State-of-the-art MBR hollow fibre UF membranes (nominal pore size usually 0.04 m and a total area 0.03 m^2) made from PVDF were sealed to a fitting on the bioreactor. A peristaltic pump is attached to the membrane fitting outside the reactor to draw filtered permeate from the bioreactor which controlled the hydraulic retention time (HRT). An automatic backwashing timer is attached to the pump to control fouling. The backwash cycle is set for 2 min in every hour. In the beginning of the experiment, the HRT is set at a definite rate for a few days such as 60 days, and then it can be at a relatively higher rate for the remaining operation period. The HRT can also be set at the definite rate throughout the operation period. The pH inside the reactor is controlled at around 7.5 ± 0.5 with an automatic pH controller that is connected to a source of dilute nitric acid (0.1 M) and sodium hydroxide (0.1 M). A

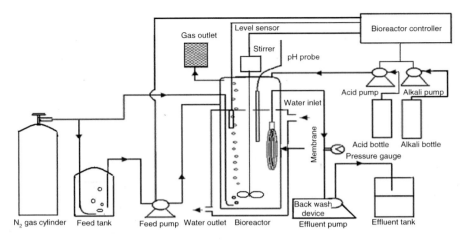

Fig. 4.1 Schematic diagram of the AnMBR

peristaltic pump is used to feed the reactor from a feed tank, which is engaged by a level sensor inside the reactor to make up for lost liquid via the membrane. Nitrogen gas is continuously sparging from the bottom of the reactor as well as the feed bottle to maintain maximum possible anaerobic conditions in the reactor (dissolved oxygen (DO) below 0.1 mg/L) measured continuously by a probe in the reactor. The reactor is to be seeded with 10% v/v anaerobic sludge collected from a food industry. The mixing between the anammox bacteria cells and the substrate is achieved through continuous stirring speed at 100 rpm during the reactor operation.

3 Synthetic Wastewater

The composition of minerals in synthetic wastewater is followed as described by van de Graaf et al. (1996) and is shown in Table 4.1. Trace element solution (1.25 mL/L) is added to the feed medium. This is prepared as described by Strous et al. (1999) and is shown in Table 4.2.

The reactor is supplied with the same synthetic wastewater makeup throughout the operation period.

Table 4.1 Composition of synthetic feed medium

Component	Concentration (mg/L)
KHCO$_3$	1250
KH$_2$PO$_4$	25
CaCl$_2 \cdot$ 2H$_2$O	300
MgSO$_4 \cdot$ 7H$_2$O	200
FeSO$_4$	6.25
EDTA	6.25

Table 4.2 The composition of trace element solution

Component	Concentration (mg/L)
EDTA	15,000
ZnSO4·7H2O	430
CoCl$_2 \cdot$ 6H$_2$O	2404
MnCl$_2 \cdot$ 4H$_2$O	990
CuSO$_4 \cdot$ 5H$_2$O	250
NaMoO$_4 \cdot$ 2H$_2$O	220
NiCl$_2 \cdot$ 2H$_2$O	190
NaSeO$_4 \cdot$ 10H$_2$O	210
H$_3$BO$_4$	14
NaWO$_4 \cdot$ 2H$_2$O	50

4 Analytical Methods

The influent and effluent samples are usually collected and analysed twice a week. The samples are analysed according to the standard methods for the analyses of water and wastewater (APHA 2012). The analysed parameters are NH_4^--N, NO_2^--N and NO_3^--N concentrations (mg/L), chemical oxygen demand (COD) (mg/L), DO (mg/L) and pH. The concentrations of nitrogen compounds and COD in the effluent and influent are measured using a spectrophotometer. The biomass concentration will be observed as total suspended solids (TSS) and volatile suspended solids (VSS). The membrane fouling rate is evaluated as development of the transmembrane pressure (TMP) during the operational time. Mixed liquor suspended solids (MLSS) and mixed liquor volatile suspended solids (MLVSS) concentrations are measured using gravimetric methods. Nitrogen removal efficiency for NH_4^--N, NO_2^--N and NO_3^-N is calculated using the following equations:

$$R_i = \frac{Ci_{inf} - Ci_{eff}}{Ci_{inf}} \times 100$$

where

R_i is the removal efficiency (%) of a parameter, i
Ci_{inf} is concentration of a parameter, i, in the influent
Ci_{eff} is concentration of a parameter, i, in the effluent

5 Start-Up Stage

A synthetic wastewater is formulated that contained mainly nitrite and ammonium to support anammox activity. Ammonium and nitrite are supplemented to mineral medium in the form of $(NH_4)_2SO_4$ and $NaNO_2$, and measured influent and effluent concentrations of nitrogen in NH_4^+ and NO_2^- are specified in the discussion section. The AnMBR is inoculated with concentrated anaerobic digester sludge biomass (10% v/v) and operated under strictly anaerobic environment. Well-flocculated microorganisms are found in the seeding biomass (Fig. 4.2), indicated by the visual observation of size flocs in the reactor vessel. The microorganisms grew rapidly under the conditions applied to the reactor, and flocs of microorganism appeared after only about 30–40 days. Microscopic observation at 1000× magnification showed high concentration of cells with a distinct orange colour being evidence of anammox (Fig. 4.3).

6 Nitrogen Removal Performance

The reactor's key performance is measured in terms of nitrogen conversion of the synthetic feed media. The results showed an average of ammonium nitrogen and nitrite nitrogen concentrations in mg/L present in the influent. Researchers suggested that the concentration of nitrite during the start-up is of crucial importance for growth. A too low amount of nitrite will lead to slower growth, while concentrations above 50–150 mg-N/L can already lead to inhibition (Strous et al. 1999; Egli et al. 2001; Dapena-Mora et al. 2007). Thus, NH_4^+-N concentration is suggesting to be taken almost double to NO_2^--N concentration to create a suitable environment for anammox growth. The effluent ammonium, nitrite and nitrate nitrogen concentrations decreased with increasing the reactor operating days the effluent ammonium nitrogen concentrations stayed usually drops less than 5 mg/L after 100 days of operation period. The nitrate concentration in the effluent indicates partial

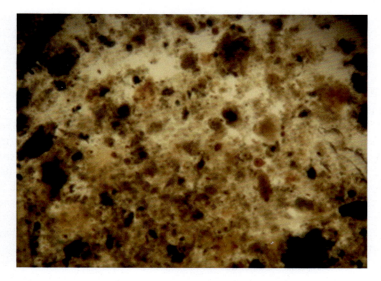

Fig. 4.2 Image of flocculating microorganisms (anammox bacterial cells) in seeding biomass (1000× magnification)

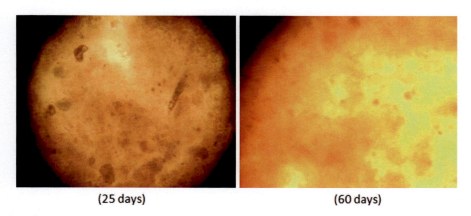

Fig. 4.3 Image of anammox bacterial cells after 25 and 60 days in an AnMBR (1000× magnification)

nitration and anammox reactions proceeded simultaneously during the first 100 days. Similar observation was found in previous reports (Abeliovich 1992; Strous et al. 1998).

The maximum ammonium nitrogen, nitrite nitrogen and total nitrogen removal is usually achieved after 140 days of inoculation. The conversion of both ammonium and nitrite nitrogen indicates that anammox start-up and growth are successfully achieved, with high ammonium conversion rates starting from day 60. A report showed that pH did not cause any influence on the anammox activity in removing nitrogen from the water and the reactor was operated even at high ammonium and nitrite concentration in an upflow anammox reactor (Mostafa et al. 2011). A further characterisation of the microbial culture by a molecular mechanism, such as PCR or FISH analysis, would have given additional conformation of the presence of an anammox culture in addition to the morphological characteristics and nitrogen utilisation pattern. Approximately 20 mg/L of the nitrate in the effluent can be attributed to anammox activity in the reactor. The presences of high levels of nitrate in the effluent if the aerobic activities inside the reactor, where ammonia is converted to nitrite and then nitrate due to air intrusion when the AnMBR is opened for repairs. However a similar trend in nitrate concentration in the effluent has been reported by others, where nitrate concentration is found higher (about 40 mg/L) in the first 200 days of inoculation and then gradually decreased with time (Tang et al. 2010a, b). The highest removal rate was achieved to be 3.12 kg-N/m^3/day under anaerobic nitrogen loading rate of 4.1 kg-N/m^3/day in a granular anammox reactor (Sen Qiao et al. 2010). The results indicated that the anammox in an anaerobic submerged bioreactor exhibited stable anaerobic ammonium oxidising performance. The nitrogen removal rate observed was higher than the reported value mentioned for other reactors (Tang et al. 2010a, b). Nitrate is the important product of the anammox reaction and thought to be an electron acceptor of this process in the earlier studies. Zhiyong Tian et al. conducted an analysis of nitrate influence on total nitrogen removal, and the results showed that the nitrate of 23.56–463.14 mg N/L had no significant effect on the nitrogen removal performance in an anammox reactor (Zhiyong Tian et al. 2015).

Thus, higher nitrate concentration did not affect the ammonium nitrogen and total nitrogen removal rates in this study. Trigo et al. (2006) illustrated that nitrogen removal rate achieved up to 710 mg l^{-1} per day with almost full nitrite removal. Another study reported conversions of NH_4^+ and NO_2^- at 0.016 kg-N/m^3/day and 0.025 kg-N/m^3/day, respectively, from 530 to 800 days (Van der Star et al. 2007). Thus, the previous report showed that a rapid anammox start-up and enrichment process was achieved compared to past work which is in part made possible by the use of membranes to retain anammox biomass. Further work by van der Star and co-workers (2008) using MBR to cultivate anammox demonstrated similar rapid conversions, reaching an order of magnitude of higher nitrite nitrogen conversion. Their faster growth compared to the previous work is possibly due to seeding of their reactor with granular sludge from a full-scale anammox reactor. Therefore, the study revealed a relatively rapid start-up seeding with sludge from an anaerobic digester instead of an operating anammox bioreactor. The AnMBR reduces biomass loss, where the growth and development of anaerobic cultures inside the reactor have been realised by several researchers (Wang et al. 2009; Huang et al. 2011; Liao et al. 2006; Stefania et al. 2012; Hongjun et al. 2011; Martinez-Sosa et al. 2012). The results from several reports indicated that the AnMBR successfully achieved anammox growth within 60 days of seeding and has the ability to remove significant amounts of nitrogen from wastewater (van der Star et al. 2008). Thus, anaerobic sludge seeded in AnMBR would be a suitable option for the application of anammox process to treat nearby ammonia-rich industrial wastewaters.

7 Influent COD/NH₄-N

The average COD removal efficiency was about 40%, and its concentration was found to be 62 mg/L, which is *below the COD effluent discharge* concentration limits of the UNEP. It is known that COD must be maintained less than 300 mg/L to ensure dominant anammox activity (Molinuevo et al. 2009). Figure 4.4 shows the relation between average rate of influent COD/NH₄-N and NH₄-N removal efficiency (%) between 60 and 145 days. The removal efficiency of ammonium nitrogen decreased with the increase of COD/NH₄-N. The figure shows that the average NH₄-N removal efficiencies achieved were 79%, 92% and 95% for the average ratio of COD/NH₄-N of 1.54, 1.44 and 1.39, respectively (Fig. 4.4). This trend is similar to anaerobic and aerobic systems; for example, in the work by Ahmed et al. (2007), removal efficiencies are 88%, 80% and 69% for the influent ratio COD/NH₄-N of 7.2, 9.9 and 14.7, respectively. It has been shown that ammonia concentration below 350 mg/L does not have an inhibitory effect on nitrification (Kim et al. 2008). Therefore, the nitrification efficiency may be increased with increasing ammonium concentration in the feed (nitrogen loading). The study suggests that the nitrification efficiency may be inhibited by substrate (ammonium nitrogen) concentration, i.e. ammonium concentration decreases in loading led to an increase in nitrification rate.

8 Membrane Performance

Microorganisms' growth inside an AnMBR can significantly affect membrane fouling (Huang et al. 2011; Haandel and Lettinga 1994; Liao et al. 2006). High mixed liquor suspended solids (MLSS) concentrations lead to membrane fouling in an anaerobic AnMBR treating wastewater (Jeison and Lier 2006b). In this study MLSS is 0.9 and 1.5 g/L after 60 and 90 days, respectively. Due to the small volume of the

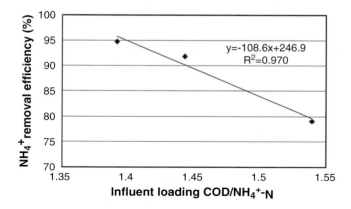

Fig. 4.4 Relationship between NH₄⁺-N removal efficiency and influent ratio of COD/NH₄⁺-N

reactor, very low average permeate flux of 0.9 L/m²/h is maintained from 1 to 60 days, and then it is set at 2 L/m²/h from 61 to 96 days. The transmembrane pressure (TMP) values immediately prior to backwash, over the operation time, are shown in Fig. 4.5. These TMP values increased with the operation, which indicates gradual membrane fouling. There is not enough data present to determine if this fouling is due to increasing anammox activity or simply gradual irreversible fouling over time. After every backwash, TMP increased rapidly in the beginning and then increased slowly with time. TMP stabilised within 20 min of each backwash suggesting more frequent backwashes could have reduced TMP build-up.

The influence of flux on membrane fouling is shown in Fig. 4.5, where TMP rate increased with decreasing HRT (increasing flux). Analyses are performed on the TMP rise rate during the observed rising periods after 20 days and before 100 days. Prior to 60 days of operation when flux is 0.9 L/m²/h, TMP rise rate is 0.07 kPa/day. Upon increasing to 2.0 L/m2/h flux, the rate nearly doubled at 0.13 kPa/day. The TMP stabilised after 90 days at 25.3 kPa, most likely due to the maximum pressure of the pump at its current setting. As a test to observe TMP rise due to fouling, HRT is temporarily set to 0.5 day by increasing pump speed. This corresponded to flux of 5.8 L/m²/h. As a result, TMP exceeded 45 kPa within an hour starting from 25 kPa, which indicates rapid fouling and thus a need to clean the membrane more rigorously.

More frequent backwashing or more membrane area (reducing flux) can help reduce TMP rise. Sustainable aerobic MBR fluxes can be achieved up to 18 L/m²/h, but the longer-term performance is sustained at lower fluxes (Le Clech et al. 2003). The higher fluxes may be due to the different chemistries of the biomass in aerobic systems, but another effect is also gas sparging that is typically adopted in MBRs. While we did not employ significant sparging, AnMBRs can use biogas to sparge the reactor, and this has a significant effect on the critical fluxes achieved, which are

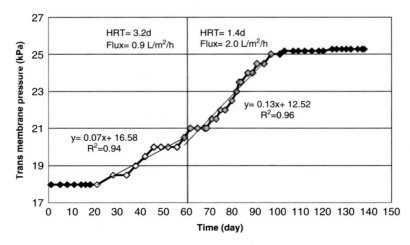

Fig. 4.5 Variation of transmembrane pressure (TMP) with time

around 20 L/m²/h with biogas sparging at 25 g of total suspended solids per litre (Jeison and Lier 2006b). The early studies of AnMBR to cultivate anammox employ fluxes that vary between 0.5 and 3.1 L/m²/h in the literature (Suneethi and Joseph 2011; Wang et al. 2009; van der Star et al. 2008), which could be considered low for aerobic MBRs and AnMBRs. A critical difference may be the gas sparging, but the viability of sparging with the anammox reactor's N_2 gas or biogas from a nearby anaerobic digester would need to be explored. So while we are able to cultivate anammox in an AnMBR at low fluxes without chemical cleaning, if higher flux is required then a critical flux study including gas sparging and varied cleaning strategies (i.e. backwash frequency and time) are needed.

We did not observe any regions/period of faster TMP rise during our 145 days of operation other than what correlated to increased flux. This may be due to the use of regular hydraulic backwashing which is a standard procedure in MBR operation.

9 Conclusions

The study observed that nitrogen removal efficiency would be higher at low COD/NH_4-N ratio. It is also shown that influent COD/NH_4-N plays an important role in nitrogen removal efficiency. The highest removal rate was achieved to be 3.12 kg-N/m³/day under anaerobic nitrogen loading rate of 4.1 kg-N/m³/day in a granular anammox reactor within 200-day operation period. The overall outcome of this study demonstrated that the submerged membrane bioreactor is an effective system to cultivate anammox from a local anaerobic sludge source within 2 months. This assists with managing the risk of operating anammox systems as sourcing biomass rapidly to start up bioreactors. Finally, cultivating anammox in the AnMBR appears to require lower than usual fluxes for MBRs, but this leads to long-term operation without significant fouling which in turn reduces the need for frequent chemical cleaning. A dedicated membrane fouling study is required to determine the types of membranes and/or required cleaning and gas sparging to achieve higher flux operation. This work demonstrated that anammox cultures can be enriched from sludges conveniently sourced from anaerobic digesters in a reasonable time frame using the AnMBR, thus enabling the next stage of utilising the biomass for treatment of industrial wastewaters. This will assist in making AnMBRs more commercially viable to rapidly cultivate anammox, supporting larger anammox bioreactors in water treatment plants.

Acknowledgements The author would like to thank Prof. Mikel Duke and Dr. Thomas Yeager at the Institute for Sustainability and Innovation, College of Engineering and Science, Victoria University, for their discussions and suggestions that contributed to this work. The author also sincerely acknowledges the Australia Endeavour Research Award 2011 for partially supporting this work at the Institute for Sustainability and Innovation, Victoria University, Melbourne, Australia.

References

Abeliovich A (1992) Transformations of ammonia and the environmental impact of nitrifying bacteria. Biodegradation 3:255–264

Ahmed M, Idris A, Adam A (2007) Combined anaerobic and aerobic system for treatment of textile wastewater. J Eng Sci Technol 2(1):55–69

Ahn YH, Kim HC (2004) Nutrient removal and microbial granulation in an anaerobic process treating inorganic and organic nitrogenous wastewater. Water Sci Technol 50(6):207–215

APHA (American Public Health Association) (2012) Standard methods for the examination of water and wastewater.22nd edn. APAH, Washington, DC

Bagchi S, Biswas R, Nandy T (2010) Start-up and stabilization of an Anammox process from a non-acclimatized sludge in CSTR. J Ind Microbiol Biotechnol 37:943–952

Casu S, Crispino NA, Farina R, Mattioli D, Ferraris M, Spagni A (2012) Wastewater treatment in a submerged anaerobic membrane bioreactor. J Environ Sci Health A: Toxic/Hazard Subst Environ Eng 47(2):204–209

Chang IS, Gander M, Jefferson B, Judd SJ (2001) Low-cost membranes for use in a submerged MBR original research article. Process Saf Environ Prot 79930:183–188

Chang IS, Clech LP, Jefferson B, Judd S (2002) Membrane fouling in membrane bioreactors for wastewater treatment. J Environ Eng-ASCE 128(11):1018–1029

Chen XL, Zheng P, Jin RC (2007) Biological nitrogen removal from monosodium glutamate-containing industrial wastewater with the anaerobic ammonium oxidation (ANAMMOX) process. Acta Sci Circumst 27(5):747–752 (in Chinese)

Chowdhary P, Yadav A, Kaithwas G, Bharagava RN (2017) Distillery wastewater: a major source of environmental pollution and its biological treatment for environmental safety. In: Singh R, Kumar S (eds) Green technologies and environmental sustainability. Springer International, Cham, pp 409–435

Ciudad G, Rubilar O, Munoz P, Ruiz G, Chamy R, Vergara C, Jeison D (2005) Partial nitrification of high ammonia concentration wastewater as a part of a shortcut biological nitrogen removal process. Process Biochem 40:1715–1719

Dapena-Mora A, Fernández I, Campos JL, Mosquera-Corral A, Méndez R, MSM J (2007) Evaluation of activity and inhibition effects on Anammox process by batch tests based on the nitrogen gas production. Enzyme Microbiol Technol 40(4):859–865

Egli K, Fanger U, Alvarez PJJ, Siegrist H, Van Der Meer JR, Zehnder AJB (2001) Enrichment and characterization of an anammox bacterium from a rotating biological contactor treating ammonium-rich leachate. Arch Microbiol 175:198–207

Fenu A, Guglielmi G, Jimenez J, Spèrandio M, Saroj D, Lesjean B, Brepols C, Thoeye C, Nopens I (2010) Activated sludge model (ASM) based modelling of membrane bioreactor (MBR) processes: a critical review with special regard to MBR specificities review article. Water Res 44(15):4272–4294

Fux C, Boehler M, Huber P, Brunner I, Siegrist H (2002) Biological treatment of ammonium-rich wastewater by partial nitritation and subsequent anaerobic ammonium oxidation (anammox) in a pilot plant. J Biotechnol 99(3):295–306

Fux C, Marchesi V, Brunner I, Siegrist H (2004) Anaerobic ammonium oxidation of ammonium-rich waste streams in fixed-bed reactors. Water Sci Technol 49(11):77–82

Haandel ACV, Lettinga G (eds) (1994) Anaerobic sewage treatment: a practical guide for regions with a hot climate. Wiley, Chichester

Hongjun L, Jianrong C, Fangyuan W, Linxian D, Huachang H (2011) Feasibility evaluation of submerged anaerobic membrane bioreactor for municipal secondary wastewater treatment. Desalination 280(1–3):120–126

Huang Z, Ong SL, Ng HY (2011) Submerged anaerobic membrane bioreactor for low-strength wastewater treatment: effect of HRT and SRT on treatment performance and membrane fouling. Water Res 45(2):705–713

Huosheng L, Shaoqi Z, Weihao M, Guotao H, Bin X (2012) Fast start-up of ANAMMOX reactor: operational strategy and some characteristics as indicators of reactor performance. Desalination 286:436–441

Hwang IS, Min KS, Choi E, Yun Z (2005) Nitrogen removal from piggery waste using the combined SHARON and ANAMMOX process. Water Sci Technol 52(10–11):487–494

Jeison D, Lier v JB (2006a) On-line cake-layer management by trans-membrane pressure steady state assessment in anaerobic membrane bioreactors for wastewater treatment. Biochem Eng J 29(3):204–209

Jeison D, Lier v JB (2006b) Cake layer formation in anaerobic submerged membrane bioreactors (AnSMBR) for wastewater treatment. J Membr Sci 284(1–2):227–236

Jetten MSM, Horn SJ, Van Loosdrecht MCM (1997) Towards a more sustainable wastewater treatment system. Water Sci Technol 35(9):171–180

Kim YM, Park D, Lee DS, Park JM (2008) Inhibitory effects of toxic compounds on nitrification process for cokes wastewater treatment. J Hazard Mater 152:915–921

Le Clech P, Jefferson B, Chang In S, Judd SJ (2003) Critical flux determination by the flux- step method in a submerged membrane bioreactor. J Membr Sci 227:81–93

Liao BQ, Kraemer JT, Bagley DM (2006) Anaerobic membrane bioreactors: applications and research directions. Crit Rev Environ Sci Technol 36(6):489–530

Lopez H, Sebastià P, Ramon G, Maël R, Balaguer MD, Jesús C (2008) Start-up and enrichment of a granular anammox SBR to treat high nitrogen load wastewaters. J Chem Technol Biotechnol 3:233–241

Martinez-Sosa D, Helmreich B, Horn H (2012) Anaerobic submerged membrane bioreactor (AnSMBR) treating low-strength wastewater under psychrophilic temperature conditions. Process Biochem 47(5):792–798

Molinuevo B, Garcia MC, Karakashev D, Angelidaki I (2009) Anammox for ammonia removal from pig manure effluents: effect of organic matter content on process performance. Bioresour Technol 100:2171–2175

Mostafa MG (2012) Recent advances in anammox bioreactors for industrial wastewater reuse. J Pet Environ Biotechnol 3(5):108. https://doi.org/10.4172/2157-7463.1000e108 (OMICS group of publications, USA)

Mostafa MG, Kawakubo Y, Furukawa K (2011) Removal of nitrogen from ammonium-rich synthetic wastewater in an upflow column type anammox reactor. Int J Water Resour Environ Eng 3(9):189–195

Mulder A, Van de Graaf AA, Robertson LA, Kuenen JG (1995) Anaerobic ammonium oxidation discovered in a denitrifying fluidized bed reactor. FEMS Microbiol Ecol 16:177–184

Qiao S, Yamamoto T, Misaka M, Isaka K, Sumino T, Bhatti Z, Furukawa K (2010) High-rate nitrogen removal from livestock manure digester liquor by combined partial nitritation–anammox process. Biodegradation 21:11–20

Reginatto V, Teixeira RM, Pereira F, Schmidell W, Furigo Jr A, Menes R, Etchebehere C, Soares HM (2005) Anaerobic ammonium oxidation in a bioreactor treating slaughterhouse wastewater. Braz J Chem Eng 22(4):593–600

Schmidt I, Sliekers O, Schmid M, Bock E, Fuerst J, Kuenen JG, Jetten MSM, Strous M (2003) New concepts of microbial treatment processes for the nitrogen removal in wastewater. FEMS Microbiol Rev 27:481–492

Strous M, Van Gerven E, Kuenen JG, Jetten MSM (1977) Appl Environ Microbiol 63:2446–2448

Strous M, Heijnen JJ, Kuenen JG, Jetten MSM (1998) The sequencing batch reactor as a powerful tool for the study of slowly growing anaerobic ammonium oxidizing microorganisms. Appl Microb Biotechnol 50:589–596

Strous M, Kuenen JG, Jetten MSM (1999) Key physiology of anaerobic ammonium oxidation. Appl Environ Microbiol 65:3248–3250

Suneethi S, Joseph K (2011) ANAMMOX process start up and stabilization with an anaerobic seed in Anaerobic Membrane Bioreactor (AnMBR). Bioresour Technol 102:8860–8867

Tang CJ, Zheng P, Wanga CH, Mahmood Q (2010a) Suppression of anaerobic ammonium oxidizers under high organic content in high-rate Anammox UASB reactor. Bioresour Technol 101:1762–1768

Tang CJ, Zheng P, Zhang L, Chen JW, Mahmood Q, Chen X-G, Hua B-L, Wang C-H, Yua Y (2010b) Enrichment features of anammox consortia from methanogenic granules loaded with high organic and methanol contents. Chemosphere 79:613–619

Tian Z, Zhang J, Song Y (2015) Several key factors influencing nitrogen removal performance of anammox process in a bio-filter at ambient temperature. Environ Earth Sci 73:5019–5026

Toh SK, Ashbolt NJ (2002) Adaptation of anaerobic ammonium- oxidising consortium to synthetic coke-ovens wastewater. Appl Microbiol Biotechnol 59(2–3):344–352

Trigo C, Campos JL, Garrido JM (2006) Start-up of the Anammox process in a membrane bioreactor. J Biotechnol 126(4):475–487

Van de Graaf AA, De Brijn P, Roberston LA, Jetten MSM, Kuenen JG (1996) Microbiology 142:2187–2196

Van der Star WRL, Abma WR, Blommers D, Mulder JW, Tokutomi T, Strous M, Picioreanu C, van Loosdrecht MCM (2007) Start-up of reactors for anoxic ammonium oxidation: experiences from the first full-scale anammox reactor in Rotterdam. Water Res 41(18):4149–4163

Van der Star WRL, Miclea AI, van Dongen UGJM, Muyzer G, Picioreanu C, van Loosdrecht MCM (2008) The membrane bioreactor: a novel tool to grow anammox bacteria as free cells. Biotechnol Bioeng 101(2):286–294

Van Dongen U, Jetten MSM, van Loosdrecht MCM (2001) The SHARON-Anammox process for treatment of ammonium rich wastewater. Water Sci Technol 44(1):153–160

Van Loosdrecht MCM, Jetten MSM (1998) Microbiological conversions in nitrogen removal. Water Sci Technol 38:1–7

Wang T, Hanmin Z, Fenglin Y, Sitong L, Zhimin F, Huihui C (2009) Start-up of the Anammox process from the conventional activated sludgein a membrane bioreactor. Bioresour Technol 100:2501–2506

Yeon KM, Jong-Sang P, Lee C-H, Kim S-M (2005) Membrane coupled high-performance compact reactor: a new MBR system for advanced wastewater treatment original research article. Water Res 39(10):1954–1961

Chapter 5
Toxicity, Beneficial Aspects and Treatment of Alcohol Industry Wastewater

Pankaj Chowdhary and Ram Naresh Bharagava

Abstract The alcohol industry is one of the most popular among all industries and a key contributor in the world's economic growth, but unfortunately, these industries are also considered one of the major sources of environmental pollution. Alcohol industry wastewater is highly toxic for the aquatic and the terrestrial eco-system. This wastewater contains a high concentration of biological oxygen demand, chemical oxygen demand, total solids, organic matter, potassium and sulphates and high acidic characteristics. Because of the high content of a toxic nature, it causes an adverse impact on soil structure and water bodies in the case of excessive amounts. In aquatic resources, it reduces the penetration power of sunlight causing a reduction in the photosynthetic activity of aquatic plants, dissolved oxygen content in water bodies and in terrestrial regions, and it causes genotoxic and phytotoxic effects on animal and plants respectively. Additionally, this wastewater may be used for the ferti-irrigation after proper dilution. In this chapter, we discuss in detail the positive and negative aspects of alcohol wastewater and miscellaneous treatment technologies for wastewater treatment. The aim of this chapter is also to provide updated information on the alternative uses of alcohol wastewater, such as energy production, ferti-irrigation, and other value-added products.

Keywords Alcohol industry · Environmental pollution · Toxicity · Genotoxicity · Energy production

1 Introduction

In the twenty-first century, human society faces serious environmental issues such as climate change, pollution and extinction of plants and animals. All these problems are interrelated and have originated from human detonation, uncontrolled use of

---

P. Chowdhary · R. N. Bharagava (✉)
Laboratory for Bioremediation and Metagenomics Research (LBMR), Department of Environmental Microbiology (DEM), Babasaheb Bhimrao Ambedkar University (A Central University), Lucknow, Uttar Pradesh, India

© Springer Nature Singapore Pte Ltd. 2019
R. N. Bharagava, P. Chowdhary (eds.), *Emerging and Eco-Friendly Approaches for Waste Management*, https://doi.org/10.1007/978-981-10-8669-4_5

natural resources, urbanisation and industrialisation. The distillery industries, belonging to one of the world's leading premium drinks companies, have turned to reed bed treatment systems to remove copper traces from the effluent produced. The demand for potable alcohol has been ever increasing with the more liberal attitude, rising middle class (disposable income) and less taboo/stigma in Indian society.

In developing countries, including India, the management of waste has a thrusting need for research, because the generated waste causes many environmental problems. It is well known that the nature and composition of wastewater varies from industry to industry. The distilleries industry is one of the most alarming dangers that faces living beings including humans (Chowdhary et al. 2017, 2018). The wastewater generated from distilleries creates a hazardous problem and causes serious environmental issues. In India, more than 319 distilleries are functional, which produce 3.25×10^9 L of alcohol and releases 40.4×10^9 L of wastewater annually (Bharagava et al. 2009). Approximately 5% of ethanol production comes from chemical synthesis worldwide. On the basis of available data, it may be concluded that 95% of ethanol is obtained from agriculture or from feedstocks related to agriculture (Tolmasquim 2008). The use of alcohol as a blend in motor fuel was not permitted in India until recently, which resulted in the under-capacity utilisation of distillation facilities.

In addition, the use of ethanol for mixing with petrol by the permission of the government is also responsible for excess demand. Distillation is an important process during alcohol production; in this step, the generation of wastewater, i.e. spent wash, has a very high pollution load (Singh and Dikshit 2012). In India, the first distillery was installed at Cawnpore (Kanpur) in 1805 by Carew & Co. Ltd., for rum manufacturing.

Overall, 52% of the alcohol produced is being utilised for potable and reaming and 48% for industrial uses. With many other uses such as beverages, pharmaceutical, flavouring, and medicinal, alcohol constitutes the feedstocks for a huge number of chemicals (organics), which are mainly used in a wide variety of intermediates such as solvents, drugs and pesticides. (Mall 1995). The spirit drinks produced are generally constructed of copper, or at the very least contain some copper, as this metal is mainly recognised as being important in flavour development (Harrison et al. 2011). Various physico-chemical methods have been previously developed for the mitigation/removal of xenobiotic compounds from the environment. These methods have potential, but require a high reagent dose and are too costly. During the treatment process, they generate huge amounts of sludge and form harmful products (Bharagava et al. 2009).

Conventional biological processes are not sufficient to treat these effluents. In addition to these techniques, bioremediation, an eco-friendly technique, has been reported, in which the removal of hazardous compounds from the environment occurs by the use of microorganisms (Bharagava and Chandra 2010a, b; Bharagava et al. 2017).

Conventional anaerobic–aerobic biological processes reduce the biochemical oxygen demand (BOD) of the molasses effluents, but are insufficient to treat melanoidins. The latter remain unaffected because they are highly resistant to microbial attack (Chandra and Pandey 2001). Modern biotechnological methods based on the ability of living organisms to accumulate and to degrade dangerous pollutants have proved to be the most effective, harmless and profitable ways of solving ecological problems (Kharayat 2012).

2 Toxicity of Distillery Wastewater

The distillery wastewater (DWW), i.e. spent wash, and its chemicals such as polysaccharides, lignin, melanoidin, and other fertiliser-based inorganic salts and wax (Sankaran et al. 2014). It is well known that in DWW melanoidins plays a key role in aquatic and edaphic pollution. It is also mainly responsible for high physico-chemical values such as BOD, chemical oxygen demand (COD), total dissolved solids (TDS), phenolics, chlorides, sulphate and phosphate values in DWW. In terrestrial region, i.e. agricultural land, it inhibits the seed germination by reducing alkalinity and manganese availability, resulting in a detrimental effect on crop production.

Treatment of spent wash or DWW is carried out by three main routes:

1. Concentration followed by incineration
2. Direct oxidation by air at a high temperature followed by incineration
3. Anaerobic digestion with biogas recovery followed by aerobic polishing (http://www.environmentalexpert.com)

In south Indian distilleries, anaerobic digestion is mainly used for economic environmental benefits.

2.1 Toxicity on Terrestrial Region

For the emission of greenhouse gases into the atmosphere, sugarcane vinasse may be a significant source. These greenhouse gas emissions may result from aerobic and anaerobic decomposition of the organic matter (OM) in vinasse that occurs during transportation, storage or even after application to soil (Oliveira et al. 2013). Carmo et al. (2012) found that when vinasse was applied to sugarcane fields in Brazil, this resulted in a large increase in greenhouse gas emissions, mainly N_2O. They also found that the presence of the straw on the soil surfaces induced the emission of carbon dioxide, N_2O and did not influence the changes in CH_4 resulting from vinasse application. Other authors observed the effect of vinasse on sugar cane crops in the city of Patrocínio Paulista, São Paulo, Brazil, which provides nutrients (Waal et al. 2009). The vinasse reduced the density and hatching rates of plant-pathogenic nematodes such as *Meloidogyne javanica* and *M. incognita* (Pedrosa et al. 2005). The fecundity rate of eggs and the fertility of *Drosophila melanogaster* were also reduced when exposed to high concentrations of vinasse (Yesilada 1999).

Various authors have observed that toxicity can arise in the study area because of soil granulation and chemical characteristics, and the high content of potassium, OM, Ca, Mn, N, P etc. present in vinasse, indicating a possible pollution in soil and ground water as a result of the leaching of N compound and ammonia (Chowdhary et al. 2017).

2.2 Toxicity in Aquatic Regions

Water is an essential constituent on earth for all living entities. Distillery industries discharge a huge amount of wastewater into bodies of water, mainly rivers located near these industries/refineries. Vinasse is harmful to flora and fauna, i.e. plants, animals, microbes and microflora from freshwater and distributed marine animals (Bharagava et al. 2009). It has also been observed by various researchers that DWW has a high pollution potential, 100 times more than household sewage, due to OM content, causing reduction of oxygen, low pH, high corrosivity, and high BOD, COD and TDS values (Freire and Cortez 2000; Kannan and Upreti 2008).

Distillery wastewater has a high organic load that is mainly responsible for the proliferation of microbes that deplete the dissolved oxygen in water bodies, kill aquatic animals and plants and make polluted water bodies more difficult to use as sources of potable water. In addition, vinasse has an unpleasant odour and is responsible for spreading endemic diseases such as malaria, amoebiasis and schistosomiasis (Laime et al. 2011). The authors also recommended that vinasse should not be discharged in water bodies because it destroys the aquatic ecosystem.

The infiltration of vinasse compromises groundwater potability, because it transfers high content of ammonia, Mn, Al, Fe, Mg, Cl, and OM to the water table. Ludovice (1997) reported that in channels used for vinasse ferti-irrigation, the contamination of the water table can reach 91.7%, polluting the groundwater. Cladocerans and fish were used for an acute toxicity test of vinasse before and after pH adjustment. Linear and quadratic regression models were adjusted to demonstrate the concentration response relationship between vinasse and the endpoints evaluated. The LC50 in 48 h of vinasse before pH adjustment for *Ceriodaphnia dubia* and *Daphnia magna* obtained by the authors were 0.67% and 0.80% respectively, and the median lethal concentration (LC50 96 h) for *Danio rerio* was 2.62%. After pH adjustment, the values increased for all organisms, demonstrating a decrease in toxicity. This study reported marked toxicity from vinasse in aquatic organisms, with toxicity reduction after pH adjustment (Botelho et al. 2012).

3 Genotoxicity of Distillery Wastewater

Industrial wastewater causes genotoxic and carcinogenic effects in living entities. Genotoxicity in organisms is induced by some chemical agents or substances that are present in the wastewater (Chowdhary et al. 2018). The genetic alteration, if not treated correctly, may cause cancer or other hereditary diseases. Yesilada (1999) has studied the genotoxicity evaluation of vinasse and its toxic effect on the fecundity and longevity of *Drosophila melanogaster* and larva survival. In this study, the somatic mutation and recombination test showed different concentrations of vinasse (0%, 25%, 50%, 75% and 100%). Authors found a decrease in egg production with an increase in vinasse concentration. Srivastava and Jain (2010) observed the effect of raw vinasse (1:5 v/v) on the cytomorphology of 11 genotypes of sugarcane (*Saccharum* species hybrids). The effect of vinasse on the mitotic index showed a wide spectrum of genotoxic effects, i.e. adherence and chromosomes delays, C-metaphases, multi-polarity, bi/multinucleated cells, and mutagenic effects such as chromosome breaks and micronuclei. According to these authors, these alterations were caused by the high concentrations of K, P, S, Fe, Mn, Zn and Cu and heavy metals such as Cd, Cr, Ni and Pb. Souza et al. (2013) observed a significant increase in chromosome aberrations in *Allium cepa* seeds (onion) exposed to soil samples from a landfarming facility with and without sugarcane vinasse. They also observed bridges, adherences and chromosome breaks in the presence of vinasse bridge adherence and chromosomes break.

4 Miscellaneous Use of Distillery Wastewater

Despite the range of potential adverse impacts, the land disposal of stillage still presents attractive possibilities that should be highlighted, such as nutrient and water recycling and the lower energy consumption associated with a reduced production of synthetic fertilisers. Further, several positive and negative impacts of DWW on environment and crops are shown in Fig. 5.1 and Table 5.1.

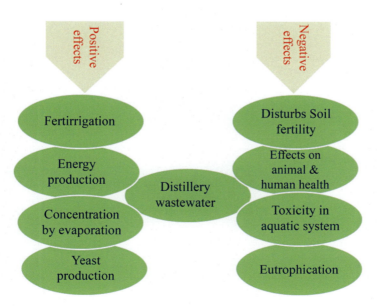

Fig. 5.1 Beneficial and adverse effects of distillery wastewater

Table 5.1 Positive and negative effects of distillery wastewater on various crops

	Types of alcohol wastewater and its concentration	Type of crops	Studied effects Positive effects	Negative effects	References
1.	Distillery wastewater (5%, 10%, 15%, 20%, 25%, 50%, 75%, 100%)	Tomato, pepper, pumpkin, cucumber and onion	Germination of seed did not inhibit (pepper, pumpkin, cucumber and onion) 10% was favourable for increasing onion seed germination	75 and 100% was more toxic for seed germination	Ramana et al. (2002)
2.	Distillery wastewater (1%, 5%, 10%, 25%)	*Oryza sativa* and *Triticum aestivum*	Seed germination increases at 1% and 5%	10 and 25% were toxic for seed germination	Ale et al. (2008)
3.	Distillery wastewater (5%, 10%, 15%, 20%, 25%, 50%, 75%, 100%)	*Zea mays* L., *Oryza sativa* L.	–	Inhibition of seed germination at an early stage, owing to non-diluted wastewater	Pandey et al. (2008)
4.	Distillery spent wash (25%, 50%, 75%, 100%)	Sugar cane (*Saccharum officinarum*) variety Co1274	Following parameter increases with an increase in the concentration of distillery spent wash up to 75% such as the height of the plant, the length of the leaves, the breadth of the leaves, the girth of the stem, the leaf area index, the number of leaves per plant, the number of tillers per plant etc.	100% was more toxic for declining parameters	Rath et al. (2010)
5.	Distillery wastewater (1%, 5%, 10%, 25%)	*Vigna angularis*, *Vigna cylindrical* and *Sorghum cernuum* (azuki bean, cowpea, and sorghum)		Decreases in seed germination with increase in wastewater concentration	Doke et al. (2011)

(continued)

Table 5.1 (continued)

	Types of alcohol wastewater and its concentration	Type of crops	Studied effects		References
			Positive effects	Negative effects	
6.	Undiluted distillery effluent (I1), one irrigation with effluent: tube-well water (1:3) at the tillering stage (I2), two irrigations with effluent: tubewell water (1:4) at tillering and 30 days after tillering stage (I3)	Sugarcane	50% recommended fertilizer dose increased the cumulative cane yields under different effluent treatments	–	Srivastava et al. (2012)

Modified from: Christofoletti et al. (2013a, b)

4.1 Ferti-Irrigation

Distillery wastewater or spent wash not only has a negative impact on the environment, but its positive aspects are also discussed by various authors. Distillery wastewater use as a fertiliser in ferti-irrigation has become most common. Literature on the technical and legal aspects still lacks reliable data relating the cons of reusing DWW in agriculture.

Camargo et al. (2009) found that ferti-irrigation consists of the infiltration of raw vinasse into the soil by irrigation of sugarcane crops. Further, Laime et al. (2011) reported the use of vinasse in ferti-irrigation to be a substitute that focuses on the rational use of natural resources, stopping the discharge of vinasse into aquatic bodies such as rivers, while fertilising agricultural land (Gianchini and Ferraz 2009). Among all the possible options for the use of vinasse established around the world, ferti-irrigation is the most popularly used, as it is a low-asset, low-maintenance, efficient application, with no complex technologies required and it increases agricultural yield (Camargo et al. 2009; Santana and Machado 2008; Fuess and Garcia 2014). The use of vinasse for agricultural practices has totally or partially replaced the use of chemical fertilisers, mainly those containing phosphorous (Corazza 1996). However, on the basis of the available literature, without proper dilution or the direct application of vinasse in the agricultural field can cause salinisation, leaching of metals present in the soil to groundwater, alteration in soil quality mainly due to improper nutrients, such as manganese, alkalinity reduction, crop losses, phytotoxicity and unpleasant odour (Navarro et al. 2000; Santana and Machado 2008). The ferti-irrigation practices give a pseudo impression of solving the problem of vinasse discharge (Santana and Machado 2008). Moreover, certain environmental factors need to be accounted for in ferti-irrigation, i.e. soil type, water retention (soil field capacity), soil salt percentage, and distance from water resources.

4.2 Energy Production

Microbial fuel cells are also becoming promising technology, producing electricity with simultaneous removal of pollutants in terms of COD, colour and TDS etc. from the wastewaters (Feng et al. 2008; Wen et al. 2010; Samsudeen et al. 2015).

Recently, biological approaches offer diverse benefits for hydrogen production such as operation under mild conditions. However, raw material cost is one of the key limitations of bio-hydrogen production. Consumption of some carbohydrate-rich, starch- or cellulose-containing solid wastes and also some food industry wastewaters is an impressive method for bio-hydrogen production. According to available data approximately 5.2 million tons of solid waste is generated per day worldwide, which can be used for the generation of useful by-products (Modak 2011). Several researchers have proposed and selected hydrogen gas as a substitute renewal source of energy and looking toward the new alternatives to H_2 gas from organic pollutants by using microorganisms (Fountoulakis and Manios 2009; Wang and Zhao 2009). Recently, many authors have reported hydrogen gas production utilising DWW as C, N, and energy sources via an anaerobic treatment process (Mishra et al. 2015). However, the main advantages of microbiological methods of hydrogen generation rely on the possibility of utilisation of industrial and municipal wastewaters, a significant reduction of costs of production and simplicity of the processes. After the bio-digestion of vinasse, it is later used in the form of fertiliser as an eco-friendly method. Moreover, biogas is chiefly used for production, because it has a high methane content. In the alcohol industry, biogas may be used to operate gas turbines combined with an electricity generator; as an alternate part of the fuels used in the agro industry during harvest time; or for use in boilers to produce vapours and to mill sugarcane (Cortez et al. 2007; Szymanski et al. 2010).

Recently, anaerobic biodigestion has gained more attention only after the progress of high-performance reactors such as the upflow anaerobic sludge blanket, which is the best adapted to vinasse. In this system, in the bottom part of the reactors, sludge adsorbs most of the OM, whereas production and removal of gas take place in two different compartments during the anaerobic process (Von Sperling 2005). According to Szymanski et al. (2010), anaerobic biodigestion is an option of great financial and environmental importance in the treatment of vinasse, as the biogas produced, once purified, has a calorific value similar to that of natural gas, with the application being a renewable and more commonly obtainable fuel. Various merits and demerits of vinasse are shown in Table 5.2.

5 Miscellaneous Approach to DWW Treatment

Because of the adverse effect of DWW on the environment and other biological systems, its proper treatment and safe management is a thrust need. Several researchers have reported the various types of biological and physiochemical

Table 5.2 Merits and demerits of vinasse application

	Application	Merits	Demerits
1.	Ferti-irrigation	Inexpensive	Costly
		Easy to implement	Transportation
			Unknown long-term effect
2.	Animal feed	Inexpensive	Little studied
		Easy to implement	
3.	Biodigestion/biogas	Energy production	Expensive
		BOD reduction	High technology
		Effluent used as fertiliser	
4.	Combustion and boilers	Complete disposal	Little studied
		Energy production	Small-scale tests
		Recovery of potassium in ashes	
5.	Production of protein	Food	Expensive
		No residue is obtained	Little studied

Adapted from: Christofoletti et al. (2013a, b)

methods for the adequate treatment of DWW, some of which are described in this section. There are many studies available in the literatures on using adsorbents, such as chemically modified bagasse, powdered activated carbon, activated charcoal, biochar, chitosan etc., for the treatment of DWW (Satyawali and Balakrishnan 2008; Agarwal et al. 2010).

The chitosan used as an adsorbent (at an optimal dosage of 10 g/L and 30 min contact time) for the effective treatment of DWW and reported 98% and 99% reduction in colour and COD respectively (Lalov et al. 2000). On the other hand, the significant colour reduction (up to 98%) from biologically treated DWW has been reported with most commonly used coagulants such as ferrous sulphate, ferric sulphate and alum under alkaline conditions (Pandey et al. 2003).

Liakos and Lazaridis (2014) reported that mitigation/removal of the colouring component (melanoidins) from stimulated and real biologically treated and untreated wastewaters was studied using the coagulation/flocculation method. In this study, the authors achieved 90% colour removal at pH 5 using a coagulation method with different concentrations of ferric ions. However, the real wastewater could be decolourised by 100 mM [Fe^{3+}] while stimulated wastewater could be decolourised by 300 mM [Fe^{3+}]. After completion of the flocculation experiment, the generated ferric hydroxide residue was washed, solubilised and re-used in a new cycle. The maximum colour reduction from the real treated, real untreated, and stimulated effluent was 95%, 90% and 45% respectively by applying 0.5 A current intensity (Liakos and Lazaridis 2014).

In addition, Satyawali and Balakrishnan (2007) have prepared 19 carbon samples by the acid and thermal activation of many agro-based by-products, i.e. bagasse, bagasse fly ash (BFA), saw dust, wood ash and rice husk ash for the colour reduction from the biomethanated DWW. The authors also observed that phosphoric acid-carbonised bagasse B (PH) has resulted in maximal colour removal (50%). Further, many commercial activated carbons ACs (ME) and AC (LB) have resulted in 80% colour removal from biomethanated DWW. Besides colour removal, these activated carbons were also found to be effective for the reduction in COD, TOC, total nitrogen content and phenol.

Apollo et al. (2013) achieved maximum colour reduction (88%) from DWW using combined treatment with an anaerobic up-flow fixed bed reactor and an annular photocatalytic reactor (as a post-treatment technique). They also found that during a single (UV photodegradation) treatment process, the colour reduction was 54% and 69% from DWW and MWW respectively. However, when UV photodegradation was applied as a pre-treatment to the anaerobic digestion process, it reduced the biogas generation and COD reduction. Farshi et al. (2013) reported 97–98% colour reduction from DWW by using electrochemical treatment at various optimised conditions, e.g. electrode distance 1 cm, pH 4, current density 2 A/dm^2 for 3 h. Arimi et al. (2015) achieved a significant reduction in colour, dissolved organic carbon and melanoidins of 92.7%, 63.3% and 48% respectively at pH 5 and a concentration of 1.6 g/L. In this experiment, the above-mentioned physico-chemical parameters were reduced by using six coagulants, of which ferric chloride was found to be more effective, resulting in 92.7% colour reduction. In another study, Arimi et al. (2015) developed an effective polishing step for the removal of colorants from melanoidin-rich DWW by using natural manganese oxides. In this process, low molecular weight colouring compounds were first removed, followed by high molecular weight colorant removal with a significant dependence on pH.

El-Dib et al. (2016) achieved 78% and 83% reduction in colour and COD respectively by using organic–inorganic nanocomposite (chitosan immobilised bentonite with chitosan content). In this study, the modified chitosan-immobilised bentonite (mCIB) and bentonite (mbent) used were prepared by intercalating cetyltrimethylammonium bromide (CTAB) as a cationic surfactant. Further, Fourier transform infrared spectroscopy (FTIR), X-ray diffraction (XRD) and scanning electron microscopy (SEM) were used to study the interlayer structure and morphology of prepared samples. Out of all the sorbents used, the modified CIB$_3$ was found to be more effective in the decolourisation of DWW.

Recently, Zhang et al. (2017) achieved ~94.0% colour reduction and ~78% reduction of dissolved OM from DWW with treatment using ferric chloride (FeCl3) as a coagulant. During the treatment process, this coagulant was found to react preferably with melanoidins (major colorant) via either surface complexation or neutralisation of an electric charge or by both mechanisms. Nure et al. (2017) reported a significant reduction in colour (64%) and COD (61%) from melanoidin solution by using activated carbon, which was produced from BFA. In this study, the surface area of used BFA was determined as 160.9 ± 2.8 m^2/g with about 90% of particles <156.8 μm in size. However, BFA was characterised by using FTIR and showed the carbonyl (R-C=O) and hydroxyl (OH-) groups, whereas XRD and SEM analysis showed an amorphous nature and heterogeneous and irregular shaped pores respectively. Sirianuntapiboon et al. (2004) isolated a strain no. WR-43-6 (*Citeromyces* sp.), which showed the highest decolourisation yield, i.e. 68.91%, from a solution containing molasses pigment in the presence of glucose 2.0%, sodium nitrate 0.1% and KH$_2$PO$_4$ 0.1% respectively at 30 °C for 8 days. Further, this bacterium was also found to be capable of removing colour (75%), BOD (76%) and COD (100%) from the stillage of an alcohol factory.

Further, Kaushik and Thakur (2009) isolated five different bacterial strains from a distillery mill site and tested for their COD and colour removal efficiency. Out of these five bacterial strains, one bacterium (*Bacillus* sp.) was found to be capable of 21 and 30% colour and COD reduction respectively from distillery spent wash. Further, under the optimised parameters such as pH, temperature aeration, carbon, nitrogen, inoculum size, and incubation time by the Taguchi approach, the same bacterium was found to be effective for 85% and 90% colour and COD reduction respectively within a 12-h incubation period.

David et al. (2015) reported that *Pseudomonas aeruginosa*, which produces polyhydroxybutyrate (PHB) in the presence of excess carbohydrate source. PHB is an intercellular polymer, which is utilised by microorganisms as an energy storage molecule when common energy sources are available in limited amounts and this bacteria in the presence of PHB resulted in 92.77% colour removal from DWW. DWW consists mainly of recalcitrant colouring compound (melanoidins) and other organic colorants that are not easily degraded in the biological treatment process. Santal et al. (2016) isolated *Paracoccus pantotrophus* and found that these bacterial strains were highly effective at decolourising melanoidins up to 81.2 ± 2.43% in the presence of carbon (glucose) and nitrogen (NH$_4$NO$_3$) sources. Further, Krzywonos et al. (2017) achieved 38% colour reduction from vinasse by using *Bacillus megaterium* ATCC 14581 and medium component (NH$_4$)$_2$SO$_4$, KH$_2$PO$_4$, yeast extract, peptone glucose and vinasse. Of these, four promising factors were chosen as follows: (NH$_4$)$_2$SO$_4$, KH$_2$PO$_4$, glucose and vinasse for further optimising the process of colour removal. Georgiou et al. (2016) reported the decolourisation of DWW by the immobilised laccase enzyme. In this study, the authors immobilised the laccase enzyme covalently on alumina or controlled pore glass-uncoated particles and achieved 71% and 74% decolourisation respectively in a 48-h incubation period. In addition, immobilised laccase on glass achieved 68% degradation of baker's wastewater in 24 h. Chen et al. (2016) achieved 97.1% colour reduction from 50% (v/v) DWW using a combined micro-electrolysis process with the help of a biological treatment method. In this study, fungal biomass and ligninolytic enzyme (LiP, MnP, and laccase) also played an important role in enhancing the decolourising efficiency of DWW. Chandra and Chowdhary (2015) also reported that laccase enzyme plays an important role in DWW decolourisation.

6 Challenges

There are many problems faced by the distillery industry, in which the key issues are wastewater treatment and safe disposal. Decolourisation or removal of DWW is a serious concern and an understanding of the chemical structure and characteristics of the colouring compounds is required to develop an appropriate approach for their removal. In a previously published article, several physico-chemical and biological treatment methods were observed (Satyawali and Balakrishnan 2007; Chowdhary et al. 2017). Among them, a physico-chemical method such as adsorption, coagulation/flocculation and oxidation have been examined in the context, but the cost of the chemicals in addition to the related sludge handling and dumping costs is a disincentive.

7 Conclusion

In this chapter, we conclude that the untreated DWW is more toxic because of the presence of undesirable organic, inorganic and other physico-chemical parameters such as BOD, COD, TDS etc. also have adverse effects on the environment. Owing to the frequent disposal of distillery waste, the physico-chemical properties of soils, lakes and rivers are changed because of waste that has low pH, electric conductivity and toxic chemicals. In addition, this complex wastewater (distillery) is also responsible for various types of phytotoxicity and genotoxicity effects on plants and animals. On the other hand, various studies prove that DWW is also used for ferti-irrigation after proper dilution. During the use of traditional practices like ferti-irrigation, environmental threats to soil and water quality may occur, and the use of more vigorous methods must be a practical solution for vinasse management. To mitigate or remove the above-mentioned problems, the treatment of wastewater is an essential step for environmental safety. Moreover, biological approaches to hydrogen production rely on the opportunity for the consumption of industrial and municipal wastewaters, a considerable reduction in the costs of production and the ease of the processes.

In this chapter, we discussed several miscellaneous (physico-chemical and biological) approaches to DWW treatment with recent information.

Acknowledgements The Rajiv Gandhi National Fellowship (Letter No. RGNF 2015-17SC-UTT-20334) awarded to Mr. Pankaj Chowdhary for Ph.D. work from the University Grant Commission (UGC), Government of India (GOI), New Delhi, is duly acknowledged.

References

Agarwal R, Lata S, Gupta M, Singh P (2010) Removal of melanoidin present in distillery effluent as a major colorant: a review. J Environ Biol 31:521–528

Ale R, Jha PK, Belbase N (2008) Effects of distillery effluent on some agricultural crops: a case of environmental injustice to local farmers in Khajura VDC. Banke Sci World 6:68–75

Apollo S, Onyango MS, Ochieng A (2013) An integrated anaerobic digestion and UV photocatalytic treatment of distillery wastewater. J Hazard Mater 261(15):435–442

Arimi MM, Zhang Y, Götz G, Geißen SU (2015) Treatment of melanoidin wastewater by anaerobic digestion and coagulation. Environ Technol 36(19):2410–2418

Bharagava RN, Chandra R (2010a) Biodegradation of the major color containing compounds in distillery wastewater by an aerobic bacterial culture and characterization of their metabolites. Biodegradation 21:703–711

Bharagava RN, Chandra R (2010b) Effect of bacteria treated and untreated post-methanated distillery effluent (PMDE) on seed germination, seedling growth and amylase activity in *Phaseolus mungo* L. J Hazard Mater 180:730–734

Bharagava RN, Chandra R, Rai V (2009) Isolation and characterization of aerobic bacteria capable of the degradation of synthetic and natural melanoidins from distillery effluent world. J Microbiol Biotechnol 25:737–744

Bharagava RN, Chowdhary P, Saxena G (2017) Bioremediation an eco-sustainable green technology, its applications and limitations. In: Bharagava RN (ed) Environmental pollutants and their bioremediation approaches. CRC Press, Taylor & Francis Group, Boca Raton, pp 1–22

Botelho RG, Tornisielo VL, Olinda RA, Maranho LA, Machado-Neto L (2012) Acute toxicity of sugarcane vinasse to aquatic organisms before and after pH adjustment. Toxicol Environ Chem 94:1–11

Camargo JA, Pereira N, Cabello PR, Teran FJC (2009) Viabilidade da aplicação do método respirométrico de Bartha para a análise da atividade microbiana de solos sob a aplicação de vinhaça. Engenharia Ambient 6:264–271

Carmo JB, Filoso S, Zotelli LC, De Sousa Neto ER, Pitombo LM, Duarte-Neto PJ, Vargas VP, Andrade CA, Gava GJC, Rossetto R, Cantarella H, Neto AE, Martinelli LA (2012) Infield greenhouse gas emissions from sugarcane soils in Brazil: effects from synthetic and organic fertilizer application and crop trash accumulation. Glob Change Biol Bioenergy 5:267–280

Chandra R, Pandey PK (2001) Decolourisation of anaerobically treated distillery effluent by activated charcoal adsorption method Ind J Environ Prot 2: 132–134

Chandra R, Chowdhary P (2015) Properties of bacterial laccases and their application in bioremediation of industrial wastes. Environ Sci: Processes Impacts 17:326–342

Chen B, Tian X, Yu L, Wu Z (2016) Removal of pigments from molasses wastewater by combining micro-electrolysis with biological treatment method. Bioprocess Biosyst Eng 39:1867–1875

Chowdhary P, Raj A, Bharagava RN (2018) Environmental pollution and health hazards from distillery wastewater and treatment approaches to combat the environmental threats: a review. Chemosphere 194:229–246

Chowdhary P, Yadav A, Kaithwas G, Bharagava RN (2017) Distillery wastewater: a major source of environmental pollution and its biological treatment for environmental safety. In: Singh R, Kumar S (eds) Green technologies and environmental sustainability. Springer International, Cham, pp 409–435

Christofoletti CA, Escher JP, Correia JE, Marinho JFU, Fontanetti CS (2013a) Sugarcane vinasse: environmental implications of its use. Waste Manag 33:2752–2761

Christofoletti CA, Pedro-Escher J, Fontanetti CS (2013b) Assessment of the genotoxicity of two agricultural residues after processing by diplopods using the *Allium cepa* assay. Water Air Soil Pollut 224:1523–1536

Corazza RI (1996) Reflexões sobre o papel das políticas ambientais e de ciência etecnologia na modelagem de opções produtivas 'mais limpas' numa perspective evolucionista: um estudo sobre disposição da vinhaça. 163f. Tese de Doutorado. Departamento de política científica e tecnológica. Instituto de Geociências, da Unicamp, Campinas-SP.

Cortez LAB, Silva A, de Lucas JJ, Jordan RA, de Castro LR (2007) Biodigestão de Efluentes. In: Cortez LAB, Lora ES (eds) Biomassa para Energia. Editora da UNICAMP, Campinas, pp 493–529

David C, Arivazhagan M, Balamurali MN, Shanmugarajan D (2015) Decolorization of distillery spent wash using biopolymer synthesized by Pseudomonas aeruginosa isolated from tannery effluent. BioMed Res Int, 2015:9 pages

Doke KM, Khan EM, Rapolu J, Shaikh A (2011) Physico-chemical analysis of sugar industry effluent and its effect on seed germination of Vigna angularis, Vigna cylindrical and Sorghum cernum. Ann Environ Sci 5:7–11

El-Dib FI, Tawfik FM, Eshaq G, Hefni HHH, ElMetwally AE (2016) Remediation of distilleries wastewater using chitosan immobilized bentonite and bentonite based organoclays. Int J Biol Macromol 86:750–755

Farshi R, Priya S, Saidutta MB (2013) Reduction of colour and COD of an aerobically treated distillery wastewater by electrochemical method. Int J Curr Eng Technol 168–171

Feng Y, Wang X, Logan BE, Lee H (2008) Brewery wastewater treatment using air-cathode microbial fuel cells. Appl Microbiol Biotechnol 78:873–880

Fountoulakis MS, Manios T (2009) Enhanced methane and hydrogen production from municipal solid waste and agro-industrial by-products co-digested with crude glycerol. Bioresour Technol 100(12):3043–3057

Freire WJ, Cortez LA (2000) Vinhaça de cana-de-açúcar. Agropecuária, Guaíba

Fuess LT, Garcia ML (2014) Implications of stillage land disposal: a critical review on the impacts of fertigation. J Environ Manag 145:210–229

Georgiou RP, Tsiakiri EP, Lazaridis NK, Pantazaki AA (2016) Decolorization of melanoidins from simulated and industrial molasses effluents by immobilized laccase. J Environ Chem Eng 4:1322–1331

Gianchini CF, Ferraz MV (2009) Benefícios da utilização de vinhaça em terras de plantio de cana-de-açúcar-Revisão de Literatura. Rev Cient Eletrônica Agron 15

Harrison B, Fagnen O, Jack F, Brosnan J (2011) The impact of copper in different parts of malt whisky pot stills on new make spirit composition and aroma. J Inst Brew 117:106–122

Kannan A, Upreti RK (2008) Influence of distillery effluent on germination and growth of mung bean (Vigna radiata) seeds. J Hazard Mater 153:609–615

Kaushik G, Thakur IS (2009) Isolation and characterization of distillery spent wash color reducing bacteria and process optimization by Taguchi approach. Int Biodeter Biodegr 63:420–426

Kharayat Y (2012) Distillery wastewater: bioremediation approaches. J Integr Environ Sci 9:69–91

Krzywonos M, Chałupniak A, Zabochnicka-Świątek M (2017) Decolorization of beet molasses vinasse by *Bacillus megaterium* ATCC 14581. Biorem J 21(2):81–88

Laime EMO, Fernandes PD, Oliveira DCS, Freire EA (2011) Possibilidades tecnológicas para a destinação da vinhaça: uma revisão. Rev Trop Ci Agric Biol 5:16–29

Lalov IG, Guerginov II, Krysteva A, Fartsov K (2000) Treatment of wastewater from distilleries with chitosan. Water Res 34:1503–1506

Liakos TI, Lazaridis NK (2014) Melanoidins removal from simulated and real wastewaters by coagulation and electro-flotation. Chem Eng J 242:269–277

Ludovice MTF (1997) Estudo do efeito poluente da vinhaça infiltrada em canal condutor de terra sobre o lençol freático. 143 p. Dissertação. Mestrado em Engenharia Civil. UNICAMP, Campinas. Available at http://libdigi.unicamp.br/document/?code=vtls000124559. Accessed 10 May 2011

Mall ID (1995) Waste utilization and management in sugar and distillery plants. Chem Eng World XXX(1):51–60

Mishra P, Roy S, Das D (2015) Comparative evaluation of the hydrogen production by mixed consortium, synthetic co-culture and pure culture using distillery effluent. Bioresour Technol 198:593–602

Modak P (2011) Waste. Investing in energy and resource efficiency. To war. a green Econ Environ. http://www.unep.org/greeneconomy/Portals/88/documents/ger/GER_8_Waste.pdf

Navarro AR, Sepúlveda MC, Rubio MC (2000) Bio-concentration of vinasse from the alcoholic fermentation of sugar cane molasses. Waste Manag 20:581–585

Nure JF, Shibeshi NT, Asfaw SL, Audenaert W, Hulle SWHV (2017) COD and colour removal from molasses spent wash using activated carbon produced from bagasse fly ash of Matahara sugar factory, Oromiya region, Ethiopia. Water SA 43(3):470–479. https://doi.org/10.4314/wsa.v43i3.12

Oliveira BG, Carvalho JLN, Cerri CEP, Cerri CC, Feigl BJ (2013) Soil greenhouse gas fluxes from vinasse application in Brazilian sugarcane areas. Geoderma 200–201:77–84

Pandey RA, Malhotra S, Tankhiwale A, Pande S, Pathe PP, Kaul SN (2003) Treatment of biologically treated distillery effluent- a case study. Int J Environ Stud 60:263–275

Pandey SN, Nautiyal BD, Sharma CP (2008) Pollution level in distillery effluent and its phytotoxic effect on seed germination and early growth of maize and rice. J Environ Biol 29:267–270

Pedrosa EMR, Rolim MM, Albuquerque PHS, Cunha AC (2005) Supressividade de nematóides em cana-de-açúcar por adição de vinhaça ao solo. Rev Bras Eng Agríc Ambient 9:197–201

Ramana A, Biswas AK, Kundu S, Saha JK, Yadava RBR (2002) Effect of distillery effluent on seed germination in some vegetable crops. Bioresour Technol 82:273–275

Rath P, Pradhan G, Mishra MK (2010) Effect of sugar factory distillery spent wash (DSW) on the growth pattern of sugarcane (*Saccharum officinarum*) crop. J Phytol 2(5):33–39

Samsudeen N, Radhakrishnan TK, Matheswaran M (2015) Bioelectricity production from microbial fuel cell using mixed bacterial culture isolated from distillery wastewater. Bioresour Technol 195:242–247

Sankaran K, Premalatha M, Vijayasekaran M, Somasundaram VT (2014) DEPHY project: distillery wastewater treatment through anaerobic digestion and phycoremediation – a green industrial approach. Renew Sustain Energy Rev 37:634–643

Santal AR, Singh NP, Saharan BS (2016) A novel application of Paracoccus pantotrophus for the decolorization of melanoidins from distillery effluent under static conditions. J Environ Manag 169:78–83

Santana VS, Machado NRCF (2008) Photocatalytic degradation of the vinasse under solar radiation. Catal Today 133:606–610

Satyawali Y, Balakrishnan M (2007) Removal of color from biomethanated distillery spentwash by treatment with activated carbons. Bioresour Technol 98:2629–2635

Satyawali Y, Balakrishnan M (2008) Wastewater treatment in molasses-based alcohol distilleries for COD and color removal: a review. J Environ Manag 86:481–497

Singh M, Chauhan Dikshit AK (2012) Indian distillery industry: problems and prospects of decolourisation of spentwash. International Conference on Future Environment and Energy IPCBEE vol. 28(2012) © (2012) IACSIT Press, Singapore

Sirianuntapiboon S, Phothilangka P, Ohmomo S (2004) Decolourization of molasses wastewater by a strain no. BP103 of acetogenic bacteria. Bioresour Technol 92:31–39

Souza TS, Hencklein FA, De Angelis DF, Fontanetti CS (2013) Clastogenicity of landfarming soil treated with sugar cane vinasse. Environ Monit Assess 185:1627–1636

Srivastava S, Jain R (2010) Effect of distillery spent wash on cytomorphological behaviour of sugarcane settlings. J Environ Biol 31:809–812

Srivastava PC, Singh RK, Srivastava P, Shrivastava M (2012) Utilization of molasses based distillery effluent for fertigation of sugarcane. Biodegradation 23:897–905

Szymanski MSE, Balbinot R, Schirmer WN (2010) Biodigestão anaeróbica da vinhaça: aproveitamento energético do biogás e obtenção de créditos de carbono – estudo de caso. Semin Ciênc Agrár 31:901–912

Tolmasquim MT (2008) Bioenergy for the future. In: Conference on biofuels: an option for a less carbon-intensive economy (4th–5th December, 2007) [cited 14 April 2008]. Available from Internet: http://www.unctad.org/sections/wcmu/docs/uxii_ditc_tedb_013_en.pdf

Von Sperling M (2005) Introdução à qualidade das águas e ao tratamento de esgotos. Departamento de Engenharia Sanitária e Ambiental – Universidade Federal de Minas Gerais, Belo Horizonte

Waal A, Jiménez-Rueda JR, Bonotto DM, Bertelli C, Hoffmann HM, Fobhag E, Santilli M (2009) Influence of the vinasse application in sugar cane fields in Patrocínio Paulista, São Paulo State, Brazil. In: Brebia CA (ed) Environmental health risk V. WIT Press, Southampton, pp 113–124

Wang H, Zhao Y (2009) A bench scale study of fermentative hydrogen and methane production from food waste in integrated two-stage process. Int J Hydrog Energy 34(1):245–254

Wen Q, Wu Y, Zhao L, Sun Q (2010) Production of electricity from the treatment of continuous brewery wastewater using a microbial fuel cell. Fuel 89:1381–1385

Yesilada E (1999) Genotoxic activity of vinasse and its effect on fecundity and longevity of Drosophila melanogaster. Bull Environ Contam Toxicol 63:560–566

Zhang M, Wang Z, Li P, Zhang H, Li X (2017) Bio-refractory dissolved organic matter and colorants in cassava distillery wastewater: characterization, coagulation treatment and mechanisms. Chemosphere 178:259–267

Chapter 6
Bioremediation Approaches for Degradation and Detoxification of Polycyclic Aromatic Hydrocarbons

Pavan Kumar Agrawal, Rahul Shrivastava, and Jyoti Verma

Abstract Waste from industry is a noteworthy risk to the earth as it contains different poisonous, mutagenic and cancer-causing substances including polycyclic aromatic hydrocarbons (PAHs). PAHs are a class of different organic compounds with two or more intertwined benzene rings in a linear, angular or cluster array. Eviction of PAHs is crucial as these are persevering toxins with ubiquitous event and adverse natural impacts. There are several remedial techniques, which are productive and financially savvy in elimination of PAHs from the affected environment. These removal approaches are not just eco-friendly; they additionally display an emerging and new strategy in mitigating the ability of PAHs to cause potential risk to living beings. Accessible physical and synthetic techniques are neither eco-accommodating nor financially viable in this way. Natural strategies such as bioremediation techniques are most appropriate for biodegradation of PAHs. Such techniques require less chemicals, less time and less contribution of energy and are cost-effective and eco-accommodating. The lethal PAH mixes can be changed into non-harmful and more straightforward ones utilizing normally occurring microorganisms like algae, bacteria and fungi in a procedure called biodegradation. This chapter mainly focuses on the enhancement in biodegradation of hazardous PAHs by using bioremedial approaches.

Keywords Polycyclic aromatic hydrocarbons · Enzymatic approach · Biodegradation · Bioremediation

---

P. K. Agrawal (✉) · J. Verma
Department of Biotechnology, G.B. Pant Engineering College, Ghurdauri, Pauri, Garhwal, Uttarakhand, India

R. Shrivastava
Department of Biotechnology & Bioinformatics, Jaypee University of Information Technology, Waknaghat, Solan, Himachal Pradesh, India

© Springer Nature Singapore Pte Ltd. 2019
R. N. Bharagava, P. Chowdhary (eds.), *Emerging and Eco-Friendly Approaches for Waste Management*, https://doi.org/10.1007/978-981-10-8669-4_6

1 Introduction

PAHs are effective environmental toxicants that consist of fused aromatic rings. PAHs are originated in unrefined petroleum, coal tar and blacktop (Ukiwe et al. 2013). PAHs are included in the US Environmental Protection Agency (EPA) and European Community (EC) Contaminant Candidate List. EPA currently regulates 16 PAH compounds as priority pollutants in water and as 'aggregate PAHs' in defiled soil and sediments (Hadibarata et al. 2009). PAHs are of big concern to humans and animals as contaminant, some even recognized as cancer causing, mutagenic or teratogenic.

PAH compounds have two- to seven-membered benzene rings. They are lacking water affinity mixes with aqueous solubility declining almost linearly with increases in molecular mass (Parrish et al. 2004). Physicochemical properties and molecular weight of PAHs vary with the number of rings in the atom. Increment in subatomic weight of PAHs leads to decrease in chemical reactivity, aqueous solubility and volatility of PAHs. High-molecular-weight PAHs have high resonance energies because of the thick mists of pi-electrons surrounding the aromatic rings making them steady in the earth and recalcitrant to degradation. The recalcitrant nature of compounds may be attributed to their low water solubility and high soil sorption (Parrish et al. 2004).

PAH degradation is controlled by several physicochemical as well as biological processes, which vary their fate and transport in the surface environment. Biodegradation of hydrocarbons is achieved either by microorganism such as bacteria (Hamamura et al. 2013), fungus (Cerniglia and Sutherland 2010) or algae (Chan et al. 2006) or by enzymatic approaches. Fungi are considered as a productive competitor for effective degradation of PAHs. In any case, filamentous growth has capacity to develop on wide spectrum of substrates by secreting extracellular hydrolytic enzymes, even equipped for growing under non-ambient environment (Juhasz and Naidu 2000). Bioremediation includes either indigenous or exogenous microbial population, which is known as not proficient degraders in contaminated site (Yadav et al. 2017; Bharagava et al. 2017). Fungi have advantages over bacteria because of their fungal hyphae and potent hydrolytic enzymes, which can enter and corrupt the hydrocarbons affected environment (Venkatesagowda et al. 2012). Fungal enzymes especially oxidoreductases, laccase and peroxidases have noticeable application in removal of PAH contaminants either in fresh, marine water or terrestrial. Nevertheless, interest on growths gets an impressive consideration for bioremediation of hydrocarbon contaminated sites associated fungi for enzyme discharge (to expel hydrocarbons from nature). The persistence, toxicity and carcinogenicity of PAH molecules draw public concern to decontaminate PAH-polluted sites.

2 Sources of Polycyclic Aromatic Hydrocarbons in the Environment

For the most part, PAH contamination happens by unprocessed and processed oil, which comes from tanker accidents, refinery effluents, metropolitan and modern release from pipelines and seaward productions and waste oil from two-wheeler and four-wheeler administration stations, which causes contamination (Uzoamaka et al. 2009). Polycyclic aromatic hydrocarbon compounds are formed, what's more, discharged into the earth through both natural and anthropogenic sources. Natural sources of PAHs include their development as exudates from trees woods and rangeland fires, fungi and bacteria (Fig. 6.1). In nature, PAHs remain prevalently distributed as parts of plant oils, cuticles of insects, components of surface waxes of leaves and lipids of microorganisms. PAHs are formed naturally during thermal geologic reactions associated with fossil fuel and mineral generation.

Anthropogenic sources like fuel ignition, vehicles, spillage of petroleum products, electric fuel generation, internal ignition motors and waste incinerators are critical sources of PAHs into the environment (Arulazhagan and Vasudevan 2011). Anthropogenic wellsprings of PAHs are the real reason for natural contamination and, hence, the focus of a lot of bioremediation programmes. PAHs remain saved in the environment through generally scattered sources covering significantly the land surface area. At such sources PAHs are observed to be consumed strongly to soil particles.

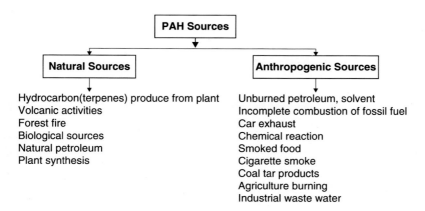

Fig. 6.1 Natural and anthropogenic sources of polycyclic aromatic hydrocarbons (PAHs)

3 Environmental Problems of PAHs

Increasing knowledge of potent adverse effects of toxicant on human health and environment has led to enhanced attention and measures for remediation and renovation of environment contaminants. Polycyclic aromatic hydrocarbons (PAHs) are omnipresent organic pollutants with serious environmental concerns. PAHs are distributed in various ecosystems and are pollutants of severe concern due to their potential toxicity, mutagenicity and carcinogenicity.

PAHs have an innate property for bioaccumulation in food chains, which makes their presence in any ecosystem alarming for human health (Morehead et al. 1986; Xue and Warshawsky 2005). Accumulation of PAHs in plants poses hazardous effect to human health because of their position in the food chain. Experimental studies have also demonstrated interference with plant carbon allocation and root symbioses by PAHs, which ultimately affect plant growth and the environment. Because of their concentration and toxicity, 16 PAHs have been enlisted as priority environmental pollutants by the US Environmental Protection Agency (US EPA).

4 Various Techniques for PAH Degradation

4.1 Chemical Degradation

The availability of PAHs in anaerobic conditions relies on certain components, which incorporate substrate interaction, pH and redox conditions (Chang et al. 2002). The most oxidation reactions in the environment are initiated by oxidants such as peroxides (H_2O_2), ozone (O_3) and hydroxyl radicals generated by photochemical processes. The degradation pathways are such that the oxidation reactions involving hydroxyl radicals or O_3 react with aromatic compounds such as PAHs at near diffusion-controlled rates by abstracting hydrogen atoms or by expansion to twofold bonds (Ukiwe et al. 2013). The reaction proceeds with complex pathways creating various intermediates. In such reactions, the last response products include a mixture of ketones, quinones, aldehydes, phenols and carboxylic acids for both oxidants (Reisen and Arey 2002). During chemical reaction PAHs are transformed into other polyaromatic hydrocarbons (they do not lose their aromatic character). Their aromaticity is conserved since considerable amounts of energy are required to change an aromatic compound into a non-aromatic compound. PAHs could be degraded through fermentative digestion system, while some studies have demonstrated that PAH degradation in anaerobic environment is much slower than in aerobic environment (Ambrosoli et al. 2005). Effects of carbon (C) and nitrogen (N) on PAH degradation have also been investigated by several authors (Quan et al. 2009). The efficiency of PAH chemical degradation is limited by their low aqueous solubility and vapour pressure (Fernando et al. 2009). However, surfactants enhance the solubility of hydrophobic compounds (Ukiwe et al. 2013). Several reports have been focused on the significance of surfactants to expand the solubility of PAHs by decreasing the interfacial surface tension amongst PAH and the dirt/water interphase (Li and Chen 2009).

4.2 Phytodegradation

It is characterized as the utilization of plants to expel contaminations from the earth to render them nontoxic (Table 6.1). Plants can take the toxicant up and accrue them in their tissues. It is an in situ, solar energy-regulated technique, which minimizes environmental disturbance and reduces costs (Haritash and Kaushik 2009). Researchers have indicated that various grasses and leguminous plants are potential candidates for phytodegradation of organics (Newman and Reynolds 2004; Ukiwe et al. 2013). Some tropical plants have also been reported to show effective degradation tendency due to inherent properties such as deep fibrous root system and tolerance to high hydrocarbon and low nutrient availability (Dzantor et al. 2000; Chandra et al. 2012). Many species of grass such as *Agropyron smithii*, *Bouteloua gracilis*, *Cynodon dactylon*, *Elymus canadensis*, *Festuca arundinacea*, *Festuca rubra*, *Melilotus officinalis*, etc. are known to degrade PAHs (McCutcheon and Schnoor 2003). Researchers are also investigating that grasses and legumes induce the removal of PAHs from affected soil. Plants also play an indirect role in the removal of PAHs by releasing of enzymes by roots. These enzymes are capable of transforming organic contaminants by catalysing chemical reactions in soil (Ndimele et al. 2010). Plant enzymes also act as causative agents in the transformation of contaminants mixed with sediment and soil. The identified enzyme systems included dehalogenase, nitroreductase, peroxidase and laccase (Thomson and Ndimele 2010). Rasmussen and Olsen studied the efficiency of orchard grass (*Dactylis glomerata*) towards PAH removal. The study reported that a soil/sand mixture vegetated with orchard grass exhibited high treatment efficiency with an input from the microbial catabolic degradation by plant exudates (Parish et al. 2004).

Table 6.1 Plants useful in phytodegradation of PAHs

S.No.	Isolate name	Compound	% Removal	Incubation	References
1.	*Festuca arundinacea*	Pyrene	38%	190 days	Chen et al. (2003)
2.	*Pannicum virgatum*	Pyrene	38%	190 days	Chen et al. (2003)
3.	*E. crassipes solani*	Naphthalene	45%	7 days	Nesterenko et al. (2012)
4.	*Scirpus Grossus*	Petroleum hydrocarbons	81.5%	72 days	Al-Baldawi et al. (2015)

4.3 *Biodegradation*

Biodegradation is a reasonable technique for degradation of natural contaminations. It is the use of microorganisms to degrade or detoxify environmental pollutants (Bamforth and Singleto 2005; Saxena and Bharagava 2017). Several additional factors affecting rate of degradation incorporate pH, temperature, nearness of oxygen and supplement accessibility. The concentration of nutrients and the state of the nutrients (organic, inorganic) are important for biodegradation. The biodegradation of PAHs is financially savvy, eco-friendly (as it prevents environmental damage during transportation of contaminants). Biological degradation is an approach that presents the possibility to remove organic pollutants with the help of natural biological activity available in the substrate (Zeyaullah et al. 2009; Bharagava and Chandra 2010). The microorganisms used for biodegradation could be indigenous to the contaminated region or site (Das and Chandran 2011). The complete mineralization products of the pollutant by biodegradation process include CO_2, H_2O and cell biomass (Gratia et al. 2006). During biodegradation process optimization involves many factors such as microbial consortia capable of degrading the pollutant, bioavailability of the pollutants to microbial attack and soil type, temperature, soil pH, oxygen level of soil, electron acceptor agents and nutrient content of soil (environmental factors) contributing to microbial growth (Gratia et al. 2006; Epelde et al. 2009; Mulla et al. 2017; Bharagava and Chandra 2010). Complete degradation of PAHs to CO_2, water, microbial carbon and other inorganic compound is the ultimate goal. Haeseler et al. (2001) showed enhanced, but incomplete, degradation of PAH compounds in a field study. When remediation was complete, final toxicity was very less because the metabolites tended to be less stable and more soluble than the parent compounds, making them more available to degraders (Haeseler et al. 2001). Microbial degradation is the most suitable alternative and effective method of removal of those toxic chemicals. Microbes (including bacteria, fungi and algae) can biologically degrade PAH compounds during direct microbial metabolism of carbon energy sources or by co-metabolism while consuming another substrate (Lundstedt et al. 2006) (Table 6.2).

5 Role of Various Microorganisms in Bioremediation

The problem linked with the PAHs can be mitigating by the use of conventional approaches, which involve degradation, modification or isolation of the toxicant. These approaches involve excavation of contaminate and its incineration or containment. These technologies are expensive and in many cases transfer the pollutant from one phase to another. On the other hand, bioremediation is the tool to transform the compounds to less hazardous/nonhazardous forms with less input of chemicals, energy and time (Ward et al. 2003; Yuan et al. 2001; Chandra et al. 2011). Microorganisms are known to be their catabolic activity in biological remediation

Table 6.2 Movement and fate of organic chemicals, such as PAHs, in the environment

S.No.	Degradation process	Consequence	Factors	Advantages/disadvantages
1.	Chemical degradation	Alteration of PAHs by chemical processes such as photochemical (i.e., UV light) and oxidation-reduction reactions	High and low pH, structure of PAH, intensity and duration of sunlight, exposure to sunlight and same factors as for microbial degradation	Limited effectiveness and can be expensive
2.	Phytodegradation	Breakdown of contaminants or pesticides through metabolic processes within the plant	Molecular weight of PAHs, sunlight, enzymes	Low-cost methods for cleaning the environment
3.	Biodegradation	Degradation of PAHs by microorganisms, biodegradation and co-metabolism	Environmental factors (pH, moisture, temperature, oxygen), nutrient status, organic matter content, PAH bioavailability, microbial community present, molecular weight of PAH (LMW or HMW)	Cost-effectiveness and complete cleanup, catabolic versatility of microorganisms, less labour-intensive, relying on solar energy, have a lower carbon footprint and have a high level of public acceptance

(Bharagava et al. 2009), but changes in microbial communities are still unpredictable, and the microbial community is still termed as a 'black box' (Dua et al. 2002). The PAH-degrading microorganism could be algae, bacteria or fungi. It involves the breakdown of organic compounds either usually by microorganism in to less complex metabolites or through mineralization into inorganic minerals, H_2O, CO_2 (aerobic) or CH_4 (anaerobic). The extent and rate of contaminant degradation depend on many factors including pH, temperature, O_2, microbial population, degree of acclimation, convenience of nutrients, compounds chemical structure, properties of cellular transport and chemical partitioning in the growth medium (Singh and Ward 2004).

5.1 Biodegradation of PAHS by Fungi

Fungus is known to have the properties of degradation of persistent organic pollutants (Table 6.3). Distinct properties differentiate filamentous fungus from other life forms to decide why they are potent biodegraders agents. First, the mycelial growth habit gives a competitive benefit over single cells such as bacteria and yeasts, especially with respect to the colonization of insoluble substrates (Bennet et al. 2002). The isolates identified as *Deuteromycetes* belonging to the genera *Cladophialophora*, *Exophiala* and *Leptodontium* and the ascomycete *Pseudeurotium zonatum* are

Table 6.3 Fungal isolates involve in degradation of PAHs

S.No	Isolate name	Compound	% Removal	Incubation	References
1.	*Phomopsis liquidambari*	Indole	41.7%	6 days	Chen et al. (2013)
2.	*Fusarium verticillioides*	Naphthalene	87.78%	8 days	Mohamed et al. (2012)
3.	*Fusarium solani*	Anthracene, benz[a]anthracene	40% and 60%, resp.	40 days	Wu et al. (2009)
4.	*Aspergillus terreus*	Naphthalene, anthracene	98.5% and 91%, resp.	4 weeks	Mohamed et al. (2012)
5.	*Fusarium* sp.	Naphthalene	42%	7 days	Ahirwae and Dehariya (2013)

toluene-degrading fungi; they use toluene as sole carbon and energy source (Francesc et al. 2001). Clemente et al. (2001) reported that degree of degradation of PAH varies with a variation of lignolytic enzymes producing deuteromycete ligninolytic fungal isolates.

Low-molecular-weight PAHs (two to three rings) were found to be degraded most extensively by *Aspergillus sp.*, *Trichocladium canadense* and *Fusarium oxysporum*. For high-molecular-weight PAHs (four to seven rings), maximum degradation has been observed by *T. canadense*, *Aspergillus sp.*, *Verticillium sp.* and *Acremonium sp.* Such studies have found that fungi have a great capability to degrade a broad range of PAHs under low-oxygen conditions. As a large and novel microbial resource, endophytic fungi have been paid more attention in their ecological functions.

The effect of microbes on litter component decomposition (Osono and Hirose 2011) but extended fungal degradation to more recalcitrant carbohydrate (Russell et al. 2011).

PAHs degradation by fungi has mostly focused on white-rot fungi. Their broad-range degradation potential is one reason of PAHs, such as *Irpex lacteus* found with a degradative ability of ANT, phenanthrene, pyrene as well as fluoranthene, and their degradative mechanisms were also investigated (Cajthaml et al. 2002). The other reason is because of their efficient production of ligninolytic enzymes; e.g. *Phanerochaete chrysosporium* could degrade ANT and phenanthrene by producing lignin peroxidase (LiP) and manganese-dependent peroxidise (MnP) (Hammel 1995). *Lentinus (Panus) tigrinus* showed out the MnP transformation ability after carrying out in vivo and in vitro degradation of PAHs (Covino et al. 2010). *Cunninghamella* sp. and *Aspergillus* sp. were reported for their potential in the transformation of benzo[a]pyrene and the conjugation mechanisms during the degradation (Wu et al. 2009). *Fusarium* spp. have shown their capability to degrade high-molecular-weight organic compounds such as coal cellulose, xylan, pectin, different hydrocarbons (Kang and Buchenauer 2000) as well as PAHs (Chulalaksananukul et al. 2006). Lignin peroxidase (LiP), manganese peroxidase (MnP) and laccase (Lac) have shown to degrade not only lignocellulose but also pollutants such as crude oil wastes, textile effluents, distillery wastewater pollutants, organochloride agrochemicals and pulp effluents which are a cause of serious environmental pollution (Mtui and Nakamura 2004; Chandra and Chowdhary 2015; Chowdhary et al. 2017a, b, 2018).

5.2 Biodegradation by Bacterial Isolates

Several reports refer to degradation ability of different bacterial isolates of environmental pollutants (Table 6.4). Many bacterial spp. are even known to nourish completely on hydrocarbons (Yakimov et al. 2007). Degradation of PAHs can occur under aerobic and anaerobic conditions, as in the case for the nitrate-reducing bacterial strains *Pseudomonas* sp. and *Brevibacillus* sp. isolated from petroleum contaminated soil (Grishchenkov et al. 2000). Several species of microorganisms have been successfully utilized in major hazardous waste cleanup processes (Levinson et al. 1994). Abd et al. (2009) reported that two- to three-ring PAHs (naphthalene, anthracene and phenanthrene) can be degraded using *Pseudomonas geniculata* and *Achromobacter xylosoxidans*. Bacterial isolates capable of chrysene metabolism include *Rhodococcus* sp. strain UW1 (Walter et al. 1991) and *Sphingomonas yanoikuyae* which oxidized chrysene (Boyd et al. 1999), while *Pseudomonas fluorescens* utilized chrysene and benz[a]anthracene as sole carbon sources (Caldini et al. 1995). The microorganisms capable of surviving in such a polluted environment are those that develop specific enzymatic and physiological responses that allow them to use hydrocarbon as a substrate. Kafilzadeh et al. (2011) reported ten genera as follows: *Bacillus*, *Corynebacterium*, *Staphylococcus*, *Streptococcus*, *Shigella*, *Alcaligenes*,

Table 6.4 Biodegradation of PAHs by various bacterial isolates

S.No	Isolate name	Compound	% Removal	Incubation	References
1.	*Pseudomonas* sp.	Naphthalene, fluorene	95%	4 days	Kumar et al. (2010)
2.	*Pseudomonas* sp.	Anthracene	74.8%	10 days	Kumar et al. (2010)
3.	*Bacillus* sp.	Anthracene	82.6%	72 h	Neelofur et al. (2014)
4.	*Mesoflavibacter zeaxanthinifaciens*	Benzo[a]pyrene	86%	42 days	Okai et al. (2015)
5.	*Mycobacterium flavescens*	Pyrene	89.4%	2 weeks	Dean-Ross et al. (2002)
6.	*Rhodococcus* sp.	Anthracene	53.0%	2 weeks	Ross et al. (2002)

Acinetobacter, Escherichia, Klebsiella and *Enterobacter* out of 80 bacterial strains. *Bacillus* was the best hydrocarbon-degrading genus. Bacterial strains that are able to degrade aromatic hydrocarbons have been repeatedly isolated, mainly from soil. The bacterial genera *Mycobacterium, Corynebacterium, Aeromonas, Rhodococcus* and *Bacillus* have been also reported for biodegradation pathways (Mrozik et al. 2003).

5.3 Biodegradation of PAHs by Algal Isolates

Algae are important microbial members in both aquatic and terrestrial ecosystems; reports are insufficient regarding their concern in hydrocarbon biodegradation (Table 6.5). (Das and Chandran 2011; Walker et al. 1975) isolated an alga, *Prototheca zopfii* which was capable of utilizing crude oil and a mixed hydrocarbon substrate and exhibited extensive degradation of n-alkanes and isoalkanes as well as aromatic hydrocarbons. Cerniglia and Gibson (1977) observed that nine cyanobacteria, five green algae, one red alga, one brown alga and two diatoms could oxidize naphthalene. Some research has demonstrated that certain fresh algae (e.g., *Chlorella vulgaris, Scenedesmus platydiscus, S. quadricauda* and *S. capricornutum*) are capable of uptaking and degrading PAHs (Wang and Zhao 2007). Warshawsky et al. (2007) found that *Selenastrum capricornutum*, a freshwater green alga, metabolizes BaP to cis-dihydrodiols using a dioxygenase enzyme system as found in heterotrophic prokaryotes. Certain algae have been reported to enhance the removal fluoranthene and pyrene when present with bacteria. Borde et al. (2003) first reported that photosynthesis enhanced degradation of toxic aromatic compounds by algal-bacterial microcosms in a one-stage treatment. *Pseudomonas migulae* and *Sphingomonas yanoikuyae* were studied for phenanthrene degradation. The green alga *Chlorella sorokiniana* was cultivated in the presence of the pollutants at different concentrations, showing increasing inhibitory effects in the order salicylate < phenol

Table 6.5 Biodegradation of PAHs by various algal isolates

S.No	Isolate name	Compound	% Removal	Incubation	References
1.	*Prototheca zopfii*	Petroleum hydrocarbons	12–41%	3 days	Kirk et al. (1999)
2.	*Selenastrum capricornutum*	Benzo[*a*]pyrene	41%	4 days	Warshawsky et al. (2007)
3.	*Selenastrum capricornutum*	Fluoranthene	99%	7 days	Ke et al. (2010)
4.	*Lyngbyala gerlerimi*	Naphthol	36.6%	5 days	Mostafa et al. (2012)
5.	*Nostoc linckia*	Catechol	56.38%	5 days	Mostafa et al. (2012)
6.	*Oscillatoria rubescens*	β-naphthol	3.04%	7 days	Mostafa et al. (2012)

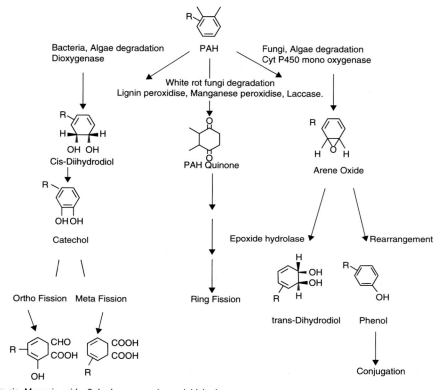

Fig. 6.2 The three main pathways for polyaromatic hydrocarbon degradation by fungi and bacteria. (Adopted by Muthuswamy et al. 2008)

<phenanthrene (Lei et al. 2007). The study of fluoranthene, pyrene and a mixture of fluoranthene and pyrene by *Chlorella vulgaris*, *Scenedesmus platydiscus*, *Scenedesmus quadricauda* and *Selenastrum capricornutum* has shown that removal is algal species-specific and toxicant-dependent. PAH removal in 7 days of treatment was 78% and 48%, respectively, by *S. capricornutum* and *C. vulgaris*. Hong et al. (2008) studied the accumulation and biodegradation of phenanthrene and fluoranthene by the algae enriched from a mangrove aquatic environment (Fig. 6.2).

6 Enzymatic Degradation of PAHs

Major enzymes useful in PAH degradation belong to oxygenase, dehydrogenase and lignolytic categories. Lignolytic enzymes secreted by majority of fungal species are lignin peroxidase, manganese peroxidase and laccase, which are extracellular in nature and are secreted for catabolism of substrate food material (Chandra and Chowdhary 2015). Spent mushroom compost (SMC) is often used as an inoculum source which enhances the rate of PAH degradation. The SMC are high in laccase and Mn-dependent peroxidase, whereas the production of ligninase is reported to be low in SMC (Haritash and Kaushik 2009). In which most of enzymes are active at many temperature and have optimum activity at mesophilic temperatures and it reduces with very high and very low temperatures. Some of the enzymes are reported to be active even at extremes of temperatures (Haritash and Kaushik 2009). Enzymes also show substrate specificity, but ligninolytic enzymes are non-specific in nature, acting on phenolic and non-phenolic organic compounds via the generation of cation radicals after one e^- oxidation (Lau et al. 2003).

6.1 Lignin Degrading Enzymes

Oxidoreductive enzymes play a key role in transformation and degradation of polymeric substances (Table 6.6). The less degraded or oxidized items can without much of a stretch be taken up by microbial cells where they are totally mineralized. A class of oxidoreductive enzymes include lignin-degrading compounds (LDEs) which have practical application in bioremediation of polluted environment (Husain 2006). LDEs belong to two classes, viz., the heme-containing peroxidases and the copper-containing laccases. A progression of redox responses are started by the laccases. LDEs degrade the lignin or lignin-derived pollutants. The LDEs oxidize the aromatic compounds until the aromatic ring structure is cleaved, which is followed by further debasement with different compounds.

Table 6.6 Biological functions of ligninolytic enzymes

Enzymes	Applications	References
Lignin peroxidase	Biodegradation of lignin	Martínez et al. (2005)
	Degradation of azo	Stolz (2001)
	Mineralization of environmental contaminants	Harms et al. (2011)
	Degradation of pharmaceuticals and their metabolites	Marco-Urrea et al. (2009)
Manganese peroxidase	Degradation of lignin	Martínez et al. (2005)
	PAH degradation	Baborová et al. (2006)
	Synthetic dyes, DDT, PCB, TNT	Hernández et al. (2008)
	Textile dye degradation and bleaching	Kalyani et al. (2008)
Laccase	Spore resistance	Lu et al. (2012)
	Rhizomorph formation	Ranocha et al. (2002)
	Pathogenesis	Langfelder et al. (2003)
	Fruit bodies formation	Nagai et al. (2003)
	Pigment synthesis	Eisenman et al. (2007)

Enzymes involved in the degradation of PAHs are oxygenase, dehydrogenase and ligninolytic enzymes. Fungal ligninolytic enzymes are lignin peroxidase, laccase and manganese peroxidase. They are extracellular and catalyse radical formation by oxidation to destabilize bonds in a molecule (Hofrichhter et al. 1999).

Peroxidases perform heme-containing degradation with other enzymes. Peroxidases are heme-containing enzymes that comprise manganese-dependant peroxidase (MnP), lignin peroxidase (LiP) and versatile peroxidase (VP). They oxidize lignin subunits using extracellular hydrogen peroxide generated by unrelated oxidases as co-substrate. Most mineralization activity of the lignin polymers to CO_2 and H_2O in terrestrial ecosystem is performed by fungal species. These fungi produce a wide range of lignin-degrading enzymes (LDEs), which in turn act on lignin and lignin-analogous compounds. PAHs are primarily degraded using extracellular oxidative enzymes, although use of laccases and peroxidases in PAH bioremediation is currently being studied (Harms et al. 2011). The white-rot fungi (WRF) belonging to the basidiomycetes produce various isoforms of extracellular ligninolytic enzymes, laccases (Lac) and different peroxidases, including lignin peroxidase (LiP), manganese peroxidase (MnP) and versatile peroxidase (VP), the latter sharing LiP and MnP catalytic properties (Martínez 2002). The natural substrate of these enzymes (lignocellulose) is degraded in the environment by the WRF, along with various xenobiotic compounds, including dyes (Wesenberg et al. 2003). Some WRF produce all the three lignin-modifying enzymes, while others produce only one or two of them. Lignin-modifying enzymes are produced by WRF during their secondary metabolism since lignin oxidation provides no net energy to the fungus (Eggert et al. 1996).

6.1.1 Lignin Peroxidases (EC 1.11.1.14)

LiPs (EC 1.11.1.14) are an extracellular hemeprotein. They are related to the family of oxidoreductases (Higuchi 2004). LiP has high redox potential and low optimum pH (Piontek et al. 2001) and is capable of oxidizing a variety of reducing substrates including polymeric substrates (Oyadomari et al. 2003). Due to their high redox potentials and enlarged substrate range, LiPs have more potential for application in several industrial processes (Erden et al. 2009). Enzymatic activity of LiP is H_2O_2 dependent; here H_2O_2 gets reduced to H_2O by picking up an electron from LiP (which itself gets oxidized). The oxidized LiP then returns to its native reduced state by picking up an e⁻ from veratryl alcohol and oxidizing into veratryl aldehyde. Veratryl aldehyde then gets reduced back to veratryl alcohol by picking up an electron from lignin or equivalent structures such as xenobiotic compounds.

6.1.2 Manganese Peroxidases (EC 1.11.1.13)

MnP (EC 1.11.1.13) belong to the family of oxidoreductases (Higuchi. 2004). Studies have shown that MnP is distributed in almost all white-rot fungi (Hofrichter 2002). Manganese peroxidases (MnP) seem to be distributed amongst white-rot fungi than LiP (Hammel and Cullen. 2008). MnP oxidizes Mn^{2+} to highly reactive Mn^{3+}, which oxidizes phenolic structures to phenoxyl radicals (Hofrichter 1999). The Mn^{3+} forms complex with chelating organic acids resulting in products such as oxalates or malates (Makela et al. 2002). The redox potential of the Mn peroxidase system is lower than that of lignin peroxidase, and it has shown capacity for preferable oxidation of phenolic substrates. On the other hand, studies indicate that contrary to LiP, MnP may oxidize Mn (II) without H_2O_2 with decomposition of acids and concomitant production of peroxyl radicals that may affect lignin structure. Versatile peroxidase (VP) enzymes produced by *Pleurotus* spp. are also able to oxidize phenolic compounds and dyes efficiently that are substrates of generic peroxidases and related peroxidases or the well-known horseradish peroxidase (HRP). VP (EC 1.11.1.16) oxidizes Mn^{2+}, similar to MnP, and have a high redox potential aromatic as LiP enzymes. Due to these qualities, interest in VP has increased during the last years (Martínez et al. 2009).

6.1.3 Laccases (EC 1.10.3.2)

Laccases (EC 1.10.3.2) belong to a multicopper oxidase family (Alcalde 2007), present in bacteria, e.g., *Azospirillum lipoferum*, *Actinomycetes* like *Streptomyces*, fungi, plants and insects (Baldrian 2006; Chandra and Chowdhary 2015). This enzyme had been reported more than a hundred years ago (Desai and Nityanand 2011), but the significance and broad studies over the role of this enzyme in wood degradation have been conducted in the last few decades. However, many laccases were reported from fungi, and most biotechnologically useful laccases also originated from fungus (Kalmis et al. 2008).

7 Challenges

In spite of considerable progress made in the study of the biodegradation of PAHs over the past few decades, removal of petroleum hydrocarbons in the environment is a daunting problem of the real world. Advancement in various approaches such as genomics, proteomics and metabolomic study has contributed immensely in understanding the PAH-degrading microorganisms and the biochemistry involved in the degradation pathways; however, challenges are posed by various aspects of PAH bioremediation which are either unknown or insufficient information is available regarding them.

Little or no information is available related to genes, enzymes and molecular mechanism of PAH degradation in high-salt environments or low-oxygen and anaerobic environments. Scarce data and research are there on the transmembrane trafficking of PAHs and their metabolites; no transporter molecule/protein has been characterized till date with specific role in the transport of PAHs into microorganisms.

Thorough understanding of genetic regulation of the pathways involved in PAH degradation by different bacteria and fungi has been used for efficient biotransformation or metabolism of PAH pollutants in recent past; a deeper understanding of the microorganism-mediated mechanisms of PAH catalysis will enable strategizing novel methods to enhance the bioremediation of PAHs in the environment.

The use of genetically modified organism in bioremediation represents a research frontier with broad implications to improve the degradation of hazardous wastes under laboratory conditions. The potential benefits of using genetically modified microorganisms are significant. Combining genetic engineering tools such as gene conversion, gene duplication and mutation, enzyme overexpression and novel strains can be produced with desirable properties for effective bioremediation applications. Ecological and environmental concerns and regulatory constraints pose major obstacles and challenges for testing genetically modified organism in the field. These problems must be solved before a genetically modified organism can provide an effective, safer and more efficient method than the present alternatives for removal process at very low-cost and eco-friendly way.

8 Conclusion

The present status of work done on biodegradation of biologically toxic PAHs using different microorganism has been reviewed. The environmental toxicity and persistence of PAHs have resulted in several laboratory-based experiments to change these substances into less unsafe/nondangerous substances with the use of microorganisms in the process called as biodegradation. Removal of PAHs from the affected environment is a tough job. Therefore, it is very essential to understand the mechanism of several degradation processes. Degradation of PAHs remains affected by numerous factors, which need to be addressed and explored. Biological approaches appear be the most efficient, cost-effective and eco-friendly method to mitigate/remove PAHs from affected area.

Acknowledgement We gratefully acknowledge TEQIP-II and G.B. Pant Engineering College, Pauri, Garhwal, for financial supports and providing other facilities.

References

Ahirwae S, Dehariya K (2013) Isolation and characterization of hydrocarbon degrading microorganism from petroleum oil contaminated soil sites. Bull Environ Sci res 2(4):5–10

Al-Baldawi IA, Abdullah SRS, Anuar N et al (2015) Phytodegradation of total petroleum hydrocarbon (TPH) in diesel-contaminated water using Scirpus grossus. Ecol Eng 74:463–473

Alcalde M (2007) Laccase: biological functions, molecular structure and industrial applications. In: Polaina J, Maccabe AP (eds) Industrial enzymes: structure, function and applications, vol 26. Springer, Netherlands, pp 461–476

Ambrosoli R, Petruzzelli L, Luis Minati J et al (2005) Anaerobic PAH degradation in soil by a mixed bacterial consortium under denitrifying conditions. Chemosphere 60(9):1231–1236

Arulazhagan P, Vasudevan N (2011) Role of nutrients in the utilization of polycyclic aromatic hydrocarbons by halotolerant bacterial strain. J Environ Sci 23(2):282–287

Baborová P, Möder M, Baldrian P et al (2006) Purification of a new manganese peroxidase of the white-rot fungus Irpex lacteus, and degradation of polycyclic aromatic hydrocarbons by the enzyme. Res Microbiol 157(3):248–253

Baldrian P (2006) Fungal laccases occurrence and properties. FEMS Microbiol Rev 30:215–242

Bamforth SM, Singleto I (2005) Bioremediation of polycyclic aromatic hydrocarbons: current knowledge and future directions. J Chem Technol Biotechnol 80:723–736

Bennet JW, Wunch KG, Faison BD (2002) Use of fungi biodegradation. Manual of environmental microbiology.2nd edn. ASM Press, Washington, DC, pp 960–971

Bharagava RN, Chandra R (2010) Biodegradation of the major color containing compounds in distillery wastewater by an aerobic bacterial culture and characterization of their metabolites. Biodegradation J 21:703–711

Bharagava RN, Chandra R, Rai V (2009) Isolation and characterization of aerobic bacteria capable of the degradation of synthetic and natural melanoidins from distillery wastewater. World J Microbiol Biotechnol 25:737–744

Bharagava RN, Chowdhary P, Saxena G (2017) Bioremediation: an eco-sustainable green technology, its applications and limitations. In: Bharagava RN (ed) Environmental pollutants and their bioremediation approaches. CRC Press, Taylor & Francis Group, Boca Raton, pp 1–22

Borde X, Guieysse B, Delgado O et al (2003) Synergistic relationships in algal-bacterial microcosms for the treatment of aromatic pollutants. Bioresour Technol 86(3):293–300

Boyd DR, Sharma ND, Hempenstall F et al (1999) Bis-cis-Dihydrodiols: a new class of metabolites from biphenyl dioxygenase catalyzed sequential asymmetric cis-dihydroxylation of polycyclic arenas and heteroarenes. J Organomet Chem 64:4005–4011

Cajthaml T, Moder M, Kacer P et al (2002) Study of fungal degradation products of polycyclic aromatic hydrocarbons using gas chromatography with ion trap mass spectrometry detection. J Chromatogr A 974:213–222

Caldini G, Cenci G, Manenti R et al (1995) The ability of an environmental isolate of Pseudomonas fluorescens to utilize chrysene and other four-ring polynuclear aromatic hydrocarbons. Appl Microbiol Biotechnol 44(1):225–229

Cerniglia CE, Gibson DT (1977) Metabolism of naphthalene by Cunninghamella elegans. Appl Environ Microbiol 34:363–370

Cerniglia CE, Sutherland JB (2010) Degradation of polycyclic aromatic hydrocarbons by fungi. In: Timmis KN (ed) Handbook of hydrocarbon and lipid microbiology. Springer, Berlin, pp 2079–2110

Chan SMN, Luan T, Wong MH et al (2006) Removal and biodegradation of polycyclic aromatic hydrocarbons by Selenastrum capricornutum. Environ Toxicol Chem 25:1772–1779

Chandra R, Chowdhary P (2015) Properties of bacterial laccases and their application in bioremediation of industrial wastes. Environ Sci Process Impacts 17:326–342

Chandra R, Bharagava RN, Kapley A, Purohit JH (2011) Bacterial diversity, organic pollutants and their metabolites in two aeration lagoons of common effluent treatment plant during the degradation and detoxification of tannery wastewater. Bioresour Technol 102:2333–2341

Chandra R, Bharagava RN, Kapley A, Purohit HJ (2012) Characterization of *Phragmites communis* rhizosphere bacterial communities and metabolic products during the two stage sequential treatment of post methanated distillery effluent by bacteria and wetland plants. Bioresour Technol 103:78–86

Chang BV, Shiung LC, Yuan SY (2002) Anaerobic biodegradation of polycyclic aromatic hydrocarbons in soil. Chemosphere 48:717–724

Chen YC, Banks MK, Schwab AP (2003) Pyrene degradation in the rhizosphere of tall fescue (*Festuca arundinacea*) and switch grass (*Panicum virgatum*). Environ Sci Technol 37(24):5778–5782

Chen Y, Xie XG, Ren CG et al (2013) Degradation of N-heterocyclic indole by a novel endophytic fungus Phomopsis liquidambari. Bioresour Technol 129:568–574

Chowdhary P, Yadav A, Kaithwas G, Bharagava RN (2017a) Distillery wastewater: a major source of environmental pollution and its biological treatment for environmental safety. Green technologies and environmental sustainability. Springer International, Cham, pp 409–435

Chowdhary P, More N, Raj A, Bharagava RN (2017b) Characterization and identification of bacterial pathogens from treated tannery wastewater. Microbiol Res Int 5:30–36

Chowdhary P, Raj A, Bharagava RN (2018) Environmental pollution and health hazards from distillery wastewater and treatment approaches to combat the environmental threats: a review. Chemosphere 194:229–246

Chulalaksananukul S, Gadd GM, Sangvanich P et al (2006) Biodegradation of benzo(a) pyrene by a newly isolated Fusarium sp. FEMS Microbiol Lett 262(1):99–106

Clemente AR, Anazawa TA, Durrant LR (2001) Biodegradation of polycyclic aromatic hydrocarbons by soil fungi. Braz J Microbiol 32(4):255–261

Covino S, Svobodova K, Kresinova Z et al (2010) In vivo and in vitro polycyclic aromatic hydrocarbons degradation by Lentinus (Panus) tigrinus CBS 577.79. Bioresour Technol 101(9):3004–3012

Das N, Chandran P (2011) Microbial degradation of petroleum hydrocarbon contaminants: an overview. Biotechnol Res Int 2011 Article ID 941810, pp 13

Dean-Ross D, Moody J, Cerniglia CE (2002) Utilization of mixtures of polycyclic aromatic hydrocarbons by bacteria isolated from contaminated sediment. FEMS Microbiol Ecol 41:1–7

Desai SS, Nityanand C (2011) Microbial laccases and their applications: a review. Asian J Biotechnol 3(2):98–124

Dua M, Singh A, Sethunathan N et al (2002) Biotechnology and bioremediation: successes and limitations. Appl Microbiol Biotechnol 59(2–3):143–152

Dzantor EK, Chekol T, Vough L (2000) Feasibility of using forage grasses and legumes for phytoremediation of organic pollutants. J Environ Sci Health A 35(9):1645–1661

Eggert C, Temp U, Eriksson KE (1996) The ligninolytic system of the white rot fungus Pycnoporus cinnabarinus: purification and characterization of the laccase. Appl Environ Microbiol 62(4):1151–1158

Eisenman HC, Mues M, Weber SE et al (2007) Cryptococcus neoformans laccase catalyses melanin synthesis from both D-and L-DOPA. Microbiology 153(12):3954–3962

El A, Haleem D, Al-Thani RF et al (2009) Isolation and characterization of polyaromatic hydrocarbons-degrading bacteria from different Qatari soils. Afr J Microbiol Res 3:761–766

Epelde L, Mijangos I, Becenil J et al (2009) Soil microbial community as bio-indicator of the recovery of soil functioning derived from metal phytoextraction with sorghum. Soil Biol Biochem 41:1788–1794

Erden E, Ucar CM, Gezer T et al (2009) Screening for ligninolytic enzymes from autochthonous fungi and applications for decolorization of remazole marine blue. Braz J Microbiol 40(2):346–353

Fernando Bautista L, Sanz R, Carmen Molina M et al (2009) Effect of different non-ionic surfactants on the biodegradation of PAHs by diverse aerobic bacteria. Int Biodeterior Biodegrad 63(7):913–922

Francesc X, Boldu P, Kuhn A et al (2001) Isolation and characterization of fungi growing on volatile aromatic hydrocarbons as their sole carbon and energy source. Mycol Res 105(4):477–484

Gratia E, Weekers F, Margesin R et al (2006) Selection of a cold-adopted bacterium for bioremediation of wastewater at low temperature. Extremophiles 13:763–768

Grishchenkov VG, Townsend RT, McDonald TJ et al (2000) Degradation of petroleum hydrocarbons by facultative anaerobic bacteria under aerobic and anaerobic conditions. Process Biochem 35(9):889–896

Hadibarata T, Tachibana S, Itoh K (2009) Biodegradation of chrysene, an aromatic hydrocarbon by Polyporus sp. S133 in liquid medium. J Hazard Mater 164(2–3):911–917

Haeseler F, Blanchet D, Werner P et al (2001) Ecotoxicological characterization of metabolites produced during PAH biodegradation in contaminated soils. In: Magar VS, Johnson G, Ong

SK, Leeson A (eds) Bioremediation of energetics phenolics and polycyclic aromatic hydrocarbons, vol 6(3). Batelle Press, San Diego, pp 227–234, 313 pp

Hamamura N, Ward DM, Inskeep WP (2013) Effects of petroleum mixture types on soil bacterial population dynamics associated with the biodegradation of hydrocarbons in soil environments. FEMS Microbiol Ecol 85:168–178

Hammel KE (1995) Mechanisms for polycyclic aromatic hydrocarbon degradation by lignolytic fungi. Environ Health Perspect 103:41–43

Hammel KE, Cullen D (2008) Role of fungal peroxidases in biological ligninolysis. Curr Opin Plant Biol 11(3):349–355

Haritash AK, Kaushik CP (2009) Biodegradation aspects of polycyclic aromatic hydrocarbons (PAHs): a review. J Hazard Mater 169(1–3):1–15

Harms H, Schlosser D, Wick LY (2011) Untapped potential: exploiting fungi in bioremediation of hazardous chemicals. Nat Rev Microbiol 9(3):177–192

Hernández RL, González-Franco AC, Crawford DL et al (2008) Review of environmental organopollutants degradation by white-rot basidiomycete mushrooms. Tecnociencia Chihuahua 2(1):32–39

Higuchi T (2004) Microbial degradation of lignin: role of lignin peroxidase, manganese peroxidase, and laccase. Proc Jpn Acad, Ser B 80(5):204–214

Hofrichhter M, Vares T, Kalsi M et al (1999) Production of manganese peroxidase and organic acids and mineralization of 14C-labelled lignin (14C-DHP) during solid state fermentation of wheat straw with the white rot fungus Nematoloma forwardii. Appl Environ Microbiol 65(5):1864–1870

Hofrichter M (2002) Review: lignin conversion by manganese peroxidase (MnP). Enzym Microb Technol 30:454–466

Hong YW, Yuan DX, Lin QM et al (2008) Accumulation and biodegradation of phenanthrene and fluoranthene by the algae enriched from a mangrove aquatic ecosystem. Mar Pollut Bull 56(8):1400–1405

Husain Q (2006) Potential applications of the oxidoreductive enzymes in the decolorization and detoxification of textile and other synthetic dyes from polluted water: a review. Crit Rev Biotechnol 26:201–221

Juhasz AL, Naidu R (2000) Bioremediation of high molecular weight polycyclic aromatic hydrocarbons: a review of the microbial degradation of benzo[a]pyrene. Int Biodeterior Biodegrad 45:57–88

Kafilzadeh F, Sahragard P, Jamali H et al (2011) Isolation and identification of hydrocarbons degrading bacteria in soil around Shiraz Refinery. Afr J Microbiol Res 4(19):3084–3089

Kalmis E, Yasa I, Kalyoncu F et al (2008) Ligninolytic enzyme activities in mycelium of some wild and commercial mushrooms. Afr J Biotechnol 7(23):4314–4320

Kalyani DC, Patil PS, Jadhav JP et al (2008) Biodegradation of reactive textile dye red BLI by an isolated bacterium Pseudomonas sp.SUK1. Bioresour Technol 99(11):4635–4641

Kang Z, Buchenauer H (2000) Ultra structural and cytochemical studies on cellulose, xylan and pectin degradation in wheat spikes infected by Fusarium culmorum. J Phytopathol 148(5):263–275

Ke L, Luo LJ, Wang P, Luan TG, Tam NFY (2010) Effects of metals on biosorption and biodegradation of mixed polycyclic aromatic hydrocarbons by a freshwater green alga Selenastrum capricornutum. Bioresour Technol 101:6950–6961

Kumar G, Singla R, Kumar R (2010) Plasmid associated anthracene degradation by pseudomonas sp. isolated from filling station site. Nat Sci 8(4):89–94

Langfelder K, Streibel M, Jahn B et al (2003) Biosynthesis of fungal melanins and their importance for human pathogenic fungi. Fungal Genet Biol 38(2):143–158

Lau KL, Tsang YY, Chiu SW (2003) Use of spent mushroom compost to bioremediate PAH-contaminated samples. Chemosphere 52(9):1539–1546

Lei AP, Hu ZL, Wong YS et al (2007) Removal of fluoranthene and pyrene by different microalgal species. Bioresour Technol 98(2):273–280

Levinson W, Stormo K, Tao H et al (1994) Hazardous waste clean-up and treatment with encapsulated or entrapped microorganisms. In: Chaudry GR (ed) Biological degradation and bioremediation of toxic chemicals. Chapman and Hall, London, pp 455–469

Li JL, Chen BH (2009) Effects of non-ionic surfactants on biodegradation of phenanthrene by marine bacteria of Neptunomnas naphthovorans. J Hazard Mater 162(1):66–73

Lu L, Zhao M, Wang T (2012) Characterization and dye decolorization ability of an alkaline resistant and organic solvents tolerant laccase from Bacillus licheniformis LS04. Bioresour Technol 115:35–40

Lundstedt S, Persson Y, Oberg LG (2006) Transformation of PAHs during ethanol- Fenton treatment of an aged gasworks soil. Chemosphere 65:1288–1294

Makela M, Galkin S, Hatakka A et al (2002) Production of organic acids and oxalate decarboxylase in lignin-degrading white rot fungi. Enzym Microb Technol 30(4):542–549

Marco-Urrea E, Pérez-Trujillo M, Vicent T et al (2009) Ability of white-rot fungi to remove selected pharmaceuticals and identification of degradation products of ibuprofen by Trametes versicolor. Chemosphere 74(6):765–772

Martínez AT (2002) Molecular biology and structure-function of lignin degrading heme peroxidases. Enzym Microb Technol 30(4):425–444

Martínez AT, Speranza M, Ruiz-Dueñas FJ et al (2005) Biodegradation of lignocellulosics: microbial chemical and enzymatic aspects of the fungal attack of lignin. Int Microbiol 8(3):195–204

Martínez AT, Ruiz-dueñas FJ, Martínez MJ et al (2009) Enzymatic delignification of plant cell wall: from nature to mill. Curr Opin Biotechnol 20(3):348–357

McCutcheon SC, Schnoor JL (2003) Phytoremediation: transformation and control of contaminants. Wiley-Inter Science, Hoboken, p 987

Mohamed I, Ali A, Khalil NM et al (2012) Biodegradation of some polycyclic aromatic hydrocarbons by Aspergillus terreus. Afr J Microbiol Res 6(16):3783–3790

Morehead NR, Eadie BJ, Lake B et al (1986) The sorption of PAH onto dissolved organic matter in Lake Michigan waters. Chemosphere 15:403–412. https://doi.org/10.1016/0045-6535(86)90534-5

Mostafa MES, Ghareib MM, Abou-EL-Souod GW (2012) Biodegradation of phenolic and polycyclic aromatic compounds by some algae and cyanobacteria. J Bioremed Biodegr 3(1):1–9

Mrozik A, Piotrowska-Seget Z, Labuzek S (2003) Bacterial degradation and bioremediation of polycyclic aromatic hydrocarbons. Pol J Environ Stud 12(1):15–25

Mtui G, Nakamura Y (2004) Lignin-degrading enzymes from mycelial cultures of basidiomycete fungi isolated in Tanzania. J Chem Eng Jpn 37(1):113–118

Mulla SI, Ameen F, Tallur PN, Bharagava RN, Bangeppagari M, Eqani SAMAS, Bagewadi ZK, Mahadevan GD, Yu CP, Ninnekar HZ (2017) Aerobic degradation of fenvalerate by a Gram-positive bacterium *Bacillus flexus* strain XJU-4. 3 Biotech 7:320–328

Muthusamy K, Gopalakrishnan S, Ravi TK, Sivachidambaram P (2008) Biosurfactants: properties, commercial production and application. Current Science 94:736–747

Nagai M, Kawata M, Watanabe H et al (2003) Important role of fungal intracellular laccase for melanin synthesis: purification and characterization of an intracellular laccase from Lentinula edodes fruit bodies. Microbiology 149(9):2455–2462

Ndimele PE, Oni AJ, Jibuike CC (2010) Comparative toxicity of crude oil-plus dispersant to Tilapia guineensis. Res J Environ Toxicol 4(1):13–22

Neelofur M, Shyam PV, Mahesh M (2014) Enhance the biodegradation of anthracene by mutation from bacillus species. BioMed Res 1(1)

Nesterenko MA, Kirzhner F, Zimmels Y et al (2012) Eichhornia crassipes capability to remove naphthalene from waste water in the absence of bacteria. Chemosphere 87(10):1186–1191

Newman L, Reynolds C (2004) Phytodegradation of organic compounds. Curr Opin Biotechnol 15:225–230

Okai M, Kihara I, Yokoyama Y et al (2015) Isolation and characterization of benzo[*a*]pyrene degrading bacteria from the Tokyo bay area and Tama river in Japan. FEMS Microbiol Lett 362 fnv143 362(18):1–7

Osono T, Hirose D (2011) Colonization and lignin decomposition of pine needle litter by Lophodermium pinastri. Forest Pathol 41:156–162

Oyadomari M, Shinohara H, Johjima T et al (2003) Electrochemical characterization of lignin peroxidase from the white-rot basidiomycete Phanerochaete chrysosporium. J Mol Catal B Enzym 21(4–6):291–297

Parrish ZD, Banks MK, Schwab AP (2004) Effectiveness of phytoremediation as a secondary treatment for polycyclic aromatic hydrocarbons (PAHs) in composted soil. Int J Phytomediation 6:119–137

Piontek K, Smith AT, Blodig W (2001) Lignin peroxidase structure and function. Biochem Soc Trans 29(2):111–116

Quan X, Tang Q, He M et al (2009) Biodegradation of polycyclic aromatic hydrocarbons in sediments from the Dalian River watershed, China. J Environ Sci 21:865–871

Ranocha P, Chabannes M, Chamayou S et al (2002) Laccase down-regulation causes alterations in phenolic metabolism and cell wall structure in poplar. Plant Physiol 129(1):145–155

Reisen F, Arey J (2002) Reactions of hydroxyl radicals and ozone with acenaphthene and acenaphthylene. Environ Sci Technol 36:4302–4311

Ross DD, Moody J, Cerniglia CE (2002) Utilization of mixtures of polycyclic aromatic hydrocarbons by bacteria isolated from contaminated sediment. FEMS Microbiol Eco 41(1):1–7

Russell JR, Huang J, Anand P et al (2011) Biodegradation of polyester polyurethane by endophytic fungi. Appl Environ Microbiol 77:6076–6084

Saxena G, Bharagava RN (2017). Organic and inorganic pollutants in industrial wastes, their ecotoxicological effects, health hazards and bioremediation approaches, Bharagava RN Environmental pollutants and their bioremediation approaches. CRC Press, Taylor & Francis Group, Boca Raton (9781138628892)

Singh A, Ward OP (2004) Biodegradation and bioremediation. Series: Soil Biology, vol 2. Springer-Verlag, New York, p 310

Stolz A (2001) Basic and applied aspects in the microbial degradation of azo dyes. Appl Microbiol Biotechnol 56(1–2):69–80

Thomson ISI, Ndimele PE (2010) A review on phytoremediation of petroleum hydrocarbon. Pak J Biol Sci 13(15):715–722

Ukiwe LN, Egereonu UU, Njoku PC et al (2013) Polycyclic aromatic hydrocarbons degradation techniques: a review. Int J Chem 5(4):43–45

Uzoamaka GO, Floretta T, Florence MO (2009) Hydrocarbon degradation potentials of indigenous fungal isolates from petroleum contaminated soils. J Phy Nat Sci 3:1–6

Venkatesagowda B, Ponugupaty E, Barbosa AM (2012) Diversity of plant oil seed-associated fungi isolated from seven oil – bearing seeds and their potential for the production of lipolytic enzymes. World J Microbiol Biotechnol 28:71–80

Walker JD, Colwell RR, Vaituzis Z et al (1975) Petroleum-degrading a chlorophyllous algae Prototheca zopfi. Nature 254:423–424

Walter U, Beyer M, Klein J et al (1991) Degradation of pyrene by Rhodococcus sp. UW1. Appl Microbiol Biotechnol 34:671–676

Wang XC, Zhao HM (2007) Uptake and biodegradation of polycyclic aromatic hydrocarbons by marine seaweed. J Coast Res 50:1056–1061

Ward OP, Singh A, Van Hamme J (2003) Accelerated biodegradation of petroleum hydrocarbon waste. J Ind Microbiol Biotechnol 30(5):260–270

Warshawsky D, Radike M, Jayasimhulu K et al (1988) Metabolism of benzo[a]pyrene by a dioxygenase system of freshwater green alga Selenastrum capricornutum. Biochem Biophys Res Commun 152:540–544

Warshawsky D, La Dow K, Schneider J (2007) Enhanced degradation of benzo[*a*]pyrene by *Mycobacterium* sp. in conjunction with green algae. Chemosphere 69(3):500–506

Wesenberg D, Kyriakides I, Aghatos SN (2003) White-rot fungi and their enzymes for the treatment of industrial dye effluents. Biotechnol Adv 22:151–187

Wu YR, He TT, Lun JS et al (2009) Removal of benzo[a]pyrene by a fungus Aspergillus sp. BAP14. World J Microbiol Biotechnol 25(8):1395–1401

Xue W, Warshawsky D (2005) Metabolic activation of polycyclic and heterocyclic aromatic hydrocarbons and DNA damage: a review. Toxicol Appl Pharmacol 206:73–93. https://doi.org/10.1016/j.taap.2004.11.006

Yadav A, Chowdhary P, Kaithwas G, Bharagava RN (2017) Toxic metals in environment, threats on ecosystem and bioremediation approaches in: handbook of metal-microbe interactions and bioremediation. In: Das S, Dash HR (eds). CRC Press, Taylor & Francis Group, Boca Raton, pp 813–841

Yakimov MM, Timmis KN, Golyshin PN (2007) Obligate oil-degrading marine bacteria. Curr Opin Biotechnol 18(3):257–266

Yuan SY, Chang JS, Yen JH et al (2001) Biodegradation of phenanthrene in river sediment. Chemosphere 43(3):273–278

Zeyaullah MD, Atif M, Islam B et al (2009) Bioremediation: a tool for environmental cleaning. Afr J Microbiol Res 36:310–314

Chapter 7
Endocrine-Disrupting Pollutants in Industrial Wastewater and Their Degradation and Detoxification Approaches

Izharul Haq and Abhay Raj

Abstract Endocrine-disrupting chemicals (EDCs), a group of chemicals that alter the normal function of the endocrine system of humans and wildlife, are a matter of great concern. These compounds are widely distributed in respective environments such as water, wastewater, sediments, soils, and atmosphere. Chemicals like pesticides, pharmaceuticals and personal care products, flame retardants, natural hormones, heavy metals, and chemicals derived from basic compounds (such as plasticizers and catalysts) are major endocrine disruptors. EDCs emerging from industries such as pulp and paper, tannery, distillery, textile, pharma, etc. have been considered as major source of contamination. Alkylphenol ethoxylates (APEOs), bisphenol A (BPA), phthalates, chlorophenols, norethindrone, triclosan, gonadotropin compounds, pesticides, etc. are generally escaped during wastewater treatment and contaminate the environment. Endocrine-disrupting activity of these compounds is well documented to have adverse effect on human-animal health. Globally, efforts are being approached for their efficient removal from sewage/wastewaters. Thus, this chapter provides updated overview on EDC generation, characteristics, and toxicity as well as removal/degradation techniques including physical, chemical, and biological methods. This chapter also reviews the current knowledge of the potential impacts of EDCs on human health so that the effects can be known and remedies applied for the problem as soon as possible.

Keywords Endocrine disruptors · Industrial wastewaters · Human-animal health · Biological treatment

---

I. Haq · A. Raj (✉)
Environmental Microbiology Laboratory, Environmental Toxicology Group, CSIR-Indian Institute of Toxicology Research (CSIR-IITR), Lucknow, Uttar Pradesh, India
e-mail: araj@iitr.res.in

© Springer Nature Singapore Pte Ltd. 2019
R. N. Bharagava, P. Chowdhary (eds.), *Emerging and Eco-Friendly Approaches for Waste Management*, https://doi.org/10.1007/978-981-10-8669-4_7

1 Introduction

Thousands of anthropogenic toxic chemicals currently are released from industries into the environment (Vandenberg et al. 2009; Eskinazi et al. 2003; Markey et al. 2003). Endocrine systems control a huge number of metabolic, developmental, and reproductive processes including embryonic development, gonadal formation, sex differentiation, growth, and digestion. Chemicals that interfere normal function of the endocrine system at certain doses are referred as endocrine-disrupting chemicals (EDCs). EDCs may affect these processes by either binding to or blocking hormone receptors, thereby triggering or preventing hormonal response (Pedersen et al. 1999; Markey et al. 2003; Witorsch 2002; Hotchkiss et al. 2008).

EDCs are highly diverse that includes synthetic chemicals used as industrial solvents/lubricants and their by-products (polychlorinated biphenyls, polybrominated biphenyls, dioxins), plastics, plasticizers, pesticides, fungicides, phytoestrogens, pharmaceutical agents, and certain industrial or commercial products (Bharagava and Mishra 2018; Diamanti-Kandarakis et al. 2009; Falconer et al. 2006). These chemicals can enter the aquatic environment through effluent discharge or storm-water runoff. The major transport of EDCs to the aquatic environment is through industrial and municipal wastewater discharge into the rivers, streams, and surface waters (Clara et al. 2005; Falconer et al. 2006; Zhang et al. 2015) (Fig. 7.1). Potable water resources, including both surface water and groundwater, can become contaminated through surface water discharge or deep-well injection of wastewater treatment plant effluent (Mompelat et al. 2009; Bharagava and Chandra 2010b; Kumari et al. 2016).

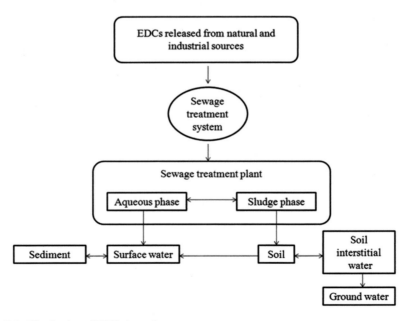

Fig. 7.1 Distribution of EDCs in environment

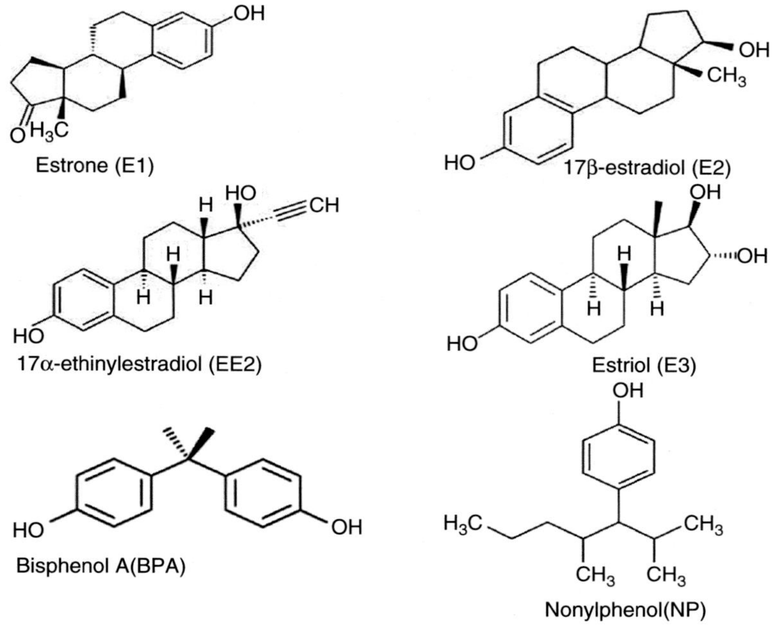

Fig. 7.2 Structure of common synthetic and industrial estrogenic compounds

Effects of EDCs on wildlife (invertebrates, fish, amphibians, reptiles, birds, and mammals) include abnormal blood hormone levels, altered gonadal development (e.g., imposex and intersex), induction of vitellogenin gene and protein expression in juveniles and males, masculinization/feminization, hermaphroditism, and decreased fertility and fecundity (Jobling et al. 2004; Kidd et al. 2007; Janex-Habibi et al. 2009; Nadzialek et al. 2010).

Estrogenic compounds specifically target estrogen signaling including steroidal estrogens, synthetic estrogens, and industrial compounds which mimic estrogen. 17β-Estradiol (E2) is the main natural estrogen, which has greatest potency than estrone (E1), a metabolite of E2 and estriol (E3), considered to be the final metabolite. 17α-Ethinylestradiol (EE2) is the synthetic steroidal estrogen component of contraceptives (Spencer et al. 2009). Bisphenol A (BPA) is a chemical used in industry to produce lacquers, food-can liners, and thermal paper (Danzl et al. 2009). Nonylphenol (NP) is the persistent and estrogenic final product of the biodegradation of the nonionic surfactant nonylphenol ethoxylate (NPEO) (Gultekin and Ince 2007; Chandra et al. 2011). β-Sitosterol, a plant sterol, emerged from pulping industry and discharged into stream waters. The structures of some commonly used synthetic and industrial EDCs are given in Fig. 7.2.

The emissions of natural and synthetic EDCs from industries and its effects in receiving aquatic flora and fauna are not well regulated. However, monitoring of chemical compounds clearly indicates that conventional wastewater treatment processes, e.g., activated sludge process (ASP) and aerated lagoon, are unable to remove most EDCs. Consequently, wastewater treatment plants are one of the most crucial point sources for the contamination of receiving waters by EDCs. Various physicochemical (ozonation, chlorination, adsorption, membrane filtration, advance oxidation process, activated carbon treatment, electrochemical method, etc.) treatment processes are currently proposed for the removal of EDCs (Hu et al. 2003; Huber et al. 2004; Nakamura et al. 2006; Bila et al. 2007). Alternatively, bioremediation is a cost-effective and environmental friendly approach which employs microorganisms for detoxification of environmental pollutants (Zhang et al. 2013; Yang et al. 2014; Bharagava et al. 2017a, b). Among the different microorganisms, fungi and bacteria have been well documented for the degradation of EDCs. During the last decade, considerable amounts of scientific and financial resources have been employed to identify the EDC sources, distribution, potential risk to human and animal health, and its remedial approaches. The present book chapter provides updated information on the EDCs including source, distribution, human and animal health risk, and their remedial strategies for environmental safety.

2 Sources of EDCs

Industrial effluents have been considered as the key source of EDCs to the aquatic environment (Ying et al. 2002; Saxena et al. 2017; Voutsa et al. 2006). The exposure of EDCs is usually diverse and widely distributed all over the environment and society of the world. But the situation is neither constant nor can be predicted easily since there is a significant usage difference of these substances among the countries. Globally, it can be seen that EDCs have notable toxicities in human as well as in wildlife (Diamanti-Kandarakis et al. 2009; Saxena and Bharagava 2017). Among the EDCs, alkylphenol ethoxylates (APEOs), BPA, phthalates, chlorophenols, norethindrone, triclosan, gonadotropin compounds, pesticides, etc. are most concerned because of their widespread in the environment from various sources (Liu et al. 2011; Staples et al. 2000). EDCs are discharges from various industries including pulp and paper, tannery, distillery, etc. Pulp and paper mill effluents are major source of various types of EDCs (Yadav et al. 2016; Chowdhary et al. 2017).

Alkylphenol polyethoxylates (APEs) and related compounds recently have been reported to be estrogenic, because they have similar effects to estradiol both in vitro and in vivo. Alkylphenols are degradation by-products of APEs, which are used in paper industry as defoamers, cleaners, and emulsifiers. β-Sitosterol, a plant sterol found in higher plants, emerged from pulp and paper industry. Phthalates are known EDCs, which originate in paper production process mostly as softeners in additives, glues, and printing inks. Besides, EDCs like NP; 4-aminobiphenyl, hexachlorobenzene, and benzidine have also been detected in effluents emerging from tannery industries (Kumar et al. 2008). Phthalates compounds like bis(2-ethylhexyl) phthalate (DHEP), dibutyl phthalate (DBP), and bis(2-methoxyethyl) phthalate have also been found in tannery effluents (Bharagava et al. 2017a, b; Alam et al. 2010).

EDCs are also present in distillery effluents such as (butanedioic acid bis(TMS) ester; 2-hydroxyisocaproic acid; di-n-octyl phthalate; dibutyl phthalate (Chowdhary et al. 2018; 2017; Chandra et al. 2012; Chandra and Kumar 2017) benzenepropanoic acid, α-[(TMS)oxy], TMS ester; vanillylpropionic acid, bis(TMS)), and other recalcitrant organic pollutants (2-furancarboxylic acid, 5-[(TMS)oxy] methyl], TMS ester; benzoic acid 3-methoxy- 4-[(TMS)oxy], TMS ester; and tricarballylic acid 3TMS) (Chandra and Kumar 2017). The major EDCs found in different industrial effluents are alkylphenol ethoxylates (APEOs), a group of nonionic surfactants, are widely used in domestic and industrial applications. APs, NPs, 4-tert-octylphenol (4-t-OP), and 4-tert-butylphenol (4-t-BP) are part of APEOs. APs are more toxic and stable than their parent compounds in the environment.

Several countries have been reduced the use of APEOs and APs due to adverse effects to humans and wildlife. BPA is a polar monomer of polycarbonate plastic, used in polycarbonate, epoxy resins, unsaturated polyester-styrene resins, flame retardants, and many other products. Some final products from the BPA are used as coatings on cans, additives in thermal paper, and antioxidants in plastics (Kang et al. 2006). Due to large-scale production and extensive use of BPA, it has become an essential part of industrial wastewater streams (Staples et al. 2002; Latorre et al. 2003). BPA has also been shown to be estrogenic via in vivo screenings (Brotons et al. 1995; Laws et al. 2000). A low-dose effect of BPA was observed in rats (Kobayashi et al. 2005), whereas in fish, it affects the progression of spermatogenesis (Sohoni and Sumpter 1998). Furthermore, BPA has also been found to have the paradoxical effect to block the beneficial effects of estradiol on neuronal synapse formation and the potential to disrupt thyroid hormone action (MacLusky et al. 2005; Zoeller et al. 2005; Mohapatra et al. 2010).

Phthalates or phthalic acid esters (PAEs) have been widely used as plasticizers for polyvinyl chloride (PVC) resin, cellulose film coating, styrene, adhesives, cosmetics, and paper manufacturing (Olujimi et al. 2010; Patnaik 2007). The PAEs such as dimethyl phthalate (DMP) and diethyl phthalate (DEP) are among the most frequently detected in diverse environmental samples including surface marine waters, freshwaters, and sediments (Jing et al. 2011; Zeng et al. 2011). The estrogenic activity of PAEs such as dibutyl phthalate (DBP), butylbenzyl phthalate (BBP), dihexyl phthalate (DHP), diisoheptyl phthalate, di-n-octyl phthalate, diisononyl phthalate, and diisodecyl phthalate is observed by Zacharewski et al. (1998). Further studies in fish have shown that both BBP and DEP induced vitellogenin (VTG) at an exposure to low concentration in the range of $\mu g\ L^{-1}$ via the water (Harries et al. 2000; Barse et al. 2007). Numerous in vivo screens and tests have demonstrated that PAEs mediated their effects through binding to the estrogen receptor (Andersen et al. 1999; Zacharewski et al. 1998).

In addition to these estrogenic effects, some PAEs are also considered to be toxic to microorganisms, aquatic life, and human beings (Bajt et al. 2001). Recent studies have indicated that phthalate metabolites such as monoethyl phthalate (MEP), mono-(2-ethylhexyl) phthalate (MEHP), mono-n-butyl phthalate (MBP), and monobenzyl phthalate (MBzP) can induce DNA damage in human sperm (Duty et al.

2003; Hauser et al. 2007). Norethindrone is one of synthetic progestogens which is used in contraceptive treatments for the promotion of menstrual cycles and correction of abnormal uterine bleeding (Chang et al. 2011). Norethindrone is released into the environment through humans' and animals' urines and feces (Fent 2015). Triclosan is an important bactericide used in various personal care and consumer products (Ying and Kookana 2007). Natural or synthetic steroids are excreted by mammals, and eventually, they occur in domestic effluents and in livestock waste (Ying et al. 2002). The previous reports revealed that the existing industrial wastewater treatment processes could not completely remove EDCs (Ifelebuegu 2011; Samaras et al. 2013) and discharge into receiving water bodies.

3   Impact of EDCs on Human and Animal Health

As mentioned in earlier sections, our daily life is surrounded by a wide range of EDCs. Thus, from a physiological perspective, natural or synthetic EDCs alter the hormonal and homeostatic systems of an organism. EDCs tend to be relatively bioaccumulative and persistent and produce adverse developmental, reproductive, neurological, hormonal, metabolic, and immune effects in both humans and wildlife (Fig. 7.3).

Studies conducted on animal, clinical observations, and also the epidemiological studies have indicated its potential role in affecting reproductive systems, prostate, breast, lung, liver, thyroid, metabolism, and obesity (Polyzos et al. 2012). EDCs can disrupt sperm production and development while exposed to the testicle of an organism (Aly et al. 2009). In the recent study, it has been suggested that high concentration of DDT/DDE has negative impact on sperm motility, morphology, count, and semen volume (Aneck-Hahn et al. 2007). Other studies also observed that exposure of DDT adversely affects sperm quality, mainly through decreased motility (Rignell-

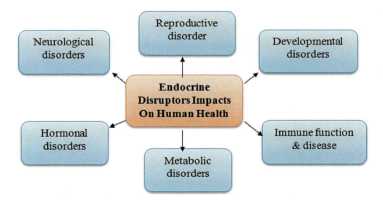

Fig. 7.3 Impact of endocrine disruptors on human health

Hydbom et al. 2004; Toft et al. 2006). Several studies have been reported carcinogenesis of environmental EDCs which leads to development of prostate cancer, breast cancer, and testicular cancer (Høyer et al. 1998; Ritchie et al. 2005; Hardell et al. 2006; McGlynn et al. 2008).

Thyroid hormones play an important role in the maintenance of body homeostasis. The unusual changes in thyroid status may lead to altered basal metabolic rate, lipid metabolism, as well as cardiovascular, gastrointestinal, neurological, and muscle function. Thyroid hormones are especially important during growth and development such as the maturation of the brain. A number of environmental agents can alter thyroid hormone levels in humans and animals. Hypothyroidism in rodents has been observed after exposure to PCB and chlorinated pesticides (Crisp et al. 1998). Developmental neurotoxicity involving cognitive and neurobehavioral disturbances has been implicated following perinatal exposure to environmental pollutants (Jacobson and Jacobson 1996). There are several studies which show that EDCs produce a wide spectrum of neurochemical and neuroendocrine effects in humans and animals (Koopman-Esseboom et al. 1996; Tilson and Kodavanti 1997; Eriksson et al. 1998).

EDCs not only affect human health but also terrestrial and aquatic organisms. There are large numbers of evidences available which indicated that endocrine disruptors are also responsible for different wildlife crises. However, wildlife is not exposed to single contaminants, instead to a mixture of chemicals, some of them acting through a common pathway. The exposure route of wildlife to EDCs is also very critical and crucial. This is because many of the endocrine disruptors do not persist in the environment and organism. Most of the EDCs are either degraded in the environment by sunlight, bacteria, and chemical processes or persist for different time ranges. The organism may follow the same uptake route as the humanlike ingestion or by absorption through the skin (Kidd et al. 2012).

For better understanding of the EDCs exposure to wildlife, studies have been conducted on animals of distant and local places. The fishes are expected to be exposed to a high level of EDCs because of their existence near the localized area. The water source as well as the low-level land is continuously being exposed to EDCs due to sewage treatment process and industrial effluent discharges. Although EDCs cannot persist in the water for longer periods of time, regular disposal of these chemicals into the water makes the aquatic wildlife be in contact with EDCs continuously. However, in spite of being in remote places, highest levels of perfluorooctanesulfonate (PFOS), polychlorinated bisphenols (PCBs), and organochlorine pesticides have been also detected in the polar bears which may be due to the long-distance transport of these chemicals into those areas (Kidd et al. 2012). However, for animal lives, primary source of EDC exposure can be thought to be the water source. The fish uptake these contaminants from water through gills, while wild birds and mammals may uptake these through drinking water. After EDC exposure to animals, it travels across the body and is metabolized. While some of the EDCs accumulate in fat tissues, or they are eliminated from the body by taking a variety of routes. They can also be end in eggs of fishes and birds exposed to them. In this way, EDCs travel from the body through lactation and pass into the offspring.

However, in the animal body, the liver is the main site of their metabolism, after which they are eliminated through urine and feces (Kidd et al. 2012). Various studies have been done previously to understand the effects of EDCs on wildlife. These studies have provided a relationship between these chemicals and their effects on animals. However, some EDCs have shorter half-lives, so sometimes the tasks are difficult to be performed. The persistent organic pollutants (POPs) have been intensively studied in wildlife including dichlorodiphenyltrichloroethane (DDT), chlordanes, dieldrin, PCBs, dioxins, polybrominated diphenyl ethers (PBDEs), etc. Therefore, it is clear that the monitoring of EDC evaluation on wildlife will take several years. Available information is still not enough to provide a visible indication of the level of EDC exposure in many areas, especially in tropical and subtropical areas. At this point, it is crucial to conduct more studies on wildlife to save our endangered wildlife as well as our precious ecological systems.

4 Wastewater Treatment Techniques and EDC Removal

4.1 Physical and Chemical Treatment

4.1.1 Membrane Technology

There are two basic types of membrane separation processes: pressure-driven and electrically driven. Microfiltration (MF), ultrafiltration (UF), nanofiltration (NF), and reverse osmosis (RO) are the pressure-driven filtration processes which use hydraulic pressure to force water molecule through the membrane (Adams et al. 2002; Walha et al. 2007). In the electrically driven membrane process, electric current is used to move ions across the membrane leaving purified water behind. Membrane treatment, applied to the end of conventional wastewater treatment system, is a viable method of achieving desired effluent quality level. Membranes are commonly used for the removal of dissolved solids, color, and hardness in drinking water. Published scientific literatures show removal of several types of EDCs present in wastewaters (Bodzek 2015). The RO and NF processes can effectively remove phytoestrogens (70–93%), PAH (85–99%), and surfactants (92–99%) from industrial wastewaters. In wastewater contaminated with greater pollutants, UF process can be used for the surfactant removal. Plant protection products such as pesticides, herbicides, and insecticides present in surface and groundwaters can be effectively removed from water during NF or by integrated systems of MF or NF and activated carbon adsorption (powdered or granulated) (Bodzek 2015; Mavrov et al. 1992).

The concentration of phthalates in different parts of the environment, especially in water, should also be controlled by these processes, as high retention of phthalates was observed during both RO and NF processes (initial concentration 40 μg/L) (Bodzek et al. 2004). Retention rates achieved for diethyl phthalate, di-n-butyl, and di2-ethylhexyl were very high and amounted from 89.7% (UF) to 99.9% (RO and NF). Phenolic xenoestrogens such as APs, NPs, BPA, and BPF can be removed from water by means of NF (Bodzek and Możliwości 2013). Recently, an increase in concentrations of the synthetic hormone such as α-ethinylestradiol, mestranol, and diethylstilbestrol which emerged from the large amounts of expired pharmaceuticals, both from households and from wastewater and hospital wastes as well as pharmaceutical plants, has been observed. It was shown that elimination of this type of pollutants from water could be performed by means of membrane processes (Bodzek and Dudziak 2006). Considering relatively low molecular weight of those pollutants RO or NF must be applied. It has been found that RO membranes totally eliminate particular hormones, while retention coefficients obtained for NF and UF membranes were lower (Bodzek and Dudziak 2006).

4.1.2 Absorption by Activated Carbon

Use of activated carbon (AC) is a well-known process for removing various organic contaminants and organic carbon in general. AC is most commonly used as a powdered form (powder activated carbon, PAC) or in a granular form (granular activated carbon, GAC) in packed bed filters. Granular activated carbon (GAC) is used at many water treatment plants in the United States and Canada. The GAC can be used as a replacement for anthracite media in conventional filters, thus providing both adsorption and filtration. Alternatively, GAC can be applied post-conventional filtration as an adsorbent bed (Snydera et al. 2007). Several authors have confirmed the effectiveness of AC (PAC and GAC) for the removal of trace organic pollutants from water (Matsui et al. 2002; Asada et al. 2004; Westerhoff et al. 2005; Zhou et al. 2007). Also, several studies have found that AC can remove broad range of EDCs for artificial and real wastewater in the laboratory and pilot and full-scale plants (Choi et al. 2005; Fukuhara et al. 2006). In a study by Abe (1999), absorbability of about 70 EDCs by AC was estimated from their chemical structures, and then their adsorption by AC was proven effective for their removal from wastewater. Representative endocrine substances such as estrone (E1) and 17β-estradiol (E2) were also used to evaluate the removal performance on absorption. Compared to artificial EDC wastewater, a great difference in EDC removal by AC adsorption between simulated wastewater and real wastewater was observed.

4.1.3 Advanced Oxidation

Advanced oxidation processes (AOPs), through the use of cheap chemicals, have been demonstrated as a viable alternative for removing trace estrogens. Few installations are in operation today because of their high operating costs on large-scale installations (Madsen et al. 2006). Ozone and hydrogen peroxide are the most diffused oxidants. Ozone has long been used for disinfection of drinking waters, and this was the origin of its application to the degradation of several organic micropollutants. Ozone can react selectively, as an oxidant in its molecular form (O_3), thus leading to typical ozonation reactions with, for example, double bonds, amines, and phenol derivatives, or nonselectively, after formation of hydroxyl radicals (HO•) (Von Gunten 2003; Haag and Yao 1992). In all cases, depending on the structure of the organic substrate, steroid derivatives can be degraded to lower molecular weight compounds (by-products) of unknown estrogenicity. The synergistic use of ozone with other oxidants (e.g., H_2O_2) or in association with physical means such as UV radiation is justified to reduce the selectivity of action in the hope of amplifying the destruction of trace organics (Von Gunten 2003). Fenton process is based on the use of ferrous ions in association with hydrogen peroxide in acidic media (pH = 3). This process couples synergistically with the oxidation process to coagulation/flocculation, with the latter process occurring later when the pH is raised to neutralize the final effluent (Petruzzelli et al. 2007). Kunde et al. (2009) explored the oxidation reaction of BPA and tetrabromobisphenol A (TBBPA), among the most heavily used polymer plasticizers and flame retardants, respectively (Haag and Yao 1992; Kunde et al. 2009). This treatment appears to be promising in the degradation of estrogens because of the self-regenerating cycle operated by MnO_2, thus proving to be cost-effective in the long run. Ferrate ion was investigated as a viable oxidant and coagulant alternative.

Although the oxidation potential of ferrate ions is greater than that of ozone, the acidic conditions of operation strongly limit its use in full-scale installations (Lee et al. 2005; Jiang et al. 2005). Chlorine and chloramines both have the potential to react with various EDCs and personal care compounds. Chlorine gas is a fast-reacting and efficient chemical for phenols and amino functional derivatives but is nearly nonreactive with ketones and alcohols, which are only partially oxidized. It is also important to note that chlorine oxidation is pH-dependent, with the best performance obtained in acidic media. In a study recently carried out by the American Water Works Association, it was confirmed that the antibiotics sulfamethoxazole, trimethoprim, and erythromycin are among the compounds that exhibit better removal efficiency upon the use of chlorine gas as compared to chloramines (Synder et al. 2011).

4.1.4 Electrochemical Methods

Electrochemical process involves chemical reactions caused by interaction between electrode and chemicals. It is an emerging technology applied in water and wastewater treatment. Most focus has been given to electrodeposition, electrocoagulation (EC), electroflotation (EF), and electrooxidation. Anodic degradation (oxidation) of EDCs, namely, 17β-estradiol (E2) and bisphenol A (BPA), by the use of a boron-doped diamond electrode (BDDE) was investigated at the laboratory scale. Cyclic voltammetry experiments were carried out to evaluate the electrochemical process response to E2 and BPA degradation as a function of the applied voltammetry cycles. Apparently, electrooxidation reaction was controlled by the applied current density, which was evaluated and discussed at three different levels. Electrolysis at high anodic potential caused quantitative oxidation of EDCs with formation of CO_2. The effects of operating conditions in the electrolytic bath [e.g., pH, background solutions (Na_2SO_4, $NaNO_3$, and $NaCl$)] were discussed in terms of electro-generated inorganic oxidants such as $S_2O_8^{2-}$, H_2O_2, and ClO^-. Better performance of the BDDE anode was found on a comparative basis with respect to, for example, Pt and amorphous graphite under similar experimental conditions (Yoshihara and Murugananthan 2009).

4.2 Biological Treatment

Physicochemical methods were shown to be effective in removing EDCs from water and wastewaters (Hu et al. 2003; Bila et al. 2007). However, use of these processes is cost-effective and energy-consuming, and the efficiency of these processes under field conditions is limited by the low concentrations of the contaminants. Biological treatment is a particularly attractive and cost-effective approach, as it represents natural and economically feasible processes for detoxification of environmental pollutants under environmental conditions. An environmental friendly process alternative for the elimination of EDCs may be the use of fungi, bacteria, algae, and plants. Among the different microorganisms, white rot fungi have been studied for their ability not only to eliminate EDCs but also to reduce their estrogenic activity. EDC-reducing ability of fungi is usually related to the production and secretion of lignin-modifying enzymes.

4.2.1 Conventional Wastewater Treatment Plants

Wastewater treatment plants receive raw wastewater from domestic and/or industrial discharges. The main objective of a wastewater treatment system is to remove only phosphorus, nitrogen, and organic substances. The wastewater treatment system mainly used activated sludge process (ASP) which is most widely used in the entire world because of its high efficiency and cost-effectiveness. In the ASP, effluent received after primary treatment is treated by the biological process using microbes for the removal of organic contaminants from wastewater. Currently, under optimized conditions, more than 90–95% of substances can be eliminated by conventional biological-based methods used in wastewater treatment plants (Li et al. 2000; Cases et al. 2011). However, ASP is energy intensive due to the high aeration requirement, and it also produces large quantity of sludge (about 0.4 g dry weight/g COD removed) that has to be treated and disposed of. As a result, the operation and maintenance cost of the ASP is considerably high. Anaerobic process for domestic wastewater treatment is an alternative that is potentially more cost-effective, particularly in the subtropical and tropical regions where the climate is warm consistently throughout the year. Anaerobic wastewater purification processes have been increasingly used in the last few decades. These processes are important because they have positive effects: removal of higher organic loading, low sludge production, high pathogen removal, methane gas production, and low energy consumption (Nykova et al. 2002). Researchers have shown that anaerobic systems such as the upflow anaerobic sludge blanket (UASB), the anaerobic sequencing batch reactor (AnSBR), and the anaerobic filter (AN) can successfully treat high-strength industrial wastewater as well as low-strength synthetic wastewater. However, the existing conventional wastewater treatment plants are not able to remove pollutants specially EDCs (Bolong et al. 2009; Liu et al. 2009; Berge´ et al. 2012).

The conventional aerobic and anaerobic treatment system cannot degrade all compounds completely or convert into biomass. For instance, the EDCs found in effluent are the products of incomplete breakdown of their respective parent compounds (Johnson and Sumpter 2001). Wintgens et al. (2002) and Gallenkemper et al. (2003), using toxicological evaluations, also indicated that wastewater treatment plants were not able to remove these novel substances sufficiently before disposing effluent into the environment. Liu et al. (2009) reported that APs, BPA, etc. are not completely removed by existing wastewater treatment plants and remained at fluctuating concentrations in effluent, so discharge of such effluent may be the main reason for the wide distribution and occurrence of EDCs in surface waters, groundwaters, and even in drinking waters.

4.2.2 Membrane Bioreactors (MBRs)

MBR technology is considered as new and promising biological approaches in wastewater treatment system, thus integrating biological degradation with membrane filtration, with specific reference to biopersistent organic substrates. The main advantages of using MBR systems are (a) improved quality of the treated wastewater, (b) more compact plant size, (c) less sludge production, and (d) higher flexibility of plant operations for improved EDC removal by the adoption of variable solid retention time (SRTs) and/or hydraulic retention time (HRTs). In these systems, the quality of the final effluent depends strongly on the settling characteristics of the sludge and the hydrodynamic conditions in the sedimentation tank. Accordingly, large-volume sedimentation tanks, offering residence times of several hours, and strict control of the biological unit are necessary to favor sludge settling (granulation), thus minimizing bulking phenomena. The final objective is to obtain adequate solid/liquid separations for optimal performance of the membrane separation to follow. Very often, site-specific and economic constraints limit such options.

4.2.3 Fungal Treatment of EDCs

The white rot fungus *Pleurotus ostreatus* HK 35 has been tested in the degradation of typical representatives of EDCs (BPA, estrone, 17β-estradiol, estriol, 17α-ethinylestradiol, triclosan and 4-nnonylphenol), and degradation efficiency under laboratory conditions was greater than 90% within 12 days (K˘resinov'a et al. 2017). Castellana and Loffredo (2014) indicated that in a period of 20 days, *Trametes versicolor* growing on the various substrates removed almost 100% of BPA, EE2,

NP, and linuron and from 59% to 97% of dimethoate and *Stereum hirsutum* showed a marked degrading activity only toward NP, which was totally removed after 20 days or less with any substrate and, to a lesser extent, linuron. Pezzella et al. (2017) have investigated degradative capabilities of *Trametes versicolor*, *Pleurotus ostreatus*, and *Phanerochaete chrysosporium* to act on five EDCs, which represent different classes of chemicals (phenols, parabens, and phthalate). *T. versicolor* was able to efficiently remove all compounds during each cycle converting up to 21 mg L^{-1} day^{-1} of the tested EDCs. In a study of Sei et al. (2007), five highly laccase producible fungal strains, *Trametes hirsute* 1674, *T. orientalis* 1071, *T. versicolor* IFO 30340, *T. versicolor* IFO 30338, and *Pycnoporus coccineus* 866, were used for the removal of various EDCs. The result found that bis(4-hydroxyphenyl) sulfone, diethylhexylphthalate (DEHP), pyrene (PY), anthracene, 3,5-dichlorophenol, and pentachlorophenol could not be removed by laccase. DEHP and PY could not be removed even with mediators. In another study of Macellaro et al. (2014), different strains of *Pleurotus ostreatus* producing laccases were also used for EDC enzymatic treatment. The use of fungi for the removal of EDCs is well documented. However, use of fungi in industrial scale is not feasible due to their reduced enzymatic activity in real effluent condition. The use of bacterial system for the removal of EDCs from wastewaters has been undertaken because the enzymatic system of bacteria is stable in harsh environmental and physiological conditions.

4.2.4 Bacterial Treatment of EDCs

Biodegradation of EDCs had been well reported by bacterial cultures in wastewater treatment plants (Federle et al. 2002; Singer et al. 2002; Bharagava et al. 2009; Thompson et al. 2005). Among biological processes, bacterial degradation is particularly an easy and prominent way to remove various EDCs present in aquatic environment (Husain and Qayyum 2012; Bharagava and Chandra 2010a). Singer et al. (2002) had reported that approximately 79% of triclosan in wastewater was biodegraded; 15% was sorbed into biosolids, and 6% was released into the receiving water bodies. Several studies have reported removal of triclosan by different biological treatment processes (Kanda et al. 2003), including activated sludge (Federle et al. 2002; Thompson et al. 2005), rotating biological contactors, and trickling filters (Thompson et al. 2005). Unlike triclosan, greater than 90% of removal has been reported for BPA (Staples et al. 1998) and ibuprofen and its metabolites (Buser et al. 1999). Recently, several estrogen-degrading bacteria were isolated from activated sludge, including *Novosphingobium tardaugens* (ARI-1) (Fujii et al. 2002), *Rhodococcus zopfii* and *Rhodococcus equi* (Yoshimoto et al. 2004), and *Achromobacter xylosoxidans* and *Ralstonia* sp. (Weber et al. 2005). *Pseudomonas nitroreducens* strain LBQSKN1, *Pseudomonas putida* strain LBQSKN2, *Stenotrophomonas* sp. LBQSKN3, *Enterobacter asburiae* strain LBQSKN4, *Pseudomonas* sp. LBQSKN5, and *Pseudomonas* sp. LBQSKN6 isolated from soil sample were able to degrade NPs (Qhanya et al. 2017).

The study of Roh et al. (2009) showed *Nitrosomonas europaea* and mixed ammonia-oxidizing bacteria could degrade triclosan, BPA, and ibuprofen in nitrifying activated sludge. In the study of De Gusseme et al. (2009), microbial consortium was used for the removal of 17α-ethinylestradiol (EE2) and was found to remove EE2 from both a synthetic minimal medium and industrial effluent with >94% removal efficiency. Villemur et al. (2013) isolated bacterial cultures from three enrichment cultures adapted to a solid-liquid two-phase partitioning system using Hytrel as the immiscible water phase and loaded with estrone, estradiol, estriol, ethinylestradiol, NP, and BPA. All molecules except ethinylestradiol were degraded in the enrichment cultures. In study of Chen et al. (2007), di-2-ethylhexyl phthalate (DEHP) degradation strain CQ0110Y was isolated from activated sludge and identified as *Microbacterium* sp. The results of this study showed the optimal pH value and temperature, which influenced the degradation rate in wastewater: pH 6.5–7.5, 25–35 °C. The efficacy of two rhizobacteria (*Sphingobium fuliginis* TIK1 and *Sphingobium* sp. IT4) of *Phragmites australis* for the sustainable treatment of water polluted with phenolic endocrine-disrupting chemicals (EDCs) was investigated (Toyama et al. 2013). Strains TIK1 and IT4 have been isolated from *Phragmites rhizosphere* and shown to degrade various 4-alkylphenols–TIK1 via phenolic ring hydroxylation and meta-cleavage and IT4 via ipso-hydroxylation.

The two strains also degraded BPA, BPB, BPE, BPF, BPP, and BPS. Yu et al. (2007) isolated 17α-estradiol-degrading bacteria (strains KC1-14) from activated sludge of a wastewater treatment plant. These isolates were widely distributed among eight different genera – *Aminobacter* (strains KC6 and KC7), *Brevundimonas* (strain KC12), *Escherichia* (strain KC13), *Flavobacterium* (strain KC1), *Microbacterium* (strain KC5), *Nocardioides* (strain KC3), *Rhodococcus* (strain KC4), and *Sphingomonas* (strains KC8, KC11, and KC14) – of three phyla, *Proteobacteria*, *Actinobacteria*, and *Bacteroidetes*. All 14 isolates were capable of converting 17α-estradiol to estrone, but only 3 strains (strains KC6, KC7, and KC8) showed the ability to degrade estrone. Only strain KC8 could use 17α-estradiol as a sole carbon source. In the study of Fernández et al. (2016), five bacteria isolated from enrichment cultures of sediments of mud volcanoes of the Gulf of Cadiz (Moroccan-Iberian margin) were identified as aerobic E2 biodegraders, which produce low amounts of biotransformed estrone (E1). An analysis of 16S rDNA gene sequences identified three of them as *Virgibacillus halotolerans*, *Bacillus flexus*, and *Bacillus licheniformis*. Among the set of strains, *Bacillus licheniformis* showed also ability to biodegrade E2 under anaerobic conditions.

5 Conclusion

This chapter provided detailed information about EDCs releasing into the environment as a result of poor treatment by industrial wastewater treatment plant which leads to environmental pollution and toxicity to human and wildlife health. The conventional wastewater treatment process such as activated sludge process is not designed to remove these micropollutants. Although available treatment technologies, such as adsorption processes, AOPs and membrane processes, and electrochemical methods as promising alternatives are efficient for efficient removal of EDCs, it is not feasible at large scale because of its high operation costs and formation of by-products. Recently, biological treatment process as a promising approach has gained popularity to remove EDCs from industrial wastewaters. The biological system which employed fungi and bacteria having ligninolytic enzymatic system is a significantly useful procedure for targeting a number of EDCs of diversified properties and structures. On the basis of available literature on the effect of EDCs on human and wildlife health and their treatment/degradation process, it seems that there is a need of attention to address the limitation in existing treatment process and provide effective solution on it. Thus, this chapter covers all EDC-associated problems and treatment technology for the sustainable development of environment.

Acknowledgment The authors are thankful to the director of CSIR-IITR, Lucknow (India), for his encouragement and support. We greatly acknowledge the Department of Biotechnology (DBT), Government of India, New Delhi, for the funding (Grant No.BT/PR20460/BCE/8/1386/2016).

References

Abe I (1999) Adsorption properties of endocrine disruptors onto activated carbon. J Water Waste 41:43–47
Adams C, Wang Y, Loftin K, Meyer M (2002) Removal of antibiotics from surface and distilled water in conventional water treatment processes. J Environ Engine 128:253–260
Alam MZ, Ahmad S, Malik A, Ahmad M (2010) Mutagenicity and genotoxicity of tannery effluents used for irrigation at Kanpur, India. Ecotoxicol Environ Saf 73:1620–1628
Aly HA, Domenech O, Abdel-Naim AB (2009) Aroclor 1254 impairs spermatogenesis and induces oxidative stress in rat testicular mitochondria. Food Chem Toxicol 47:1733–1738
Andersen HR, Andersson AM, Arnold SF et al (1999) Comparison of short-term estrogenicity tests for identification of hormone-disrupting chemicals. Environ Health Perspect 107:89–108
Aneck-Hahn NH, Schulenburg GW, Bornman MS, Farias P, de Jager C (2007) Impaired semen quality associated with environmental DDT exposure in young men living in a malaria area in the Limpopo Province, South Africa. J Androl 28:423–434
Asada T, Oikawa K, Kawata K, Ishihara S, Iyobe T (2004) Study of removal effect of bisphenol-A and β-estradiol by porous carbon. J Health Sci 50:588–593
Bajt O, Mailhot G, Bolte M (2001) Degradation of dibutyl phthalate by homogeneous photocatalysis with Fe(III) in aqueous solution. Appl Catal B 33:239–248
Barse AV, Chakrabarti T, Ghosh TK, Pal AK, Jadhao SB (2007) Endocrine disruption and metabolic changes following exposure of *Cyprinus carpio* to diethyl phthalate. Pestic Biochem Physiol 88:36–42
Berge' A, Cladie're M, Gasperi J, Coursimault A, Tassin B, Moilleron R (2012) Meta-analysis of environmental contamination by alkylphenol. Environ Sci Pollut Res 19:3798–3819
Bharagava RN, Chandra R (2010a) Biodegradation of the major color containing compounds in distillery wastewater by an aerobic bacterial culture and characterization of their metabolites. Biodegradation J 21:703–711
Bharagava RN, Chandra R (2010b) Effect of bacteria treated and untreated post-methanated distillery effluent (PMDE) on seed germination, seedling growth and amylase activity in *Phaseolus mungo* L. J Hazard Mater 180:730–734

Bharagava RN, Mishra S (2018) Hexavalent chromium reduction potential of *Cellulosimicrobium* sp. isolated from common effluent treatment plant of tannery industries. Ecotoxicol Environ Saf 147:102–109

Bharagava RN, Chandra R, Rai V (2009) Isolation and characterization of aerobic bacteria capable of the degradation of synthetic and natural melanoidins from distillery wastewater. World J Microbiol Biotechnol 25:737–744

Bharagava RN, Chowdhary P, Saxena G (2017a) Bioremediation: an eco-sustainable green technology, it's applications and limitations. Bharagava RN Environmental pollutants and their bioremediation approaches. CRC Press, Taylor & Francis Group Boca Raton 9781138628892

Bharagava RN, Saxena G, Mulla SI, Patel DK (2017b) Characterization and identification of recalcitrant organic pollutants (ROPs) in tannery wastewater and its phytotoxicity evaluation for environmental safety. Arch Environ Contam Toxicol 14:1–14. https://doi.org/10.1007/s00244-017-0490-x

Bila D, Montalvão AF, Azevedo DA, Dezotti M (2007) Estrogenic activity removal of 17β-estradiol by ozonation and identification of by-products. Chemosphere 69:736–746

Bodzek M (2015) Application of membrane techniques for the removal of micropollutants from water and wastewater. Copernican Lett 6:24–33

Bodzek M, Dudziak M (2006) Elimination of steroidal sex hormones by conventional water treatment and membrane processes. Desalination 198:24–32

Bodzek M, Możliwości P (2013) wykorzystania technik membranowych w usuwaniu mikroorganizmów i zanieczyszczeń organicznych ze środowiska. Inżynieria Ochrona Środowiska 16:5–37

Bodzek M, Dudziak M, Luks–Betlej K (2004) Application of membrane techniques to water purification. Removal of phthalates. Desalination 162:121–128

Bolong N, Ismail AF, Salim MR, Matsuura T (2009) A review of the effects of emerging contaminants in wastewater and options for their removal. Desalination 239:229–246

Brotons JA, Olea-Serrano MF, Villalobos M, Pedraza V, Olea N (1995) Xenoestrogens released from lacquer coatings in food cans. Environ Health Perspect 103:608–612

Cases V, Argandona AV, Rodriguez M, Prats D (2011) Endocrine disrupting compounds: a comparison of removal between conventional activated sludge and membrane reactors. Desalination 272:240–245

Castellana G, Loffredo E (2014) Water Air Soil Pollut 225:1872. https://doi.org/10.1007/s11270-014-1872-6

Chandra R, Kumar V (2017) Detection of androgenic-mutagenic compounds and potential autochthonous bacterial communities during in situ bioremediation of post-methanated distillery sludge. Front Microbiol 8:887

Chandra R, Bharagava RN, Kapley A, Purohit JH (2011) Bacterial diversity, organic pollutants and their metabolites in two aeration lagoons of common effluent treatment plant during the degradation and detoxification of tannery wastewater. Bioresour Technol 102:2333–2341

Chandra R, Bharagava RN, Kapley A, Purohit HJ (2012) Characterization of *Phragmites cummunis* rhizosphere bacterial communities and metabolic products during the two stage sequential treatment of post methanated distillery effluent by bacteria and wetland plants. Bioresour Technol 103:78–86

Chang H, Wan Y, Wu S, Fan Z, Hu J (2011) Occurrence of androgens and progestogens in wastewater treatment plants and receiving river waters: comparison to estrogens. Water Res 45:732–740

Chen J, Li X, Li J et al (2007) Degradation of environmental endocrine disruptor di-2-ethylhexyl phthalate by a newly discovered bacterium, *Microbacterium* sp. strain CQ0110Y. Appl Microbiol Biotechnol 74:676

Choi KJ, Kim SG, Kim CW, Kim SH (2005) Effects of activated carbon types and service life on the re-moval of endocrine disrupting chemicals: amitrol, nonylphenol and bisphenol-A. Chemosphere 58:1535–1545

Chowdhary P, Yadav A, Kaithwas G, Bharagava R N (2017) Distillery wastewater: a major source of environmental pollution and it's biological treatment for environmental safety. Singh R &

Kumar S, Green technology and environmental sustainability. Springer International, Cham 978-3-319-50653-1

Chowdhary P, Raj A, Bharagava RN (2018) Environmental pollution and health hazards from distillery wastewater and treatment approaches to combat the environmental threats: A review. Chemosphere 194:229–246

Clara M, Strenn B, Gans O, Martinez E, Kreuzinger N, Kroiss (2005) Removal of selected pharmaceuticals, fragrances and endocrine disrupting compounds in a membrane bioreactor and conventional wastewater treatment plants. Water Res 39:4797–4807

Crisp TM, Clegg ED, Cooper RL, Wood WP, Anderson DG, Baetcke KP, Hoffmann JL, Morrow MS, Rodier DJ, Schaeffer JE, Touart LW, Zeeman MG, Patel YM (1998) Environmental endocrine disruption: an effects assessment and analysis. Environ Health Perspect 106(Suppl. 1):11–56

Danzl E, Sei K, Soda S, Ike M, Fujita M (2009) Biodegradation of bisphenol A, bisphenol F and bisphenol S in seawater. Int J Environ Res Public Health 6:1472–1484

De Gusseme B, Pycke B, Hennebel T, Marcoen A, Vlaeminck SE, Noppe H et al (2009) Biological removal of 17α-ethinylestradiol by a nitrifier enrichment culture in a membrane bioreactor. Water Res 43:2493–2503

Diamanti-Kandarakis E, Bourguignon JP, Giudice LC, Hauser R, Prins GS, Soto AM, Zoeller RT, Gore AC (2009) Endocrine-disrupting chemicals: an Endocrine Society scientific statement. Endocr Rev 30:293–342

Duty SM, Singh NP, Silva MJ et al (2003) The relationship between environmental exposures to phthalates and DNA damage in human sperm using the neutral comet assay. Environ Health Perspect 111:1164–1169

Eriksson P, Jakobsson E, Fredriksson A (1998) Developmental neurotoxicity of brominated flame retardants, polybrominated diphenyl ethers and tetrabromo-bis-phenol A. Organohalogen Compd 35:375

Eskinazi B, Mocarelli P, Warner M, Chee WY, Gerthoux PM, Samuels S, Needham LL, Patterson Jr DG (2003) Maternal serum dioxin levels and birth outcomes in women of Seveso, Italy. Environ Health Perspect 111:947–953

Falconer IR, Chapman HF, Moore MR, Ranmuthugala G (2006) Endocrine-disrupting compounds: a review of their challenge to sustainable and safe water supply and water reuse. Environ Toxicol 21:181–191

Federle TW, Kaiser SK, Nuck BA (2002) Fate and effects of triclosan in activated sludge. Environ Toxicol Chem 21:1330–1337

Fent K (2015) Progestins as endocrine disrupters in aquatic ecosystems: concentrations, effects and risk assessment. Environ Int 84:115–130

Fujii K, Kikuchi S, Satomi M, Ushio-Sata N, Morita N (2002) Degradation of 17β-estradiol by a gram-negative bacterium isolated from activated sludge in a sewage treatment plant in Tokyo, Japan. Appl Environ Microbiol 68:2057–2060

Fukuhara T, Iwasaki S, Kawashima M, Shinohara O, Abe I (2006) Adsorbability of estrone and 17β -estradiol in water onto activated carbon. Water Res 40:241–248

Gallenkemper M, Wintgens T, Melin T (2003) Nanofiltration of endocrine disrupting compounds. Membr Drinking Ind Water Prod 3:321–327

Gultekin I, Ince NH (2007) Synthetic endocrine disruptors in the environment and water remediation by advanced oxidation processes. J Environ Manag 85:816–832

Haag WR, Yao CCD (1992) Rate constants for reaction of hydroxyl radicals with several drinking water contaminants. Environ Sci Technol 26:1005–1013

Hardell L, Andersson SO, Carlberg M, Bohr L, van Bavel B, Lindstorm G et al (2006) Adipose tissue concentrations of persistent organic pollutants and the risk of prostate cancer. J Occup Environ Med 48:700–707

Harries JE, Runnalls T, Hill E et al (2000) Development of a reproductive performance test for endocrine disrupting chemicals using pair-breeding fathead minnows (*Pimephales promelas*). Environ Sci Technol 34:3003–3011

Hauser R, Meeker JD, Singh NP et al (2007) DNA damage in human sperm is related to urinary levels of phthalate monoester and oxidative metabolites. Hum Reprod 22:688–695

Hotchkiss AK, Rider CV, Blystone CR, Wilson VS, Hartig PC, Ankley GT, Foster PM, Gray CL, Gray LE (2008) Fifteen years after "wingspread" environmental endocrine disrupters and human and wildlife health: where we are today and where we need to go. Toxicol Sci 105:235–259

Høyer AE, Grandlean P, Jørgensen T, Brock JW, Hartvig HB (1998) Organochlorine exposure and risk of breast cancer. Lancet 352:1816–1820

Hu J, Chen S, Aizawa T, Terao Y, Kunikane S (2003) Products of aqueous chlorination of 17β-estradiol and their estrogenic activities. Environ Sci Technol 37:5665–5670

Huber MM, Ternes TA, Gunten UV (2004) Removal of estrogenic activity and formation of oxidation products during ozonation of 17a-ethinylestradiol. Environ Sci Technol 38:177–5186

Husain Q, Qayyum S (2012) Biological and enzymatic treatment of bisphenol A and other endocrine disrupting compounds: a review. Crit Rev Biotechnol 3:260–292

Ifelebuegu AO (2011) The fate and behavior of selected endocrine disrupting chemicals in full scale wastewater and sludge treatment unit processes. Int J Environ Sci Technol 8:245–254

Jacobson JL, Jacobson SW (1996) Intellectual impairment in children exposed to polychlorinated biphenyls in utero. New Eng J Med 335:783

Janex-Habibi ML, Huyard A, Esperanza M, Bruchet A (2009) Reduction of endocrine disruptor emissions in the environment: the benefit of wastewater treatment. Water Res 43:1565–1576

Jiang JQ, Yin Q, Zhou JL, Pearce P (2005) Occurrence and treatment trials of endocrine disrupting chemicals (EDCs) in wastewaters. Chemosphere 61:544–550

Jing Y, Li LS, Zhang QY, Lu PP, Liu H, Lu XH (2011) Photocatalytic ozonation of dimethyl phthalate with TiO2 prepared by a hydrothermal method. J Hazard Mater 189:40–47

Jobling S, Casey D, Rogers-Gray T, Oehlmann J, Schulte-Oehlmann U, Pawlowski S, Baunbeck T, Turner AP, Tyler CR (2004) Comparative responses of molluscs and fish to environmental estrogens and an estrogenic effluent. Aquat Toxicol 66:207–222

Johnson AC, Sumpter JP (2001) Removal of endocrine-disrupting chemicals in activated sludge treatment works. Environ Sci Technol 35:4697–4703

Kanda R, Griffin P, James HA, Fothergill J (2003) Pharmaceutical and personal care products in sewage treatment works. J Environ Monit 5:823–830

Kang JH, Kondo F, Katayama Y (2006) Human exposure to bisphenol A. Toxicology 226:79–89

Kidd KA, Blanchfield PJ, Mills KH, Palace VP, Evans RE, Lazorchak JM, Flick RW (2007) Collapse of a fish population after exposure to a synthetic estrogen. Proc Natl Acad Sci U S A 104:8897–8901

Kidd KA, Becher G, Bergman A, Muir DCG, Woodruff TJ (2012) Human and wildlife exposures to EDC's; Chapter 3; State of the science of endocrine disrupting chemicals. UNEP, Geneva, pp 189–250

Kobayashi K, Miyagawa M, Wang RS, Suda M, Sekiguchi S, Honma T (2005) Effects of in utero and lactational exposure to bisphenol a on thyroid status in F1 rat offspring. Ind Health 43:685–690

Koopman-Esseboom C, Weisglas-Kuperus N, de Ridder MA, Van der Paauw CG, Tuinstra LG, Sauer PJ (1996) Effects of polychlorinated biphenyl/dioxin exposure and feeding type on infants' mental and psychomotor development. Pediatrics 97:700–706

Kumar V, Majumdar C, Roy P (2008) Effects of endocrine disrupting chemicals from leather industry effluents on male reproductive system. J Steroid Biochem Mol Biol 111:208–216

Kumari V, Yadav A, Haq I, Kumar S, Bharagava RN, Singh SK, Raj A (2016) Genotoxicity evaluation of tannery effluent treated with newly isolated hexavalent chromium reducing *Bacillus cereus*. J Environ Manag 183:204–211

Kunde L, Weiping L, Gan J (2009) Oxidative removal of bisphenol A by manganese dioxide: efficacy, products, and pathways. Environ Sci Technol 43:3860–3864

Latorre A, Lacorte S, Barcel'o D (2003) Presence of nonylphenol, octyphenol and bisphenol a in two aquifers close to agricultural, industrial and urban areas. Chromatographia 57:111–116

Laws SC, Carey SA, Ferrell JM, Bodman GJ, Cooper RL (2000) Estrogenic activity of octylphenol, nonylphenol, bisphenol A andmethoxychlor in rats. Toxicol Sci 54:154–167

Lee Y, Yoon J, Von Gunten U (2005) Kinetics of the oxidation of phenols and phenolic endocrine disruptors during water treatment with ferrate (Fe (VI)). Environ Sci Technol 39:8978–8984

Li HQ, Jiku F, Schroder HF (2000) Assessment of the pollutant elimination efficiency by gas chromatography/mass spectrometry, liquid chromatography–mass spectrometry and tandem mass spectrometry-comparison of conventional and membrane-assisted biological wastewater treatment processes. J Chromatogr 889:155–176

Liu ZH, Kanjo Y, Mizutami S (2009) Removal mechanisms for endocrine disrupting compounds (EDCs) in wastewater treatment-physical means, biodegradation, and chemical advanced oxidation: a review. Sci Total Environ 407:731–748

Liu J, Wang R, Huang B, Lin C, Wang Y, Pan X (2011) Distribution and bioaccumulation of steroidal and phenolic endocrine disrupting chemicals in wild fish species from Dianchi Lake, China. Environ Pollut 159:2815–2822

Macellaro G, Pezzella C, Cicatiello P, Sannia G, Piscitelli A (2014) Fungal laccases degradation of endocrine disrupting compounds. Bio Med Res Int 2014:614038

MacLusky NJ, Hajszan T, Leranth C (2005) The environmental estrogen bisphenol A inhibits estradiol-induced hippocampal synaptogenesis. Environ Health Perspec 113:675–679

Madsen PB, Johansen NH, Andersen HR, Kaas P (2006) Removal of endocrine disruptors and pathogens. Advanced photo oxidation processes at Hørsholm WWTP. Presented at the IWA World Water Congress, Beijing, China

Markey CM, Rubin BS, Soto AM, Sonnenschein C (2003) Endocrine disruptors: from wingspread to environmental developmental biology. J Steroid Biochem Mol Biol 83:235–244

Matsui Y, Knappe DRU, Iwaki K, Ohira H (2002) Pesticide adsorption by granular activated carbon adsorbers 2. Effects of pesticide and natural organic matter characteristics on pesticide breakthrough curves. Environ Sci Technol 36:3432–3438

Mavrov V, Nikolov ND, Islam MA, Nikolova JD (1992) An investigation on the configuration of inserts in tubular ultrafiltration module to control concentration polarization. J Membr Sci 75:197–201

McGlynn KA, Quraishi SM, Graubard BI, Weber JP, Rubertone MV, Erickson RL (2008) Persistent organochlorine pesticides and risk of testicular germ cell tumors. J Natl Cancer Inst 100:663–671

Mohapatra DP, Brar SK, Tyagi RD, Surampalli RY (2010) Physico-chemical pre-treatment and biotransformation of wastewater and wastewater sludge-fate of bisphenol A. Chemosphere 78:923–941

Mompelat S, Le Bot B, Thomas O (2009) Occurrence and fate of pharmaceutical products and by-products, from resource to drinking water. Environ Int 35:803–814

Nadzialek S, Vanparys C, Van der Heiden E, Michaux C, Brose F, Scippo ML, De Coen W, Kestemont P (2010) Understanding the gap between the estrogenicity of an effluent and its real impact into the wild. Sci Total Environ 408:812–821

Nakamura H, Shiozawa T, Terao Y, Shiraishi F, Fukazawa H (2006) By-products produced by the reaction of estrogens with hypochlorous acid and their estrogen activities. J Health Sci 52:124–131

Nykova N, Muller TG, Gyllenberg M, Timmer J (2002) Quantitative analyses of anaerobic wastewater treatment processes: identifiability and parameter estimation. Biotechnol Bioeng 78:89–103

Olujimi OO, Fatoki O, Odendaal SJP, Okonkwo JO (2010) Endocrine disrupting chemicals (phenol and phthalates) in the South African environment: a need for more monitoring. Water SA 36:671–682

Patnaik P (2007) A comprehensive guide to the hazardous properties of chemical substances.3rd edn. Wiley Interscience, Hoboken

Pedersen SN, Christiansen LB, Pedersen KL, Korsgaard B, Bjerregaard P (1999) In vivo estrogenic activity of branched and linear alkylphenols in rainbow trout (Oncorhynchus mykiss). Sci Total Environ 233:89–96

Petruzzelli D, Boghetich G, Petrella M, Dell'Erba AL, Abbate P, Sanarica S (2007) Advanced oxidation as a pretreatment of industrial landfill leachate. Global NEST J 9:51–56

Pezzella C, Macellaro G, Sannia G, Raganati F, Olivieri G, Marzocchella A, Schlosser D, Piscitelli A (2017) Exploitation of Trametes versicolor for bioremediation of endocrine disrupting chemicals in bioreactors. PLoS One 12:e0178758

Polyzos SA, Kountouras J, Deretzi G, Zavos C, Mantzoros CS (2012) The emerging role of endocrine disruptors in pathogenesis of insulin resistant: a concept implicating nonalcoholic fatty liver disease. Cur Mol Med 12:68–82

Qhanya Lehlohonolo B et al (2017) Isolation and characterisation of endocrine disruptor nonylphenol-using bacteria from South Africa. S Afr J Sci 113:1–7

Rignell-Hydbom A, Rylander L, Giwercman A, Jönsson BAG, Nilsson-Ehle P, Hagmar L (2004) Exposure to CB-153 and p, p0-DDE and male reproductive function. Hum Reprod 19:2066–2075

Ritchie JM, Vial SL, Fuortes LJ, Robertson LW, Guo H, Reedy VE et al (2005) Comparison of proposed frameworks for grouping polychlorinated biphenyl congener data applied to a case-control pilot study of prostate cancer. Environ Res 98:104–113

Roh H, Subramanya N, Zhao F, Yu CP, Sandt J, Chu KH (2009) Biodegradation potential of wastewater micropollutants by ammonia-oxidizing bacteria. Chemosphere 77:1084–1089

Samaras VG, Stasinakis AS, Mamais D, Thomaidis NS, Lekkas TD (2013) Fate of selected pharmaceuticals and synthetic endocrine disrupting compounds during wastewater treatment and sludge anaerobic digestion. J Hazard Mater 244:259–267

Saxena G, Bharagava RN (2017). Organic and inorganic pollutants in industrial wastes, their ecotoxicological effects, health hazards and bioremediation approaches, Bharagava RN Environmental pollutants and their bioremediation approaches. CRC Press, Taylor & Francis Group, Boca Raton 9781138628892

Saxena G, Chandra R, Bharagava RN (2017) Environmental pollution, toxicity profile and treatment approaches for tannery wastewater and its chemical pollutants. Rev Environ Contam Toxicol 240:31–69

Sei K, Takeda T, Soda SO, Fujita M, Ike M (2007) Removal characteristics of endocrine-disrupting chemicals by laccase from white-rot fungi. J Environ Sci Health A 43:53–60

Singer H, Muller S, Tixier C, Pillonel L (2002) Triclosan: occurrence and fate of a widely used biocide in the aquatic environment: field measurements in wastewater treatment plants, surface waters, and lake sediments. Environ Sci Technol 36:4998–5004

Snyder S, Westerhoff P, Song R, Levine B, Long B (2011) American Water Works Association Research Foundation (AWWARF) Project #2758: Evaluation of conventional and advanced treatment processes to remove endocrine disruptors and pharmaceutically active compounds. http://enpub.fulton.asu.edu/pwest/awwarf_project_EDC.htm

Snydera SA, Adhamb S, Reddingc AM, Cannonc FS, DeCarolisb J, Oppenheimerb J, Werta EC, Yoond Y (2007) Role of membranes and activated carbon in the removal of endocrine disruptors and pharmaceuticals. Desalination 202:156–181

Sohoni P, Sumpter JP (1998) Several environmental oestrogens are also anti-androgens. J Endocrinol 158:327–339

Spencer AL, Bonnema R, McNamara MC (2009) Helping women choose appropriate hormonal contraception: update on risks, benefits, and indications. Am J Med 122:497–506

Staples CA, Dome PB, Klecka GM, Oblock ST, Harris LR (1998) A review of the environmental fate, effects, and exposures of bisphenol A. Chemosphere 36:2149–2173

Staples CA, Woodburn KN, Hall AT, Klecka GM (2002) A weight of evidence approach to the aquatic hazard assessment of bisphenol A. Human Ecol Risk Assess 8:1083–1105

Thompson A, Griffin P, Stuetz R, Cartmell E (2005) The fate and removal of triclosan during wastewater treatment. Water Environ Res 77:63–67

Tilson HA, Kodavanti PR (1997) Neurochemical effects of polychlorinated biphenyls: an overview and identification of research needs. Neurotoxicology 18:727–743

Toft G, Rignell-Hydbom A, Tyrkiel E, Shvets M, Giwercman A, Lindh CH et al (2006) Semen quality and exposure to persistent organochlorine pollutants. Epidemiology 17:450–458

Toyama T, Ojima T, Tanaka Y, Mori K, Morikawa M (2013) Sustainable biodegradation of phenolic endocrine-disrupting chemicals by *Phragmites australis*-rhizosphere bacteria association. Water Sci Technol 68:522–529

Vandenberg LN, Maffini MV, Sonnenschein C, Rubin BS, Soto AM (2009) Bisphenol-A and the great divide: a review of controversies in the field of endocrine disruption. Endocrinol Rev 30:75–95

Villemur R, dos Santos SCC, Ouellette J, Juteau P, Lepine F, Deziel E (2013) Biodegradation of endocrine disruptors in solid-liquid two-phase partitioning systems by enrichment cultures. Appl Environ Microbiol 79:4701–4711

Von Gunten U (2003) Ozonation of drinking water: part I. Oxidation kinetics and product formation. Water Res 37:1443–1467

Voutsa et al (2006) Benzotriazoles, alkylphenols and bisphenol A in municipal wastewaters and in the Glatt River, Switzerland. Environ Sci Pollut Res 13:333–341

Walha K, Amar RB, Firdaous L, Quem'eneur F, Jaouen P (2007) Brackish groundwater treatment by nanofiltration, reverse osmosis and electrodialysis in Tunisia: performance and cost comparison. Desalination 207:95–106

Weber S, Leuschner P, Kampfer P, Dott W, Hollender J (2005) Degradation of estradiol and ethinyl estradiol by activated sludge and by a defined mixed culture. Appl Microbiol Biotechnol 67:106–112

Westerhoff P, Yoon Y, Snyder S, Wert E (2005) Fate of endocrine-disruptor, pharmaceutical, and personal care product chemicals during simulated drinking water treatment processes. Environ Sci Technol 39:6649–6663

Wintgens T, Gallenkemper M, Melin T (2002) Endocrine disrupter removal from wastewater using membrane bioreactor and nanofiltration technology. Desalination 146:387–391

Witorsch RJ (2002) Endocrine disruptors: can biological effects and environmental risks be predicted? Regul Toxicol Pharmacol 36:118–130

Yadav A, Mishra S, Kaithwas G, Raj A, Bharagava RN (2016) Organic pollutants and pathogenic bacteria in tannery wastewater and their removal strategies. In: Singh JS, Singh DP (eds) Microbes and environmental management. Studium Press (India) Pvt. Ltd., New Delhi, 2015, pp 101–127

Yang YY, Wang Z, Xie SG (2014) Aerobic biodegradation of bisphenol A in river sediment and associated bacterial community change. Sci Total Environ 470:1184–1188

Ying GG, Kookana RS (2007) Triclosan in wastewaters and biosolids from Australian wastewater treatment plants. Environ Int 33:199–205

Ying GG, Williams B, Kookana R (2002) Environmental fate of alkylphenols and alkylphenol ethoxylates-a review. Environ Int 28:215–226

Yoshihara S, Murugananthan M (2009) Decomposition of various endocrine-disrupting chemicals at boron-doped diamond electrode. Electrochim Acta 54:2031–2038

Yoshimoto T, Nagai F, Fujimoto J, Watanabe K, Mizukoshi H, Makino T, Kimura K, Saino H, Sawada H, Omura H (2004) Degradation of estrogens by *Rhodococcus zopfii* and *Rhodococcus equi* isolates from activated sludge in wastewater treatment plants. Appl Environ Microbiol 70:5283–5289

Yu CP, Roh H, Chu KH (2007) 17β-estradiol-degrading bacteria isolated from activated sludge. Environ Sci Technol 41:486–492

Zacharewski TR, Meek MD, Clemons JH, Wu ZF, Fielden MR, Matthews JB (1998) Examination of the in vitro and in vivo estrogenic activities of eight commercial phthalate esters. Toxicol Sci 46:282–293

Zeng F, Liu W, Jiang H, Yu HQ, Zeng RJ, Guo Q (2011) Separation of phthalate esters from bio-oil derived from rice husk by a basification-acidification process and column chromatography. Bioresour Technol 102:1982–1987

Zhang WW, Yin K, Chen LX (2013) Bacteria-mediated bisphenol A degradation. Appl Microbiol Biotechnol 97:5681–5689

Zhang C, Li Y, Wang C, Niu L, Cai W (2015) Occurrence of endocrine disrupting compounds in aqueous environment and their bacterial degradation: a review. Crit Rev Environ Sci Technol 46:1–59

Zhou JH, Sui ZJ, Zhu J, Li P, Chen D, Dai YC et al (2007) Characterization of surface oxygen complexes on carbon nanofibers by TPD, XPS and FT-IR. Carbon 45:785–796

Zoeller RT, Bansal R, Parris C (2005) Bisphenol-A, an environmental contaminant that acts as a thyroid hormone receptor antagonist in vitro, increases serum thyroxine, and alters RC3/neurogranin expression in the developing rat brain. Endocrinology 146:607–612

Chapter 8
Arsenic Toxicity and Its Remediation Strategies for Fighting the Environmental Threat

Vishvas Hare, Pankaj Chowdhary, Bhanu Kumar, D. C. Sharma, and Vinay Singh Baghel

Abstract Arsenic (As) is an abundant element found ubiquitously in nature, primarily in the earth's crust and also in the environment. Arsenic is necessary for living beings; however, it is also an emerging issue by virtue of the toxicity it causes in living beings, including humans and animals. Basically, the ground water is badly affected by As contamination, coming from sources including As-affected aquifers, and has severely threatened humanity around the world. Arsenic poisoning is worse in Bangladesh and Uttar Pradesh, where As(III) is found in higher concentrations in ground water, which is used by people. The dissolution process caused by oxidation and reduction reactions leads to the natural occurrence of As in groundwater. There are several review articles on As toxicity and exposure, but with scattered information and no systematic knowledge in a combined way. However, in this chapter we try to compile all the information in systematic manner, which will be helpful for people who are working for As mitigation and removal from environment for sustainable development. This chapter will be helpful in providing detailed knowledge on As occurrence, speciation, factors affecting As toxicity arising because of its biogeochemistry, and various physico-chemical and biological strategies for combating the environmental threats.

Keywords Arsenic toxicity · Accumulation · Arsenic speciation · Health problems · Remediation approaches

---

V. Hare (✉) · P. Chowdhary · V. S. Baghel
Department of Environmental Microbiology (DEM), Babasaheb Bhimrao Ambedkar University (A Central University), Lucknow, Uttar Pradesh, India

B. Kumar
Pharmacognosy and Ethnopharmacology Division, CSIR-National Botanical Research Institute, Lucknow, Uttar Pradesh, India

D. C. Sharma
Department of Microbiology, Dr Shakuntala Mishra National Rehabilitation University, Lucknow, Uttar Pradesh, India

1 Introduction

Arsenic is a harmful metal, found everywhere, i.e. in soil, water, air, and is highly toxic to all living beings. Its distribution among geochemical sources is irregular and is commonly found in the earth's crust. It comes from both anthropogenic and geogenic sources (Smith et al. 1999; Juhasz et al. 2003). However, the worst contamination conditions have been encountered in West Bengal (India) and neighbouring countries such as Bangladesh, and have been created because of natural process (Tripathi et al. 2007). Arsenic is also reported to be invading the food chain, for instance rice, most probably due to contaminated ground water (Meharg et al. 2009; Hare et al. 2017). Arsenic (As) persists in nature for a long time.

Arsenic is found in two forms (organic and inorganic) in nature, but the inorganic As shows higher toxicity than the organic form. Inorganic As is represented by two biological forms {Arsenate As(V) and arsenite As(III)} which can interconvert regulated by the environment, especially redox conditions. Arsenate As(V) interferes with necessary cellular processes such as oxidative phosphorylation and ATP synthesis as it is a phosphate analogue and thus, the main route of arsenate uptake by the roots may be through the phosphate transport mechanism. A little is exported to the shoot through the xylem as the ox anions As(V) and As(III). The toxicity of arsenite is governed by its tendency to bind to sulfhydryl groups, with resultant harmful effects on protein functioning, and As(III) is transported in the neutral As(OH)$_3$ form through aquaglyceroporins (Meharg 2004). Arsenic occurs as oxyanions in soils, predominantly as arsenite As(III) and arsenate As(V). In oxygen-rich environments and well-drained soils, arsenate species dominate (H$_2$AsO$_4^-$ in acidic soils and HAsO$_4^{2-}$ in alkaline ones), whereas in a reducing environment, for example, regularly flooded soils, As(III) is more stable (Gomez-Caminero et al. 2001). The relative prevalence of the various forms of As in soils depends on the type and amount of adsorbing component of the soil, pH and redox potential (Buschmann et al. 2006). Materials with iron and aluminium oxide surfaces are capable of adsorbing arsenate and arsenite. Humic acids and fulvic acids in soil have been reported to show great affinity for As (Meunier et al. 2011). These may interfere strongly with As adsorption in some situations and As mobility may be increased in soil when they are present (Jackson and Miller 1999). This relationship is postulated to involve bridging metals and deprotonated functional groups within the humic acid (Warwick et al. 2005). Increased As concentration retards usual growth and development of plants if present in irrigation water or in soil with toxicity symptoms such as biomass reduction and yield decreases. Arsenic interferes with plant metabolic system and can inhibit growth, often leading to death when present in higher concentrations (Jiang and Singh 1994). There are several reports on the loss of fresh and dry biomass of roots and shoots, loss of yield and fruit production, morphological changes if the plants were grown in soils treated with As (Mokgalaka-Matlala et al. 2008). Miteva (2002) reported a decrease in growth of the aerial parts and the root system of tomato plants at higher As concentrations. A disease known as "straight head" is a physiological disorder of rice (*Oryzasativa* L.) characterised by the sterility of florets/spikelets leading to a decreased crop yield.

Singh et al. (2010) worked on the effect of As on rice crops in the Indo-Gangetic plains of north-western India and found a positive correlation between the rice growth and As in the irrigation water and soil. There are several reports revealing the positive correlation between rice crops and the fields versus As-contaminated irrigation water (Khan et al. 2009).

Arsenic that has accumulated in plants finally reaches human beings, causing potential health risks such as skin cancer as it become an integral part of the food chain. Arsenic is regarded as a group A human carcinogen by the United States (US) Environmental Protection Agency (EPA). The major health hazards include skin, lung, bladder, and liver cancers, and many other cardiovascular, neurological, haematological, renal, and respiratory diseases (Halim et al. 2009; Johnson et al. 2010; Yadav et al. 2017), mostly ascribed to intake along with contaminated fresh drinking water. Arsenic toxicity in severe stages is marked by symptoms such as dermal lesions (e.g. hyperkeratosis, hyper pigmentation, desquamation and loss of hair (Zaloga et al. 1985), peripheral neuropathy, skin cancer and peripheral vascular disease. The most affected parts are those directly involved in absorption, accumulation and excretion of As, especially the gastrointestinal tract and liver (Yadav et al. 2017). Apart from this, the vascular system and other soft organs such as the heart and kidneys are very sensitive to As. The skin tissues are secondarily affected. The intestinal epithelium is the first barrier against such exogenous inorganic As toxicity.

2 Occurrence/Sources of Arsenic

Arsenic is the 20th most frequently occurring trace element in the soil, 14th in the seawater, and 12th in living systems (humans). Arsenic is abundantly distributed in nature and its derivatives are mobile in the environment. In the environment, the major sources of As include natural/geogenic and anthropogenic sources (Fig. 8.1).

2.1 Natural/Geogenic Sources

Arsenic is ubiquitous in nature, and is distributed throughout the earth's crust, soil, sediments, water, atmosphere and living organisms. Arsenic is reported to be found in soil in higher concentrations than in rocks. Arsenic is naturally found in more than 200 different mineral forms, which contain arsenates (60%), sulphides (20%), sulfosalts and the rest consist of arsenides, arsenites, oxides, silicates and elemental As (20%). On the land, the concentration of As present is around 1.5–3.0 mg kg^{-1}. The concentrations of As in the earth crust of different nations are said to vary within the range from 0.1 to 40 mg kg^{-1} (mean 6 mg kg^{-1}), 1–50 mg kg^{-1} (mean 6 mg kg^{-1}) and mean 5 mg kg^{-1}, but varies considerably among geographic regions (Kabata and Pendias 1984). In natural waters, As is reported to be present at low concentrations. The seawater ordinarily contains 0.001–0.008 mgl^{-1} of As. The

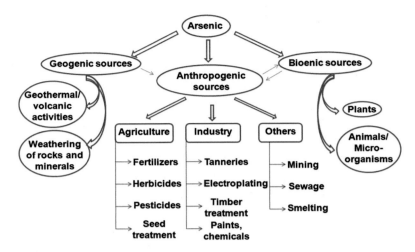

Fig. 8.1 Sources of arsenic contamination in the environment

maximum permissible limits of As in potable water is 50 μgl^{-1} and the recommended value by the EPA and the WHO is 10 μgl^{-1} (WHO Arsenic Compounds 2001; U.S. Environmental Protection Agency 1975). In air, As is present primarily absorbed on particulate matters, is found most of the time as a mixture of As(III) and As(V), with the organic species being of little significance except in areas where the application of As pesticides is frequent. The As content present in air is quite low and ranges from 0.4 to 30 ng m^{-3}; therefore, through air humans generally encounter very low exposure. USEPA estimate that the general public is exposed to a range of approximately 40–90 ng per day by inhalation.

2.2 Anthropogenic Sources

For As, the principal anthropogenic contributions to soils come from the ways of managing municipal solid waste, i.e. primarily combustion, the application of different agricultural support substances, including pesticides such as herbicides, fungicides and insecticides, which contain As. Application of solid waste/sewage sludge into land, river and irrigation waters (Kabata-Pendias and Adriano 1995), the mining and smelting of As-containing ores, combustion of fossil fuels (especially coal), land filling of industrial wastes (pulp and paper, tannery, textile and distillery), the release or disposal of chemical warfare agents (Chowdhary et al. 2017a, b, c; Goh and Lim 2005), manufacturing of metals and alloys, petroleum refining and pharmaceutical manufacturing (Ning 2002; Mishra and Bharagava 2016). Nowadays, for preserving wood chromated copper arsenate (CCA) has become popular, and is another potential source of As in soil (Bharagava and Mishra 2018).

Arsenic is used in various kinds of pesticides, such as lead arsenate, Ca_3AsO_4, copper acetoarsenite, Paris green (copper-acetoarsenite), H_3AsO_4, monosodium methanearsonate (MSMA), disodium methanearsonate (DSMA) and cacodylic acid are used in the production of cotton as pesticides. Arsenic is also used for wood preservatives such as fluor-chrome-arsenic-phenol (FCAP), CCA and ammoniacal copper arsenate (ACA) (U.S. Department of Agriculture 1974). Many As compounds are used for feed additives, such as H_3AsO_4, 3-nitro-4-hydroxy phenylarsonic acid and 4-nitrophenylarsonic acid.

3 Accumulation and Transformation of Arsenic in Edible Crops

Plants vary significantly in their ability to accumulate As and the rice plant is more inclined to As accumulation than other cereals, as it is generally grown under flood conditions where As mobility is elevated. However, wide range of As-resistant microorganisms are present in nature; relatively few of the microorganisms are known to hyper-accumulate As (non-genetically engineered microorganisms) (Xie et al. 2013). Bioaccumulation refers to a process in which As is accumulated in cell membranes and cytoplasm, different from biosorption, in which accumulation takes place at the cell surface (Joshi et al. 2009; Xie et al. 2013). Arsenic transformation in the environment is predominantly biotic (Meng et al. 2011), abiotic transformation of As has been shown to be considerably slower and is believed to be less important than microbially mediated reduction (Jones et al. 2000). The transformation reactions of As include microbial reduction, oxidation, methylation and demethylation, which play a detrimental role in the environmental behaviour of As, because different types of mobility are shown by different chemical forms of As [Methyl As(III)>Methyl As(V)>As(III)>As(V)] (Lafferty and Loeppert 2005), toxicity [Methyl As(III)>As(III)>As(V)>Methyl As(V)] and susceptibility to plant uptake [e.g. uptake by the rice root As(III)>Monomethyl As(V)>Dimethyl As(V)].

3.1 Conversion and Internalisation of Arsenate

In arsenate uptake, the phosphate transporters present in the plant roots play a key role. Confirmation is shown in different physiological and electrophysiological studies exhibiting that arsenate uptake causes an effective inhibition of phosphate, and recent reports show that *A. thaliana* mutants flawed in phosphate transport are more tolerant to arsenate. Reduction of As(V) usually demonstrates an enhancement in the mobility of As in the natural environment, as As(III) has greater mobility than As(V) (Smedley and Kinniburgh 2002). Reduction of As (V) through microbes may occur via respiratory reduction, as microorganisms use arsenate as the terminal electron acceptor in anaerobic respiration (Lloyd and Oremland 2006; Mukhopadhyay et al. 2002; Stolz et al. 2006). Detoxification is the other mechanism of reduction of arsenate to arsenite through microbes (Langner and Inskeep 2000; Stolz et al. 2002). Arsenic detoxification/mitigation has been well documented in *E. coli*, *Staphylococcus aureus* and *Staphylococcus xylosis* and is controlled by "ars" genes that encode for arsenate (Cervantes et al. 1994; Tamaki and Frankenberger 1992). Arsenates detoxifying reducing bacteria were found to play an important role in As mobilisation under oxic conditions. Under flooding conditions, in soil amended with citrate, strong As mobilisation was observed at the commencement of incubation when oxic conditions prevailed (Eh >250 mV) (Corsini et al. 2010).

3.2 Conversion and Uptake of Arsenite

Uptake of arsenite by plant roots is mainly in the form of the neutral molecule $As(OH)_3$. Arsenite enters through plant root cells via aquaglyceroporin channels. In higher plants, the nodulin 26-like intrinsic proteins (NIPs) are the structural and functional equivalents of the microbial and mammalian aqua-glyceroporins (Wallace et al. 2006). NIPs are a subfamily of the plant major intrinsic proteins, collectively known as water channels or aquaporins, and arsenite permeability is widespread in different subclasses of NIPs (Zhao et al. 2009; Maurel et al. 2008). Microorganisms tolerate higher As(III) levels by carrying out microbial As(III) oxidation as an effective detoxification process (Paez-Espino et al. 2009; Tamaki and Frankenberger 1992). As(III) oxidation is catalysed by a wide range of microorganisms, e.g. *Alcaligenes faecalis*, *Hydrogenophaga* sp., *A. ferrooxidans*, *T. aquaticus*, *T. thermophilus*. Stolz et al. 2006; Wang and Zhao 2009). The major impact of microbial oxidation of As(III) to As(V) is to reduce As mobility in the environment, as the resemblance of As(V) to mineral solids is usually higher than that of As(III) (Dixit and Hering 2003; Huang et al. 2011; Smedley and Kinniburgh 2002).

3.3 Methylation

Several aerobic and anaerobic microorganisms (Kuehnelt and Goessler 2003) established As methylation. Microbial methylation allows the conversion of aqueous or solid associated inorganic As into gaseous arsines and leaves from the living medium, which is usually regarded as detoxification (Jia et al. 2013). The gaseous arsines endure long distances as they are highly mobile in comparison to aqueous As and may easily travel in the environment (Mukai et al. 1986). The lower adsorption capacity of methylated As than inorganic As has been reported and was considered to be mobilisation owing to the formation of aqueous trivalent and pentavalent methyl As (Huang and Matzner 2006; Lafferty and Loeppert 2005). Low redox potentials (i.e. reducing conditions) encourage the production and mobilisation of methylated As (Frohne et al. 2011). Under reducing conditions, the reduction of As(V) to As(III) may enhance microbial methylation of As followed by enhanced levels of dissolved As in soils and sediments (Bennett et al. 2012; Du Laing et al. 2009).

3.4 Demethylation

Demethylation may occur under oxic and anoxic conditions, but is usually faster under oxic conditions (Huang et al. 2007). Removal of the organic moieties not only induces the general toxicity of As, but also reduces its mobility. Arsenic demethylation usually refers to the degradation of aqueous methylated As. In the atmosphere, the gaseous methylated arsines undergo rapid photo oxidative degradation (Mestrot et al. 2013). Whether the microorganisms found in the environment are able to perform As demethylation is still an open question.

3.5 Nitrate Reduction

Nitrate reduction may not only affect As(V) reduction, but can also influence As cycling under anoxic conditions. For example, nitrate-respiring sediments could reduce As(V) to As(III) once all of the nitrate has been removed (Gibney and Nusslein 2007; Dowdle et al. 1996). In urban lakes, microbial oxidation of Fe(II) and As(III) helped by nitrate may be an important process, leading to the arrangement of particulate ferric-oxide and As(V); an important consequence of enriched nitrate is therefore the presence of As(V) associated with hydrous ferric oxide colloids (Senn and Hemond 2002). Introducing nitrate may support the anoxic oxidation of Fe(II) and arsenite in the subsurface as a means of immobilising As in the form of As(V) adsorbed onto biogenic Fe(III) (hydr)oxides (Sun et al. 2009).

4 Toxicity Profile of Arsenic

West Bengal is most affected by As and it was also first reported here in the 1990s when people started to build up arseniosis, starting with skin rashes and leading to fatal problems, with a toxic effect on organs, for instance, the lungs, kidneys, and bladder (Chowdhury et al. 2000). Related issues were detected in Bangladesh, which is in close proximity to West Bengal and has a similar land pattern based on alluvial and deltaic sediments. Arsenic is not reported to act as nutrient and it is hazardous even at very minimal concentrations to plants and animals (Smedley and Kinniburgh 2002; Buschmann et al. 2008). Toxicity of As could affect different varieties of organisms, including humans (Cervantes et al. 1994). As accumulation in plants affects not only the growth, but also, over time, it becomes a part of the food chain, which eventually causes adverse health problems to human beings such as skin cancer. Arsenic toxicity is largely manifested in the cytoplasm and a common mechanism of detoxifying cytoplasmic metals and metalloids is complexation via sulphur bonds (Tripathi et al. 2007; Rosen 2002).

The chemical forms and oxidation states of As are more significant with regard to toxicity. The mode of toxicity differs between As species. As(V) interferes with phosphate metabolism such as phosphorylation and ATP synthesis, whereas As(III) binds to the vicinal sulfhydryl groups of proteins, affecting their structures or catalytic functions. Because arsenate decreases to arsenite, much of the toxic effect of As(V) may actually be due to its reduction product arsenite. It has been reported that iAs(V) is less toxic than iAs(III) to both animal and humans, but can be toxic to plants. An element is most toxic when it inactivates the enzyme systems, which serves as a biological catalyst. iAs(V) does not react openly with the active sites of enzymes (Johnstone 1963). It first reduces to iAs(III) in vivo before exerting its toxic effect. The enzymes, which generated cellular energy in the citric acid cycle, are harmfully affected. The inhibitory action is based on inactivation of pyruvate dehydrogenase by complexation with iAs(III), whereby the generation of adenosine-5-triphosphate (ATP) is prohibited. Arsenic can inhibit many enzymes, for example the pyruvate oxidase, *S*-aminoacid oxidase, choline oxidase, and transaminase. Although iAs(III) is regarded as the more toxic form of the element, iAs(V) as arsenate can be disruptive as it competes with phosphate. Arsenate is also capable of replacing the phosphorous group in the DNA molecule and this appears to inhibit the DNA repair mechanism. In environments where phosphate content is high, arsenate toxicity to flora and fauna is generally reduced. Arsenic seems to have antagonistic relation to selenium in the body and each counteracts the toxicity of the other. However, As may also interfere with the essential role in metabolism. Arsenic prevents the biological role of selenium, which results in the apparent deficiency of the glutathione peroxidise system (selenium-dependent enzyme).

The Chronic effects of As are well documented. The worst affected organs are those that take part in As absorption, accumulation, and excretion. Chronic As toxicity causes symptoms such as dermal lesions, and loss of hair (Zaloga et al. 1985), peripheral neuropathy, skin cancer and peripheral vascular disease. These symptoms have been identified mostly in populations, used As-contaminated water (Tseng 1977; Zaldivar 1980; Cebrian et al. 1983; Smith et al. 2010). The skin is known to localise and store As because of its high keratin content, which contains several sulfhydryl groups to which As^{3+} may bind (Kitchin 2001). The hazardous health effects of As exposure to different internal and external organs are summarised in Table 8.1.

8 Arsenic Toxicity and Its Remediation Strategies for Fighting the Environmental Threat

Table 8.1 Toxicological effects of arsenic in living organisms

Effect	Toxicity	References
Respiratory effects	Tracheae, bronchitis, rhinitis, pharyngitis, shortness of breath, chest sounds, nasal congestion and perforation of the nasal septum	Morton et al. (1994)
Pulmonary effects	1. Non-malignant pulmonary disease	Borgono et al. (1997)
	2. Abnormal skin pigmentation, complained of chronic cough	
	3. Restrictive lung disease	
Cardiovascular effects	1. Cardiovascular abnormalities, Raynaud's disease, myocardial infarction, myocardial depolarisation, cardiac arrhythmias and thickening of blood vessels	WHO Arsenic Compounds (2001)
Gastrointestinal effects	1. Sub-acute arsenic poisoning as dry mouth and throat, heartburn, nausea, abdominal pains and cramps, and moderate diarrhoea	Nagvi et al. (1994)
	2. Chronic low-dose arsenic produces a mild oesophagitis, gastritis, or colitis with respective upper and lower abdominal discomfort	
Haematological effects	1. Anaemia and leukopaenia is reported to result from acute, intermediate and chronic oral exposures	Franzblau and Lilis (1989)
	2. High doses of arsenic are reported to cause bone marrow depression in humans	Environmental Protection Agency (1984)
Hepatic effects	Chronic arsenic induced hepatic changes, including bleeding oesophageal varices, ascites, jaundice, or simply an enlarged tender liver, mitochondrial damage, impaired mitochondrial functions, and porphyrin metabolism, congestion, fatty infiltration, cholangitis, cholecystitis and acute yellow atrophy, and swollen and tender liver	Guha Mazumder (2001)
		Santra et al. (1999)
		Chakraborty and Saha (1987)
Renal effects	1. Sites of arsenic damage in the kidney include capillaries, tubules and glomeruli, which lead to haematuria and proteinuria, oliguria, shock and dehydration with a real risk of renal failure, cortical necrosis and cancer	
Dermal effects	1. Chronic exposure to arsenic diffused and spotted melanosis, leukomelanosis, keratosis, hyperkeratosis, dorsum, Bowen's disease and cancer	Southwick et al. (1981)
	2. Chronic doses of 0.003–0.01 mg As kg^{-1} per day	Valentine et al. (1985)
Neurological effects	1. Acute high exposure (1 mg As kg^{-1} per day) often causes encephalopathy with symptoms such as headache, lethargy, mental confusion, hallucination, seizures and coma	Grantham and Jones (1977)
Developmental effects	1. Babies born to women exposed to arsenic dusts during pregnancy have a higher than expected incidence of congenital malformations, below average birth weight	Nordstrom et al. (1979)

(continued)

Table 8.1 (continued)

Effect	Toxicity	References
Reproductive effects	1. Inorganic arsenic readily crosses the placental barrier and affects fetal development	Squibb and Fowler (1983)
	2. Organic arsenicals do not seem to cross the placenta so readily and are stored in the placenta and cause elevations in low birth weights, spontaneous abortions, still-birth, pre-eclampsia and congenital malformations	
Immunological effects	1. Low doses of arsenite (2×10–6 M) and arsenate (5×10–6 M), phytohemagglutinin (PHA)-induced stimulation of cultured human lymphocytes is increased 49% with arsenite and 19% with arsenate	McCabe et al. (1983)
	2. High doses of arsenite (1.9×10–5 M) and arsenate (6×10–4 M), PHA-induced stimulation is completely inhibited with an impairment of immune response	
Genotoxic effects	1. Trivalent forms are far more potent and genotoxic than the pentavalent forms	Barrett et al. (1989)
	2. Dimesthylarsinic acid (DMA) is more toxic than monomethylarsonous acid (MMA) in assays using mammalian and human cells and trimethylarsine oxide (TMAO) is more potent at inducing both mitotic arrest and tetraploids	Moore et al. (1997)
Mutagenetic effects	Arsenic promoted genetic damage in large part by inhibiting DNA repair and may cause cancer or problems in the exposed generation	Hoffman (1991)
Carcinogenic effects	1. Patients who received chronic treatment with arsenical medications have greatly increased the incidence of both basal cell and squamous cell carcinomas of the skin	Hutchinson (1887)
Diabetes mellitus	1. Drinking water arsenic exposure	Rahman et al. (1998–1999)
	2. The presence of keratosis as an indicator of arsenic exposure showed elevated risks for diabetes in those exposed to arsenic in their drinking water	

5 Removal Strategies of Arsenic from a Contaminated Environment

Several strategies of As remediation are reported. In-situ treatments are preferred at large sites and are given at the place of contamination itself, whereas ex-situ remediation is done at another location, which involves different treatments of contaminated soil or water (Table 8.2) (Pierzynski et al. 2005).

Table 8.2 Arsenic bioremediation proposed in the literature to date

Microbial process	Comments	References
Biosorption	Fe(III) treated *Baccilus subtulis* has 11 times higher As(V) sorption capacity than that of the native bacteria	Yang et al. (2012)
	The maximum biosorption capacity of living cells of *Bacillus cereus* for As(III) was found to be 32.42 mg g^{-1} at pH 7.5, at optimal conditions of contact time of 30 min, biomass dosage of 6 g L^{-1}, and temperature of 30 °C	Giri et al. (2013)
	Bacillus cereus strain W2 retained As(III) and As(V) up to 1.87 mg As g^{-1} of dry cell weight and dry cell removal capacity up to 0.18 mg As g^{-1}	Miyatake and Hayashi (2011) and Prasad et al. (2011)
	The biosorption capacity of the *Rhodococcus* sp. WB-12 for As(III) was 77.3 mg g^{-1} at pH 7.0 using 1 g L^{-1} biomass with the contact time of 30 min at 30 °C	
Bioaccumulation	Engineering of phytochelatin producing, As transporter GlpF co-expressing and an As efflux deletion *Escherichia coli* showed a 80-fold more As accumulation than a control strain, achieving an accumulation level of 16.8 μmol g^{-1} (dry cell weight)	Singh et al. (2010)
	Saccharomyces cerevisiae was engineered for 3- to 4-fold greater As(III) uptake and accumulation by overexpression of transporters genes FPS1 and HXT7 responsible for the influx of the contaminant coupled with and without high-level production of cytosolic As sequestors (phytochelatins or bacterial ArsRp)	Shah et al. (2010)
	Engineered *Escherichia coli* expressing ArsR accumulated 50–60 times higher As(III) and As(V) than control	Kostal et al. (2004)
Bioreduction	The co-presence of anthraquinone-2,6-disulfonate with As(V) respiratory-reducing bacteria (*Bacillus selenatarsenatis* SF-1) improved the removal efficiency and can be an effective strategy for remediation of As-contaminated soils	Yamamura et al. (2008)
Biomethylation	Engineering the soil bacterium *Pseudomonas* putida expressing the As(III) S-adenosylmethionine methyltransferase gene has the potential for bioremediation of environmental As	Chen et al. (2013)
	Soil microorganism, e.g. *Trichoderma* sp., sterile mycelial strain, *Neocosmospora* sp. and *Rhizopus* sp. Fungal strains could be used for soil As bioremediation via biovolatilisation	Srivastava et al. (2011)

(continued)

Table 8.2 (continued)

Microbial process	Comments	References
Biomineralisation	The nitrate- and sulphate-plus-lactate-amended microcosms with sediment from an aquifer with naturally elevated As levels decreased effective soluble As levels from 3.9 to 0.01 and 0.41 µM via sorption onto freshly formed hydrous ferric oxide and iron sulphide	Omoregie et al. (2013)
	The biogenic Mn oxides generated by *Marinobacter* sp. MnI7-9 oxidised the highly toxic As(III) to As(V) and decreased the concentration of As(III) from 55.02 to 5.55 µM	Liao et al. (2013)
	Arsenic immobilisation by biogenic Fe mineral formed by *Acidovorax* sp. BoFeN1, an anaerobic nitrate-reducing Fe(II)-oxidising ß-proteobacteria	Hitchcock et al. (2012)
	Microbial calcite precipitated by an As(III) tolerant bacterium *Sporosarcina ginsengisoli* CR5 to retain As	Achal et al. (2012)
	Bioremediation strategy based on injecting nitrate to support the anoxic oxidation of Fe(II) and As(III) in the subsurface as a means of immobilising As in the form of As(V) adsorbed onto biogenic Fe(III) (hydr)oxides	Sun et al. (2009)
Other process	**Comments**	
Phytoextraction	This uses pollutant-accumulating plants to extract and translocate pollutants to the harvestable parts. It can be sub-divided into phytoextraction using hyperaccumulator plants and chemically induced phytoextraction for the accumulation of metals to plants. Induced phytoextraction, however, has not yet been applied to As	Fitz and Wenzel (2002) and Salt et al. (1998)
Phytostabilisation	Uses pollutant-tolerant plants to mechanically stabilise polluted land to prevent bulk erosion, reduce air-borne transport and leaching of pollutants. In contrast to phytoextraction, plants are required to take up only small amounts of As and other metals to prevent transfer into the wild-life food chain	Fitz and Wenzel (2002)
Phyto-immobilisation	The use of plants to decrease the mobility and bioavailability of pollutants by altering soil factors that lower pollutant mobility by formation of precipitates and insoluble compounds and by sorption on roots	Fitz and Wenzel (2002)
Volatilisation	The use of plants to volatilise pollutants. Volatilisation of As is known to occur in the natural environments	Frankenberger Jr and Arshad (2002)
Rhizofiltration	The use of plants with extensive root systems and high accumulation capacity for contaminants, to absorb and adsorb pollutants, mainly metals, from water and streams	Salt et al. (1998)
Phytofiltration	The use of plant uptake contaminants into the biomass, thus removing the pollutant	Raskin and Kumar (1994)
Rhizodegradation	Transformation of the contaminant in the rhizosphere can occur in soil organisms such as fungi or bacteria or via enzymes exuded from microorganisms or plants	Schultz et al. (2001)

5.1 Physico-Chemical Strategies for Arsenic Remediation

5.1.1 pH

Arsenic (V) adsorption reduces with increasing pH, and highest absorption occurs at above pH 8.5, adsorption of As(III) increases with increasing pH. The degree of influence of pH on As adsorption varies from soil to soil. The highest adsorption of As(V) is at pH 4, whereas for As(III) the adsorption maxima are found at approximately pH 7–8.5 (Fitz and Wenzel 2002; Mahimairaja et al. 2005).

5.1.2 Phosphate

Phosphate (PO_4^{-3}), an analogue of As(V), acts as an essential factor for the nature of As in oxygen-rich soils (Mahimairaja et al. 2005; Williams et al. 2006). The effect of PO_4^{-3} additions to aerobic soils on the uptake of As therefore dependS on the balance between competition for sorption sites and competition for uptake. As(III) is not an analogue of PO_4^{-3}; hence, making the presence of PO_4^{-3} most likely less relevant to the behaviour of As under flooded conditions (Takahashi et al. 2004). The role of PO_4^{-3} in the rhizosphere is not known, where aerobic conditions may prevail under flooded situations. There are many factors that affect the biogeochemistry of As (Table 8.3).

5.1.3 The Effect of Iron Hydroxides

As(V) and As(III) adsorb primarily to iron-hydroxides (FeOOH) of the soil and As(V) has stronger bonds compared with As(III). The behaviour of FeOOH primarily depends on its redox conditions; thus, iron redox chemistry play as an important parameter in regulating As behaviour (Fitz and Wenzel 2002; Takahashi et al. 2004). Under anaerobic conditions, the dissolution of FeOOH readily occurs, which results in the release of As in the soil. Such As occurs chiefly as As(III) (Takahashi et al. 2004). FeOOH serves as a sink for As and is quietly insoluble under aerobic conditions. Fe and As behaviour is closely related and very dynamic in lowland paddy fields. In comparison with sandy soils, FeOOH more frequently occurs in clayey soil (Fitz and Wenzel 2002; Mahimairaja et al. 2005). At the same concentration, the clayey soils are found to be less toxic in comparison with sandy soils owing to the strong bond of As in clayey soils.

5.2 Microbial-Based Arsenic Remediation

Arsenic is very toxic to the biological system, although microorganisms have evolved several mechanisms to deal with this problem. The key modes of microbial remediation of As-affected soils are: oxidation and reduction, biosorption,

Table 8.3 Factors affecting the biogeochemistry of arsenic

	Factor	Effects
1.	pH	Arsenic mobility increases at a very low pH (pH <5). As(III) solubility increases as the pH decreases within the range (pH 3–9). In contrast, the pattern is reversed in the case of As(V)
2.	Speciation	As speciation in soil is essential to assess the As toxicity in plants. Among all species toxicity order of As: As(III) > As(V) > MMA > DMA
3.	Redox condition	Increased solubility of As in reducing environment
4.	Fe-plaque	Fe-plaque decrease As uptake by plants
5.	Organic matter	High amount of OM can reduce As solubility and lead to less As availability for plants
6.	Soil texture	Arsenic is five times more in sandy and loamy soils than in clay soils
7.	Fe-Mn oxides	Under aerobic conditions, the Fe-Mn phases decrease As mobility. However, under flooding conditions, these phases can release As mobility, leading to more available As for plant uptake
8.	Heavy metals	Heavy metals can form ternary complexes with arsenate on Fe and Al oxide surfaces, thus lowering the bioavailability of arsenic
9.	Phosphate	Phosphate PO_4 is a chemical analogue of arsenate and its increasing concentration decreases As content in Fe-plaques
10.	Silica	High silica availability in soil reduces the arsenite uptake
11.	Sulphate	The application of sulphur significantly reduces As accumulation in rice
12.	Nitrate	Nitrate (NO_3) is a strong oxidiser and its reduces As uptake
13.	Irrigation practice	Under aerobic water management practices, rice takes up less As (0.23–0.26 ppm) than under anaerobic practices (0.60–0.67 ppm)
14.	Seasonal variation	Winter season crops have a high accumulation of As in comparison with monsoon season crops
15.	Genotype	The genotype variation in As uptake may be due to differences in root anatomy, which controls root aeration and Fe-plaque formation on the root surfaces or differences in the arsenic tolerance gene

complexation, biomethylation, sequestration, solubilisation, and microbe-mediated phytoremediation. Microbial oxidation and reduction involve various metal ions that drastically affect As mobility. Microbes evolved their biochemical pathways to utilise As oxyanions, either as an electron acceptor [e.g. As(V)] for anaerobic respiration, or as an electron donor [e.g. As(III)] to support chemoautotrophic fixation of CO_2 into cell carbon (Santini et al. 2000; Rhine et al. 2006).

5.2.1 Arsenic Remediation by Bacteria

Bacterially mediated remediation involved the reduction of As oxyanions. Stolz and Oremland (1999) decode the stepwise reduction of As oxyanions that selected bacteria perform. Up to 80% As removal was reported by researchers. Many Gram-negative and Gram-positive bacteria employ common ars operon (typically ars RDABC)-based mechanisms for resistance against As. The operon may be encoded either on chromosomes or on plasmids (Xu et al. 1998).

5.2.1.1 Oxidation/Reduction

The energy for microbial growth is derived from the oxidation process. Arsenite oxidase is the enzyme that plays a pivotal role in arsenite oxidation. The enzyme was identified and sequenced and found to belong to the DMSO reductase family. Like many arsenite oxidases, the one (AoxAB) isolated from *Hydrogenophaga* sp. strain NT-14 is made up of a heterodimer (from the gene aoxAB) containing iron and molybdenum as a part of their catalytic unit. The bio-oxidation contributes to As non-availability by immobilising it, alleviating its toxicity in flowing acidic water and neutralising it by environmental alkalinity.

The reduction or conversion of As(V) to As(III) in an anaerobic environment is mediated by mixed populations of anaerobes such as methanogens, fermentative, sulphate- and iron-reducing bacteria. Microbial reduction of As(V) could even occur when As is found bounded with iron hydroxides (Langner and Inskeep 2000). Zobrist et al. (2000) reported that the reduction of As(V) may lead to its mobilisation, without dissolving the sorbent phase. The anaerobically incubating As(V) is co-precipitated with Al hydroxide in the presence of *Sulfurospirillum barnesii*. Moreover, the microbial reduction of As(V) to As(III) under aerobic conditions in As-contaminated soils may occur relatively quickly, resulting in excessive As mobilisation and transport from contaminated soils to groundwater (Macur et al. 2001). Arsenate reduction is a part of anaerobic arsenate respiration in some bacteria (e.g. *Shewanella* sp. strain ANA-3) (Krafft and Macy 1998), where arsenates serve as a terminal electron acceptor.

5.2.1.2 Methylation/Demethylation

The methylation of As occurs via the reduction of pentavalent As to trivalent As and the addition of a methyl group. The conversion of As (V) to small amounts of volatile methylarsines was first described in *Methanobacterium bryantii* (McBride and Wolfe 1971). Methylation is considered to be detoxification. However, not all methylated As products are less toxic (Bentley and Chasteen 2002); As(V) can be rehabilitated to mono- and di-methylarsine by *Achromobacter* sp. and *Enterobacter* sp., and to mono-, di-, and trimethylarsine by *Aeromonas* sp. and *Nocardia* sp. (Cullen 1989). Methylation and demethylation also play critical roles in the toxicity and availability of As in soils and groundwater (Wang and Mulligan 2006). Methylation of As(III) and As(V) sometimes forms volatile species, which lead to the escape of As from water and soil surfaces. The oxidation of methylated forms of As may convert them back to the oxidised As(V); however, demethylation of mono- and microbes could use dimethyl As compounds and methylated arsenicals as possible carbon sources (Maki et al. 2004).

5.2.1.3 Biofilm and Biosorption

There are several microorganisms that can form biofilms and in a biofilm, microniches could coexist with variable physiological needs, enabling the coexistence of spatially separated conflicting redox processes simultaneously (Labrenz et al. 2000; Van Hullebusch et al. 2003). This is evident by the enrichment of As in biofilms. The As concentrations in rock biofilm may reach up to 60 mg kg^{-1} of dry weight (Drewniak et al. 2008).

5.2.1.4 Biostimulation and Bioaugmentation

Bioaugmentation is the process that exploits metal-immobilising microbial population(s) to convert highly toxic metal into a less toxic state at the site itself. In biostimulation, a stimulus is provided to microorganisms pre-existing at the site. The stimulus may be a nutrient, growth substrates with or without electron acceptors or donors. This could be achieved by delivering nutrients (biostimulation) or microbial populations (bioaugmentation) into the soil using biosurfactant foam technology (Hug et al. 2011). The biosurfactants enhance the availability of metal ion [Fe(III), As etc.] to the microorganisms by reducing the interfacial tension and formation of micelles.

5.2.1.5 Biomineralisation

In biomineralisation, the As is immobilised by its precipitation, e.g. scorodite and As sulphide. More than 300 compounds of As are known to occur in various environments (Drahota and Filippi 2009). In As-contaminated suspended aquifer, the availability of free Ca^{2+} control the As mobility by digenetic precipitation of calcium arsenates (Martinez-Villegas et al. 2013).

5.2.2 Arsenic Remediation by Algae

Certain species of algae take up copper, cadmium, chromium, lead, and nickel (Qiming et al. 1999) from aqueous solutions by releasing a protein called metallothioneins. Metallothioneins bind the metal as a defence mechanism to remove/mitigate the toxic effect of metal by its regular cellular activity. The biosorption by algae is a viable mechanism for the treatment of metal waste and could work efficiently for multi-component metal systems. Algae respond to metal ions by synthesising low molecular weight compounds such as glutathione and carotenoids, and initiate the synthesis of several antioxidants and enzymes (superoxide dismutase, catalase, glutathione peroxidise and ascorbate peroxidise). *Chlorella* and *Scenedesmus* are the two most frequently exploited species used for metal uptake. *Scenedesmus* sp. has much greater metal binding potentials than *Chlorella* sp.

Suhendrayatna et al. (1999) exposed *Chlorella* sp. to various concentrations of arsenite ranging from 0 to 100 μg As cm^{-3}. They found that the cell growth of *Chlorella* sp. was affected at concentrations higher than 50 μg As cm^{-3} At concentrations greater than 50 μg As cm^{-3} The cell growth of *Chlorella* was suppressed. There is a lack of studies on wastewater treatment plants where mixed bacteria culture or the alga *Scenedesmus abundans* was used for the removal of As. Available literature indicates that arsenite As(III) is a more toxic form and thus more difficult to remove using conventional treatment methods (Zouboulis and Katsoyiannis 2005).

5.2.3 Arsenic Remediation by Yeast

In yeast, As(V) enters in the cells through high-affinity phosphate transporters such as Pho84, whereas As(III) influx occurs through the aquaglyceroporins Fps1 (Wysocki et al. 2001). In addition, glucose permeases are also involved in As uptake by yeasts (Liu et al. 2004). Expression of yeast hexose carriers in mutant *Dfps1* restored As sensitivity, which was As(III)-tolerant earlier (Liu et al. 2006). Arsenic tolerance in yeast is encoded by the gene cluster *ACR1*, *ACR2* and *ACR3*. *ACR1* encodes a putative transcription factor that regulates the transcription of *ACR2* and *ACR3*, possibly by directly sensing cellular As levels. *ACR2* encodes an arsenate reductase and *ACR3* encodes a plasma membrane-expressed As(III) efflux transporter. This gene cluster provides a process for the sensing, reduction and efflux of As. In addition, a second pathway is also present in yeasts, in the form of vacuoles. Cytosolic As(III) complex with glutathione can be sequestered through an ABC-type transporter, *YCF1*, that also transports conjugates of other toxic compounds (Ghosh et al. 1999).

5.2.4 Arsenic Remediation by Fungi

Fungi present in polluted soils play an important role in the maintenance of indigenous diversity and protect them against the uptake of toxic heavy metals (HMs) by plants. Sharples et al. (2000) showed that the ericoid mycorrhizal fungus *Hymenoscyphus ericae* significantly reduces the uptake of As by *Calluna vulgaris* when grown in As-affected soil. Various mycorrhiza-based phosphate transporters have been reported to increase expression in arbuscular mycorrhizae (AM). They are StPT3 in *Solanum tuberosum* LePT3 and LePT4 in *Lycopersicon esculentum* (Nagy et al. 2005; Xu et al. 2007), and MtPT4 in *Medicago truncatula* (Javot et al. 2007). Contradictorily, mycorrhizal association increases As accumulation in the fern *P. vittata* (Liu et al. 2005). The study by Uppanan (2000) exposed the *Alcaligenes* spp. from soil that could oxidise arsenite to arsenate and taken up by the plant via phosphate transporter (Meharg and Hartley-Whitaker 2002). In AM-mediated phytoremediation of As, contradictory results have been reported on As uptake by plants, e.g. the addition of rhizo-fungi in As-contaminated soil, and the uptake and the accumulation of As were induced in *P. vittata* (Leung et al. 2006).

5.2.5 Arsenic Remediation by Genetic Engineered Microbes

The genetically engineered microbes may serve as selective biosorbents and lay the foundation of green technology for the eco-friendly, low-cost, efficient removal of HMs such as (Singh et al. 2008a). For the development of As-metabolising/accumulating microbe, the microbe should comprise the ability to modify its naturally occurring defence mechanisms and be able to develop a novel pathway. The bacterial enzymes arsenate reductase (ArsC) and GSH synthase (g-ECS) were successfully expressed in the plant *Arabidopsis thaliana*. These enzyme systems confer the ability to accumulate arsenate as GSH–As complexes. In the same way, YCF1 protein of yeast expressed in *A. thaliana* enhanced As storage in the vacuole (Song et al. 2003). These reports open up the path of genetically engineered microbes and their metabolic pathways for successful As sequestration. The phytochelatin (PC) synthase of *A. thaliana* was expressed in *E. coli* (Sauge-Merle et al. 2003) and produced PC, leading to moderate levels of As accumulation. However, the level of GSH, a key precursor of PC, acts as a limiting factor for high PC production and As accumulation. The PC synthase of *S. pombe* (SpPCS) when expressed in *E. coli* resulted in higher As accumulation than the wild type (Singh et al. 2008b). Significant levels of PC were obtained by co-expressing As transporter GlpF; this led to a 1.5-fold higher accumulation of As. On deletion of As efflux in the *E. coli* strain, the highest As accumulation of 16.8 mmol g^{-1} cells of *E. coli* was recorded.

Sulphur-reducing bacteria naturally precipitate As(V) by the formation of an insoluble sulphide complex with H_2S (Rittle et al. 1995). Recently, a yeast strain was engineered that was coexpressing AtPCS and cysteine desulfhydrase, to elevate the accumulation of As by the formation of PC metal–sulphide complexes (Tsai et al. 2009). The approach to exploiting the resting cells of high-affinity biosorbent property for As had also been exploited. To achieve this, AtPCS were expressed in *S. cerevisiae,* which naturally has a higher level of GSH; the engineered yeast strain accumulated high levels of As in resting cell cultures (Singh et al. 2008b). Arsenic accumulation in *E. coli* was achieved by over-expressing the As-specific regulatory protein ArsR. These engineered resting cells were able to remove 50 ppb of As(III) per hour (Kostal et al. 2004). The concept has been extended to using a naturally occurring As-binding MT (Singh et al. 2004). These resting cells were able to remove 35 ppb of As(III) in 20 min. New irrational approaches such as directed evolution, genome shuffling and metagenomic studies could also be used to develop new As-removing pathways suitable for As remediation. It was demonstrated by the modification of As resistance operon by DNA shuffling (Crameri et al. 1997). The cells with optimised operon can grow in 0.5 M arsenate, showing 40-fold enhancement in As resistance. Meanwhile, Chauhan and coworkers constructed a metagenomic library of sludge obtained from industrial effluent treatment plants, and identified a novel As(V) resistance gene (*arsN*) encoding a protein similar in action to acetyltransferases. Its over-expression resulted in higher As resistance in *E. coli* (Chauhan et al. 2009).

5.3 Plant-Mediated Arsenic Remediation

Phytoremediation, the plant-mediated bioremediation of polluted soil, water and air, is an emerging cost-effective, non-invasive and widely accepted method for removing/mitigating environmental pollutants (Boyajian and Carreira 1997; Singh et al. 2003; Bharagava et al. 2017). Plants have a natural ability to accumulate inorganic and organic contaminants. The organic pollutants are metabolised and microbial degradation of organic pollutants is promoted in the rhizosphere. Phytoremediation involved the cultivation of metal-tolerant plants able to concentrate the metal in their tissues. The plant biomass thus produced is harvested and dried and is deposited in a landfill or added to smelter feed (Kramer 2005). The exceptional ability of *Pteris vittata* to accumulate As could be explored to design an efficient phytoremediation strategy, although greenhouse studies demonstrate the promising potential of As extraction by *P. vittata*. However, the results of two small-scale field trials are less promising owing to low biomass production (<1 tonne dry biomass ha^{-1}) (Salido et al. 2003). In 2 years of growth, the total As removal by *P. vittata* was about 1% of As in top soil (30 cm in depth) (Kertulis-Tartar et al. 2006).

6 Challenges

- Chronic As poisoning is a major threat to large sections of the global population and food consumption is one of the biggest contributors to human As exposure.
- Contamination of As in paddy soils is a widespread problem because of the irrigation of As-laden groundwater in the south southeast, including India.
- Anaerobic conditions in flooded paddy soils are conducive to the mobilisation of As, leading to much enhanced As bioavailability to rice plants (Lee et al. 2008; Xu et al. 1998). Rice is a major source of inorganic As for populations based on a rice diet and not exposed to high As in drinking water (Kile et al. 2007; Meharg et al. 2009).
- Accumulation of As in paddy soil can cause phyto-toxicity to rice plants and a significant reduction in grain yield, thus threatening the long-term sustainability of the rice cropping system in the affected areas (Khan et al. 2009).
- Research concerning microbial As is still in its infancy; therefore, a thorough understanding of the true As behaviour in the surface and subsurface environments under the influence of microbial activities is still very challenging.

7 Conclusion

The following points have been concluded in this study:

- This chapter has attempted to summarise As content in edible crops, its relationships with soil and irrigated groundwater and the factors controlling As mobilisation and uptake in edible crops.
- Arsenic accumulation in plants is largely influenced by a variety of factors, including soil physicochemical parameters; other elements such as iron, phosphorus, sulphur and silicon concentrations; and environmental conditions that control As availability and uptake in the soil–rhizosphere–plant system.
- Environmental conditions can be managed by changing irrigation practices. For example, the flooding of the paddy soil mobilises As in the soil solution and can increase As accumulation in rice.
- Therefore, changing agricultural practices to aerobic rice cultivation throughout the entire season may be a viable strategy for mitigating this problem. However, there are arguments in certain cases because of flood conditions.
- Arsenic is present as arsenite, which cannot compete with phosphate; furthermore, phosphate increases As mobility because it competes with arsenate for the adsorption site of Fe oxides/hydroxides.
- The use of vegetation directly or indirectly to remove contaminants from water or soil is an important innovative remediation technology potentially applicable to a variety of contaminated sites.
- Selection of the appropriate plant species is a critical process for the success of this technology.
- Approaches to reducing As uptake in crops, especially in the edible parts, would provide a viable alternative.
- Many natural substances are expected to exhibit substantial effects on the microbial processes and subsequently change the environmental behaviour of As, either directly or indirectly.
- Researching microorganism–As interactions also provides the opportunity to study As remediation taking advantage of microbial activities.
- Arsenic concentration in the agricultural field soil was below 20.0 mg kg^{-1} (the maximum acceptable limit for agricultural soil, recommended by the European Community (EC)) and 1.0 mg kg^{-1} dry weight of As (the permissible limit of As in rice according to WHO recommendations).
- The total amount of As in raw rice is not taken into the human body because of its distribution in the following order: root>straw>husk>grain. An appreciably high efficiency in the translocation of As from shoot to grain was observed compared with the translocation of As from root to shoot.

Thus, this chapter covers all As-oriented problems and their fate in the environment and treatment technology for the sustainable development of the environment and environmental safety.

Acknowledgements The fellowship awarded from the University Grant Commission (UGC), Government of India (GOI), New Delhi, to Mr. Vishvas Hare for his Ph.D. work is duly acknowledged.

References

Achal V, Pan XL, Fu QL, Zhang DY (2012) Biomineralization based remediation of As(III) contaminated soil by *Sporosarcina ginsengisoli*. J Hazard Mater 201:178–184

Barrett JC, Lamb PW, Wang TC, Lee TC (1989) Mechanisms of arsenic-induced cell transformation. Biol Trace Elem Res 21:421

Bennett WW, Teasdale PR, Panther JG, Welsh DT, Zhao HJ, Jolley DF (2012) Investigating arsenic speciation and mobilization in sediments with DGT and DET: a mesocosm evaluation of oxic-anoxic transitions. Environ Sci Technol 46(7):3981–3989

Bentley R, Chasteen TG (2002) Microbial methylation of metalloids: arsenic, antimony, and bismuth. Microbiol Mol Biol Rev 66:250–271

Bharagava RN, Mishra S (2018) Hexavalent chromium reduction potential of *Cellulosimicrobium* sp. isolated from common effluent treatment industries. Ecotoxicol Environ Saf 147:102–109

Bharagava RN, Chowdhary P, Saxena G (2017) Bioremediation an eco-sustainable green technology, its applications and limitations Environmental pollutants and their bioremediation approaches, Bharagava, RN (ed)CRC Press, Taylor & Francis Group, Boca Raton, p. 1–22

Borgono JM, Vicent P, Venturino H et al (1997) Arsenic in the drinking water of the city of Antofagasta: epidemiological and clinical study before and after the installation of a treatment plant. Environ Health Perspect 19:103–105

Boyajian GE, Carreira LH (1997) Phytoremediation: a clean transition from laboratory to marketplace? Nat Biotechnol 15:127–128

Buschmann J, Kappeler A, Lindauer U, Kistler D, Berg M, Sigg L (2006) Arsenite and arsenate binding to dissolved humic acids: influence of pH, type of humic acid, and aluminum. Environ Sci Technol 40:6015–6020

Buschmann J, Berg M, Stengel C, Winkel L, Sampson ML, Trang PTK, Viet PH (2008) Contamination of drinking water resources in the Mekong delta floodplains: arsenic and other trace metals pose serious health risks to population. Environ Int 34:756–764

Cebrian ME, Albores A, Aguilar M, Blakely E (1983) Chronic arsenic poisoning in the North of Mexico. Hum Toxicol 2:121–133

Cervantes C, Ji GY, Ramirez JL, Silver S (1994) Resistance to arsenic compounds in microorganisms. FEMS Microbiol Rev 15(4):355–367

Chakraborty AK, Saha KC (1987) Arsenical dermatitis from tube well water in West Bengal. Indian J Med Res 85:326–334

Chauhan NS, Ranjan R, Purohit HJ, Kalia VC, Sharma R (2009) Identification of genes conferring arsenic resistance to Escherichia coli from an effluent treatment plant sludge. FEMS Microbiol Ecol 67:130–139

Chen J, Qin J, Zhu YG, de Lorenzo V, Rosen BP (2013) Engineering the soil bacterium *Pseudomonas putida* for arsenic methylation. Appl Environ Microbiol 79(14):4493–4495

Chowdhary P, Yadav A, Kaithwas G, Bharagava RN (2017a) Distillery wastewater: a major source of environmental pollution and its biological treatment for environmental safety. In: Singh R, Kumar S (eds) Green technologies and environmental sustainability. Springer International, Cham, pp 409–435

Chowdhary P, More N, Raj A, Bharagava RN (2017b) Characterization and identification of bacterial pathogens from treated tannery wastewater. Microbiol Res Int 5(3):30–36

Chowdhury UK, Biswas BK, Chowdhury TR, Samanta G, Mandal BK, Bas GC, Chanda CR, Lodh D, Saha KC, Mukherjee SK, Kabir S, Quamruzzaman Q, Chakraborti D (2000) Groundwater arsenic contamination in Bangladesh, West Bengal, and India. Environ Health Perspect 108(5):393–397

Corsini A, Cavalca L, Crippa L, Zaccheo P, Andreoni V (2010) Impact of glucose on microbial community of a soil containing pyrite cinders: role of bacteria in arsenic mobilization under submerged condition. Soil Biol Biochem 42(5):699–707

Crameri A, Dawes G, Rodriguez E, Silver S, Stemmer WPC (1997) Molecular evolution of an arsenate detoxification pathway DNA shuffling. Nat Biotechnol 15:436

Cullen WR (1989) The metabolism of methyl arsine oxide and sulphide. Appl Organomet Chem 3:71–78

Dixit S, Hering JG (2003) Comparison of arsenic(V) and arsenic(III) sorption onto iron oxide minerals: implications for arsenic mobility. Environ Sci Technol 37(18):4182–4189

Dowdle PR, Laverman AM, Oremland RS (1996) Bacterial dissimilatory reduction of arsenic(V) to arsenic(III) in anoxic sediments. Appl Environ Microbiol 62(5):1664–1669

Drahota P, Filippi M (2009) Secondary arsenic minerals in the environment: a review. Environ Int 35(8):1243–1255

Drewniak L, Styczek A, Majder-Lopatka M, Sklodowska A (2008) Bacteria, hyper tolerant to arsenic in the rocks of an ancient gold mine, and their potential role in dissemination of arsenic pollution. Environ Pollut 156(3):1069–1074

Du Laing G, Chapagain SK, Dewispelaere M, Meers E, Kazama F, Tack FMG, Rinklebe J, Verloo MG (2009) Presence and mobility of arsenic in estuarine wetland soils of the Scheldt estuary (Belgium). J Environ Monit 11(4):873–881

Environmental Protection Agency (EPA) (1984) Health assessment document for inorganic arsenic, final report, EPA 600/8-83-021F. USEPA, Environmental Criteria and Assessment Office, Research Triangle Park

Fitz WJ, Wenzel WW (2002) Arsenic transformations in the soil-rhizosphere-plant system: fundamentals and potential application to phytoremediation. J Biotechnol 99:259–278

Frankenberger Jr WT, Arshad M (2002) Volatilisation of arsenic. In: Frankenberger Jr WT (ed) Environmental chemistry of arsenic. Marcel Dekker, New York, pp 363–380

Franzblau A, Lilis R (1989) Acute arsenic intoxication from environmental arsenic exposure. Arch Environ Health 44:385–390

Frohne T, Rinklebe J, Diaz-Bone RA, Du Laing G (2011) Controlled variation of redox conditions in a floodplain soil: impact on metal mobilization and biomethylation of arsenic and antimony. Geoderma 160(3–4):414–424

Ghosh M, Shen J, Rosen BP (1999) Pathways of As-III detoxification in *Saccharomyces cerevisiae*. Proc Natl Acad Sci U S A 96:5001–5006

Gibney BP, Nusslein K (2007) Arsenic sequestration by nitrate respiring microbial communities in urban lake sediments. Chemosphere 70(2):329–336

Giri AK, Patel RK, Mahapatra SS, Mishra PC (2013) Biosorption of arsenic (III) from aqueous solution by living cells of Bacillus cereus. Environ Sci Pollut Res 20(3):1281–1291

Goh KH, Lim TT (2005) Arsenic fractionation in a fine soil fraction and influence of various anions on its mobility in the subsurface environment. Appl Geochem 20:229–239

Gomez-Caminero A, Howe P, Hughes M, Kenyon E, Lewis DR, Moore M, Ng J, Aitio A, Beecking G (2001) Arsenic and arsenic compounds. The Environmental Health Criteria.2nd edn. World Health Organisation, Finland

Grantham DA, Jones JF (1977) Arsenic contamination of water wells in Nova Scotia. J Am Water Works Assoc 69:653–657

Guha Mazumder DN (2001) Arsenic and liver disease. J Indian Med Assoc 99(6):311–320

Halim MA, Majumder RK, Nessa SA, Hiroshiro Y, Uddin MJ, Shimada J, Jinno K (2009) Hydrogeochemistry and arsenic contamination of groundwater in the Ganges Delta plain. Bangladesh J Hazard Mater 164:1335–1345

Hare V, Chowdhary P, Baghel VS (2017) Influence of bacterial strains on Oryza sativa grown under arsenic tainted soil: accumulation and detoxification response. Plant Physiol Biochem 119:93–102

Hitchcock AP, Obst M, Wang J, Lu YS, Tyliszczak T (2012) Advances in the detection of As in environmental samples using low energy X-ray fluorescence in a scanning transmission X-ray microscope: arsenic immobilization by an Fe(II)-oxidizing freshwater bacteria. Environ Sci Technol 46(5):2821–2829

Hoffman GR (1991) Genetic toxicology. In: Amdur MO, Doull J, Klaassen CD (eds) Toxicology. Pergamon Press, New York, pp 201–225

Huang JH, Matzner E (2006) Dynamics of organic and inorganic arsenic in the solution phase of an acidic fen in Germany. Geochim Cosmochim Acta 70(8):2023–2033

Huang JH, Scherr F, Matzner E (2007) Demethylation of dimethylarsinic acid and arsenobetaine in different organic soils. Water Air Soil Pollut 182(1–4):31–41

Huang JH, Voegelin A, Pombo SA, Lazzaro A, Zeyer J, Kretzschmar R (2011) Influence of arsenate adsorption to ferrihydrite, goethite, and boehmite on the kinetics of arsenate reduction by *Shewanella putrefaciens* strain CN-32. Environ Sci Technol 45(18):7701–7709

Hug SMI, Sultana S, Chakraborty G, Chowdhury MTA (2011) A mitigation approach to alleviate arsenic accumulation in rice through balance fertilization. Appl Environ Soil Sci. https://doi.org/10.1155/2011/835627

Hutchinson J (1887) Arsenic cance. Br Med J 2:1280

Jackson BP, Miller WP (1999) Soluble arsenic and selenium species in fly ash/organic waste-amended soils using ion chromatography inductively coupled plasma spectrometry. Environ Sci Technol 33:270–275

Javot H, Penmetsa RV, Terzaghi N, Cook DR, Harrison MJ (2007) A *Medicago truncatula* phosphate transporter indispensable for the arbuscular mycorrhizal symbiosis. Proc Natl Acad Sci U S A 104:1720–1725

Jia Y, Huang H, Zhong M, Wang FH, Zhang LM, Zhu YG (2013) Microbial arsenic methylation in soil and rice rhizosphere. Environ Sci Technol 47(7):3141–3148

Jiang QQ, Singh BR (1994) Effect of different forms and sources of arsenic on crop yield and arsenic concentration. Water Air Soil Pollut 74:321–343

Johnnson MO, Cohly HHP, Isokpehi RD, Awofolu OR (2010) The case for visual analytic of arsenic concentrations in foods. Int J Environ Res Public Health 7:1970–1983

Johnstone RM (1963) Sulfhydryl agents: arsenicals. In: Hochster RM, Quastel JH (eds) Metabolic inhibitors: a comprehensive treatise, vol 2. Academic, New York, pp 99–118

Jones CA, Langner HW, Anderson K, McDermott TR, Inskeep WP (2000) Rates of microbially mediated arsenate reduction and solubilization. Soil Sci Soc Am J 64(2):600–608

Joshi DN, Flora SJS, Kalia K (2009) *Bacillus sp.* strain DJ-1, potent arsenic hyper tolerant bacterium isolated from the industrial effluent of India. J Hazard Mater 166(2–3):1500–1505

Juhasz AL, Naidu R, Zhu YG, Wang LS, Jiang JY, Cao ZH (2003) Toxicity tissues associated with geogenic arsenic in the groundwater-soil-plant-human continuum. Bull Environ Contam Toxicol 71:1100–1107

Kabata-Pendias A, Adriano DC (1995) Trace metals. In: Rechcigl JE (ed) Soil amendments and environmental quality. CRC Press, Boca Raton, pp 139–167

Kabata-Pendias A, Pendias H (1984) Trace elements in soils and plants, vol 315. CRC Press, Boca Raton

Kertulis-Tartar GM, Ma LQ, Tu C, Chirenje T (2006) Phytoremediation of an arsenic-contaminated site using *Pteris vitrata* L.: a two-year study. Int J Phytoremediation 8:311–322

Khan MA, Islam MR, Panaullah GM, Duxbury JM, Jahiruddin M, Loeppert RH (2009) Fate of irrigation-water arsenic in rice soils of Bangladesh. Plant Soil 322:263–277

Kile ML, Houseman EA, Breton CV, Smith T, Quamruizzaman Q, Rahman M, Mahiuddin G, Christini DC (2007) Dietary arsenic exposure in Bangladesh. Environ Health Perspect 115:889–893

Kitchin KT (2001) Recent advances in arsenic carcinogenesis: modes of action, animal model systems, and methylated arsenic metabolites. Toxicol Appl Pharmacol 172:249–261

Kostal J, Yang R, Wu CH, Mulchandani A, Chen W (2004) Enhanced arsenic accumulation in engineered bacterial cells expressing ArsR. Appl Environ Microbiol 70:4582–4587

Krafft T, Macy JM (1998) Purification and characterization of the respiratory arsenate reductase of Chrysiogenes arsenatis. Eur J Biochem 255:647–653

Kramer U (2005) Phytoremediation: novel approaches to cleaning up polluted soils. Curr Opin Biotechnol 16:133–141

Kuehnelt D, Goessler W (2003) Organ arsenic compounds in the terrestrial environment. In: Craig PJ (ed) Organometallic compounds in the environment. Wiley, Heidelberg, pp 223–275

Labrenz M, Druschel GK, Thomsen-Ebert T, Gilbert B, Welch SA, Kemner KM, Logan GA, Summons RE, De Stasio G, Bond PL, Lai B, Kelly SD, Banfield JF (2000) Formation of sphalerite (ZnS) deposits in natural biofilms of sulfate-reducing bacteria. Science 290(5497):1744–1747

Lafferty BJ, Loeppert RH (2005) Methyl arsenic adsorption and desorption behavior on iron oxides. Environ Sci Technol 39(7):2120–2127

Langner HW, Inskeep WP (2000) Microbial reduction of arsenate in the presence of ferrihydrite. Environ Sci Technol 34:3131–3136

Lee JS, Lee SW, Chon HT, Kim KW (2008) Evaluation of human exposure to arsenic due to rice ingestion in the vicinity abandoned Myungbong Au-Ag mine site. Korea J Geochem Explor 96:231–235

Leung HM, Ye ZH, Wong MH (2006) Interactions of mycorrhizal fungi with *Pteris vittata* (As hyperaccumulator) in As-contaminated soils. Environ Pollut 139:1–8

Liao SJ, Zhou JX, Wang H, Chen X, Wang HF, Wang GJ (2013) Arsenite oxidation using biogenic manganese oxides produced by a deep-sea manganese-oxidizing bacterium, Marinobacter sp MnI7-9. Geomicrobiol J 30(2):150–159

Liu WJ, Zhu YG, Smith FA, Smith SE (2004) Do phosphorus nutrition and iron plaque alter arsenate (As) uptake by rice seedlings in hydroponic culture? New Phytol 162:481–488

Liu Y, Zhu YG, Chen BD, Christie P, Li XL (2005) Influence of the arbuscular mycorrhizal fungus Glomus mosseae on uptake of arsenate by the As hyperaccumulator fern *Pteris vittata* L. Mycorrhiza 15:187–192

Liu WJ, Zhu YG, Hu Y, Williams PN, Gault AG, Meharg AA, Charnock JM, Smith FA (2006) Arsenic sequestration in iron plaque, its accumulation and speciation in mature rice plants (*Oryza sativa* L.) Environ Sci Technol 40:5730–5736

Lloyd JR, Oremland RS (2006) Microbial transformations of arsenic in the environment: from soda lakes to aquifers. Elements 2(2):85–90

Macur PE, Wheeler JT, McDermott TR, Inskeep WP (2001) Microbial population associated with the reduction and enhanced mobilization of arsenic in mine tailings. Environ Sci Technol 35:3676–3682

Mahimairaja S, Bolan NS, Adriano DC, Robinson B (2005) Arsenic contamination and its risk management in complex environmental settings. Adv Agron 86:1–82

Maki T, Hasegawa H, Watarai H, Ueda K (2004) Classification for dimethylarsenate-decomposing bacteria using a restrict fragment length polymorphism analysis of 16S rRNA genes. Anal Sci 20:61–68

Martinez-Villegas N, Briones-Gallardo R, Ramos-Leal JA, Avalos-Borja M, Castanon-Sandoval AD, Razo-Flores E, Villalobos M (2013) Arsenic mobility controlled by solid calcium arsenates: a case study in Mexico showcasing a potentially widespread environmental problem. Environ Pollut 176:114–122

Maurel C, Verdoucq L, Luu DT, Santoni V (2008) Plant aquaporins: membrane channels with multiple integrated functions. Annu Rev Plant Biol 59:595–624

McBride BC, Wolfe RS (1971) Biosynthesis of dimethylasrine by a methanobacterium. Biochemistry 10:4312–4317

McCabe M, Maguire D, Nowak M (1983) The effects of arsenic compounds on human and bovine lymphocyte mitogenesis in vitro. Environ Res 31:323

Meharg AA (2004) Arsenic in rice-understanding a new disaster for South-East Asia. Trends Plant Sci 9:415–417

Meharg AA, Hartley-Whitaker J (2002) Arsenic uptake and metabolism in arsenic resistant and nonresistant plant species. New Phytol 154:29–43

Meharg AA, Williams PN, Adomako E, Lawgali YY, Deacon C, Villada A, Cambell RCJ, Sun G, Zhu YG, Feldmann J, Raab A, Zhao FJ, Islam R, Hossain S, Yanai J (2009) Geographical variation in total and inorganic arsenic content of polished (white) rice. Environ Sci Technol 43:1612–1617

Meng XY, Qin J, Wang LH, Duan GL, Sun GX, Wu HL, Chu CC, Ling HQ, Rosen BP, Zhu YG (2011) Arsenic biotransformation and volatilization in transgenic rice. New Phytol 191:49–56

Mestrot A, Planer-Friedrich B, Feldmann J (2013) Biovolatilisation: a poorly studied pathway of the arsenic biogeochemical cycle. Environ Sci: Processes Impacts 15(9):1639–1651

Meunier L, Koch I, Reimer KJ (2011) Effects of organic matter and ageing on the bioaccessibility of arsenic. Environ Pollut 159:2530–2536

Mishra S, Bharagava RN (2016) Toxic and genotoxic effects of hexavalent chromium in environment and its bioremediation strategies. J Environ Sci Health C 34(1):1–34

Miteva E (2002) Accumulation and effect of arsenic on tomatoes. Commun Soil Sci Plant Anal 33(11):1917–1926

Miyatake M, Hayashi S (2011) Characteristics of arsenic removal by *Bacillus cereus* strain W2. Resour Process 58(3):101–107

Mokgalaka-Matlala NS, Flores-Tavizon E, Castillo-Michel H, Peralta-Videa JR, Gardea-Torresdey JL (2008) Toxicity of Arsenic(III) and (V) on plant growth, element uptake, and total amylolytic activity of Mesquite (*Prosopis Juliflora* 9 P. Velutina). Int J Phytoremed 10(1):47–60

Moore MM, Harrington-Brock K, Doerr CL (1997) Relative genotoxic potency of arsenic and its methylated metabolites. Mutat Resuscitation 386:279

Morton WE, Dunnette DA, Nriagu JO (1994) Arsenic in the environment. II. human health and ecosystem effects. Wiley, New York, pp 17–34

Mukai H, Ambe Y, Muku T, Takeshita K, Fukuma T (1986) Seasonal-variation of methylarsenic compounds in airborne particulate matter. Nature 324(6094):239–241

Mukhopadhyay R, Rosen BP, Pung LT, Silver S (2002) Microbial arsenic: from geocycles to genes and enzymes. FEMS Microbiol Rev 26(3):311–325

Nagvi SM, Vaishnavi C, Singh H (1994) Toxicity and metabolism of arsenic in vertebrates. In: Nriagu JO (ed) Arsenic in the environment. Part II: human health and ecosystem effects. Wiley, New York, pp 55–91

Nagy R, Karandashov V, Chague V, Kalinkevich K, Tamasloukht M, Xu G, Jakobsen I, Levy AA, Amrhein N, Bucher M (2005) The characterization of novel mycorrhizaspecific phosphate transporters from *Lycopersicon esculentum* and *Solanum tuberosum* uncovers functional redundancy in symbiotic phosphate transporter in solanaceous species. Plant J 42:236–250

Ning RY (2002) Arsenic ermoval by reverse osmosis. Desalinisation 143:237–241

Nordstrom S, Beckman L, Nordenson I (1979) Occupational and environmental risks in and around a smelter in northern Sweden. Hereditas 90:297

Omoregie EO, Couture RM, Van Cappellen P, Corkhill CL, Charnock JM, Polya DA, Vaughan D, Vanbroekhoven K, Lloyd JR (2013) Arsenic bioremediation by biogenic iron oxides and sulfides. Appl Environ Microbiol 79(14):4325–4335

Paez-Espino D, Tamames J, de Lorenzo V, Canovas D (2009) Microbial responses to environmental arsenic. Biometals 22(1):117–130

Pierzynski GM, Sims JT, Vance GF (2005) Soils and environmental quality.3rd edn. CRC Press, Boca Raton, p 569

Prasad KS, Srivastava P, Subramanian V, Paul J (2011) Biosorption of As (III) ion on *Rhodococcus sp*. WB-12: biomass characterization and kinetic studies. Sep Sci Technol 46(16):2517–2525

Qiming Y, Matheickal Jose T, Yin P, Kaewsarn P (1999) Heavy metal uptake capacities of common marine macro algal biomas. Water Res 36(6):1534–1537

Raskin I, Kumar PBAN (1994) Bioconcentration of heavy metals by plants. Curr Opin Biotechnol 5:285–290

Rhine ED, Phelps CD, Young LY (2006) Anaerobic arsenite oxidation by novel denitrifying isolates. Environ Microbiol 8:889–908

Rittle KA, Drever JI, Colberg PJS (1995) Precipitation of arsenic during bacterial sulfate reduction. Geomicrobiol J 13:1–11

Rosen B (2002) Biochemistry of arsenic detoxification. FEBS Lett 529:86–92

Salido AL, Hasty KL, Lim JM, Butcher DJ (2003) Phytoremediation of arsenic and lead in contaminated soil using Chinese Brake Ferns (*Pteris vittata*) and Indian mustard (*Brassica juncea*). Int J Phytoremediation 5:89–103

Salt DE, Smith RD, Raskin I (1998) Phytoremediation. Annu Rev Plant Physiol Plant Mol Biol 49:643–668

Santini JM, Sly LI, Schnagl RD, Macy JM (2000) A new chemolitoautotrophic arsenite oxidizing bacterium isolated from a gold-mine: phylogenetic, physiological, and preliminary biochemical studies. Appl Environ Microbiol 66:92–97

Santra A, Das Gupta J, De BK et al (1999) Hepatic damage caused by chronic arsenic toxicity in experimental animals. Ind Soc Gastroenterol 18:152

Sauge-Merle S, Cuine S, Carrier P, Lecomte-Pradines C, Luu DT, Peltier G (2003) Enhanced toxic metal accumulation in engineered bacterial cells expressing Arabidopsis thaliana phytochelatin synthase. Appl Environ Microbiol 69:490–494

Schultz A, Jonas U, Hammer E, Schauer F (2001) Dehalogenation of chlorinated hydroxybiphenyls by fungal laccase. Appl Environ Microbiol 67:4377–4381

Senn DB, Hemond HF (2002) Nitrate controls on iron and arsenic in an urban lake. Science 296(5577):2373–2376

Shah D, Shen MWY, Chen W, Da Silva NA (2010) Enhanced arsenic accumulation in *Saccharomyces cerevisiae* overexpressing transporters Fps1p or Hxt7p. J Biotechnol 150(1):101–107

Sharples JM, Meharg AA, Chambers SM, Cairney JWG (2000) Symbiotic solution to arsenic contamination. Nature 404:951–952

Singh OV, Labana S, Pandey G, Budhiraja R, Jain RK (2003) Phytoremediation: an overview of metallic ion decontamination from soil. Appl Microbiol Biotechnol 61:405–412

Singh SK, Hawkins C, Clarke ID, Squire JA, Bayani J, Hide T (2004) Identification of human brain tumour initiating cells. Nature 432:396–401

Singh S, Lee W, DaSilva NA, Mulchandani A, Chen W (2008a) Enhanced arsenic accumulation by engineered yeast cells expressing *Arabidopsis thaliana Phytochelatin synthase*. Biotechnol Bioeng 99:333–340

Singh S, Mulchandani A, Chen W (2008b) Highly selective and rapid arsenic removal by metabolically engineered Escherichia coli cells expressing *Fucus vesiculosus* metallothionein. Appl Environ Microbiol 74:2924–2927

Singh S, Kang SH, Lee W, Mulchandani A, Chen W (2010) Systematic engineering of phytochelatin synthesis and arsenic transport for enhanced arsenic accumulation in *E. coli*. Biotechnol Bioeng 105(4):780–785

Smedley PL, Kinniburgh DG (2002) A review of the source, behaviour and distribution of arsenic in natural waters. Appl Geochem 17:517–568

Smith E, Naidu R, Alston AM (1999) Chemistry of As in soil: I. Sorption of arsenate and arsenite by four Australian soils. J Environ Qual 28:1719–1726

Smith SE, Christophersen HM, Pope S, Smith FA (2010) Arsenic uptake and toxicity in plants: integrating mycorrhizal influences. Plant Soil 327:1–21

Song WY, Sohn EJ, Martinoia E, Lee YJ, Yang YY, Jasinski M, Forestier C, Hwang I, Lee Y (2003) Engineering tolerance and accumulation of lead and cadmium in transgenic plants. Nat Biotechnol 21:914–919

Southwick JW, Western AE, Beck MM (1981) Community health associated with arsenic in drinking water in Millard Country, Utah, EPA-600/1-81-064, NTIS No. PB82-108374. USEPA, Health Effects Laboratory, Cincinnati

Squibb KS, Fowler BA (1983) The toxicity of arsenic and its compounds. In: Fowler BA (ed) Biological and environmental effects of arsenic. Elsevier, New York, pp 233–269

Srivastava PK, Vaish A, Dwivedi S, Chakrabarty D, Singh N, Tripathi RD (2011) Biological removal of arsenic pollution by soil fungi. Sci Total Environ 409(12):2430–2442

Stolz JF, Oremland RS (1999) Bacterial respiration of arsenic and selenium. FEMS Microbiol Rev 23:615–627
Stolz JF, Basu P, Oremland RS (2002) Microbial transformation of elements: the case of arsenic and selenium. Int Microbiol 5:201–207
Stolz JF, Basu P, Santini JM, Oremland RS (2006) Arsenic and selenium in microbial metabolism. Annu Rev Microbiol 60:107–130
Suhendrayatna A, Ohki TK, Maeda S (1999) Arsenic compounds in the freshwater green microalga *Chlorella vulgaris* after exposure to arsenite. Appl Organomet Chem 13:127–133
Sun WJ, Sierra-Alvarez R, Milner L, Oremland R, Field JA (2009) Arsenite and ferrous iron oxidation linked to chemo lithotrophic denitrification for the immobilization of arsenic in anoxic environments. Environ Sci Technol 43(17):6585–6591
Takahashi Y, Minamikawa R, Hattori KH, Kurishima K, Kiho N, Yuita K (2004) Arsenic behaviour in paddy elds during the cycle of ooded and non-ooded periods. Environ Sci Technol 38:1038–1044
Tamaki S, Frankenberger WT (1992) Environmental biochemistry of arsenic. Rev Environ Contam Toxicol 124:79–110
Tripathi RD, Srivastava S, Mishra S, Singh N, Tuli R, Gupta DK, Maathuis FJM (2007) Arsenic hazards: strategies for tolerance and remediation by plants. Trends Biotechnol 25:158–165
Tsai SL, Singh S, Chen W (2009) Arsenic metabolism by microbes in nature and the impact on arsenic remediation. Curr Opin Biotechnol 20:659–667
Tseng WP (1977) Effects and dose–response relationships of skin cancer and Blackfoot disease with arsenic. Environ Health Perspect 19:109–119
Uppanan P (2000) Screening and characterization of bacteria capable of biotransformation of toxic arsenic compound in soil, M.Sc. Thesis, Mahidol University, Thailand
US Department of Agriculture (1974) Wood preservatives. In: The pesticide review. Washington, DC, p. 21
US Environmental Protection Agency (1975) Interim primary drinking water standards. Fed Regist 40(11):990
Valentine JL, Reisbord LS, Kang HK, Schluchter MD (1985) Arsenic effects of population health histories. In: Mills CF, Bremner IM, Chesters KJ (eds) Trace elements in man and animals. Commonwealth Agricultural Bureau, Slough, pp 289–294
Van Hullebusch ED, Zandvoort MH, Lens PNL (2003) Metal immobilisation by biofilms: mechanisms and analytical tools. Rev Environ Sci Biotechnol 2:9–33
Wallace IS, Choi WG, Roberts DM (2006) The structure, function and regulation of the nodulin 26-like intrinsic protein family of plant aquaglyceroporins. Biochim Biophys Acta 1758:1165–1175
Wang S, Mulligan CN (2006) Effect of natural organic matter on arsenic release from soil and sediments into groundwater. Environ Geochem Health 28:197–214
Wang SL, Zhao XY (2009) On the potential of biological treatment for arsenic contaminated soils and groundwater. J Environ Manag 90(8):2367–2376
Warwick P, Inam E, Evans N (2005) Arsenic's interactions with humic acid. Environ Chem 2:119–124
WHO Arsenic Compounds (2001) Environmental health criteria 224.2nd edn. World Health Organisation, Geneva
Williams PN, Islam MR, Adomako EE, Raab A, Hossain SA, Zhu YG, Feldmann J, Meharg AA (2006) Increase in rice grain arsenic for regions of Bangladesh irrigating paddies with elevated arsenic in ground waters. Environ Sci Technol 40:4903–4908
Wysocki R, Che'ry C, Wawrzycka D, Hulle VM, Cornelis R, Thevelein J, Tama's M (2001) The glycerol channel Fps1p mediates the uptake of arsenite and antimonite in *Saccharomyces cerevisiae*. Mol Microbiol 40(6):1391–1401
Xie ZM, Luo Y, Wang YX, Xie XJ, Su CL (2013) Arsenic resistance and bioaccumulation of an indigenous bacterium isolated from aquifer sediments of Datong Basin, northern China. Geomicrobiol J 30(6):549–556

Xu C, Zhou TQ, Kuroda M, Rosen BP (1998) Metalloid resistance mechanisms in prokaryotes. J Biochem 123:16–23

Xu GH, Chague V, Melamed-Bessudo C, Kapulnik Y, Jain A, Raghothama KG, Levy AA, Silber A (2007) Functional characterization of LePT4: a phosphate transporter in tomato with mycorrhiza-enhanced expression. J Exp Bot 58:2491–2501

Yadav A, Chowdhary P, Kaithwas G, Bharagava RN (2017) Toxic metals in the environment, their threats on ecosystem and bioremediation approaches. In: Das S, Singh HR (eds) Handbook of metal-microbe interaction and bioremediation. CRC Press, Taylor & Francis Group, Boca Raton, pp 128–141

Yamamura S, Watanabe M, Kanzaki M, Soda S, Ike M (2008) Removal of arsenic from contaminated soils by microbial reduction of arsenate and quinone. Environ Sci Technol 42(16):6154–6159

Yang T, Chen ML, Liu LH, Wang JH, Dasgupta PK (2012) Iron(III) modification of Bacillus subtilis membranes provides record sorption capacity for arsenic and endows unusual selectivity for As(V). Environ Sci Technol 46(4):2251–2256

Zaldivar R (1980) A morbid condition involving cardiovascular, brochopulmonary, digestive and neural lesions in children and young adults after dietary arsenic exposure. Zentralbl Bacteriologie 1 Abt Originale B: Hyg Krankenhaushygiene Betriebshygiene Preventive Med 170:44–56

Zaloga GP, Deal J, Spurling T, Richter J, Chernow B (1985) Case report: unusual manifestations of arsenic intoxication. Am J Med Sci 289:210–214

Zhao FJ, Ma JF, Meharg AA, McGrath SP (2009) Arsenic uptake and metabolism in plants. New Phytol 181, 777–7794

Zobrist J, Dowdle PR, Davis JA, Oremland RS (2000) Mobilization of arsenite by dissimilatory reduction of arsenate. Environ Sci Technol 34:4747–4753

Zouboulis AI, Katsoyiannis IA (2005) Recent advances in the bioremediation of arsenic- contaminated ground waters. Environ Int 31:213–219

Chapter 9
Microalgal Treatment of Alcohol Distillery Wastewater

Alexei Solovchenko

Abstract Microalgae possess unique metabolic mechanisms making them highly efficient at removal of the nutrients from and decomposition of organic components of different wastewater types including alcohol distillery wastewater. Microalgae generate dissolved oxygen increasing the efficiency of the alcohol distillery wastewater treatment. Microalgal cells take up and store large amounts of nitrogen (N) and, especially, phosphorus (P). This process induces the nutrient removal efficiency and generates the valuable algal biomass enriched in N and P. Nevertheless, the microalgae-based alcohol distillery wastewater treatment processes are much less established in comparison with conventional anaerobic and aerobic approaches based on bacteria and other heterotrophic organisms. It is not likely that microalgal treatment, either in the open ponds or in the closed systems, can replace completely the conventional alcohol distillery wastewater treatment methods but will complement the latter.

Keywords Microalgae · Biodegradation · Distillery wastewater · Photosynthetic oxygenation · Organic pollutant · Nutrient bioremoval

1 Introduction

Production of ethanol from agricultural raw materials such as sugarcane, sugar beet, corn, and derived materials displays a steady growth which is supported by its consumption in the form of industrial solvent, pharmaceuticals, food, perfumery, and a beverage component as well as an alternative CO_2-neutral fuel (Scott et al. 2010; Simate et al. 2011). The latter usage attracts increasing worldwide attention due to the shortage of nonrenewable energy resources and variability of oil and natural gas prices (Kondili and Kaldellis 2007; Scott et al. 2010).

Annual production of ethyl alcohol is estimated as several billion liter; it is accompanied by generation of huge volumes of assorted types of wastewater

---

A. Solovchenko (✉)
Biological Faculty of Lomonosov Moscow State University, Moscow, Russia
e-mail: solovchenko@mail.bio.msu.ru

Table 9.1 Typical alcohol distillery wastewater generation rate and characteristics according to Satyawali and Balakrishnan (2008)

Wastewater type	Generation rate (L/L alcohol)	Color	pH	Suspended solids (mg/L)	BOD (mg/L)	COD (mg/L)
Spent wash	14.4	Dark brown	4.6	615	36,500	82,080
Fermenter cleaning	0.6	Yellow	3.5	3000	4000	16,500
Fermenter cooling	0.4	Colorless	6.3	220	105	750
Condenser cooling	2.88	Colorless	9.2	400	45	425
Floor wash	0.8	Colorless	7.3	175	100	200
Bottle plant	14	Hazy	7.6	150	10	250
Other	0.8	Pale yellow	8.1	100	30	250

collectively referred to below as alcohol distillery wastewater, ADW; see Table 9.1 (Mohana et al. 2009; Pant and Adholeya 2007; Chowdhary et al. 2017). On an average, 8–15 L of ADW is generated for every liter of alcohol produced (Saha et al. 2005). Increasingly stringent environmental regulations are forcing distilleries to improve existing treatment and to explore alternative methods of the ADW management. As a result, space and funds required for building the ADW treatment plants are turning to be the most serious obstacles for implementation of most advanced and efficient technologies of ADW treatment.

Conventional well-established methods of ADW bioremediation include different combinations of anaerobic and aerobic steps based on the use of heterotrophic bacteria and/or fungi (Mohana et al. 2009). However, a novel approach to ADW treatment based on cultivation of phototrophic microorganisms (collectively referred to as microalgae) became increasingly widespread. In particular, microalgae-based ADW treatment is believed to be more environment-friendly and conductive for the implementation of "circular economy"-based approach to integrated wastewater treatment and biofuel production which currently is showing an upward trend. In this chapter, we review the possibilities, limitations, and perspectives of the ADW treatment with microalgae.

2 Generation and Characteristics of Alcohol Distillery Wastewater

Alcohol distilleries generate several types of wastewater originating from different stages of the process (Table 9.1). These wastewater streams differ considerably in their composition and amount of wastewater generated per L alcohol produced; typical characteristics of the waste streams are presented in Table 9.1. Actual pollutant composition and load of ADW depend on the quality of raw material, details of its processing, and recovery of alcohols (Pandey et al. 2003; Satyawali and Balakrishnan 2008). The most abundant and most problematic in terms of treatment is spent wash—the residue of the fermented mash which comes out as liquid waste (Singh et al. 2004). The spent wash usually features a high content of organic pollutants (expressed as chemical oxygen demand (COD) and biological oxygen demand (BOD)) represented mostly by carbonic acids, sugar decomposition products, dextrans, etc. as well as a considerable amount of inorganic pollutants (mainly nitrate, ammonia, and phosphate ions).

This type of waste stream is characterized by strong odor, dark-brown color, and low pH, making it, unless properly treated, a serious environment threat to water bodies and soils (see next section). Actual generation rate and composition of the spent wash are highly variable and dependent on the raw material used and various aspects of the ethanol production process (Satyawali and Balakrishnan 2008). Washwater which is used to clean the fermenters, as well as cooling water blow down and broiler water blow down further contribute to its variability (see Table 9.1). The composition and environmental hazards of ADW are reviewed in detail by Mohana et al. (2009) and Singh et al. (2004).

3 Environmental Impact and Difficulties of the Alcohol Distillery Wastewater

Discharge of ADW into the environment is highly hazardous. Thus, its high organic content as well as total nitrogen and total phosphate concentration is conductive for eutrophication of natural water bodies (Kumar et al. 1997). This risk is exacerbated by the presence of colored components attenuating the sunlight in water column hence further reducing the photosynthetic activity and dissolved oxygen concentration in the water. The ADW toxicity also manifests itself as adverse effects, e.g., on the physiology, including respiration, and behavior of fish and other aquatic organisms (see, e.g., Kumar and Gopal 2001; Kumar et al. 1995). Dangerous effects of pollution with ADW were also observed in vegetation exposed to the ADW. These effects include inhibition of seed germination and growth and decline in essential microelement bioavailability (Kannan and Upreti 2008; Kumar et al. 1997). The later effects might result, in particular, from the alteration of the soil pH, (nitrogen-fixing) microflora, and other parameters (Juwarkar and Dutta 1990).

Recalcitrance and environmental hazards of ADW stem from its composition that presents a considerable difficulty for conventional treatment approaches. Thus, ADW often possesses high sulfate level, brown-colored polymers, and melanoidins, which are resistant to biodegradation and tend to escape to the environment. Of particular concern are the products of sugar decomposition, e.g., caramels, plant-derived phenolics, and xenobiotics (Pandey et al. 2003). Anaerobic treatment, which is a widespread technique of treatment of ADW possessing a high COD, has a higher potential for destruction of such resilient compounds. Still, operation of an anaerobic digesters can suffer if carbon dioxide-reducing methanogens fail to keep

Table 9.2 Typical composition of anaerobically treated ADW (Mohana et al. 2009)

Parameters (units or mg L^{-1})	Values of anaerobically treated effluent
pH	7.5–8
BOD	8000–10,000
COD	45,000–52,000
Total solid (TS)	70,000–75,000
Total volatile solid (TVS)	68,000–70,000
Total suspended solid (TSS)	38,000–42,000
Total dissolved solids (TDS)	30,000–32,000
Chlorides	7000–9000
Phenols	7000–8000
Sulfate	3000–5000
Phosphate	1500–1700
Total nitrogen	4000–4200

pace with hydrogen production (Nagamani and Ramasamy 1999). Other commonly mentioned shortcomings of the anaerobic processes include sensitivity to shock loadings of BOD/COD and acidic pH. Although the anaerobic treatment reduces significantly the environmental hazard of ADW (Table 9.2), the effluent from the anaerobic digesters is still unsafe to discharge into the environment. Therefore, it is subjected to the second stage of treatment represented by aerobic processes (aerated basins and tanks) (Mohana et al. 2009; Pant and Adholeya 2007). At the same time, the anaerobically treated ADW contains ample nutrient, and its pH is close to neutral making it a suitable medium for cultivation of microalgae (Travieso et al. 1999).

4 Microalgae: An Emerging Solution for ADW Treatment

Wastewater treatment with photosynthetic microorganisms in open ponds has been studied for more than 50 years (Oswald 1988; Oswald and Gotaas 1957), although the history of the microalgal treatment of wastewater in closed systems (photobioreactors, PBRs) is somewhat shorter. Accordingly, alternative approaches for remediation of ADW with microalgae started to emerge over the last two decades (Pant and Adholeya 2007; Valderrama et al. 2002). These approaches involved mostly combinations of microalgae with higher plants (Mata et al. 2012) and/or cyanobacteria and heterotrophic bacteria (Satyawali and Balakrishnan 2008). Treatments of wastewater, including ADW, with microalgae have distinct advantages. An important one is a lower aeration need (in the case of the aerobic treatment) since the oxygen generated by the microalgae during photosynthesis readily oxidizes organic molecules and supports growth of heterotrophic bacteria which also consume organic substances from the wastewater (Muñoz and Guieysse 2006). The lower need of aeration naturally translates into electric energy savings due to reduced need of mechanical mixing and/or air blowing. Another important advantages are simultaneous removal of nutrients such as nitrogen and phosphorus (Aslan and Kapdan 2006) with possibility of their recycling (Solovchenko et al. 2016) as well as biosequestration of CO_2, the most abundant greenhouse gas (Van Den Hende et al. 2012). Treatment with photosynthetic microorganisms is also believed to be more environment-friendly because it does not generate large amounts of secondary wastes such as sludge.

In spite of numerous advantages, the microalgae-based systems for ADW remediation are so far much less widespread in comparison with the traditional systems based on heterotrophic bacteria. Reports on efficient microalgal processes specifically aimed to ADW bioremediation are scarce and mostly limited to laboratory-scale experiments (Mata et al. 2012; Raposo et al. 2010). Thus, Travieso et al. (2008b, c) evaluated the performance of a laboratory-scale microalgae pond treating effluent from an anaerobic fixed-bed reactor digesting ADW. The microalgae pond operated with an effluent recycling attained the global solid removal efficiency of 92.6%. The removal efficiencies for organic nitrogen and ammonia were 90.2% and 84.1%, respectively. The removal for total phosphorus and orthophosphate was 85.5% and 87.3%, respectively.

Most of the currently tested methods of ADW remediation are based on cultivation of green microalgae (Chlorophyta), particularly representatives of the genus *Chlorella*, due to its remarkable stress tolerance and capability of mixotrophic growth (Alcántara et al. 2014; Perez-Garcia et al. 2011). Apart from the green eukaryotic microalgae, a marine cyanobacterium *Oscillatoria boryana* was used for decolorization of distillery spent wash where the microalgae consumed the recalcitrant biopolymer melanoidin as a nitrogen and carbon source (Kalavathi et al. 2001). The cyanobacteria *Oscillatoria* sp., *Lyngbya* sp., and *Synechocystis* sp. exhibited 96%, 81%, and 26% decolonization of the ADW, respectively (Patel et al. 2001).

Valderrama et al. (2002) combined a microalga *Chlorella vulgaris* and an aquatic plant *Lemna minuscule* for the treatment of ADW. In this system, microalgae removed the bulk of organic matter, and the higher plant was used for the "polishing" treatment of the effluent. More recently, Yang et al. (2008) cultivated *C. pyrenoidosa* in cycle tubular photobioreactor (PBR) in undiluted wastewater from ethanol fermentation using cassava powder as raw material. High rates of organic pollutant removal were achieved with a fixed-bed reactor coupled with a laboratory-scale microalgal pond (Travieso et al. 2008a).

It should be noted in addition that the upscaling of microalgal photobiotechnology for wastewater remediation requires fast and reliable, preferably nondestructive, techniques for online monitoring of the state of algal culture (Havlik et al. 2013). This is necessary for timely and informed decisions on adjustment of illumination conditions, the rate of wastewater inflow, and on the time for biomass harvesting. Often, the decisions must be taken within hours, and mistakes may lead to a significant reduction in productivity or in a total culture loss. In line with this reasoning, Solovchenko et al. (2014) used a high-density culture of *Chlorella sorokiniana* monitored in real time via chlorophyll fluorescence measurements for treatment of ADW in semi-batch mode (Fig. 9.1).

An important aspect of wastewater treatment is the removal of chemically bound nitrogen (N) and phosphorus (Olguín 2003). Failing to do so leads to eutrophication,

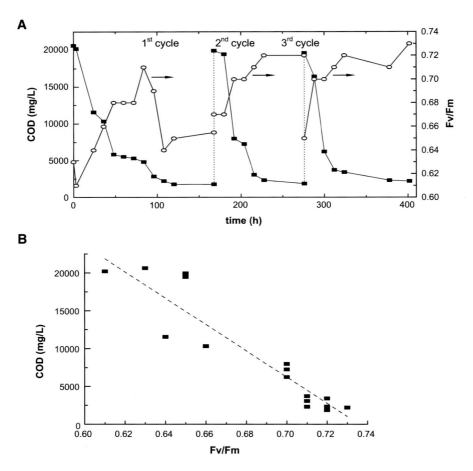

Fig. 9.1 An example of using variable chlorophyll fluorescence for real-time monitoring of ADW treatment with a microalgal culture in a photobioreactor. (**a**) The time course of changes in COD (closed squares) and maximum PSII efficiency (open squares, right scale) was determined as variable chlorophyll fluorescence, Fv/Fm. The data for three sequential cycles of semi-batch cultivation are shown. The points recorded just before and after resupply with fresh ADW are connected with vertical dashed lines. (**b**) Relationships between Fv/Fm and COD decline during ADW treatment with *Chlorella sorokiniana* in the photobioreactor. (Reprinted from Solovchenko et al. 2014 with permission)

hypoxia, harmful algae blooms, and loss of biodiversity (Anderson et al. 2002; Smil 2000). Microalgae are naturally equipped for a rapid uptake of nutrients from the medium taking up more nutrients than they currently need to support their cell cycle; this mechanism is known as luxury uptake (de Mazancourt and Schwartz 2012). This capability makes cultivation of microalgae an efficient tool for nutrient bioremoval from wastewater including ADW (Coppens et al. 2014; Solovchenko et al. 2016) (Table 9.3).

Table 9.3 Bioremoval of sulfates, nitrates, and phosphates by *Chlorella sorokiniana* cells from alcohol distillery wastewater (Solovchenko et al. 2014)

Cultivation time (h)	Concentration (mg/L)		
	NO_3^-	SO_4^{2-}	PO_4^{3-}
0 (at pH 7)	15 ± 1.3	178 ± 18	773 ± 73
96	ND[a]	116 ± 15	176 ± 18
	Removal extent (%)		
	>95	35	77

[a]*ND* not detected

On the other hand, N and P are the macronutrients essential for plant growth. It is important that MA assimilate N (in the form of ammonium and nitrate), while denitrifying bacteria reduce these forms of N to N_2, which escapes to the atmosphere. Accordingly, the nutrients captured by microalgae can be returned to agroecosystems, e.g., in the form of biofertilizer made from the algal biomass (Mulbry et al. 2005). The use of the biofertilizer from microalgae (e.g., grown in ADW) also reduces the consumption of synthetic N and P fertilizers further declining the impact on the environment and energy demand.

Efficient aerobic ADW treatment requires aeration providing oxygen for direct oxidation of organic ADW components as well for the heterotrophic aerobic bacteria further reducing BOD and COD. In conventional bioreactors, the aeration is accomplished by mechanical mixing and/or sparging (blowing) of the air through the reactor volume. During the retreatment of the wastewater with microalgae, the oxygen demand is satisfied, to a considerable extent, by photosynthesis reducing considerably the need for aeration.

The treatment of ADW with microalgae looks especially promising since this type of wastewater does not contain significant amounts of dangerous low-molecular organic pollutants and/or heavy metals (Mohana et al. 2009; Raposo et al. 2010). Microalgae hold great potential as a feedstock for biodiesel and other kinds of biofuels since their oil yield is an order of magnitude higher than that of oleaginous higher plants (Chisti 2010; Hu et al. 2008). Furthermore, ADW can be, in principle, used to grow microalgae accumulating value-added products: essential polyunsaturated fatty acids, e.g., arachidonic (Crawford et al. 2003), linolenic (Wang et al. 2012), eicosapentaenoic acid, and other long-chain fatty acids (Cohen and Khozin-Goldberg 2010). Another important line of bioproducts from microalgae is constituted by carotenoid antioxidants such as astaxanthin (Dhankhar et al. 2012) and β-carotene (Takaichi 2011).

Biofuel production is a promising way of utilization of the microalgal biomass grown in wastewater (Hu et al. 2008; Pittman et al. 2011). Solid (pelleted dry biomass), liquid (biodiesel and hydrocarbons), and gaseous (biogas obtained via anaerobic digestion) forms of fuels can be produced from the microalgal biomass which is enriched with lipids and carbohydrates, and it could be converted to biofuel such as bio-oil or biogas (methane) (Georgianna and Mayfield 2012). It is accepted that the biomass of microalgae grown in wastewater is the most cost-effective feed-

Table 9.4 A typical pattern of fatty acid content and composition of a green microalga (*Chlorella sorokiniana*) biomass grown on the alcohol distillery wastewater and potentially suitable for conversion into biodiesel (Solovchenko et al. 2014).

Fatty acid	mg L^{-1} 0 h	96 h	% TFA 0 h	96 h
14:0	2.8	136.5	2.0	5.8
16:0	16.9	508.0	11.8	21.5
16:1	14.2	228.5	9.9	9.7
16:2	5.6	282.7	3.9	12.0
16:3	5.6	3.2	3.9	0.1
18:0	10.0	62.8	7.0	2.7
18:1	67.1	739.7	40.7	31.3
18:2	12.3	396.4	8.6	16.8
18:3	8.6	ND	6.0	ND
Sum of fatty acids	**143.5**	**2357.8**	–	–
% DW	10	20	–	–
Unsaturation index	–	–	111.5	98.9
Saturated/unsaturated fatty acid ratio	–	–	0.27	0.43

stock, so the biofuel produced form this biomass could compete successfully with fossil fuels (Pittman et al. 2011).

The fatty acid composition of the microalgal biomass grown in wastewater is crucial for the preferred way of its utilization (see Table 9.3). In particular, a good-quality biodiesel has high cetane number, which is associated with saturated fatty acids such as palmitic (16:0) and stearic (18:0) acids; on the contrary, a low cetane number was observed with highly unsaturated fatty acids such as linolenic (18:3) acid (Knothe et al. 2003). The high proportion of saturated fatty acids in algal biomass may provide increased energy yield, superior oxidative stability, and higher cetane numbers that cause less problems in fuel polymerization during combustion (Canakci and Sanli 2008; Mallick et al. 2012; Walsh et al. 2015) (Table 9.4).

It should be taken into account, however, that cultivation of microalgae under eutrophic conditions does not necessarily induce the synthesis of reserve lipids as it happens during cultivation in media lacking N. As a result, the biomass of the microalgae grown in nutrient-rich wastewater rarely contains more than 20% of lipids. Anaerobic digestion for biogas production seems to be a better way of the utilization of the wastewater-grown biomass; production of feed additives and fertilizers is a viable alternative as well. The microalgae accumulating the essential long-chain fatty acids in the chloroplast membrane lipids such as *Nannochloropsis* are preferred for feed additive production (Benemann 1992; El-Sayed 1998) although *Chlorella* strains which are especially suitable for ADW treatment also yield the biomass of a quality acceptable for this purpose (Raposo et al. 2010).

Another promising way of utilization of the ADW-grown biomass is its application to soil as a biofertilizer. In particular, the performance of the microalgal biomass-based biofertilizer was commensurate to this of traditional chemical fertilizers (Mulbry et al. 2005). An added benefit of algal biomass is that it does not need to be tilled into soil, which is necessary for mineral P fertilizers. The algal biomass could be side-dressed into growing crops, thereby saving much labor and energy. Sludge from waste stabilization ponds used for wastewater treatment with microorganism consortia including microalgae has potential as a biofertilizer as well (Powell et al. 2011).

Finally, many microalgal species are capable of mixotrophic growth and formation of stable algal-bacterial consortia, including the consortia with participation of cyanobacteria, taking up organic matter and inorganic nutrients from the wastewater. These capabilities significantly enhance the COD and nutrient removal capacities of microalgae (de-Bashan et al. 2004; Meza et al. 2015; Raposo et al. 2010). It is difficult to determine precisely the contributions of the autotroph (the microalga) and the heterotroph (the bacteria) components of the consortia to the destruction of organic pollutants in ADW although some reports demonstrate that microalgae are essential for efficient ADW treatment (Muñoz and Guieysse 2006; Solovchenko et al. 2014). Accordingly, isolation of autochthonous algal-(cyano)bacterial consortia from the facilities of ADW treatment is a promising method for bioprospecting of strains and consortia suitable for microalgal remediation of ADW.

5 Cultivation Conditions for Microalgae During ADW Treatment

Efficient treatment of ADW with photosynthetic microorganisms such as microalgae requires sufficient and uniform illumination of the microalgal cell suspension. On one hand, high-density cultures are able to cope with higher nutrient and pollutant loads in ADW (Lau et al. 1995; Lavoie and de la Noüe 1985; Muñoz and Guieysse 2006). On the other hand, in thick suspension layers, photosynthesis and hence the efficiency of treatment are limited by light intensity due to self-shading of the cells. As a result, only the cells in the surface (>10 mm thick) layer get sufficient light for photosynthesis (Richmond 2004). This problem is exacerbated by a significant turbidity and/or dark color of ADW. However, it is not possible to solve the problem of insufficient illumination of dark inner part of the suspensions of relatively high-density microalgal cultures (>1 g/L dry cell weight) just by increasing incident illumination intensity. Apart from significant increase of electricity costs, excessive illumination of the surface layer of the suspensions can damage or even kill the cells, while the cells in deeper suspension layers will suffer from light energy starvation (Richmond 2004, see Fig. 9.2).

Viable strategies for achieving uniform illumination of microalgal cells during wastewater treatment include decreasing of the suspension layer thickness (increasing the surface-to-volume ratio of the treatment system) and/or vigorous mixing of the cell suspension. Either strategy has its limitations: the former increases the footprint of the treatment system, whereas the latter imposes the risk of damage to the microalgal cell because of sheer deformation if the mixing rate is too high, let alone the increase of electricity consumption (Richmond 2004; Zittelli et al. 2013).

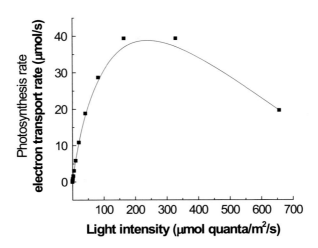

Fig. 9.2 Selection of optimal illumination intensity is based in the light response curves like one presented in this figure for *Chlorella sorokiniana* culture (OD_{678} = 1.0) grown in alcohol distillery wastewater (Reprinted from Solovchenko et al. 2014 with permission). Note that the use of more dense cultures for ADW treatment will require higher illumination intensity and a more efficient mixing (see text for explanation)

The rate and the efficiency of wastewater treatment depend on the concentration of inorganic (nutrients) and organic (COD/BOD) pollutants per cell density or biomass content of the culture used for the treatment which is designated as "biomass load" (Boelee et al. 2014). Of special importance is the initial cell density of microalgae. The reason is that the inoculation of ADW with microalgal culture is, to a certain degree, stressful for the cell per se (Lavoie and de la Noüe 1985). If the concentration of the ADW components is in the range optimal for the microalgae, the treatment (uptake of nutrients and the destruction of the organic pollutants) proceeds at the maximum rate (see Figs. 9.3 and 9.4), so it is important to keep the COD loading rate within the capacity of the efficient decomposition by the microalgal culture. It should be noted that, depending on the agricultural raw material and the variation of technological process used, COD load of ADW could exceed this capacity. In this case, dilution of ADW is necessary; a practical solution to this problem (dilution with the washwater originating from the same distillery, see Table 9.1) was suggested by Valderrama et al. (2002). Another option is to operate PBR in fed-batch mode at a suitable dilution rate, e.g., as described by Yang et al. (2008); the added benefit of the fed-batch cultivation is the possibility to operate at the native ADW pH.

Under excessive biomass load i.e. in the situations when the concentration of the ADW components per unit culture density is too high, the microalgae can suffer from the stress leading to a decline of cell division rate (a longer lag period), inhibition of the treatment process or even in a culture crash. On the other hand, excessive culture density results in light or other kind of limitation of the microalgae, causing a decline in the efficiency of ADW treatment as well. Therefore, finding the

Fig. 9.3 A correct determination of optimal biomass load (ratio of microalgal culture density and ADW concentration) is crucial for the efficiency of the ADW treatment. Here, the effect of starting culture density of *Chlorella sorokiniana* (indicated near the curves) on its growth (**a**) and (**b**) photosynthetic activity in alcohol distillery wastewater is demonstrated. (Reprinted from Solovchenko et al. 2014 with permission)

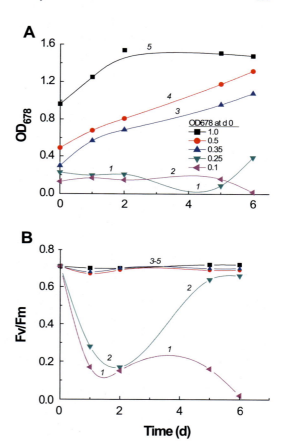

optimal biomass load is of paramount importance for the development of a viable microalgae-based wastewater treatment process (Lau et al. 1995).

In principle, the optimal biomass load can be attained in the case of ADW by blending waste streams with different pollutant content and composition. It is also important to understand that the biomass load is limited by a component or parameter of the ADW, which is most toxic/inhibitory to the microalgae. Thus, a low pH (high acidity) typical of ADW (Table 9.1) severely inhibits microalgal growth, so the rate of ADW treatment will be limited by the overall pH of the inoculated ADW (Solovchenko et al. 2014). On the contrary, steady increase in the pH of the ADW treated with microalgae (Fig. 9.5) is characteristic of robust and balanced treatment system.

Currently, microalgae are cultivated predominantly in traditional waste stabilization (oxidation) ponds which are shallow man-made basins accumulating wastewater. The wastewater is treated there over a long time with a combination of aerobic and anaerobic bacteria as well as microalgae (Mara and Pearson 1986; Von Sperling 2007). WSP are well established and widespread for the treatment of domestic and industrial wastewater in regions with ample land and sunshine.

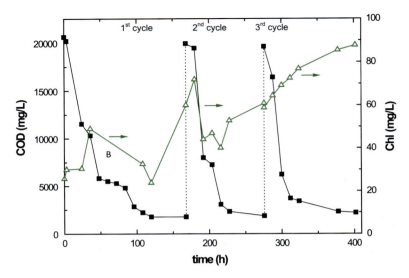

Fig. 9.4 An optimal biomass load results in a steady decline of COD and increase in microalgal cell density (apparent as accumulation in chlorophyll, right scale) as exemplified by treatment of alcohol distillery wastewater in the photobioreactor with *Chlorella sorokiniana*. The data for three representative sequential cycles of semi-batch cultivation are shown. The points recorded just before and after resupply with fresh ADW are connected with vertical dashed lines. (Reprinted from Solovchenko et al. 2014 with permission)

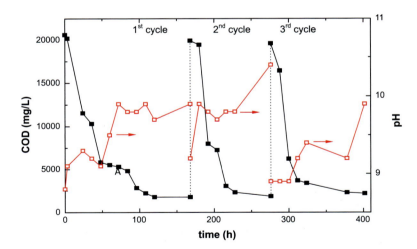

Fig. 9.5 The changes in COD (closed squares) and pH (open squares, right square) in *Chlorella sorokiniana* culture grown on alcohol distillery wastewater in a photobioreactor. The data for three representative sequential cycles of semi-batch cultivation are shown. The points recorded just before and after resupply with fresh ADW are connected with vertical dashed lines. (Reprinted from Solovchenko et al. 2014 with permission)

Waste stabilization ponds are further divided into anaerobic, facultative, and maturation/oxidation ponds that are often set up in a series to achieve a reasonable efficiency of wastewater treatment (Mara and Pearson 1986). Microalgae exert a sizeable contribution to nutrient (P and N) removal in the upper (photic) layer of facultative ponds, which can be primary or secondary. A substantial drawback of such systems is the dependence of treatment efficiency on climatic conditions (efficiency decline in cold seasons, high water evaporation in summer). Such systems have a large footprint and high construction and maintenance costs.

A more advanced modification of open pond is high-rate algal pond (HRAP) (Nurdogan and Oswald 1996; Park et al. 2011). HRAPs were the first systems for intensive algal cultivation adapted for wastewater treatment and involving mixing and CO_2 addition. HRAPs further evolved into the concept of "luxury uptake ponds," facilitating uptake of nutrients such as P by microalgal cells from treated wastewater (Powell et al. 2009). The characteristic features of high-rate algal pond and luxury uptake ponds include (1) relatively high P load rates, (2) low pond depth (typically <30 cm) and hence high light intensity, and (3) aeration and/or mixing of the suspension in the pond.

The open pond-based treatment systems are widespread because they are inexpensive and simple to construct. PBRs are the closed systems for cultivation of microalgae essentially free of the abovementioned drawbacks of open ponds excepting high costs. The closed cultivation systems can be considered for special situations in which, for example, a wastewater effluent is used as an inexpensive nutrient source in emerging large-scale applications for third-generation biofuels or CO_2 sequestration. In addition, PBRs may be considered when high costs of harvesting microalgae from diluted suspensions, water evaporation, dependence on climatic conditions, large footprint, and unstable yield and biomass quality in open-pond systems outweigh their lower cost.

Vertical (see Fig. 9.6), horizontal, tubular, circular, and flat-panel reactors are currently the most widespread PBR types. The specific advantages and disadvantages of these FBR are reviewed in details elsewhere (Zittelli et al. 2013). The key factors of successful use of microalgae for nutrient sequestration from wastewater including ADW are (Solovchenko et al. 2013) the choice of rapidly growing strains that are tolerant to eutrophic conditions and a particular wastewater composition, an energy-efficient PBR design supporting intensive cultivation of microalgae, and fine-tuning of the illumination intensity, mixing rate, and temperature based on a deep understanding of the physiology of luxury nutrient uptake by microalgal cells. Microalgae grown in PBR possess a high potential for nutrient and COD removal, and PBR allows, as a closed system, the use of engineered algal strains, so we believe that the application of the closed cultivation systems holds promise for P recovery from wastewater, especially in regions with temperate and cold climates or high land cost.

An interesting alternative to suspension culture is the use of immobilized cells of microalgae for wastewater treatment. Cells can be immobilized by trapping in alginate gelsor by forming biofilms (Abe et al. 2008; Boelee et al. 2011; de-Bashan and Bashan 2010).This method has many advantages, including steady growth, high purification efficiency, and higher cell retention in PBR (Vasilieva et al. 2016).

Fig. 9.6 Operating annular PBR unit with the *Chlorella sorokiniana* culture on alcohol distillery wastewater (filled to 10% of nominal volume, left) and its 3D model showing the mixing paddles (right). (Reprinted from Solovchenko et al. 2014 with permission)

6 Conclusions and Outlook

Microalgae possess unique metabolic mechanisms making them highly efficient at removal of the nutrients from and decomposition of organic components of different wastewater types including ADW. As photosynthetic organisms, microalgae generate dissolved oxygen increasing the efficiency of the ADW treatment. Thanks to the luxury uptake capability, microalgal cells take up and store large amounts of nitrogen and, especially, phosphorus. This process enhances the nutrient removal efficiency and generates the valuable algal biomass enriched in N and P. Nevertheless, the microalgae-based ADW treatment processes are much less established in comparison with conventional anaerobic and aerobic approaches based on bacteria and other heterotrophic organisms. It is not likely that microalgal treatment, either in the open ponds or in the closed systems, can replace completely the conventional ADW treatment methods. At the same time, the microalgae-based stages can be easily integrated into the existing wastewater treatment plants, e.g., at the final (tertiary) stage. Still, microalgae are capable of growth in anaerobically treated ADW or even in untreated or minimally pretreated ADW. There is a ground to believe that a full-cycle treatment of ADW with microalgae, in principle, is feasible (Ip et al. 1982). A benefit of ADW remediation with microalgae is constituted by diverse applications of the resulting biomass. This biomass, depending on its composition, can be converted into assorted biofuels, biofertilizers, and feed additives.

The crucial factors of efficient treatment of ADW with microalgae include (i) the rapidly growing strain tolerant to the eutrophic conditions and to specific ADW composition, (ii) energy-efficient design of the cultivation system supporting high-density cultures, and (iii) knowledge-based optimization of the cultivation regime based on deep understanding of the physiology and, preferably, online monitoring of the culture condition. A promising solution to this problem is based in optical and/or luminescent (based on fluorescence of chlorophyll) methods. This feature is important since an ADW treatment method has to be flexible enough to overcome the constant fluctuations of the organic and inorganic ion load by timely adjusting of dilution rate, light intensity, etc. The use of optical and Chl fluorescence-based sensors is a plausible option for automation of microalgae-based ADW treatment.

Further development of the microalgae-based ADW treatment will depend on development of energy-efficient cultivation systems (photobioreactors) suitable for integration with current ADW treatment processes. A separate goal is bioprospecting of tolerant strains capable of growth in ADW under harsh eutrophic conditions. Finally, a cost-effective and practical method of biomass harvesting is essential for development of the comprehensive solution for microalgal treatment of ADW.

Acknowledgments Financial support of the Russian Ministry of Science and Education (grant 14.616.21.0080) is gratefully acknowledged.

References

Abe K, Takahashi E, Hirano M (2008) Development of laboratory-scale photobioreactor for water purification by use of a biofilter composed of the aerial microalga *Trentepohlia aurea* (Chlorophyta). J Appl Phycol 20:283–288. https://doi.org/10.1007/s10811-007-9245-9

Alcántara C, Fernández C, García-Encina P, Muñoz R (2014) Mixotrophic metabolism of *Chlorella sorokiniana* and algal-bacterial consortia under extended dark-light periods and nutrient starvation. Appl Microbiol Biotechnol 1–12:6125–6125. https://doi.org/10.1007/s00253-014

Anderson DM, Glibert PM, Burkholder JM (2002) Harmful algal blooms and eutrophication: nutrient sources, composition, and consequences. Estuaries 25:704–726

Aslan S, Kapdan IK (2006) Batch kinetics of nitrogen and phosphorus removal from synthetic wastewater by algae. Ecol Eng 28:64–70

Benemann J (1992) Microalgae aquaculture feeds. J Appl Phycol 4:233–245. https://doi.org/10.1007/bf02161209

Boelee NC, Temmink H, Janssen M, Buisman CJ, Wijffels RH (2011) Nitrogen and phosphorus removal from municipal wastewater effluent using microalgal biofilms. Water Res 45:5925–5933. https://doi.org/10.1016/j.watres.2011.08.044

Boelee N, Temmink H, Janssen M, Buisman C, Wijffels R (2014) Balancing the organic load and light supply in symbiotic microalgal–bacterial biofilm reactors treating synthetic municipal wastewater. Ecol Eng 64:213–221

Canakci M, Sanli H (2008) Biodiesel production from various feedstocks and their effects on the fuel properties. J Ind Microbiol Biotechnol 35:431–441

Chisti Y (2010) Fuels from microalgae. Biofuels 1:233–235

Chowdhary P, Yadav A, Kaithwas G, Bharagava RN (2017) Distillery wastewater: a major source of environmental pollution and its biological treatment for environmental safety. In: Singh R, Kumar S (eds) Green technologies and environmental sustainability. Springer International, Cham, pp 409–435

Cohen Z, Khozin-Goldberg I (2010) Searching for PUFA-rich microalgae. In: Cohen Z, Ratledge C (eds) Single cell oils2nd edn. American Oil Chemists' Society, Champaign, pp 201–224

Coppens J, Decostere B, Van Hulle S et al (2014) Kinetic exploration of nitrate-accumulating microalgae for nutrient recovery. Appl Microbiol Biotechnol 98:8377–8387. https://doi.org/10.1007/s00253-014-5854-9

Crawford M et al (2003) The potential role for arachidonic and docosahexaenoic acids in protection against some central nervous system injuries in preterm infants. Lipids 38:303–315

de Bashan LE, Bashan Y (2010) Immobilized microalgae for removing pollutants: review of practical aspects. Bioresour Technol 101:1611–1627

de Bashan LE, Hernandez J-P, Morey T, Bashan Y (2004) Microalgae growth-promoting bacteria as "helpers" for microalgae: a novel approach for removing ammonium and phosphorus from municipal wastewater. Water Res 38:466–474. https://doi.org/10.1016/j.watres.2003.09.022

de Mazancourt C, Schwartz MW (2012) Starve a competitor: evolution of luxury consumption as a competitive strategy. Theor Ecol 5:37–49

Dhankhar J, Kadian SS, Sharma A (2012) Astaxanthin: a potential carotenoid. Int J Pharm Sci Res 3:1246–1259

El-Sayed AF (1998) Total replacement of fish meal with animal protein sources in Nile tilapia, *Oreochromis niloticus* (L.), feeds. Aquac Res 29:275–280

Georgianna DR, Mayfield SP (2012) Exploiting diversity and synthetic biology for the production of algal biofuels. Nature 488:329–335

Havlik I, Lindner P, Scheper T et al (2013) On-line monitoring of large cultivations of microalgae and cyanobacteria. Trends Biotechnol 31(7):406–414. https://doi.org/10.1016/j.tibtech.2013.04.005

Hu Q, Sommerfeld M, Jarvis E et al (2008) Microalgal triacylglycerols as feedstocks for biofuel production: perspectives and advances. Plant J 54:621–639

Ip S, Bridger J, Chin C et al (1982) Algal growth in primary settled sewage: the effects of five key variables. Water Res 16:621–632

Juwarkar A, Dutta S (1990) Impact of distillery effluent application to land on soil microflora. Environ Monit Assess 15:201–210

Kalavathi DF, Uma L, Subramanian G (2001) Degradation and metabolization of the pigment—melanoidin in distillery effluent by the marine cyanobacterium Oscillatoria boryana BDU 92181. Enzym Microb Technol 29:246–251

Kannan A, Upreti RK (2008) Influence of distillery effluent on germination and growth of mung bean (Vigna radiata) seeds. J Hazard Mater 153:609–615

Knothe G, Matheaus AC, Ryan III TW (2003) Cetane numbers of branched and straight-chain fatty esters determined in an ignition quality tester. Fuel 82:971–975

Kondili EM, Kaldellis JK (2007) Biofuel implementation in East Europe: current status and future prospects. Renew Sust Energ Rev 11:2137–2151

Kumar S, Gopal K (2001) Impact of distillery effluent on physiological consequences in the freshwater teleost Channa punctatus. Bull Environ Contam Toxicol 66:617–622

Kumar S, Sahay S, Sinha M (1995) Bioassay of distillery effluent on common guppy, Lebistes reticulatus (Peter). Bull Environ Contam Toxicol 54:309–316

Kumar V, Wati L, FitzGibbon F et al (1997) Bioremediation and decolorization of anaerobically digested distillery spent wash. Biotechnol Lett 19:311–314

Lau P, Tam N, Wong Y (1995) Effect of algal density on nutrient removal from primary settled wastewater. Environ Pollut 89:59–66

Lavoie A, de la Noüe J (1985) Hyperconcentrated cultures of *Scenedesmus obliquus*: a new approach for wastewater biological tertiary treatment. Water Res 19:1437–1442. https://doi.org/10.1016/0043-1354(85)90311-2

Mallick N, Mandal S, Singh AK et al (2012) Green microalga Chlorella vulgaris as a potential feedstock for biodiesel. J Chem Technol Biotechnol 87:137–145. https://doi.org/10.1002/jctb.2694

Mara D, Pearson H (1986) Artificial freshwater environment: waste stabilization ponds. In: Rehm H, Reed G (eds) Biotechnology. A comprehensive treatise. Verlagsgesellschaft, Weinheim, pp 177–206

Mata TM, Melo AC, Simões M et al (2012) Parametric study of a brewery effluent treatment by microalgae Scenedesmus obliquus. Bioresour Technol 107:151–158. https://doi.org/10.1016/j.biortech.2011.12.109

Meza B, de Bashan LE, Hernandez JP et al (2015) Accumulation of intra-cellular polyphosphate in Chlorella vulgaris cells is related to indole-3-acetic acid produced by Azospirillum brasilense. Res Microbiol 166:399–407. https://doi.org/10.1016/j.resmic.2015.03.001

Mohana S, Acharya BK, Madamwar D (2009) Distillery spent wash: treatment technologies and potential applications. J Hazard Mater 163:12–25

Mulbry W, Westhead EK, Pizarro C et al (2005) Recycling of manure nutrients: use of algal biomass from dairy manure treatment as a slow release fertilizer. Bioresour Technol 96:451–458

Muñoz R, Guieysse B (2006) Algal-bacterial processes for the treatment of hazardous contaminants: a review. Water Res 40:2799–2815. https://doi.org/10.1016/j.watres.2006.06.011

Nagamani B, Ramasamy K (1999) Biogas production technology: an Indian perspective. Curr Sci 77:44–55

Nurdogan Y, Oswald WJ (1996) Tube settling of high-rate pond algae. Water Sci Technol 33:229–241

Olguín EJ (2003) Phycoremediation: key issues for cost-effective nutrient removal processes. Biotechnol Adv 22:81–91

Oswald WJ (1988) Micro-algae and waste-water treatment

Oswald WJ, Gotaas HB (1957) Photosynthesis in sewage treatment. Trans Am Soc Civ Eng 122:73–105

Pandey R, Malhotra S, Tankhiwale A et al (2003) Treatment of biologically treated distillery effluent-a case study. Int J Environ Stud 60:263–275

Pant D, Adholeya A (2007) Biological approaches for treatment of distillery wastewater: a review. Bioresour Technol 98:2321–2334. https://doi.org/10.1016/j.biortech.2006.09.027

Park JBK, Craggs RJ, Shilton AN (2011) Wastewater treatment high rate algal ponds for biofuel production. Bioresour Technol 102:35–42. https://doi.org/10.1016/j.biortech.2010.06.158

Patel A, Pawar R, Mishra S, Tewari A (2001) Exploitation of marine cyanobacterial for removal of colour from distillery effluent. Indian J Environ Protect 21:1118–1121

Perez-Garcia O, Escalante FME, de Bashan LE et al (2011) Heterotrophic cultures of microalgae: metabolism and potential products. Water Res 45:11–36. https://doi.org/10.1016/j.watres.2010.08.037

Pittman JK, Dean AP, Osundeko O (2011) The potential of sustainable algal biofuel production using wastewater resources. Bioresour Technol 102:17–25. https://doi.org/10.1016/j.biortech.2010.06.035

Powell N, Shilton A, Chisti Y et al (2009) Towards a luxury uptake process via microalgae – defining the polyphosphate dynamics. Water Res 43:4207–4213. https://doi.org/10.1016/j.watres.2009.06.011

Powell N, Shilton A, Pratt S et al (2011) Phosphate release from waste stabilisation pond sludge: significance and fate of polyphosphate. Water Sci Technol 63:1689

Raposo M, Oliveira SE, Castro PM, Bandarra NM, Morais RM (2010) On the utilization of microalgae for brewery effluent treatment and possible applications of the produced biomass. J Inst Brew 116:285–292

Richmond A (2004) Principles for attaining maximal microalgal productivity in photobioreactors: an overview. Hydrobiologia 512:33–37

Saha N, Balakrishnan M, Batra V (2005) Improving industrial water use: case study for an Indian distillery resources. Conserv Recycl 43:163–174

Satyawali Y, Balakrishnan M (2008) Wastewater treatment in molasses-based alcohol distilleries for COD and color removal: a review. J Environ Manag 86:481–497. https://doi.org/10.1016/j.jenvman.2006.12.024

Scott EL, Kootstra AMJ, Sanders JPM (2010) Perspectives on bioenergy and biofuels. Sustainable biotechnology, pp 179–194

Simate GS, Cluett J, Iyuke SE et al (2011) The treatment of brewery wastewater for reuse: state of the art. Desalination 273:235–247

Singh PN, Robinson T, Singh D et al (2004) Treatment of industrial effluents. Distillery effluent. In: Pandey A (ed) Concise encyclopedia of bioresource technology. Food Products Press, New York, pp 135–142

Smil V (2000) Phosphorus in the environment: natural flows and human interferences. Annu Rev Energy Environ 25:53–88

Solovchenko A, Lukyanov A, Vasilieva S et al (2013) Possibilities of bioconversion of agricultural waste with the use of microalgae. Moscow Univ Biol Sci Bull 68:206–215

Solovchenko A et al (2014) Phycoremediation of alcohol distillery wastewater with a novel Chlorella sorokiniana strain cultivated in a photobioreactor monitored on-line via chlorophyll fluorescence. Algal Res-Biomass Biofuels Bioprod 6:234–241. https://doi.org/10.1016/j.algal.2014.01.002

Solovchenko A, Verschoor AM, Jablonowski ND, Nedbal L (2016) Phosphorus from wastewater to crops: an alternative path involving microalgae. Biotechnol Adv 34:550–564

Takaichi S (2011) Carotenoids in algae: distributions. Biosyntheses Funct Mar Drugs 9:1101–1118

Travieso L, Benitez F, Dupeyron R (1999) Algae growth potential measurement in distillery wastes. Bull Environ Contam Toxicol 62:483–489

Travieso L, Benítez F, Sanchez E et al (2008a) Assessment of a microalgae pond for post-treatment of the effluent from an anaerobic fixed bed reactor treating distillery. Wastewater Environ Technol 29:985–992

Travieso L, Benítez F, Sánchez E et al (2008b) Performance of a laboratory-scale microalgae pond for secondary treatment of distillery wastewaters. Chem Biochem Eng Q 22:467–473

Travieso L, Benítez F, Sánchez E et al (2008c) Assessment of a microalgae pond for post-treatment of the effluent from an anaerobic fixed bed reactor treating distillery. Wastewater Environ Technol 29:985–992. https://doi.org/10.1080/09593330802166228

Valderrama LT, Del Campo CM, Rodriguez CM et al (2002) Treatment of recalcitrant wastewater from ethanol and citric acid production using the microalga Chlorella vulgaris and the macrophyte Lemna minuscula. Water Res 36:4185–4192

Van Den Hende S, Vervaeren H, Boon N (2012) Flue gas compounds and microalgae: (bio-) chemical interactions leading to biotechnological opportunities. Biotechnol Adv 30:1405–1424. https://doi.org/10.1016/j.biotechadv.2012.02.015

Vasilieva S, Lobakova E, Lukyanov A et al (2016) Immobilized microalgae in biotechnology Moscow University. Biol Sci Bull 71:170–176

Von Sperling M (2007) Waste stabilisation ponds. IWA publishing, London

Walsh BJ et al (2015) New feed sources key to ambitious climate targets. Carbon Balance Manag 10:26. https://doi.org/10.1186/s13021-015-0040-7

Wang X, Lin H, Gu Y (2012) Multiple roles of dihomo-gamma-linolenic acid against proliferation diseases. Lipids Health Dis 11:25

Yang C-f, Ding Z-y, Zhang K-c (2008) Growth of Chlorella pyrenoidosa in wastewater from cassava ethanol fermentation. World J Microbiol Biotechnol 24:2919–2925. https://doi.org/10.1007/s11274-008-9833-0

Zittelli G, Biondi N, Rodolfi L et al (2013) Photobioreactors for mass production of microalgae. In: Richmond A, Hu Q (eds) Handbook of microalgal culture: applied phycology and biotechnology2nd edn. Blackwell, Oxford, pp 225–266

Chapter 10
Endophytes: Emerging Tools for the Bioremediation of Pollutants

Carrie Siew Fang Sim, Si Hui Chen, and Adeline Su Yien Ting

Abstract Pollutants are toxic to living organisms and the environment. Removal of these pollutants using biological agents has been attempted, with many of these successfully performed by a variety of bacteria and fungi. In recent years, a group of microorganisms known as endophytes have been explored for their bioremediation potential. Endophytes are microorganisms that exist in the tissues of the host plant and have traditionally been studied for their plant growth-promoting properties, biocontrol activities, and production of bioactive compounds. Their bioremediation potential is new and has tremendous room for research and development. Endophytes are, therefore, interesting microorganisms in our effort to discover new tools for the bioremediation of pollutants. In this chapter, the nature of endophytes, their tolerance to pollutants, and their application and mechanisms in removing pollutants such as toxic metals and triphenylmethane dyes are discussed. Examples of known endophytic species are also highlighted, and the methods in bioprospecting for these endophytic isolates are also discussed.

Keywords Bioremediation · Endophytes · Toxic metals · Triphenylmethane dyes

1 Introduction

Our environment is increasingly contaminated with pollutants such as metals and dyes. Metals, such as aluminum (Al), copper (Cu), lead (Pb), zinc (Zn), cadmium (Cd), and chromium (Cr), are introduced into the environment through industrialization (mining, electroplating, and wielding) and urbanization (Wang and Chen 2009; Mishra and Bharagava 2016; Chowdhary et al. 2017). In addition, metal-based fertilizers and pesticides from agricultural usage also contribute to the release of metals into the environment. These metals are a cause of concern as they are toxic

C. S. F. Sim · S. H. Chen · A. S. Y. Ting (✉)
School of Science, Monash University Malaysia, Bandar Sunway, Selangor Darul Ehsan, Malaysia
e-mail: adeline.ting@monash.edu

© Springer Nature Singapore Pte Ltd. 2019
R. N. Bharagava, P. Chowdhary (eds.), *Emerging and Eco-Friendly Approaches for Waste Management*, https://doi.org/10.1007/978-981-10-8669-4_10

at low concentrations, with potential carcinogenic, teratogenic, and mutagenic risks (Lee et al. 2012; Bilal et al. 2013). The toxic metals leach, migrate, deposit, and typically accumulate in water bodies. This may result in biomagnification of the toxic metals through food chains, which are eventually integrated into living organisms (Abdel-Baki et al. 2011; Barakat 2011). Metal toxicity also affects biodiversity, evidenced by the endemic threat toward *Acacia holosericea* and *Eucalyptus crebra* plants in Australia (Lamb et al. 2012).

For synthetic dyes, it is estimated that there are over 10,000 different synthetic dyes available in the market, with 700,000–1,000,000 tons produced annually (Daneshvar et al. 2007). These synthetic dyes are classified based on their chemical constitutions and industrial applications. Dyes that are commonly used include azo, triphenylmethane, and anthraquinone dyes. The applications of these dyes are diverse, ranging from dyeing of textiles and paper to manufacturing of printing ink, plastic, cosmetic, and pharmaceuticals and to food production (Ghaedi et al. 2013). Among these, the textile industry accounts for two third of the dye market, making it the largest consumer of synthetic dyes (Nath and Ray 2015). The imperfect binding of dyes to fabrics during the dyeing processes consequently lead to 10–15% of dyes being released into the environment.

Untreated dye effluents have negative impact on the environment. Low concentrations (<1 ppm) of synthetic dyes in the effluent can impart color to water bodies, affecting the aesthetic merit, impede photosynthesis, and disrupt aquatic life (Fallah and Barani 2014). In addition, these effluents have high content of chemicals (e.g., heavy metal salts, surfactants, bleaches, grease, solvents, chlorinated compounds, acids, and alkalis) and other undesirable properties, such as high temperature, poor turbidity, and extreme pH (both acidic and alkaline), that makes the environment toxic to various living organisms (Shehzadi et al. 2014; Dadi et al. 2017).

There are generally two methods to remove pollutants, the conventional physicochemical and the biological method (bioremediation). Throughout the years, physicochemical methods such as ion exchange, reverse osmosis, and chemical precipitation have been widely used (Mohee and Mudhoo 2012; Cid et al. 2015). These conventional methods are effective, but long-term application revealed formation of toxic by-products (sludge), which are costly to remove/dispose and toxic to the environment (Volesky 2001). This prompts the search for a greener and more environmentally friendly alternative, bioremediation. Bioremediation is a biological-based method involving the use of plants and microorganisms such as bacteria, fungi, and algae to remove pollutants from the environment. Bacteria and fungi are the most useful as they are not only ubiquitous but are capable in withstanding different environmental conditions, hence a broader range of application (Radha et al. 2005; Das et al. 2008; Eman 2012; Kinoshita et al. 2013).

The use of endophytes for bioremediation is relatively new. Endophytes, particularly endophytic fungi, are fungi living in tissues of plants without causing profound symptoms to the plant. Endophytes are found in various host species, which include wild herbs to cultivated plants (vegetables, fruits) (Xiao et al. 2010; Deng et al. 2011). The presence of endophytes benefits the hosts as they protect the hosts

10 Endophytes: Emerging Tools for the Bioremediation of Pollutants

against pathogens and herbivores (Young et al. 2006). Endophytes also ensure and enhance the survivability of their host plants in unfavorable environments, especially in areas with heavily polluted soils. Endophytes residing in such harsh environments are hypothesized to have adapted and evolved to acquire increased tolerance toward the adverse conditions.

In this chapter, the potential of endophytes in removing dyes and metals is discussed. This is presented in comparison with existing technologies of removing the pollutants using other bioremediation techniques. The effectiveness of endophytes and non-endophytic strains in removing pollutants will further be compared. Lastly, the advantages of using endophytes as the emerging tool for bioremediation of pollutants will also be discussed.

2 Pollutants in the Environment

Two types of pollutants, namely, organic and inorganic pollutants, can be found in the environment. Synthetic dyes, pharmaceutical products, pesticides and phenol-based compounds are examples of organic pollutants, while toxic metals are inorganic pollutants (Iram et al. 2015; Yadav et al. 2017). Organic pollutants originate from the excessive use of organic chemicals in the production of food, cosmetics, fertilizers, and pharmaceuticals. These organic compounds have been used since the twentieth century (Chinalia et al. 2007; Hamid et al. 2011). Exposure to the organic pollutants at high concentrations can result in an array of damages to the heart, eyes, and neurons (Seth et al. 2000). Absorbing, inhaling, or ingesting phenol-based organic pollutants may cause poisoning and irritation (Bravo 1998). Inorganic pollutants such as toxic metals are also commonly found in the environment. Similar to organic contaminants, inorganic pollutants pose serious health impacts to living organisms. Toxic metals are comparatively more persistent due to their nondegradable characteristics. Among the various types of pollutants, this chapter mainly focuses on dyes and toxic metals.

2.1 Toxic Metals

Metals comprise of (i) toxic metals such as Cu, Pb, and Zn; (ii) valuable elements including platinum (Pt), silver (Ag), and gold (Au); and (iii) radionuclides such as uranium (U) and radium (Ra) (Wang and Chen 2006). Metals are primarily discharged from anthropogenic activities (Table 10.1), although natural occurrences, such as mineral weathering, soil erosion as well as volcanic activities, are known to release metals into the environment as well.

Metals can dissolve in water and appear to possess multiple oxidation states (Lim et al. 2014). Cu and Cr, for example, exist in oxidation states of +1, +2 (for Cu)

Table 10.1 Characteristics of common toxic metals and their permitted threshold concentrations in drinking water

Metals	Essentiality	Toxicity	Common sources	Adverse effects	Concentrations in drinking water (mg/L)	Threshold concentrations (mg/L)
Cu^{2+}	Essential	Average	Metal finishing, Plating	Accumulates and damages important organs such as the brain, liver, and gastro intestine	≤0.005 to>30	2.0
Zn^{2+}	Essential	Average	Metal finishing, Plating	Harms nervous system	<0.1	3.0
Pb^{2+}	Nonessential	High	Mining, Welding, Plumbing, Discharge of petrol	Heavily affects cardiovascular, nervous, immune, and reproductive systems	<0.005	0.01
Cd^{2+}	Nonessential	High	Steel industries	Carcinogenic and primarily damages kidneys	<0.001	0.003
			Batteries	Destroys red blood cells		

Data is tabulated as a summary from findings by Cheremisinoff (1995), Davis et al. (2000), Järup (2003), Thakur (2011), and World Health Organization (2011)

and +3 and +6 (for Cr). The different oxidation states pose different toxicities, such as in the case for hexavalent Cr which was revealed to be more toxic than the trivalent state (Lopez-Luna et al. 2009). The numerous oxidation states of metals are a result of interactions among the metals and various environmental factors. Their water-soluble characteristics easily allow the transport of the elements in water bodies. Upon discharged into the environment, they are transported and concentrated in waterlogged areas such as lakes, estuaries, and peat swamps (Cozma et al. 2010; Cvijovic et al. 2010). Some of these metals are not required by living organisms (Al^{3+}, Pb^{2+}, Cd^{2+}), but some are essential metals (Cu^{2+}, Zn^{2+}, Fe^{3+}, Mn^{2+}) required in trace amounts to support physiological and enzymatic activities in living organisms. In high concentrations, essential metals render toxic effect (Table 10.1) leading to debilitating health concerns. The persistence and nondegradable features of metals further harms the environment. It is imperative that metal concentrations should be at a balance, as expressed in the following equation (Lombi and Gerzabek 1998):

$$M_{total} = \left(M_p + M_a + M_f + M_{ag} + M_{ow} + M_{ip}\right) - \left(M_{cr} + M_l\right)$$

whereby:

M metal and the subscripts represent different sources, p parental matter, a atmosphere, f fertilizer, ag agrochemical, ow organic waste, ip inorganic pollutant, cr removal of crops, l leached losses

Therefore, it is crucial that metals in the environment are kept to a minimal tolerable range. The World Health Organization (WHO) (2011) has proposed a guideline on the permissible levels of common metals in drinking water (Table 10.1). This guideline serves as a reference to various water management practices.

2.2 Triphenylmethane Dyes

Triphenylmethane (TPM) dyes are water-soluble organic compounds, where the molecular structures are based on hydrocarbon (triphenylmethane) and tertiary alcohol (triphenylcarbinol), with the quinonoid group as the chromophore and amino/hydroxyl groups as auxochromes (Geethakrishnan et al. 2015). This class of dyes often comes in shades of violet, blue, green, and red, with examples such as Crystal Violet, Ethyl Violet, Methyl Violet, Bromophenol Blue, Brilliant Blue FCF, Coomassie Brilliant Blue, Methyl Blue, Patent Blue V, Brilliant Green, Malachite Green, Methyl Green, and Cresol Red (Lucova et al. 2013; Chen and Ting 2015a; Kus and Eroglu 2015; Zhuo et al. 2015; Kristanti et al. 2016; Mekhalif et al. 2016). The focus of this review is on four TPM dyes: Crystal Violet, Methyl Violet, Methyl Blue, and Malachite Green. These four dyes are extensively used in industrial processes such as dyeing of textiles (e.g., wool, silk, cotton, leather, nylon, and polyacrylonitrile), paper, plastic, printing inks, food, cosmetic, and drugs (Table 10.2). TPM dyes are favored for applications as they are relatively cheaper than other dyes, have intense colors, and have high tinctorial strength. Certain TPM dyes (i.e., Malachite Green) are also utilized as therapeutic agents in aquaculture and poultry, while others are used as pH indicators and biological staining agents (Table 10.2).

TPM dyes or their metabolites (e.g., Leucomalachite Green, Leucocrystal Violet) have adverse effects on living organisms. Acute or short-term exposure to certain TPM dyes, such as Crystal Violet, Methyl Violet, and Malachite Green, causes irritation and sensitization of the eyes, skin, respiratory, and gastrointestinal tract (Velpandian et al. 2007; Lucova et al. 2013; Mani and Bharagava 2016) (Table 10.2). Longer exposures to the dyes damage cells and organs due to their carcinogenic and mutagenic effects (Kus and Eroglu 2015). TPM dyes are also reportedly phytotoxic toward plants (e.g., *Sorghum bicolor*, *Triticum aestivum*, *Vigna radiata*, *Lemna minor*, and *Zea mays*) (Ayed et al. 2008; Przystas et al. 2012; Bera et al. 2016) and fishes in fish farms (Fallah and Barani 2014).

Table 10.2 Summary of common triphenylmethane dyes used for various industrial applications and their toxic effects on human

TPM dyes	Other known names	Applications	Toxic effects on human	References
Crystal violet	Basic violet 3, gentian violet, methyl violet 10b	Dyeing of cotton and silk, paint and printing ink manufacturing, biological stain, antimicrobial agent, and skin disinfectant	Irritation to the eyes, skin, and digestive tract; permanent injury to cornea and conjunctiva; may cause respiratory and kidney failures	Fu et al. (2013), and Mani and Bharagava (2016)
Methyl violet	Methyl violet 2b	Purple colorant for textiles, printing ink, leather, rubber, adhesives, petroleum products; pH indicator; cell viability assay	Irritation to the eyes, skin, respiratory and gastrointestinal systems	Lim et al. (2016)
Malachite green	Basic green 4, aniline green, diamond green B	Antifungal and antiparasitic agents; dyeing of cotton, wool, jute, silk, leather, acrylic fibers, ceramics, and paper	Eye irritant; potential hazard to reproductive system	Velpandian et al. (2007) and Koçer and Acemioğlu (2015)
Methyl blue	Acid blue 93, cotton blue	Staining agent for histological samples and fungi; dyeing of cotton, cotton-based fibers and leather	Irritation to the eyes, skin, respiratory, and gastrointestinal systems	Arunarani et al. (2013)

3 Bioremediation Approaches in Removing Pollutants

Bioremediation is the use of living organisms, such as microorganisms and plants, to remove pollutants (Radha et al. 2005; Bharagava et al. 2017). In the early days, bioremediation of pollutants (i.e., synthetic dyes, toxic metals, polycyclic aromatic hydrocarbons, pesticides, crude oil, petroleum-based products) was achieved using natural decomposers of the environment – bacteria, fungi, and yeasts (Du et al. 2013; Jasińska et al. 2015; Xin et al. 2017). Subsequent studies explored plants and their associated microorganisms (rhizosphere, endosphere) for removal of pollutants in a process called phytoremediation (Fu et al. 2013). In all bioremediation strategies, the process is often aided by favorable conditions (temperature, pH, and presence of nutrients and oxygen) (Przystas et al. 2012; Shi et al. 2015).

Bioremediation is typically achieved by two main mechanisms: biosorption and bioaccumulation (Sathish et al. 2012; Martins et al. 2017). The former involves both live and dead cells, whereas the latter only occurs for live cells. Biosorption is a form of sequestration in which pollutants bind onto surfaces, primarily the cell wall, which has functional groups such as carboxyl, hydroxyl, amine, and phosphoryl groups that act as binding sites (Kousha et al. 2013). The bioaccumulation process involves transport and accumulation of pollutants in the cells, where chemical transformation of pollutants into less harmful compounds can occur with possible complete mineralization of pollutants into carbon dioxide and water (Daneshvar et al. 2007). Unlike bioaccumulation, biosorption of pollutants can occur within a shorter operation time (often within hours rather than days), and the biosorbents can be regenerated for reuse, have longer shelf life, and are unaffected by the toxicity of pollutants and lack of nutrient supply (Nath and Ray 2015). Hence, removal of pollutants via biosorption is an attractive approach. Nevertheless, this process does not breakdown the pollutants. Instead, the pollutants are entrapped in the biomass matrix without fragmentation.

Various materials can be used as biosorbents for the removal of toxic metals and TPM dyes. They include inert materials (e.g., charcoal and agricultural wastes) and microbial biomass (algae, bacteria, yeast, and filamentous fungi) (Daneshvar et al. 2007; Bahafid et al. 2013; Erto et al. 2013; Broadhurst et al. 2015; Nath and Ray 2015; Sim et al. 2016; Martins et al. 2017). Each type of biosorbents has their own advantages and disadvantages that will be further discussed in the following sections.

3.1 Activated Charcoal

Activated charcoal, or activated carbon (AC), is an amphoteric, nonpolar solid commercially available in granular and powder forms (Erto et al. 2013). It has been heralded as one of the best adsorbent for pollutants due to its high sorption capacity, diverse functional groups, microporous nature, and large surface area (500–1500 m^2/g) (Gorzin and Ghoreyshi 2013; Kavand et al. 2014). Studies on commercial activated charcoal revealed high maximum adsorption capacities (q_m) of 151.52 and 200 mg/g for TPM dyes Crystal Violet and Basic Green 4 (Malachite Green), respectively (Nagda and Ghole 2008; Koçer and Acemioğlu 2015). Activated charcoal has also been documented to effectively remove toxic metals such as Cd^{2+}, Cr^{6+}, Cu^{2+}, Ni^{2+}, Pb^{2+}, and Zn^{2+} (Erto et al. 2013; Gorzin and Ghoreyshi 2013; Owamah 2013; Gomez et al. 2014; Kavand et al. 2014). Despite the excellent biosorption properties, the use of commercially activated charcoal for actual cleanup of polluted environments has not gain traction. This is due to the energy-consuming processes required to produce activated charcoal and their poor amenability for reuse (Gomez et al. 2014).

3.2 Agricultural Wastes

The agricultural sector generates large volumes of by-products from fruits, vegetables, and plants (Martins et al. 2017). These wastes are often discarded or utilized as animal feeds or organic fertilizers. In recent years, agricultural wastes have shown potential for removal of various pollutants (e.g., synthetic dyes, toxic metals, phenolic compounds, and radionuclides) (Tseng et al. 2011; Neupane et al. 2014; Uçar et al. 2014). This includes removal of TPM dyes and toxic metals via biosorption by a variety of agricultural wastes, namely, sugarcane bagasse (Martins et al. 2017), *Miscanthus sacchariflorus* (grass) biochar (Kim et al. 2013), walnut shell (Song et al. 2016), *Pinus brutia* (Calabrian pine) cones (Mekhalif et al. 2016), *Ananas comosus* (pineapple) leaves (Neupane et al. 2014), *Artocarpus odoratissimus* (tarap) leaves (Lim et al. 2016), *Psidium guajava* (guava) leaves (Rehman et al. 2015), spent tea leaves (Akar et al. 2013), spent ground coffee (Jutakridsada et al. 2015), orange peel (Santos et al. 2015), olive pomace (Koçer and Acemioğlu 2015), rice husk (Leng et al. 2015), potato stems and leaves (Gupta et al. 2016), and oak charcoal (Hamad et al. 2016). The adsorption capabilities of these wastes are attributed to their composition, which primarily includes cellulose, hemicellulose, and lignin (Leng et al. 2015). These lignocellulosic materials have large quantities of functional groups, particularly hydroxyl groups, that provide binding sites for pollutants (Martins et al. 2017). Agricultural wastes for remediation purposes can be utilized in their raw form, modified, or converted to activated carbon (Akar et al. 2013; Uçar et al. 2014). Modified forms (via esterification, etherification, oxidizing agents) have additional functional groups such as carboxylic groups to increase their biosorption capacity (Martins et al. 2017). However, the use of these wastes as adsorbents is limited by the production of toxic sludge and potential clogging problems of bioreactors. Consequently, adsorbents from agricultural wastes are often used in the event where low concentrations of pollutants are detected (Koçer and Acemioğlu 2015).

3.3 Algae

Algae comprise of a diverse group of ubiquitous, photosynthetic organisms that may be single-celled or as large as kelps (Cheng et al. 2016). Brown, green, and red algae have been documented to have bioremediation potential for toxic metals, particularly brown algae for metal biosorption (Carro et al. 2015; Cid et al. 2015). In contrast, studies on TPM dye removal by algae are limited (Kousha et al. 2013; Jegan et al. 2016). Species of algae with bioremediation potential include *Aphanocapsa roseana* de Bary (Nandi et al. 2016), *Chlorella vulgaris* (Cheng et al. 2016), *Cosmarium* sp. (Daneshvar et al. 2007), *Chlamydomonas* sp., *Euglena* sp., *Cystoseira compressa* (Benfares et al. 2015), *Durvillaea antarctica* (Cid et al. 2015), *Gracilaria edulis* (Jegan et al. 2016), *Kappaphycus alvarezii* (El Nemr et al. 2015), *Pterocladia capillacea* (El Nemr et al. 2015), *Scenedesmus quadricauda* (Kousha et al. 2013), and *Sargassum muticum* (Carro et al. 2015). Algae are abundant in nature and fast-growing; thus they are extremely cost-effective. They have large surface area-to-volume ratio, high tolerance, and uptake capacities for pollutants (Daneshvar et al. 2007; Cid et al. 2015). Functional groups (primarily carboxylic) for metal- or dye-binding are found in the alginate polymers of algae (Kousha et al. 2013). Nevertheless, more studies are required to better understand the biochemistry of algae in response to pollutants and environmental parameters.

3.4 Yeast

Yeasts are single-celled organisms with diverse industrial applications, varying from production of enzymes, proteins, and oils to use for bioremediation purposes (Singh et al. 2012; Zhuo et al. 2015). The attractiveness of yeast lies in the fact that they can grow rapidly and thrive under unfavorable environmental conditions (Bahafid et al. 2013). Yeast is also available in abundance, especially as waste from fermentation, brewing, and distillation processes (Singh et al. 2012; Xin et al. 2017). Species of this microorganism (e.g., *Candida tropicalis*, *Saccharomyces cerevisiae*, *Pichia acacia*, *Kluyveromyces lactis*, *Cyberlindnera fabianii*, *Wickerhamomyces anomalus*, *Cryptococcus* sp.) are known for having high binding capacities for toxic metals such as Cd^{2+}, Cr^{3+}, Cu^{2+}, Hg^{2+}, and Pb^{2+} (Bahafid et al. 2013; Kuang et al. 2015; Honfi et al. 2016; Liang et al. 2016; Xin et al. 2017). Nevertheless, TPM dye removal by yeast has not been fully explored. Notable research this far documented *Saccharomyces cerevisiae* and *Yarrowia lipolytica* as effective decolorizers of Malachite Green, Brilliant Green, Methyl Violet, Aniline Blue, and Crystal Violet (Safarikova et al. 2005; Singh et al. 2012; Ghaedi et al. 2013; Asfaram et al. 2016). The use of yeasts to remove pollutants is, however, limited by their poor reusability, leading to disposal problems (Xin et al. 2017).

3.5 Bacteria

Bacteria have excellent bioremediation potential. They are fast-growing, are adaptable to harsh conditions, and have demonstrated rapid removal of dyes and metals (Zhou et al. 2014; Limcharoensuk et al. 2015; Sun et al. 2017; Bharagava and Mishra 2017). Examples of bacteria known to remove metals and TPM dyes include species of *Bacillus*, *Klebsiella*, *Shewanella*, *Burkholderia*, *Pseudomonas*, and *Sphingomonas*. These bacteria are isolated from soil and water samples of polluted rivers, textile effluents, effluent disposal sites, activated sludge, and metal mines (Ayed et al. 2008; Arunarani et al. 2013; Wu et al. 2013; Zhou et al. 2014; Limcharoensuk et al. 2015; Nath and Ray 2015; Zablocka-Godlewska et al. 2015; Yang et al. 2016). Bacteria remove dyes through biosorption onto the cell wall and/or through intracellular uptake for further degradation (Du et al. 2011; Nath and Ray 2015). The biosorption process is aided by the presence of negatively charged functional groups such as carboxyl, amine, hydroxyl, and phosphonate on the cell wall, as well as electrostatic interactions (Arunarani et al. 2013; Bera et al. 2016). Bacterial degradation of dyes may involve enzymes such as laccase, lignin peroxidase, manganese peroxidase, and NADH-DCIP reductase (Chaturvedi et al. 2013; Du et al. 2013). Tolerance and removal of metals by bacteria involves additional mechanisms, including active removal from the cells, precipitation, as well as intracellular and extracellular accumulation of metals (Kiran et al. 2017).

3.6 Filamentous Fungi

Filamentous fungi are a diverse group of multicellular organisms characterized by the presence of microscopic fruiting bodies and hyphae showing extreme polar growth (Kück et al. 2009). These organisms are capable of converting a wide range of recalcitrant pollutants, such as toxic metals, synthetic dyes, polycyclic aromatic hydrocarbons, pesticides, crude oil, and products of petroleum, into less toxic compounds (Adekunle and Oluyode 2005; Iram et al. 2015). The bioremediation of toxic metals and TPM dyes is well studied for species of *Phanerochaete*, *Coriolopsis*, *Irpex*, *Trichoderma*, *Aspergillus*, *Penicillium*, and *Rhizopus* (Radha et al. 2005; Saratale et al. 2006; Chen and Ting 2015a, b; Chew and Ting 2015; Iram et al. 2015). The effectiveness of fungi for bioremediation processes is attributed to the secretion of non-specific, extracellular enzymes to catalyze the breakdown of pollutants (Radha et al. 2005; Chen and Ting 2015b). Fungi also have defense mechanisms to exclude, adsorb, and sequester metals (Rhee et al. 2016). Fungi also remove pollutants via biosorption, which is aided by the presence of various functional groups and large cell-to-surface ratio of the biomass (Chew and Ting 2015).

4 Bioprospecting Endophytes for Removal of Metals

Endophytes are a group of microorganisms found inhabiting plant tissues without causing any visible symptoms to the host plant. Every plant on Earth has the potential to harbor one or more endophytes, which suggests a large endophyte community as more than 300,000 plant species exist on Earth (Strobel and Daisy 2003; Huang et al. 2007). Endophytes remained ubiquitous despite the heterogeneity in the environment (soil, farming practices) (Seghers et al. 2004; Conrath et al. 2006; Hardoim et al. 2008; Singh et al. 2009). Endophytes form a mutualistic relationship with host plant as summarized in Fig. 10.1. They protect host plants against

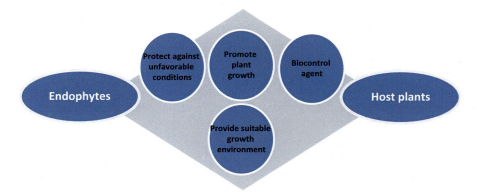

Fig. 10.1 The mutualistic relationship between endophytes and host plants

unfavorable biotic or abiotic stresses, promote plant growth, and produce an array of bioactive compounds for various applications (Young et al. 2006; Waqas et al. 2012). In return, endophytes derive sufficient nutrient sources from their host plant for growth and survival.

In recent years, endophytes were found to have a prominent role in removing pollutants (Gai et al. 2009; Ho et al. 2012; Kim et al. 2012). This bioremediation potential of endophytes was conceived from the use of plants in phytoremediation. Plants naturally harbor endophytes, and it has been hypothesized that endophytes in these plants may have natural tolerance and adaptation toward the pollutants (Rodriguez et al. 2008). This "habitat-adapted symbiosis" is suspected to render endophytes with adaptability and enhanced tolerance to the pollutants.

5 Endophytes for Metal Removal

Endophytes with metal removal potential have been discovered from a variety of plants. They include cultivated plants such as *Brassica napus* (rapeseed), *Nicotiana tabacum* (cultivated tobacco), *Solanum nigrum* (black nightshade) (Sheng et al. 2008; Mastretta et al. 2009; Guo et al. 2010; Luo et al. 2011), ornamentals (*Rosa longicuspis*, *Commelina communis*), shrubs (*Acacia decurrens*), and hyperaccumulators (*Phragmites* sp.) (Sheng et al. 2008; Mastretta et al. 2009; Guo et al. 2010; Luo et al. 2011; Sim et al. 2016). The endophyte communities from these plants are diverse and include both endophytic bacteria and fungi (*Pseudomonas fluorescens*, *Stenotrophomonas* sp., *Clostridium aminovalericum*, *Microbacterium* sp., *Flavobacterium* sp., *Bacillus* sp., *Acinetobacter* sp., *Trichoderma asperellum*, *Phomopsis* sp., *Saccharicola bicolor*, *Phoma* sp., *Aspergillus* sp., *Mucor* sp.) (Sheng et al. 2008; Mastretta et al. 2009; Guo et al. 2010; Deng et al. 2011; Li et al. 2011; Luo et al. 2011; Zhang et al. 2011; Sim et al. 2016). Studies revealed that endophytes from both hyperaccumulator and non-hyperaccumulator plants have similar potential in removing metals. Endophytic *T. asperellum* (from *Phragmites* sp.) and *Microsphaeropsis* sp. (from *S. nigrum*) removed Al^{3+}, Pb^{2+}, Zn^{2+}, and Cd^{2+} efficiently (Xiao et al. 2010; Sim et al. 2016). Govarthanan et al. (2016) demonstrated similar outcomes using five bacterial endophytes from the non-hyperaccumulator *Tridax procumbens* (coatbuttons daisy).

5.1 Mechanisms of Metal Removal by Endophytes

Endophytes remove metals by employing either the biosorption or the bioaccumulation mechanisms. Biosorption is a process that involves chemical and physical interaction between adsorbate (biosorbent) and pollutant (metal). Endophytes known for their biosorption activities include *Trichoderma asperellum*, *Microsphaeropsis* sp., and *Penicillium lilacinum* (Xiao et al. 2010; El-Gendy et al. 2011; Sim et al. 2016). They remove metals at varying capacity; *T. asperellum* successfully removed Zn^{2+} (18 mg/g), Cu^{2+} (17.26 mg/g), Pb^{2+} (19.24 mg/g), Cd^{2+} (19.78 mg/g), and Cr^{3+} (16.75 mg/g) (Sim et al. 2016); *Microsphaeropsis* sp. LSE10 adsorbed a maximum of 247.5 mg/g Cd^{2+} (Xiao et al. 2010); and *P. lilacinum* removed Cu^{2+} and Cd^{2+} at 85.4% and 31.43%, respectively (El-Gendy et al. 2011). Effective biosorption is achieved when metal cations bind successfully to the functional groups of the cell walls and cell membranes (Khalil et al. 2016). This can be detected via Fourier-transform infrared spectroscopy (FTIR), based on changes in peaks of functional groups.

Bioaccumulation is a removal process, which involves the movement of metal cations across cellular structures such as the cell membrane. Live cells typically perform bioaccumulation, and this energy-dependent process is achieved via precipitation, covalent bonding, redox reactions, and crystallization (Wilde and Benemann 1993; Malekzadeh et al. 2002). In bioaccumulation, the metal cations are transported across the cell membrane and are further compartmentalized or detoxified. Bioaccumulation in living cells is relatively slower than biosorption as the viable cells require a period of adaptation to the metal stress. Nevertheless, the amount of metal removed is higher than biosorption. Deng et al. (2014) characterized a metal-tolerant endophytic *Lasiodiplodia* sp. MXSF31 from hyperaccumulator *Portulaca olercea*, capable of bioaccumulating Pb^{2+} (5.6×10^5 mg kg^{-1}), Cd^{2+} (4.6×10^4 mg kg^{-1}), and Zn^{2+} (7.0×10^4 mg kg^{-1}) effectively. Endophytic *Mucor* sp. from *Brassica chinensis* also demonstrated higher Cd^{2+} removal by live (173 mg g^{-1}) than dead (108 mg g^{-1}) cells (Deng et al. 2011).

5.2 Biosourcing Endophytes for Metal Removal

Endophytic isolates are first isolated from plant tissues after a vigorous triple surface sterilization procedure with a series of ethanol, sodium hypochlorite, and sterile distilled water (Taylor et al. 1999). Sterilized tissues are injured and plated onto agar plate to allow the growth of endophytes. The sterile distilled water (used as final rinsing water) is also inoculated via spread-plate, where negative microbial growth indicates successful sterilization and subsequently confirming the isolates obtained were strictly endophytic (Ting 2014). Screening for tolerance to metals is performed via plating on agar supplemented with metals. Growth and tolerance of endophytes are measured based on tolerance index (TI) and/or minimum inhibitory concentration (MIC) (Zafar et al. 2007; Ezzouhri et al. 2009). Metal removal by endophytes is primarily attributed to the functional groups in the cell walls. For bacterial endophytes, the peptidoglycan of their cell wall are rich with *N*-acetylglucosamine (NAG) and *N*-acetylmuramic acid (NAM), which assist in metal binding (Das et al. 2008). Additionally, capsules or slime layer produced by some bacteria further supports metal binding as numerous functional groups such as sulfate and phosphate are present to chelate and bind metals (Moat et al. 2003; Deng

Table 10.3 A summary of metal-tolerant endophytes isolated from various host plants with their tolerable range toward Al^{3+}, Cu^{2+}, Pb^{2+}, Zn^{2+}, and Cd^{2+}

Endophyte	Host plant	Range of metal tolerance	References
Penicillium sp. FT2G59[a]	Dysphania ambrosioides	30–50 mmol/L Pb^{2+}; >680 mmol/L Zn^{2+}; 20–30 mmol/L Cd^{2+}	Li et al. (2016)
Phomopsis columnaris FT2G7[a]	Dysphania ambrosioides	30–50 mmol/L Cd^{2+}	Li et al. (2016)
Trichoderma asperellum T2[a]	Musa sp.	500 mg/L Al^{3+}; 300 mg/L Pb^{2+}; 100 mg/L Zn^{2+}	Ting and Jioe (2016)
Penicillium citrinum BTF08[a]	Musa sp.	50 mg/L Al^{3+}, Cu^{2+}, Pb^{2+}, Cd^{2+}, Zn^{2+}	Ting and Jioe (2016)
Diaporthe phaseolorum WAA02[a]	Portulaca sp.	50 mg/L Al^{3+}, Cu^{2+}, Pb^{2+}, Cd^{2+}, Zn^{2+}	Ting and Jioe (2016)
Exophiala pisciphila H93[a]	Arundinella bengalensis	2.3 mg/mL Pb^{2+}; 0.51 mg/mL Cd^{2+}; 3.1 mg/mL Zn^{2+}	Zhang et al. (2008)
Thysanorea papuana H125[a]	Artemisia carvifolia	1.2 mg/mL Pb^{2+}; 0.015 mg/mL Cd^{2+}; 0.1 mg/mL Zn^{2+}	Zhang et al. (2008)
Thysanorea papuana H114[a]	Artemisia carvifolia	1.0 mg/mL Pb^{2+}; 0.01 mg/mL Cd^{2+}; 0.08 mg/mL Zn^{2+}	Zhang et al. (2008)
Thysanorea papuana B3b[a]	Curculigo capitulata	0.9 mg/mL Pb^{2+}; 0.025 mg/mL Cd^{2+}; 0.1 mg/mL Zn^{2+}	Zhang et al. (2008)
Rahnella sp. JN6[b]	Polygonum pubescens	150 mg/L Cu^{2+}; 150 mg/L Ni^{2+}; 1250 mg/L Pb^{2+}; 1550 mg/L Cd^{2+}; 3000 mg/L Zn^{2+}	He et al. (2013)
Staphylococcus saprophyticus PJSI1[b]	Prosopis juliflora	2000 mg/L Cr^{3+}; <100 mg/L Cd^{2+}, Pb^{2+}, Zn^{2+}; 100 mg/L Cu^{2+}	Khan et al. (2015)
Pseudomonas aeruginosa PJRS20[b]	Prosopis juliflora	3000 mg/L Cr^{3+}; 100 mg/L Cd^{2+}, Cu^{2+}, Pb^{2+}; <100 mg/L Zn^{2+}	Khan et al. (2015)
Pantoea stewartii ASI11[b]	Prosopis juliflora	3000 mg/L Cr^{3+}; 100 mg/L Cd^{2+}, Cu^{2+}, Pb^{2+}, Zn^{2+}	Khan et al. (2015)
Microbacterium arborescens HU33[b]	Prosopis juliflora	3000 mg/L Cr^{3+}; 100 mg/L Cd^{2+}, Cu^{2+}, Pb^{2+}, Zn^{2+}	Khan et al. (2015)

[a]Endophytic fungi
[b]Endophytic bacteria

and Wang 2012). Similarly, metal complexation is also observed in fungal endophytes (Dursun et al. 2003). A summary of metal-tolerant endophytes from various host plants is shown in Table 10.3.

The metal removal efficacy by endophytes can be improved by providing the following conditions at optimal: pH, adsorbent dosages, initial metal concentrations, and temperature. Among these, pH is the most important factor as it affects precipitation, availability of functional groups, and influences competitiveness between metal cations and hydrogen groups (Bayramoglu et al. 2002). The optimum pH for metal removal is between pH 4 and 8 (Wang and Chen 2006; Congeevaram et al. 2007). Hydrogen and metal cations are highly competitive in lower pH due to the presence of more hydrogen ions in acidic environment (Kalyani et al. 2004). As a result, hydrogen ions bind to the functional groups, repelling metal ions leading to decrease in metal biosorption. Other factors such as initial metal concentration, adsorbent dosages, and temperature are also important but are dependent on the type of endophytes used and metal ions.

5.3 *Innovations to Endophytes for Enhanced Metal Removal*

Various pretreatment methods could be employed to improve and enhance metal removal. These methods, typically used for non-endophytic isolates, could be translated to modify endophytes in a similar manner. They include physical and chemical pretreatments, grafting to increase the endophytes' sorption capacity, immobilizing endophytic cells, the use of endophytic consortium, and trials in single- and multi-metal solutions.

Physical and chemical pretreatments are implemented to modify functional groups present on the cell walls to allow more ligands for metal binding (Vieira and Volesky 2000). Examples of physical pretreatments include exposure to heat such as boiling or autoclaving and to cold conditions, while examples of chemical pretreatments comprise of treatment with acids, alkalis, or detergents. Of these, pretreatment with acids (hydrochloric acid, nitric acid) or alkalis (sodium carbonate) is typically conducted and has been known to yield good metal removal efficacy. For example, Khalil et al. (2016) treated their fungal endophyte STRI: ICBG-Panama: TK1285 with 0.5 M sodium carbonate, which enhanced Cr^{3+} removal from 27% to 91.2% for samples without and with alkali pretreatment, respectively. Similarly, Yang et al. (2013) reported the effectiveness of pretreating endophytic *Fusarium* sp. with citric acid, which led to increased thorium (IV) removal from 11.35 to 75.47 mg/g for endophyte without and with citric acid pretreatment, respectively. Clearly, pretreatments enhanced the performance of endophytic biosorbents in metal bioremediation.

Another innovation is the introduction of grafting to raise the sorption ability of endophytes (Luo et al. 2014). Polymerization and layer-by-layer grafting by simple cross-linking formations allow surface modification, which adds functional groups (amine groups) and surface area for biosorption activities. This method has been successful for endophytic *Pseudomonas* sp. (from Cd-hyperaccumulator *Solanum nigrum* L.) to enhance removal of Pb^{2+}, Cd^{2+}, and Cu^{2+}. Grafting was performed to add poly(allylamine hydrochloride) (PAA) layers onto the bacterial surface using glutaraldehyde (cross-linking agent) (Luo et al. 2014). This led to a metal removal that was ten times higher than the non-modified bacteria. There is definite potential in grafting endophytes to develop excellent-performing biosorbents for metal bioremediation.

Immobilization is also another method widely adopted to enhance metal removal. Immobilization is achieved by encapsulating or entrapping cells within a natural matrix (alginate) or synthetic polymers (polyacrylamide, polyurethane) (Mehta and Gaur 2005). Endophytic isolate STRI: ICBG-Panama: TK1285 by Khalil et al. (2016) showed improved Cr^{3+} removal when alginate-immobilized forms (98.5%) were used compared to free-cells (27%). This enhanced removal was attributed to the role of alginate as endophyte-free alginate removed 98.6% of Cr^{3+}. Evidently, alginate has superior surface binding or functional groups, which allowed for more efficient biosorption of metals (Chew and Ting 2015). Application of immobilized forms benefited large-scale industries as well. There are lesser issues with blockage and separation of biosorbents from wastewaters, unlike application with free-cell forms. Other advantages of applying immobilized forms over free-cells include superior biocompatibility, inexpensive, high biomass loading with standardized preferred sizes, and reusability (Chew and Ting 2015).

Another innovative application is the use of microbial consortium rather than single inoculum. This is attributed to the fact that microbial species hardly exist as a single species to thrive in a multifaceted environment but as communities (Collins and Stotzky 1989). Therefore, microbial consortium is expected to have better metal removal rate. Improvements are also introduced to the testing approach. Instead of the typical single-metal solution systems, many researchers have incorporated the multi-metal solutions into their assessments to further validate the efficacy of the endophytes. In single-metal systems, the efficiency of endophytes in removing a particular metal is easily determined, but this interaction does not include their possible interaction with multiple metals in the environment. The multi-metal systems are therefore conceived to mimic natural wastewaters and are used to estimate their metal removal efficacies in the environment (Göksungur et al. 2005; Sim and Ting 2017a, b). Sim and Ting (2017a) evaluated single (using endophytic *Saccharicola bicolor* from *Phragmites* plant and an environmental bacterium, *Stenotrophomonas maltophilia*) and microbial consortium (mixed biosorbent-MB, consisting of *S. bicolor* and *S. maltophilia*) in both single and multi-metal systems. Results revealed the improved removal of Cu^{2+}, Pb^{2+}, Zn^{2+}, and Cd^{2+} by consortium (MB) than the single isolate *S. bicolor*. Similarly, Mastretta et al. (2009) found the positive effect of endophytic bacterial consortium (consisting of *Pseudomonas* sp., *Sanguibacter* sp., *Stenotrophomonas* sp., *Enterobacter* sp.), which enhanced Cd^{2+} accumulation by threefold.

6 Bioprospecting Endophytes for Removal of Triphenylmethane Dyes

6.1 Endophytes for TPM Dye Removal

Limited studies are available on the decolorization potential of endophytic fungi and bacteria for TPM dyes. These microorganisms were isolated from plants growing in unpolluted environments, with the majority of the isolates being endophytic fungi rather than bacteria (Gayathri et al. 2010; Przystas et al. 2012). These endophytes have decolorization efficiencies similar to or higher than non-endophytic strains (Deng et al. 2008; Jasińska et al. 2015). The removal of TPM dyes by endophytic bacteria has been reported in two studies (Gayathri et al. 2010; Pundir et al. 2014). Gayathri et al. (2010) discovered that 12 of the 36 endophytic bacteria from plants in mangrove and salt-marsh were able to decolorize Malachite Green (initial concentration at 0.01%) via plate assay. Similarly, Pundir et al. (2014) found 16 endophytic bacteria from several indigenous plants of India (Ambala, Haryana) with dye-degrading potential. The isolates were identified as species of *Escherichia*, *Proteus*, *Bacillus*, and *Salmonella* (Işik and Sponza 2003; Deng et al. 2008).

There are more reports on endophytic fungi with TPM dye removal potential. They include *Pleurotus ostreatus*, *Polyporus picipes*, *Gloeophyllum odoratum*, *Colletotrichum gloeosporioides*, and *Diaporthe* sp. (Przystas et al. 2012; Gangadevi and Muthumary 2014; Ting et al. 2016). Endophytic fungus *Diaporthe* sp. WAA02 was isolated from *Portulaca* weed and tested for removal of 50 mg/L Cotton Blue and 100 mg/L Crystal Violet, Methyl Violet, and Malachite Green (Ting et al. 2016). High decolorization efficiency (85.55–85.73%) was achieved for Cotton Blue, regardless of whether live or dead biomass was used. In contrast, decolorization of Malachite Green, Methyl Violet, and Crystal Violet was more effective in the presence of live biomass (78.81–87.80%) rather than dead biomass (18.82–48.32%). Przystas et al. (2012) made similar observations using endophytic fungi (*Pleurotus ostreatus*, *Polyporus picipes*, and *Gloeophyllum odoratum*) isolated from the woods in Poland. These endophytes removed Brilliant Green effectively (73.81–95.00%). *Eucalyptus globulus* and *E. citriodora* also harbored endophytes, which decolorized 50 mg/L Aniline Blue within 21 days (almost 100%) (Sathish et al. 2012).

6.2 Mechanisms of Dye Removal by Endophytes

Bioremediation of TPM dyes by microorganisms involves two mechanisms: biodegradation and/or biosorption (Przystas et al. 2012). Most of the dye removal studies on endophytic fungi and bacteria revealed that biodegradation is the main removal mechanism, especially in live cells (Gayathri et al. 2010; Przystas et al. 2012; Sathish et al. 2012; Pundir et al. 2014; Ting et al. 2016). This process involves non-specific ligninolytic enzymes that catalyze the breakdown of dyes, particularly their chromophoric centers, into simpler compounds that are less harmful (Przystas et al. 2012). This has been proven via toxicity assays of treated dye samples on zooplankton (e.g., *Daphnia magna*) (Przystas et al. 2012), bacteria (e.g., *Pseudomonas aeruginosa, Escherichia coli,* and *Staphylococcus aureus*) (Jasińska et al. 2015), and plants (e.g., *Lemna minor, Triticum aestivum,* and *Ervum lens* Linn) (Lopez-Luna et al. 2009; Przystas et al. 2012). The pathway of dye biodegradation by live cells generally begins with secretion of extracellular enzymes to catalyze degradation of dyes that were adsorbed onto the cell wall (Sathish et al. 2012; Ting et al. 2016). Dyes, accumulated within the cells, are further degraded with the aid of intracellular enzymes. For endophytes, dye degradation is primarily attributed to the enzyme laccase (Lac), although other ligninolytic enzymes such as lignin peroxidase (LiP), manganese peroxidase (MnP), reductases, and tyrosinase are also known to degrade TPM dyes (Urairuj et al. 2003; Chandra and Chowdhary 2015; Patil et al. 2016; Sing et al. 2017). Laccase (EC 1.10.3.2) is a multi-copper oxidase that is widely found in microorganisms and higher plants (Shi et al. 2015). In the presence of oxygen, laccase oxidizes dye compounds to generate free radicals for further dye degradation. Sathish et al. (2012) have also demonstrated that laccase-producing endophytic fungi were capable of decolorizing the TPM dye Aniline Blue and Azo dye Methylene Blue. Laccase was also identified as the primary enzyme produced by endophytic *Colletotrichum gloeosporioides* (isolated from medicinal plant *Piper betle*) in decolorizing Malachite Green (100%) and the Azo dye Bismarck Brown (95.31%) (Sidhu et al. 2014).

Biosorption is another removal mechanism demonstrated by endophytic fungi (Przystas et al. 2012; Ting et al. 2016). This process is dependent on the functional groups (e.g., carboxyl, hydroxyl, amino, and phosphate groups) that are present in the lipid, chitin, glucan, and chitosan components of the cell wall (Patil et al. 2016). The involvement of biosorption in dye removal can be deduced from the proportional decrease in major absorption peaks for ultraviolet-visible spectrum of dye supernatant as well as stained biomass after treatment (Ting et al. 2016). The biosorption of Malachite Green, Crystal Violet, and Methyl Violet by dead cells of endophytic *Diaporthe* sp. documented decolorization at 18.82%, 39.88%, and 48.32%, respectively (Ting et al. 2016). Similar results were obtained by Przystas et al. (2012) for the biosorption of Brilliant Green by dead endophytic *Pleurotus ostreatus, Polyporus picipes,* and *Gloeophyllum odoratum*.

6.3 Developing Endophytes As Bio-Agents for TPM Dye Removal

Endophytes with potential for bioremediation of TPM dyes are firstly screened for dye removal. This involved inoculating pure cultures of endophytic bacteria and fungi on solid agar (e.g., screening medium, minimal base agar, potato dextrose agar, malt extract agar) supplemented with the dye of interest (Gayathri et al. 2010;

Sathish et al. 2012; Sidhu et al. 2014; Shilpa and Shikha 2015). The presence of clear zone around the inoculum or change in color of the agar suggests dye degradation by the respective isolates. The next step is to investigate the capability of endophytes for removing TPM dyes in aqueous solutions (Sathish et al. 2012; Sidhu et al. 2014; Ting et al. 2016). This experiment is conducted over a period of time to determine the number of hours or days at which complete decolorization is achieved (Przystas et al. 2012; Ting et al. 2016). The use of dead cells allows for rapid decolorization activity (within 3 h), but the removal efficiency is often lesser than the amount achieved by live cells (Ting et al. 2016).

The mechanisms of dye removal by endophytic microorganisms can be investigated through several methods. Enzymes involved in dye degradation may be assayed qualitatively (on agar plate) and quantitatively (in liquid solution) (Sathish et al. 2012; Sidhu et al. 2014; Shi et al. 2015). The former demonstrates the presence or absence of a particular enzyme, whereas the latter measures the precise amount/activity of enzyme present in the sample. Further confirmation on the breakdown of parent dye compound into simpler metabolites (as conducted for other classes of dyes) may involve analyses such as Fourier-transform infrared spectroscopy (FTIR), high-performance liquid chromatography (HPLC), and gas chromatography-mass spectrometry (GC-MS) (Shilpa and Shikha 2015; Patil et al. 2016). FTIR spectra show changes in infrared absorption peaks (shifts, disappearance, and/or broadening); HPLC detects the absence and/or presence of peaks with different retention times; while GC-MS enables identification of degradation products based on mass fragmentation patterns and *m/z* values. Based on these results, the possible pathways involved in dye degradation by endophytes may be elucidated (Patil et al. 2016).

The removal efficiency of promising isolates are then further optimized in terms of initial dye concentrations and incubation conditions (agitation, aeration) (Przystas et al. 2012, 2015). Increasing the initial concentrations of dye mixture generally results in decreased decolorization efficiency. This phenomenon has also been reported for non-endophytic fungi. *Phanerochaete chrysosporium* decolorized 0.05 g/L Methyl Violet at approximately 90%, which decreased to around 30% when initial concentration was increased to 0.4 g/L (Radha et al. 2005). The decolorization of Crystal Violet, Methyl Violet, and Malachite Green by *Penicillium simplicissimum* and *Coriolopsis* sp. also showed similar observations (Chen and Ting 2015a, b). High initial dye concentrations may exert toxic effects on the metabolism of living cells and inhibit secretion of enzymes involved in degrading dye compounds (Chen and Ting 2015a). Decolorization of TPM dyes was also better when incubated with agitation as demonstrated by Przystas et al. (2012). Nevertheless, the rate of agitation differs according to fungal species. Higher decolorization efficiencies for Brilliant Green were observed for *Pleurotus ostreatus* (BWPH and MB strains), *Polyporus picipes* (RWP17), and *Gloeophyllum odoratum* (DCA) when agitated (73.81–95.00%) as compared to static incubation (56.74–79.30%). These observations were similar to TPM dye removal by non-endophytic fungi. Decolorization of Cresol Red by *Absidia spinosa* M15 after 30 days was almost complete (97.1%) when agitated at 120 rpm, whereas static conditions resulted in lower removal (44.5%) (Kristanti et al. 2016). Malachite Green was removed by *Aspergillus ochraceus* at a higher rate (96%) with 150 rpm agitation compared to without agitation (72%) (Saratale et al. 2006). This efficiency was attributed to enhanced contact between cells and dye compounds for oxidative metabolism.

6.4 Innovations to Endophytes for Enhanced TPM Dye Removal

Several innovations have been attempted to enhance dye bioremediation potential of endophytes. They include the use of genetic engineering, solid-state fermentation (SSF), and immobilization. With advancements in molecular biology techniques and bioinformatics analysis, genetic engineering has been utilized to produce recombinant dye-degrading enzymes and the over-expressing of these enzymes in hosts such as bacteria (Shi et al. 2015). This enables large-scale production of stable enzymes at low costs and high consistency in terms of quality. A laccase gene (Lac4) from endophytic bacterium *Pantoea ananatis* Sd-1 was successfully cloned and expressed by Shi et al. (2015) in *E. coli*. The recombinant laccase was stable in acidic pH (1–3), a wide range of temperatures (30–70 °C), and metal ions (Fe^{2+} and Cu^{2+}), with decolorization potential for TPM dye Aniline Blue (47%) as well as Azo dyes Remazol Brilliant Blue R (35%) and Congo Red (89%).

Other innovations such as solid-state fermentation allow endophytes to grow in an environment (solid media without free liquid) that mimics their natural conditions (Muthezhilan et al. 2014). This will induce secretion of enzymes in large quantities that can be purified for dye degradation applications. Endophytic fungus *Fusarium* sp. AEF17 (isolated from coastal sand dune plants of India) produced maximum amount of laccase (15.04 U/mL) when cultured on minimal salt media supplemented with wheat bran (Muthezhilan et al. 2014). The researchers subsequently immobilized or entrapped the purified laccase in 3% sodium alginate to facilitate decolorization of nine textile dyes, which include Black B, Orange M2R, and Blue M2R (66–89%) (Muthezhilan et al. 2014).

7 Conclusions and Future Prospects

The use of endophytes for metal and dye removal is fairly new but is gaining progress rapidly. Endophytes are clearly an excellent group of microorganisms, with potential to remove metals and TPM dyes. The use of endophytes is an environmentally friendly and sustainable approach in bioremediation. Endophytes could be sampled from a variety of hosts and has shown good tolerance and removal potential. Endophytes are also amenable to innovations and improvements such as optimizing the factors influencing bioremediation (pH, adsorbent dosages, initial metal concentrations, and temperature), pretreating, grafting, and immobilizing as well as utilizing endophytes in the form of a consortium with other microorganisms to improve their biosorption and biodegradation activities.

In the near future, improvements can be made to the endophytes to improve their metal and dye removal efficacy. Further studying and understanding of the mechanisms used by endophytes for bioremediation can enhance biosorption and bioaccumulation of metals and dyes. Future studies may also focus on the use of a combination of biological and physicochemical approaches as a potential solution for more effective cleanup of the environment.

Acknowledgments The authors are thankful for the research funding provided by the Malaysian Ministry of Higher Education under the FRGS grant scheme (FRGS/2/2013/STWN01/MUSM/02/2) and to the Monash University Malaysia for the funds and research facilities.

References

Abdel-Baki AS, Dkhil MA, Al-Quraishy S (2011) African bioaccumulation of some heavy metals in tilapia fish relevant to their concentration in water and sediment of Wadi Hanifah. J Biotechnol 10:2541–2547

Adekunle A, Oluyode T (2005) Biodegradation of crude petroleum and petroleum products by fungi isolated from two oil seeds (melon and soybean). J Environ Biol 26:37–42

Akar E, Altinişik A, Seki Y (2013) Using of activated carbon produced from spent tea leaves for the removal of malachite green from aqueous solution. Ecol Eng 52:19–27. https://doi.org/10.1016/j.ecoleng.2012.12.032

Arunarani A, Chandran P, Ranganathan BV et al (2013) Bioremoval of basic violet 3 and acid blue 93 by pseudomonas putida and its adsorption isotherms and kinetics. Colloids Surf B Biointerfaces 102:379–384. https://doi.org/10.1016/j.colsurfb.2012.08.049

Asfaram A, Ghaedi M, Ghezelbash GR et al (2016) Biosorption of malachite green by novel biosorbent Yarrowia lipolytica isf7: application of response surface methodology. J Mol Liq 214:249–258. https://doi.org/10.1016/j.molliq.2015.12.075

Ayed L, Chaieb K, Cheref A et al (2008) Biodegradation of triphenylmethane dye Malachite Green by Sphingomonas paucimobilis. World J Microbiol Biotechnol 25:705–711. https://doi.org/10.1007/s11274-008-9941-x

Bahafid W, Joutey NT, Sayel H et al (2013) Chromium adsorption by three yeast strains isolated from sediments in Morocco. Geomicrobiol J 30:422–429. https://doi.org/10.1080/01490451.2012.705228

Barakat MA (2011) New trends in removing heavy metals from industrial wastewater. Arab J Chem 4:361–337

Bayramoglu G, Denizli A, Bektas S et al (2002) Entrapment of *Lentinussajorcaju* into Ca-alginate gel beads for removal of Cd (II) ions from aqueous solution: preparation and biosorption kinetics analysis. Microchem J 72:63–76

Benfares R, Seridi H, Belkacem Y et al (2015) Heavy metal bioaccumulation in brown algae *Cystoseira compressa* in Algerian Coasts, Mediterranean Sea. Environ Process 2:429–439. https://doi.org/10.1007/s40710-015-0075-5

Bera S, Sharma VP, Dutta S et al (2016) Biological decolorization and detoxification of malachite green from aqueous solution by Dietzia maris NIT-D. J Taiwan Inst Chem Eng 67:271–284. https://doi.org/10.1016/j.jtice.2016.07.028

Bharagava RN, Mishra S (2017) Hexavalent chromium reduction potential of *Cellulosimicrobium sp.* isolated from common effluent treatment industries. Ecotoxicol Environ Saf 147:102–109

Bharagava RN, Chowdhary P, Saxena G (2017) Bioremediation an eco-sustainable green technology, its applications and limitations. In: Bharagava RN (ed) Environmental pollutants and their bioremediation approaches. CRC Press/Taylor & Francis Group, Boca Raton, pp 1–22

Bilal M, Shah JA, Ashfaq TQ et al (2013) Waste biomass adsorbents for copper removal from industrial wastewater- a review. J Hazard Mater 263:322–333

Bravo L (1998) Polyphenols: chemistry, dietary sources, metabolism, and nutritional significance. Nutr Rev 56:317–333

Broadhurst CL, Chaney RL, Davis AP et al (2015) Growth and cadmium phytoextraction by Swiss chard, maize, rice, Noccaea caerulescens, and Alyssum murale in pH adjusted biosolids amended soils. Int J Phytoremediation 17:25–39. https://doi.org/10.1080/15226514.2013.828 015

Carro L, Barriada JL, Herrero R, Sastre de Vicente ME (2015) Interaction of heavy metals with Ca-pretreated Sargassum muticum algal biomass: characterization as a cation exchange process. Chem Eng J 264:181–187. https://doi.org/10.1016/j.cej.2014.11.079

Chandra R, Chowdhary P (2015) Properties of bacterial laccases and their application in bioremediation of industrial wastes. Environ Sci Process Imp 17:326–342

Chaturvedi V, Bhange K, Bhatt R et al (2013) Biodetoxification of high amounts of malachite green by a multifunctional strain of Pseudomonas mendocina and its ability to metabolize dye adsorbed chicken feathers. J Environ Chem Eng 1:1205–1213. https://doi.org/10.1016/j.jece.2013.09.009

Chen SH, Ting ASY (2015a) Biodecolorization and biodegradation potential of recalcitrant triphenylmethane dyes by Coriolopsis sp. isolated from compost. J Environ Manag 150:274–280. https://doi.org/10.1016/j.jenvman.2014.09.014

Chen SH, Ting ASY (2015b) Biosorption and biodegradation potential of triphenylmethane dyes by newly discovered Penicillium simplicissimum isolated from indoor wastewater sample. Int Biodeterior Biodegrad 103:1–7. https://doi.org/10.1016/j.ibiod.2015.04.004

Cheng J, Qiu H, Chang Z et al (2016) The effect of cadmium on the growth and antioxidant response for freshwater algae Chlorella vulgaris. Springerplus 5:1290. https://doi.org/10.1186/s40064-016-2963-1

Cheremisinoff PN (1995) Handbook of water and wastewater treatment technology. Marcel Dekker, New York

Chew SY, Ting ASY (2015) Common filamentous *Trichoderma asperellum* for effective removal of triphenylmethane dyes. Desalin Water Treat 57:13534–13539. https://doi.org/10.1080/194 43994.2015.1060173

Chinalia FA, Reghali-Seleghin MH, Correa EM (2007) 2,4-D toxicity, cause, effect and control. Terr Aquat Environ Toxicol 1:24–33

Chowdhary P, Yadav A, Kaithwas G, Bharagava RN (2017) Distillery wastewater: a major source of environmental pollution and its biological treatment for environmental safety. In: Singh R, Kumar S (eds) Green technologies and environmental sustainability. Springer, Cham, pp 409–435

Cid H, Ortiz C, Pizarro J et al (2015) Characterization of copper (II) biosorption by brown algae Durvillaea antarctica dead biomass. Adsorption 21:645–658. https://doi.org/10.1007/s10450-015-9715-3

Collins JS, Stotzky G (1989) Factors affecting the toxicity of heavy metals to microbes. In: Beveridge TJ, Doyle RJ (eds) Metal ions and bacteria. Wiley, Toronto, p 31

Congeevaram S, Dhanarani S, Park J et al (2007) Biosorption of chromium and nickel by heavy metal resistant fungal and bacterial isolates. J Hazard Mater 146:270–277

Conrath U, Beckers GJM, Flors V et al (2006) Priming: getting ready for battle. Mol Plant-Microbe Interact 19:1062–1071

Cozma D, Tanase C, Tunsu C et al (2010) Statistical study of heavy metal distribution in the specific mushrooms from the sterile dumps Calimani area. Environ Eng Manag J 9:659–665

Cvijovic M, Djurdjevic P, Cvetkovic S et al (2010) A case study of industrial water polluted with chromium (VI) and its impact to river recipient in western Serbia. Environ Eng Manag J 9:45–49

Dadi D, Stellmacher T, Senbeta F et al (2017) Environmental and health impacts of effluents from textile industries in Ethiopia: the case of Gelan and Dukem, Oromia Regional State. Environ Monit Assess 189:1. https://doi.org/10.1007/s10661-016-5694-4

Daneshvar N, Khataee AR, Rasoulifard MH et al (2007) Biodegradation of dye solution containing Malachite Green: optimization of effective parameters using Taguchi method. J Hazard Mater 143:214–219. https://doi.org/10.1016/j.jhazmat.2006.09.016

Das N, Vimala R, Karthika P (2008) Biosorption of heavy metals- a review. Indian J Biotechnol 7:159–169

Davis TA, Volesky B, Mucci A (2000) A review of the biochemistry of heavy metal biosorption by brown algae. Water Resour 37:4311–4330

Deng X, Wang P (2012) Isolation of marine bacteria highly resistant to mercury and their bioaccumulation process. Bioresour Technol 121:342–347

Deng D, Guo J, Zeng G et al (2008) Decolorization of anthraquinone, triphenylmethane and azo dyes by a new isolated Bacillus cereus strain DC11. Int Biodeterior Biodegrad 62:263–269. https://doi.org/10.1016/j.ibiod.2008.01.017

Deng Z, Cao L, Huang H et al (2011) Characterization of Cd- and Pb-resistant fungal endophyte Mucor sp. CBRF59 isolated from rapes (Brassica chinensis) in a metal-contaminated soil. J Hazard Mater 185:717–724

Deng Z, Zhang R, Shi Y et al (2014) Characterization of Cd-, Pb-, Zn-resistant endophytic Lasiodiplodia sp. MXSF31 from metal accumulating Portulaca oleracea and its potential in promoting the growth of rape in metal-contaminated soils. Environ Sci Pollut Res 31:2346–2357

Du LN, Wang S, Li G et al (2011) Biodegradation of malachite green by Pseudomonas sp. strain DY1 under aerobic condition: characteristics, degradation products, enzyme analysis and phytotoxicity. Ecotoxicology 20:438–446. https://doi.org/10.1007/s10646-011-0595-3

Du L-N, Zhao M, Li G et al (2013) Biodegradation of malachite green by Micrococcus sp. strain BD15: biodegradation pathway and enzyme analysis. Int Biodeterior Biodegrad 78:108–116. https://doi.org/10.1016/j.ibiod.2012.12.011

Dursun AY, Uslu G, Tepe O et al (2003) A comparative investigation on the bioaccumulation of heavy metal ions by growing Rhizopus arrhizus and Aspergillus niger. Biochem Eng J 15:87–92

El Nemr A, El-Sikaily A, Khaled A et al (2015) Removal of toxic chromium from aqueous solution, wastewater and saline water by marine red alga Pterocladia capillacea and its activated carbon. Arab J Chem 8:105–117. https://doi.org/10.1016/j.arabjc.2011.01.016

El-Gendy MMA, Hassanein NM, El-Hay IHA et al (2011) Evaluation of some fungal endophytes of plant potentiality as low-cost adsorbents for heavy metals uptake from aqueous solution. Aust J Basic Appl Sci 5:466–473

Eman ZG (2012) Production and characteristics of a heavy metals removing bioflocculant produced by Pseudomonas aeruginosa. Pol J Microbiol 61:281–289

Erto A, Giraldo L, Lancia A et al (2013) A comparison between a low-cost sorbent and an activated carbon for the adsorption of heavy metals from water. Water Air Soil Pollut 224:1531. https://doi.org/10.1007/s11270-013-1531-3

Ezzouhri L, Castro E, Moya M et al (2009) Heavy metal tolerance of filamentous fungi isolated from polluted sites in Tangier, Morocco. Afr J Microbiol Res 3:35–48

Fallah AA, Barani A (2014) Determination of malachite green residues in farmed rainbow trout in Iran. Food Control 40:100–105. https://doi.org/10.1016/j.foodcont.2013.11.045

Fu X-Y, Zhao W, Xiong A-S et al (2013) Phytoremediation of triphenylmethane dyes by over-expressing a Citrobacter sp. triphenylmethane reductase in transgenic Arabidopsis. Appl Microbiol Biotechnol 97:1799–1806. https://doi.org/10.1007/s00253-012-4106-0

Gai CS, Lacava PT, Quecine MC et al (2009) Transmission of Methylobacterium mesophilicum by Bucephalogonia xanthophis for paratransgenic control strategy of citrus variegated chlorosis. J Microbiol 47:448–454

Gangadevi V, Muthumary J (2014) Isolation of Colletotrichum gloeosporioides, a novel endophytic taxol-producing fungus from the leaves of a medicinal plant, Justicia gendarussa. IJSER 5:1087–1094

Gayathri S, Saravanan D, Radhakrishnan M et al (2010) Bioprospecting potential of fast growing endophytic bacteria from leaves of mangrove and salt-marsh plant species. Indian J Biotechnol 9:397–402

Geethakrishnan T, Sakthivel P, Palanisamy PK (2015) Triphenylmethane dye-doped gelatin films for low-power optical phase-conjugation. Opt Commun 335:218–223. https://doi.org/10.1016/j.optcom.2014.09.033

Ghaedi M, Hajati S, Barazesh B et al (2013) Saccharomyces cerevisiae for the biosorption of basic dyes from binary component systems and the high order derivative spectrophotometric method for simultaneous analysis of brilliant green and methylene blue. J Ind Eng Chem 19:227–233. https://doi.org/10.1016/j.jiec.2012.08.006

Göksungur Y, Dagbagli S, Ucan A et al (2005) Optimization of pullulan production from synthetic medium by Aureobasidium pullulans in a stirred tank reactor by response surface methodology. J Chem Technol Biotechnol 80:819–827

Gomez LM, Colpas-Castillo F, Fernandez-Maestre R (2014) Cation exchange for mercury and cadmium of xanthated, sulfonated, activated and non-treated subbituminous coal, commercial activated carbon and commercial synthetic resin: effect of pre-oxidation on xanthation of subbituminous coal. Int J Coal Sci Technol 1:235–240. https://doi.org/10.1007/s40789-014-0033-2

Gorzin F, Ghoreyshi AA (2013) Synthesis of a new low-cost activated carbon from activated sludge for the removal of Cr (VI) from aqueous solution: equilibrium, kinetics, thermodynamics and desorption studies. Korean J Chem Eng 30:1594–1602. https://doi.org/10.1007/s11814-013-0079-7

Govarthanan M, Mythili R, Selvankumar T et al (2016) Bioremediation of heavy metals using an endophytic bacterium Paenibacillus sp. RM isolated from the roots of Tridax procumbens. 3 Biotech 6:242. https://doi.org/10.1007/s13205-016-0560-1

Guo H, Luo S, Chen L et al (2010) Bioremediation of heavy metals by growing hyper accumulator endophytic bacterium Bacillus sp. L14. Bioresour Technol 101:8599–8605

Gupta N, Kushwaha AK, Chattopadhyaya MC (2016) Application of potato (Solanum tuberosum) plant wastes for the removal of methylene blue and malachite green dye from aqueous solution. Arab J Chem 9:S707–S716. https://doi.org/10.1016/j.arabjc.2011.07.021

Hamad H, Ezzeddine Z, Kanaan S et al (2016) A novel modification and selective route for the adsorption of Pb^{2+} by oak charcoal functionalized with glutaraldehyde. Adv Powder Technol 27:631–637. https://doi.org/10.1016/j.apt.2016.02.019

Hamid AA, Aiyelaagbe OO, Balogun GA (2011) Herbicides and its application. Adv Nat Appl Sci 5:201–213

Hardoim PR, van Overbeek LS, van Elsas JD (2008) Properties of bacterial endophytes and their proposed role in plant growth. Trends Microbiol 16:463–471

He H, Ye Z, Yang D et al (2013) Characterization of endophytic Rahnella sp. JN6 from Polygonum pubescens and potential in promoting growth and Cd, Pb, Zn uptake by Brassica napus. Chemosphere 90:1960–1965

Ho YN, Mathew DC, Hsiao SC et al (2012) Selection and application of endophytic bacterium Achromobacter xylosoxidans strain F3B for improving phytoremediation of phenolic pollutants. J Hazard Mater 15:43–49

Honfi K, Tálos K, Kőnig-Péter A et al (2016) Copper(II) and phenol adsorption by cell surface treated Candida tropicalis cells in aqueous suspension. Water Air Soil Pollut 227:61. https://doi.org/10.1007/s11270-016-2751-0

Huang WY, Cai YZ, Xing J et al (2007) A potential antioxidant resource: endophytic fungi isolated from traditional Chinese medicinal plants. Econ Bot 61:14–30

Iram S, Shabbir R, Zafar H et al (2015) Biosorption and bioaccumulation of copper and lead by heavy metal-resistant fungal isolates. Arab J Sci Eng 40:1867–1873. https://doi.org/10.1007/s13369-015-1702-1

Işik M, Sponza DT (2003) Effect of oxygen on decolorization of azo dyes by Escherichia coli and Pseudomonas sp. and fate of aromatic amines. Process Biochem 38:1183–1192. https://doi.org/10.1016/S0032-9592(02)00282-0

Järup L (2003) Hazards of heavy metal contamination. Br Med Bull 68:167–182

Jasińska A, Paraszkiewicz K, Sip A et al (2015) Malachite green decolorization by the filamentous fungus Myrothecium roridum – mechanistic study and process optimization. Bioresour Technol 194:43–48. https://doi.org/10.1016/j.biortech.2015.07.008

Jegan J, Vijayaraghavan J, Bhagavathi PT et al (2016) Application of seaweeds for the removal of cationic dye from aqueous solution. Desalin Water Treat 57:25812–25821. https://doi.org/10.1080/19443994.2016.1151835

Jutakridsada P, Prajaksud C, Kuboonya-Aruk L et al (2015) Adsorption characteristics of activated carbon prepared from spent ground coffee. Clean Techn Environ Policy 18:639–645. https://doi.org/10.1007/s10098-015-1083-x

Kalyani S, Rao PS, Krishnaiah A (2004) Removal of nickel (II) from aqueous solutions using marine macroalgae as the sorbing biomass. Chemosphere 57:1225–1229

Kavand M, Kaghazchi T, Soleimani M (2014) Optimization of parameters for competitive adsorption of heavy metal ions (Pb^{+2}, Ni^{+2}, Cd^{+2}) onto activated carbon. Korean J Chem Eng 31:692–700. https://doi.org/10.1007/s11814-013-0280-8

Khalil MMH, Abou-Shanab RAI, Salem ANM et al (2016) Biosorption of trivalent chromium using Ca-alginate immobilized and alkali-treated biomass. J Chem Sci Technol. https://doi.org/10.5963/JCST0501001

Khan MU, Sessitsch A, Harris M et al (2015) Cr-resistant rhizo- and endophytic bacteria associated with *Prosopis juliflora* and their potential as phytoremediation enhancing agents in metal-degraded soils. Front Plant Sci. https://doi.org/10.3389/fpls.2014.00755

Kim TU, Cho SH, Han JH et al (2012) Diversity and physiological properties of root endophytic Actinobacteria in native herbaceous plants of Korea. J Microbiol 50:50–57

Kim WK, Shim T, Kim YS et al (2013) Characterization of cadmium removal from aqueous solution by biochar produced from a giant Miscanthus at different pyrolytic temperatures. Bioresour Technol 138:266–270. https://doi.org/10.1016/j.biortech.2013.03.186

Kinoshita KF, Bolleman J, Campbell MP et al (2013) Introducing glycomics data into the semantic web. J Biomed Semant 3:39–43

Kiran MG, Pakshirajan K, Das G (2017) Heavy metal removal from multicomponent system by sulfate reducing bacteria: mechanism and cell surface characterization. J Hazard Mater 324:62–70. https://doi.org/10.1016/j.jhazmat.2015.12.042

Koçer O, Acemioğlu B (2015) Adsorption of basic green 4 from aqueous solution by olive pomace and commercial activated carbon: process design, isotherm, kinetic and thermodynamic studies. Desalin Water Treat 57:16653–16669. https://doi.org/10.1080/19443994.2015.1080194

Kousha M, Farhadian O, Dorafshan S et al (2013) Optimization of malachite green biosorption by green microalgae-Scenedesmus quadricauda and Chlorella vulgaris: application of response surface methodology. J Taiwan Inst Chem Eng 44:291–294. https://doi.org/10.1016/j.jtice.2012.10.009

Kristanti RA, Fikri Ahmad Zubir MM, Hadibarata T (2016) Biotransformation studies of cresol red by Absidia spinosa M15. J Environ Manag 172:107–111. https://doi.org/10.1016/j.jenvman.2015.11.017

Kuang X, Fang Z, Wang S et al (2015) Effects of cadmium on intracellular cation homoeostasis in the yeast Saccharomyces cerevisiae. Toxicol Environ Chem 97:922–930. https://doi.org/10.1080/02772248.2015.1074689

Kück U, Pöggeler S, Nowrousian M et al (2009) Sordaria macrospora, a model system for fungal development. In: Anke T, Weber D (eds) Physiology and genetics: selected basic and applied aspects. Springer, Berlin/Heidelberg, pp 17–39. https://doi.org/10.1007/978-3-642-00286-1_2

Kus E, Eroglu H (2015) Genotoxic and cytotoxic effects of sunset yellow and brilliant blue, colorant food additives, on human blood lymphocytes. Pak J Pharm Sci 28:227–230

Lamb DT, Naidu R, Ming H et al (2012) Copper phytotoxicity in native and agronomical plant species. Ecotoxicol Environ Saf 85:23–239

Lee JC, Son YO, Pratheeshkumar P et al (2012) Oxidative stress and metal carcinogenesis. Free Radic Bil Med 53:742–757

Leng L, Yuan X, Zeng G et al (2015) Surface characterization of rice husk bio-char produced by liquefaction and application for cationic dye (Malachite green) adsorption. Fuel 155:77–85. https://doi.org/10.1016/j.fuel.2015.04.019

Li HY, Li DW, He CM et al (2011) Diversity and heavy metal tolerance of endophytic fungi from six dominant plant species in a Pb-Zn mine wasteland in China. Fungal Ecol. https://doi.org/10.1016/j.funeco.2011.06.002

Li X, Li W, Chu L et al (2016) Diversity and heavy metal tolerance of endophytic fungi from *Dysphania ambrosioides*: a hyperaccumulator from Pb-Zn contaminated soils. J Plant Interact 11:186–192

Liang X, Csetenyi L, Gadd GM (2016) Lead bioprecipitation by yeasts utilizing organic phosphorus substrates. Geomicrobiol J 33:294–307. https://doi.org/10.1080/01490451.2015.1051639

Lim KT, Shukor MY, Wasoh H (2014) Physical, chemical, and biological methods for the removal of arsenic compounds. Biomed Res Int. https://doi.org/10.1155/2014/503784

Lim LBL, Priyantha N, Mohamad Zaidi NAH (2016) A superb modified new adsorbent, Artocarpus odoratissimus leaves, for removal of cationic methyl violet 2B dye. Environ Earth Sci 75. https://doi.org/10.1007/s12665-016-5969-7

Limcharoensuk T, Sooksawat N, Sumarnrote A et al (2015) Bioaccumulation and biosorption of Cd(2+) and Zn(2+) by bacteria isolated from a zinc mine in Thailand. Ecotoxicol Environ Saf 122:322–330. https://doi.org/10.1016/j.ecoenv.2015.08.013

Lombi E, Gerzabek MH (1998) Determination of mobile heavy metal fraction in soil: results of a pot experiment with sewage sludge. Commun Soil Sci Plant Anal 29:2545–2556

Lopez-Luna J, Gonzalez-Chavez MC, Esparza-Garcia FJ et al (2009) Toxicity assessment of soil amended with tannery sludge, trivalent chromium and hexavalent chromium, using wheat, oat and sorghum plants. J Hazard Mater 163:829–834. https://doi.org/10.1016/j.jhazmat.2008.07.034

Lucova M, Hojerova J, Pazourekova S et al (2013) Absorption of triphenylmethane dyes brilliant blue and patent blue through intact skin, shaven skin and lingual mucosa from daily life products. Food Chem Toxicol 52:19–27. https://doi.org/10.1016/j.fct.2012.10.027

Luo S, Chen L, Chen J et al (2011) Analysis and characterization of cultivable heavy metal-resistant bacterial endophytes isolated from Cd hyper accumulator Solanumnigrum L. and their potential use for phytoremediation. Chemosphere 85:1130–1138

Luo S, Li X, Chen L et al (2014) Layer-by-layer strategy for adsorption capacity fattening of endophytic bacterial biomass for highly effective removal of heavy metals. Chem Eng J 239:312–321

Malekzadeh F, Latifi AM, Shahamat M et al (2002) Effects of selected physical and chemical parameters on uranium uptake by the bacterium Chryseomonas MGF-48. World J Microbiol Biotechnol 18:599–602

Mani S, Bharagava RN (2016) Exposure to crystal violet, its toxic, genotoxic and carcinogenic effects on environment and its degradation and detoxification for environmental safety. In: de Voogt WP (ed) Reviews of environmental contamination and toxicology, vol 237. Springer, Cham, pp 71–104. https://doi.org/10.1007/978-3-319-23573-8_4

Martins LR, Rodrigues JAV, Adarme OFH et al (2017) Optimization of cellulose and sugarcane bagasse oxidation: application for adsorptive removal of crystal violet and auramine-O from aqueous solution. J Colloid Interface Sci 494:223–241. https://doi.org/10.1016/j.jcis.2017.01.085

Mastretta C, Taghavi S, Van der Lelie D et al (2009) Endophytic bacteria from seeds of *Nicotiana tabacum* can reduce cadmium phytotoxicity. Int J Phytoremediation 11:251–267

Mehta SK, Gaur JP (2005) Use of algae for removing heavy metal ions from wastewater: progress and prospects. Crit Rev Biotechnol 25:113–152

Mekhalif T, Guediri K, Reffas A et al (2016) Effect of acid and alkali treatments of a forest waste, Pinus brutia cones, on adsorption efficiency of methyl green. J Dispers Sci Technol 38:463–471. https://doi.org/10.1080/01932691.2016.1178585

Mishra S, Bharagava RN (2016) Toxic and genotoxic effects of hexavalent chromium in environment and its bioremediation strategies. J Environ Sci Health Part C 34(1):1–34

Moat AG, Foster JW, Spector MP (2003) Microbial physiology. Wiley, Canada

Mohee R, Mudhoo A (eds) (2012) Bioremediation and sustainability: research and applications. Wiley, Canada

Muthezhilan R, Vinoth S, Gopi K et al (2014) Dye degrading potential of immobilized laccase from endophytic fungi of coastal sand dune plants. Int J ChemTech Res 6:4154–4160

Nagda G, Ghole V (2008) Utilization of lignocellulosic waste from bidi industry for removal of dye from aqueous solution. Int J Environ Res Public Health 2:385–390

Nandi R, Laskar S, Saha B (2016) Surfactant-promoted enhancement in bioremediation of hexavalent chromium to trivalent chromium by naturally occurring wall algae. Res Chem Intermed. https://doi.org/10.1007/s11164-016-2719-0

Nath J, Ray L (2015) Biosorption of Malachite green from aqueous solution by dry cells of *Bacillus cereus* M116 (MTCC 5521). J Environ Chem Eng 3:386–394. https://doi.org/10.1016/j.jece.2014.12.022

Neupane S, Ramesh ST, Gandhimathi R et al (2014) Pineapple leaf (Ananas comosus) powder as a biosorbent for the removal of crystal violet from aqueous solution. Desalin Water Treat 54:2041–2054. https://doi.org/10.1080/19443994.2014.903867

Owamah HI (2013) Biosorptive removal of Pb(II) and Cu(II) from wastewater using activated carbon from cassava peels. J Mater Cycles Waste 16:347–358. https://doi.org/10.1007/s10163-013-0192-z

Patil SM, Chandanshive VV, Rane NR et al (2016) Bioreactor with Ipomoea hederifolia adventitious roots and its endophyte Cladosporium cladosporioides for textile dye degradation. Environ Res 146:340–349. https://doi.org/10.1016/j.envres.2016.01.019

Przystas W, Zablocka-Godlewska E, Grabinska-Sota E (2012) Biological removal of azo and triphenylmethane dyes and toxicity of process by-products. Water Air Soil Pollut 223:1581–1592. https://doi.org/10.1007/s11270-011-0966-7

Przystas W, Zablocka-Godlewska E, Grabinska-Sota E (2015) Efficacy of fungal decolorization of a mixture of dyes belonging to different classes. Braz J Microbiol 46:415–424. https://doi.org/10.1590/S1517-838246246220140167

Pundir R, Rana S, Kaur A et al (2014) Bioprospecting potential of endophytic bacteria isolated from indigenous plants of Ambala (Haryana, India). Int J Pharma Sci Res 5. https://doi.org/10.13040/ijpsr.0975-8232.5(6).2309-19

Radha KV, Regupathi I, Arunagiri A et al (2005) Decolorization studies of synthetic dyes using Phanerochaete chrysosporium and their kinetics. Process Biochem 40:3337–3345. https://doi.org/10.1016/j.procbio.2005.03.033

Rehman R, Mahmud T, Irum M (2015) Brilliant green dye elimination from water using Psidium guajava leaves and Solanum tuberosum peels as adsorbents in environmentally benign way. J Chem 2015:1–8. https://doi.org/10.1155/2015/126036

Rhee YJ, Hillier S, Gadd GM (2016) A new lead hydroxycarbonate produced during transformation of lead metal by the soil fungus Paecilomyces javanicus. Geomicrobiol J 33:250–260. https://doi.org/10.1080/01490451.2015.1076544

Rodriguez RJ, Henson J, Van VE et al (2008) Stress tolerance in plants via habitat-adapted symbiosis. ISME J 2:404–416

Safarikova M, Ptackova L, Kibrikova I et al (2005) Biosorption of water-soluble dyes on magnetically modified Saccharomyces cerevisiae subsp. uvarum cells. Chemosphere 59:831–835. https://doi.org/10.1016/j.chemosphere.2004.10.062

Santos CM, Dweck J, Viotto RS et al (2015) Application of orange peel waste in the production of solid biofuels and biosorbents. Bioresour Technol 196:469–479. https://doi.org/10.1016/j.biortech.2015.07.114

Saratale G, Kalme S, Govindwar S (2006) Decolorisation of textile dyes by Aspergillus ochraceus (NCIM-1146). Indian J Biotechnol 5:407–410

Sathish L, Pavithra N, Ananda K (2012) Antimicrobial activity and biodegrading enzymes of endophytic fungi from eucalyptus. Int J Pharm Sci Res 3:2574–2583

Seghers D, Wittebolle L, Top EM et al (2004) Impact of agricultural practices on the Zea mays L. endophytic community. Appl Environ Microbiol 70:1475–1482

Seth PK, Jaffery FN, Khanna VK (2000) Toxicology. Indian J Pharm 32:134–151

Shehzadi M, Afzal M, Khan MU et al (2014) Enhanced degradation of textile effluent in constructed wetland system using Typha domingensis and textile effluent-degrading endophytic bacteria. Water Res 58:152–159. https://doi.org/10.1016/j.watres.2014.03.064

Sheng X, Xia J, Jiang C et al (2008) Characterization of heavy metal-resistant endophytic bacteria from rape (Brassica napus) roots and their potential in promoting the growth and lead accumulation of rape. Environ Pollut 156:1164–1170

Shi X, Liu Q, Ma J et al (2015) An acid-stable bacterial laccase identified from the endophyte Pantoea ananatis Sd-1 genome exhibiting lignin degradation and dye decolorization abilities. Biotechnol Lett 37:2279–2288. https://doi.org/10.1007/s10529-015-1914-1

Shilpa S, Shikha R (2015) Biodegradation of dye reactive black-5 by a novel bacterial endophyte. Int Res J Env Sci 4:44–53

Sidhu A, Agrawal S, Sable V et al (2014) Isolation of Colletotrichum gloeosporioides gr., a novel endophytic laccase producing fungus from the leaves of a medicinal plant, Piper betle. Int J Sci Eng Res 5:1087–1096

Sim CSF, Ting ASY (2017a) Metal biosorption in single- and multi-metal solutions by biosorbents: indicators of efficacy in natural wastewater. Clean Soil Air Water. https://doi.org/10.1002/clen.201600049

Sim CSF, Ting ASY (2017b) FTIR and kinetic modelling of fungal biosorbents *Trichoderma asperellum* for the removal of Pb(II), Cu(II), Zn(II) and Cd(II) from multi-metal solutions. Desalination Water Treat 63:167–171

Sim CSF, Tan WS, Ting ASY (2016) Endophytes from Phragmites for metal removal: evaluating their metal tolerance, adaptive tolerance behaviour and biosorption efficacy. Desalin Water Treat 57:6959–6966. https://doi.org/10.1080/19443994.2015.1013507

Sing NN, Husaini A, Zulkharnain A et al (2017) Decolourisation capabilities of Ligninolytic enzymes produced by Marasmius cladophyllus UMAS MS8 on Remazol brilliant blue R and other azo dyes. Biomed Res Int 2017:1325754. https://doi.org/10.1155/2017/1325754

Singh G, Singh N, Marwaha TS (2009) Crop genotype and a novel symbiotic fungus influences the root endophytic colonization potential of plant growth promoting rhizobacteria. Physiol Mol Biol Plants 15:87–92

Singh A, Manju, Rani S et al (2012) Malachite green dye decolorization on immobilized dead yeast cells employing sequential design of experiments. Ecol Eng 47:291–296. https://doi.org/10.1016/j.ecoleng.2012.07.001

Song Y, Fang H, Xu H et al (2016) Treatment of wastewater containing crystal violet using walnut Shell. J Residuals Sci Technol 13:243–249. https://doi.org/10.12783/issn.1544-8053/13/4/1

Strobel G, Daisy B (2003) Bioprospecting for microbial endophytes and their natural products. Microbiol Mol Biol Rev 67:491–502

Sun J, Zheng M, Lu Z et al (2017) Heterologous production of a temperature and pH-stable laccase from Bacillus vallismortis fmb-103 in Escherichia coli and its application. Process Biochem. https://doi.org/10.1016/j.procbio.2017.01.030

Taylor JE, Hyde KD, Jones EBG (1999) Endophytic fungi associated with the temperate palm, Trachycarpus fortune, within and outside its natural geographic range. New Phytol 142:335–346

Thakur IS (2011) Environmental biotechnology: basic concepts and applications. I.K. International, New Delhi

Ting ASY (2014) Biosourcing endophytes as biocontrol agents of wilt diseases. In: Verma VC, Gange AC (eds) Advances in endophytic research. Springer, New Delhi, pp 283–300

Ting ASY, Jioe E (2016) In vitro assessment of antifungal activities of antagonistic fungi towards pathogenic Ganoderma boninense under metal stress. Biol Control 96:57–63

Ting ASY, Lee MVJ, Chow YY et al (2016) Novel exploration of endophytic Diaporthe sp. for the biosorption and biodegradation of Triphenylmethane dyes. Water Air Soil Pollut 227:109. https://doi.org/10.1007/s11270-016-2810-6

Tseng R-L, Wu P-H, Wu F-C et al (2011) Half-life and half-capacity concentration approach for the adsorption of 2,4-dichlorophenol and methylene blue from water on activated carbons. J Taiwan Inst Chem Eng 42:312–319. https://doi.org/10.1016/j.jtice.2010.07.002

Uçar G, Bakircioglu D, Kurtulus YB (2014) Determination of metal ions in water and tea samples by flame-AAS after preconcentration using sorghum in nature form and chemically activated. J Anal Chem 69:420–425. https://doi.org/10.1134/s1061934814050098

Urairuj C, Khanongnuch C, Lumyong S (2003) Ligninolytic enzymes from tropical endophytic *Xylariaceae*. Fungal Divers 13:209–219

Velpandian T, Saha K, Ravi AK et al (2007) Ocular hazards of the colors used during the festival-of-colors (Holi) in India – malachite green toxicity. J Hazard Mater 139:204–208. https://doi.org/10.1016/j.jhazmat.2006.06.046

Vieira RHSF, Volesky B (2000) Biosorption: a solution to pollution? Int Microbiol 3:17–24

Volesky B (2001) Detoxification of metal-bearing effluents: biosorption for the next century. Hydrometall 59:203–216

Wang JL, Chen C (2006) Biosorption of heavy metals by *Saccharomyces cerevisiae*: a review. Biotechnol Adv 24:427–451

Wang J, Chen C (2009) Biosorbents for heavy metals removal and their future. Biotechnol Adv 27:195–226

Waqas M, Khan AL, Kamran M et al (2012) Endophytic fungi produce gibberellins and indole acetic acid and promotes host-plant growth during stress. Molecules 17:10754–10773

Wilde EW, Benemann JP (1993) Bioremoval of heavy metals by the use of microalgae. Biotechnol Adv 11:781–812

World Health Organization (2011) Guidelines for drinking-water quality. http://www.who.int/water_sanitation_health/publications/2011/dwq_guidelines/en/. Accessed 9 Feb 2017

Wu Y, Xiao X, Xu C et al (2013) Decolorization and detoxification of a sulfonated triphenyl-methane dye aniline blue by Shewanella oneidensis MR-1 under anaerobic conditions. Appl Microbiol Biotechnol 97:7439–7446. https://doi.org/10.1007/s00253-012-4476-3

Xiao X, Luo S, Zeng G et al (2010) Biosorption of cadmium by endophytic fungus (EF) *Micro sphaeropsis sp.* LSE10 isolated from cadmium hyper accumulator *Solanum nigrum* L. Bioresour Technol 101:1668–1674

Xin S, Zeng Z, Zhou X et al (2017) Recyclable Saccharomyces cerevisiae loaded nanofibrous mats with sandwich structure constructing via bio-electrospraying for heavy metal removal. J Hazard Mater 324:365–372. https://doi.org/10.1016/j.jhazmat.2016.10.070

Yadav A, Chowdhary P, Kaithwas G, Bharagava RN (2017) Toxic metals in environment, threats on ecosystem and bioremediation approaches. In: Das S, Dash HR (eds) Handbook of metal-microbe interactions and bioremediation. CRC Press/Taylor & Francis Group, Boca Raton, p 813

Yang SK, Tan N, Yan XM et al (2013) Thorium (IV) removal from aqueous medium by citric acid treated mangrove endophytic fungus Fusarium sp. #ZZF51. Mar Pollut Bull 74:213–219

Yang J, Pan X, Zhao C, Mou S, Achal V, Al-Misned FA, Mortuza MG, Gadd GM (2016) Bioimmobilization of heavy metals in acidic copper mine tailings soil. Geomicrobiol J 33:261–266. https://doi.org/10.1080/01490451.2015.1068889

Young CA, Felitti S, Shields K (2006) A complex gene cluster for indole-diterpene biosynthesis in the grass endophyte *Neotyphodium lolii*. Fungal Genet Biol 43:679–693

Zablocka-Godlewska E, Przystas W, Grabinska-Sota E (2015) Dye decolourisation using two Klebsiella strains. Water Air Soil Pollut 226:2249. https://doi.org/10.1007/s11270-014-2249-6

Zafar S, Aqil F, Ahmad I (2007) Metal tolerance and biosorption potential of filamentous fungi isolated from metal contaminated agricultural soil. Bioresour Technol 98:2557–2561

Zhang Y, Zhang Y, Liu M et al (2008) Dark septate endophyte (DSE) fungi isolated from metal polluted soils: their taxonomic position, tolerance, and accumulation of heavy metals in vitro. J Microbiol 46:624–632

Zhang Y, He L, Chen Z et al (2011) Characterization of lead-resistant and ACC deaminase-producing endophytic bacteria and their potential in promoting lead accumulation of rape. J Hazard Mater 186:1720–1725

Zhou F, Cheng Y, Gan L et al (2014) Burkholderia vietnamiensis C09V as the functional biomaterial used to remove crystal violet and Cu(II). Ecotoxicol Environ Saf 105:1–6. https://doi.org/10.1016/j.ecoenv.2014.03.028

Zhuo R, He F, Zhang X et al (2015) Characterization of a yeast recombinant laccase rLAC-EN3-1 and its application in decolorizing synthetic dye with the coexistence of metal ions and organic solvents. Biochem Eng J 93:63–72. https://doi.org/10.1016/j.bej.2014.09.004

Chapter 11
Textile Wastewater Dyes: Toxicity Profile and Treatment Approaches

Sujata Mani, Pankaj Chowdhary, and Ram Naresh Bharagava

Abstract Textile industry is one of the major industries in the world that play a major role in the economy of many countries. Wastewater discharges from textile industries are highly colored which contains various polluting substances including synthetic dyes, chemicals, etc., causing severe health hazards to humans, animals, plants, as well as microorganisms. This highly colored textile wastewater severely affects photosynthetic function in plant as well as aquatic life by eutrophication. So, this textile wastewater must be treated before their discharge. In this chapter, different treatment methods to treat the textile wastewater have been presented such as physical methods (adsorption, ion exchange, and membrane filtration), chemical methods (chemical precipitation, coagulation and flocculation, chemical oxidation), and biological methods (aerobic and anaerobic). This chapter also recommends the possible eco-friendly approaches for treating diverse types of effluent generated from textile operation.

Keywords Textile industry · Dyes · Environmental pollution · Aquatic toxicity · Treatment approaches

1 Introduction

Industrialization is considered as the key factor to development of countries in economic terms. However, improper discarding of industrial wastes such as textile, distillery, pulp, and paper, tannery is the root of environmental damages (Kumari et al. 2016; Chowdhary et al. 2017a, b; Bharagava and Mishra 2017). The recognition that environmental pollution is a worldwide menace to public health has given rise to innovative initiatives for environmental restitution for both economic and ecological reasons (Mishra and Bharagava 2016). With the increase in demands for

S. Mani · P. Chowdhary · R. N. Bharagava (✉)
Laboratory for Bioremediation and Metagenomics Research (LBMR), Department of Environmental Microbiology (DEM), Babasaheb Bhimrao Ambedkar University (A Central University), Lucknow, Uttar Pradesh, India

© Springer Nature Singapore Pte Ltd. 2019
R. N. Bharagava, P. Chowdhary (eds.), *Emerging and Eco-Friendly Approaches for Waste Management*, https://doi.org/10.1007/978-981-10-8669-4_11

textile products, there is amplifying in the number of textile industries and, simultaneously, its wastewater proportion, making it one of the core sources of severe pollution problems worldwide. Until recently there is no data accessible on the exact amount of annual production of organic dyes in the world. However, more than 100,000 types of commercially accessible dyes exist, and an annual worldwide production of 700,000–1,000,000 tons has been reported previously (Gupta and Suhas 2009). It is anticipated that 90% of the total dyes produced will end up in fabrics, while the remaining portion will be used in leather, paper, plastic, and chemical industry (Hameed et al. 2007). It is also estimated that 280,000 tons of textile dyes are discharged as industrial effluent every year worldwide (Ali 2010; Zainith et al. 2016).

Textile fibers are categorized into two principal groups, i.e., natural fibers and man-made or manufactured fibers. The types of natural fibers include cotton, linen, wool, and silk. Man-made fibers are either derived from natural polymers such as viscose and cellulose acetate or synthetic polymers such as polyester, polyamide, polyacrylonitrile, polyurethane, and polypropylene (Bechtold et al. 2006). Further, the textile industries are classified into two groups such as dry (solids wastes are generated) and wet fabric industry (liquid wastes from various processing operations are generated). During different wet processing operations, large amount of water is consumed at different processing step, thus releasing chemicals such as acids, alkalis, dyes, hydrogen peroxide, starch, surfactants, dispersing agents, and soaps of metals containing large amount of wastewater (Paul et al. 2012). Hence, in terms of environmental impacts, the textile industry is expected to utilize more water as compared to other industries and is likely to produce highly polluted wastewater.

Aromatic and heterocyclic dyes with complicated and stable structures used in textile industries are posing greater threat in degradation when present in textile wastewaters (Ding et al. 2010; Mani and Bharagava 2017). Hence, the toxicity of the textile effluent produced by different textile industries and dye manufacturing industries is a main challenge and also an ecological distress. Therefore, the key aim of this chapter is to provide a widespread knowledge about different wet processing operations in cotton textile industry, the toxicological effects of emerging wastewaters, factors affecting during their degradation, and the cost of methods instigated for during the treatment of the dyes in textile wastewater. This chapter also explains the critical study of the generally used methods such as chemical, physical, biological, and some ecological friendly approaches applied for the removal of dyes from textile industrial effluents.

2 Textile Wastewater: Generation and Characteristics

The characteristics of textile wastewater generally depend on the textiles manufactured and chemicals used in it. The textile wastewater has high BOD, COD, total solids, chemicals, odor, and color which causes threat to environment as well as human health. The 1:4 BOD and COD ratio of the wastewater indicates the

Table 11.1 Characteristics of typical untreated textile wastewater

S. No.	Parameters	Range
1	pH	6–10
2	Temperature	35–45
3	TDS	8000–12,000
4	BOD	80–6000
5	COD	150–12,000
6	Chlorine	1000–6000
7	Sodium	70%
8	Trace elements (mg/L)	<10
9	Oil and grease (mg/L)	10–30
10	Free ammonia	<10
11	SO4	600–1000
12	Silica (mg/L)	<15
13	TNK	10–30
14	Total nitrogen (mg/L)	70–80
15	Color (Pt-Co)	50–2500

Adopted from Ghaly et al. (2014)

existence of non-biodegradable substances into it (Arya and Kohli 2009). The textile wastewater also contains trace metals like Cr, As, Cu, and Zn which harm the environment. The distinctive characteristics of textile effluent are shown in Table 11.1.

The processing in the textile operations are one of the key factors in the successful trading of textile products. At first, the textile industries prepare the fibers which are further transformed into yarn and finally turned to fabrics. These fabrics are run through several stages of wet processing. In textile dyeing process, a portion of unfixed dyes get washed along with the water. In dry process, much solid wastes are generated as compared to wet process. The possible pollutant and the nature of effluent released from each step of the wet process are listed in Fig. 11.1.

2.1 Sizing and Desizing

Sizing is one of the frequent substances that is applied or included into other materials, which act as defensive protective material or glaze in order to modify the absorption and characteristics of those materials. Sizing of yarn reduces its breakage, which ultimately strengthens and scuffs its resistance. Sizing agents are selected on the basis of fabric types, environmental friendliness, easy removal, cost-effective, effluent treatments, etc. Sizing agents can be natural substances, such as starch and its derivatives, cellulosic derivatives, protein-based starch, etc., or synthetic agents such as polyesters, PVA, polyacrylates, etc.

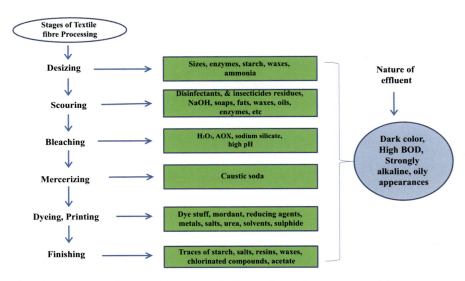

Fig. 11.1 Possible pollutant and the nature of effluent released from each step of the wet process

Desizing is the process of removal of size materials from yarn, which can be done by the processes like enzyme desizing, oxidative desizing, acid desizing, or removing water-soluble sizes. The effluents released after desizing have pH around 4–5 with high BOD range (300–450 ppm) (Magdum et al. 2013). During sizing process, starch is used which can be degraded into CO_2 and H_2O by H_2O_2 oxidation, and through enzymes desizing it can be converted into ethanol which can be further used as an energy, thus reducing the BOD load on treatment (Sarayu and Sandhya 2012).

2.2 Scouring

Scouring is a chemical washing process, which is carried out to remove impurities or any added soil or dirt particles from fibers or yarns through hydrodynamic scour, bridge scour, ice scour, or tidal scour.

2.3 Bleaching

Creamy look of the fabrics is due to the natural color substances present in it. It becomes essential to remove natural colors from fabrics in order to get white and bright fabrics, and this process of eradicating color is known as bleaching. Earlier, chlorine was used to bleach fabrics, and further, hypochlorites were made due to its disinfecting ability, but later it was replaced by hydrogen peroxide and peracetic acid. One of the major benefits of using peracetic acid is the less destruction of yarn as well as elevated sheen (Abdel-Halim and Al-Deyab 2013; Liang and Wang 2015).

2.4 Mercerization

The treatment of cotton fabrics with strong caustic alkaline solution (about 18–24% by weight) in order to improve the luster or shine is known as mercerization. In this process, cotton fabrics are first dipped in high concentration of alkaline solution, and then the swollen fabrics are placed into water to provide stress conditions for removing alkali which in result provides permanent silk-like luster to fibers. Due to mercerization process, the fabrics also gain increased ability to absorb dye, improved reaction with a variety of chemicals, improved stability of form, improved smoothness, and improved strength (Fu et al. 2013; Lee et al. 2014).

2.5 Dyeing and Printing

The process of adding color to fabrics or yarns by treating with dyes is known as dyeing. The key factors in dyeing are temperature and controlled time. The dyes responsible for color have chromophore groups or auxochrome groups (Waring and Hallas 2013). Dyes of textile goods are applied by dyeing from dye solution and by printing from thick paste form of the dye to prevent its spread. The effluents from both dyeing and printing consist of waste components (Ratthore et al. 2014).

2.6 Finishing

In textile manufacturing, finishing is a process that converts the woven or knitted cloth in a usable material with improved definite properties like softness, waterproof, antibacterial, and UV protected. This final process contributes to water pollution (Kant 2012).

3 Toxicological Effects of Textile Wastewater Dyes in Environment

During processing stages, a large volume of fresh water and chemicals are consumed by textile industries which deliver a considerable quantity of coloring agents along with other chemicals. The color of the dyes being dichromatically stronger is visible at a very low concentration, i.e., 1 ppm in water. Release of water-insoluble and soluble dyes in the wastewater is due to the improper uptake of dye as well as degree of fixation on the substrate, which is bridled by factors such as depth of the shade, application method, material to liquor ratio and pH, etc. Exhaustion and degree of fixation of dye are indirectly proportional to the depth of the shade in almost all dye-fiber combinations. Due to its high color intensity and high composition variability, it becomes very difficult to treat the textile wastewater. Further, it is predicted that approx 2% of dyes are directly discharged into the aqueous effluent and 10% is consequently lost during the coloration process. It is practically assumed that approx 20% of colorants are discharged into the environment through textile effluents, thus creating serious environmental problems like toxicity to aquatic life and mutagenicity to humans. Since, the water quality is influenced by the observation of its color, thus making the removal of color much more important than the removal of soluble colorless organic substances from the wastewater.

Because of high thermal and photostability, dyes can remain in environment for an extended period of time. Under aerobic conditions, the dyes (particularly azo dyes) are resistant to biodegradation, but under anaerobic conditions these undergo reductive splitting of the azo bond easily and hence release aromatic amines (Blaise et al. 1988). The determination of BOD and COD is an excellent clue of the pollution load of water; in spite of this, these measures are inadequate to gather information regarding the harmful effects of chemicals (APHA 2012). The evaluation of toxicity is done by acute and enrichment toxicity tests (Schowanck et al. 2001). For the identification and evaluation of toxicity, a protocol was proposed by the US Environment Protection (USEPA 1988) which consists of a series of fractional procedures followed by a bioassay which determines the source of effluent toxicity. The acute and/or chronic effects of dyes depend on the dye concentration and exposure time on the exposed organisms (Sponza and Isik 2006).

The utmost environmental fear is the absorption and reflection of sunlight into the water containing dyes. Light absorption affects the photosynthetic activity of algae and hence manipulates the food chain. Some dyes and their by-products are reported to be carcinogenic, mutagenic, and/or poisonous to life causing cancers of the kidney, urinary bladder, and liver of the workers in dye-using textile companies. Mathur et al. (2005) studied the mutagenicity of the dyes used by their trade names only in textile industries without knowing its chemical nature and natal hazards and its effect on the health of the workers as well as on the environment. They used the dyes in their crude form without following any purification process for testing the potential threat of the dyes in its actual form and found that the dyes used were highly mutagenic. Also Pagga and Brown (1986) published an article on the possibility to foresee the toxicity of new azo dyes. In the European Union, the dyes used in textile companies were tested for mutagenicity, and the dyes found mutagenic were replaced with fewer risky substances. The dyes used in textiles can also cause contact dermatitis, allergenic reaction in eyes, skin irritation, respiratory diseases, and irritation to the mucous membrane and upper respiratory tract. Thus, the textile effluents require an appropriate treatment before discharge into the environment for safety purposes.

3.1 Case Study

A report on "Toxic Sludge Irrigating Fields for 20 years" was published in The Tribune on 7th April, 2009 which stated about the farmers of this cancer-prone district have to irrigate their field with toxic sludge flowing into a drain due to unavailability of canal water and subsoil water, which passes through their village. The effect on crops can be determined by the symptoms appearing in villagers like plaque in teeth, joint pain, and gray hairs (Kant 2012).

4 Factors Affecting the Degradation of Dyes

The bacterial degradation is controlled by various physicochemical parameters such as oxygen, temperature, pH, concentration of dye, structure of dye, concentration of carbon, and nitrogen sources. Thus, to get a more effective and faster bacterial degradation, it is necessary to conclude the consequence of and every parameter on the biodegradation. Table 11.2 summarizes possible range of equipped parameters for a better biodegradation.

5 Treatment Processes for Textile Wastewater

The effluent generated from cotton dyeing textile industries is extremely polluted with high color, BOD, COD, and total solids due to the existence of dyes and other chemicals. The color present in the wastewater directly hinders with the photosynthetic activity and development of the aquatic organisms, resulting to the imbalance in the environment. In order to avoid treatment cost problems of the river, water used for drinking should be colorless and free from toxic compounds. Therefore, textile wastewater should go through many treatment processes including physical, chemical, and biological methods before discharging into any freshwater body (Fig. 11.2).

5.1 Physical Treatment Methods

The removal of substances from wastewater through ordinarily taking place forces such as electrical attraction, gravity, van der Waal forces or by physical barriers is a known physical treatment method. These methods do not result in the change of chemical structure of the substances present in water. Sometimes, physical state is changed or coagulation of dispersed substances occurs. Some of these methods have been discussed here.

Table 11.2 Various factors affecting the degradation and decolorization of dyes

Factors	Descriptions
pH	The pH has a major effect on the dye decolorization efficiency; the optimal pH for color removal in bacteria is often between 6.0 and 10.0. The tolerance to high pH is important in particular for industrial processes using reactive azo dyes, which are usually performed under alkaline conditions. The pH has a major effect on the efficiency of dye decolorization; the optimal pH for color removal in bacteria is often between 6.0 and 10.0 (Chen et al. 2003; Guo et al. 2007)
Temperature	Temperature is also a very important factor for all processes associated with microbial vitality, including the remediation of water and soil. It was also observed that the decolorization rate of azo dyes increases up to the optimal temperature, and afterward there is a marginal reduction in the decolorization activity
Dye concentration	Earlier reports show that increasing the dye concentration gradually decreases the decolorization rate, probably due to the toxic effect of dyes with regard to the individual bacteria and/or inadequate biomass concentration, as well as blockage of active sites of azo reductase by dye molecules with different structures
Carbon and nitrogen sources	Dyes are deficient in carbon and nitrogen sources, and the biodegradation of dyes without any supplement of these sources is very difficult. Microbial cultures generally require complex organic sources, such as yeast extract, peptone, or a combination of complex organic sources and carbohydrates for dye decolorization and degradation
Oxygen and agitation	Environmental conditions can affect the azo dyes degradation and decolorization process directly, depending on the reductive or oxidative status of the environment and, indirectly, influencing then microbial metabolism. It is assumed that under anaerobic conditions reductive enzyme activities are higher; however a small amount of oxygen is also required for the oxidative enzymes which are involved in the degradation of azo dyes
Dye structure	Dyes with simpler structures and low molecular weights exhibit higher rates of color removal, whereas the removal rate is lower in the case of dyes with substitution of electron withdrawing groups such as SO_3H and $-SO_2NH_2$ in the para position of phenyl ring, relative to the azo bond and high molecular weight dyes
Electron donor	It has been observed that the addition of electron donors, such as glucose or acetate ions, apparently induces the reductive cleavage of azo bonds. The type and availability of electron donors are important in achieving good color removal in bioreactors operated under anaerobic conditions
Redox mediator	Redox mediators (RM) can enhance many reductive processes under anaerobic conditions, including azo dye reduction

Adopted from Sujata and Bharagava (2016) and Khan et al. (2012)

5.1.1 Adsorption

This is the most commonly equilibrium separation physicochemical method used in the potential treatment of wastewater. Due to higher efficiency in the removal of pollutants, adsorption technique has gained more favor recently as compared to other methods. The most commonly used adsorbents are activated carbon, silicon

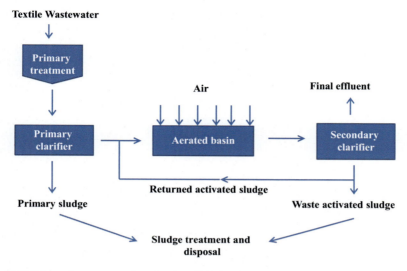

Fig. 11.2 Schematic representation of activated sludge process

polymers, and kaolin which are capable of adsorbing different dyes with high adsorption capacity (Jadhav and Srivastava 2013). In this process, ions or molecules present in one phase (either gas or liquid) tend to accumulate and concentrate on the surface of another phase (usually solids). When weak interspecies bond exists between the adsorbate and adsorbents, then physical adsorption takes place, but when strong interspecies bonds exists due to exchange of electrons, then chemical adsorption occurs (Bizuneh 2012). In order to modernize the treatment methods, several new adsorbents has been developed such as eggshells, sugarcane bagasse, hen feathers, almond peel, etc. (Ahmad and Mondal 2009; Chakraborty et al. 2012a, b, c, d; Chowdhury et al. 2013a, b, c) (Table 11.3).

5.1.2 Ion Exchange

Ions are replaced among two electrolytes or linking an electrolyte solution and a complex in ion exchange method. In some cases, ion exchange method is applied as a technique of purification, separation, and decontamination of ion and aqueous solution. Some typical ion exchangers are resins, zeolites, montmorillonite, clay, and soil humus. Since the ion exchangers cannot accommodate a wide range of dyes, thus this method has not been widely used for the treatment of textile wastewater effluents. It can be only used to remove the undesirable cationic (basic dyes) or anionic (acid, direct, and reactive dyes) from wastewater. There are some amphoteric exchangers which are able to exchange both cations and anions simultaneously that can be efficiently used in mixed beds containing mixture of cation and anion

Table 11.3 Various adsorbent for adsorption of dye

S.No.	Adsorbent	Dye	References
1	Eggshells	Crystal violet	Chowdhury et al. (2013a, b, c)
2	Citric acid-modified rice straw	Crystal violet	Chowdhury et al. (2013a, b, c) and Chakraborty et al. (2013)
3	NaOH-modified rice husk	Crystal violet	Chowdhury et al. (2013a, b, c)
4	Sugarcane bagasse, H_2SO_4 modified sugarcane bagasse, fish scales, hen feathers	Crystal violet, methylene blue and CV, Congo red and CV	Chakraborty et al. (2012a, b, c, d)
5	*Artocarpus heterophyllus* leaf powder	Crystal violet	Saha et al. (2012)
6	Almond peel	Brilliant green	Ahmad and Mondal (2009)
7	Water nut carbon	Congo red and malachite green	Ahmad and Mondal (2010)
8	Ginger waste	Malachite green	Kumar and Ahmad (2010)
9	Activated carbon/iron oxide nanocomposite, conducting polyaniline/iron oxide composite, bael shell carbon	Brilliant green, Amido Black 10 B, Congo red, respectively	Ahmad and Kumar (2010a, b, c)
10	Alumina reinforced polystyrene	Amaranth dye	Ahmad and Kumar (2011)
11	PAni/TiO_2 nanocomposite	Methylene blue	Ahmad and Mondal (2012a, b)

exchange resins. Meanwhile, the advantage of ion exchange treatment method is the recovery of adsorbent, retrieval of solvent after use, and the effective removal of soluble dyes (Mani and Bharagava 2016).

5.1.3 Membrane Filtration

Membrane filtration method has emerged as a feasible alternative method used for the removal of dyes from effluent effectively and has proven to be cost-effective and has less water consumption (Koyuncu 2002). Thus, the method simultaneously reduces the coloration and BOD or COD of wastewaters and has special features like resistance to temperature and adverse chemical effects. The advantage of membrane filtration method is its quick processing with low requirements, and its drench can be reused, but due to its high cost, clogging possibility and replacement of membrane affect its applicability (Bizuneh 2012).

5.1.3.1 Ultrafiltration

Ultrafiltration is a process of separation of macromolecules and particles of size about 1 nm–0.05 μm, but it does not eliminate some dissolved polluting substances such as dyes, thus prohibiting the reuse of treated textile effluent for some sensitive processes such as dyeing of textile fabrics. Rott and Minke (1999) highlighted that 40% of the ultrafiltration treated wastewater can be reused for minor processes such as washing or rinsing in textile industry. This method can be used in combination with biological reactor or as a pretreatment for reverse osmosis (Ramesh Babu et al. 2007).

5.1.3.2 Nanofiltration

This filtration method is used for the treatment of tinted textile effluents from the textile industry of nanometers size with retention weight of about 80–1000 da. For the treatment of textile effluents, adsorption and nanofiltration combination is adopted in which nanofiltration process is followed after adsorption method in order to decrease the polarization concentration which occurs during filtration processes. The nanofiltration membranes preserve the low molecular divalent ions, large monovalent ions, hydrolyzed reactive dyes, weight organic compounds, and dyeing auxiliaries. Treatment of textile effluents from nanofiltration is one of the possible solutions for management of extremely intense and intricate solutions (Ramesh Babu et al. 2007). It is an effective and potential cure of textile effluent and also favorable in terms of environmental regulations.

5.1.3.3 Microfiltration

Microfiltration is used for the treatment of dyebaths containing dye pigments with aperture about 0.1–1 μm. This method is also used as a pretreatment process for reverse osmosis or nanofiltration (Ramesh Babu et al. 2007).

5.1.3.4 Reverse Osmosis

Decolorization as well as elimination of chemical compounds from dye house wastewater can be carried out in a single step of reverse osmosis process. For most types of ionic compounds, reverse osmosis membrane has 90% or more of retention rate. Reverse osmosis helps in the exclusion of all mineral salts, hydrolyzed reactive dyes, and chemical compounds. The osmotic pressure is directly proportional to the concentration of dissolved salts and thus to the energy required for the separation (Ramesh Babu et al. 2007).

5.2 Chemical Treatment Methods

In an array of processes to accelerate disinfection, chemicals are used during wastewater treatment. These chemical processes including chemical reactions are known as chemical unit processes which are used alongside with physical and biological processes. These include various processes including chemical precipitation, coagulation and flocculation, chemical oxidation, Fenton oxidation, etc., which are applied during wastewater treatment.

5.2.1 Chemical Precipitation

The most common method used for the treatment of textile effluent for removing dissolved toxic metals from it is chemical precipitation. In this process, the dissolved metals in wastewater are converted into solid particle forms by adding precipitation reagent which triggers a chemical reaction causing dissolved metals to form solid particles which is further removed through filtration method. The probability of the method depends on the kind of metals present, its concentration, and the kind of reagents used. In hydroxide precipitation, sodium or calcium hydroxides are used as a reagent to convert dissolved metals into solid particles, but it is very difficult to create hydroxides since wastewater consists of mixed metals.

5.2.2 Coagulation and Flocculation

These processes are generally used for removing organic materials by partly removing BOD, COD, TDS, and color from effluent (Aguilar et al. 2005). This method basically depends on the law of addition of coagulants which associates with pollutants forming coagulate or flock and later precipitate which is removed either by flotation, settling, filtration, or other physical technology to form sludge which is further treated for reducing its toxicity (Golob and Ojstrsek 2005; Mishra and Bajpai 2005). The high cost for treating sludge and disposal restrictions into the environment is the major disadvantage of this process (Bizuneh 2012).

5.2.3 Chemical Oxidation

Chemical oxidation is totally a chemical operation based on strict chemical reactions. Chemical treatment depends on the chemical interactions of the desired contaminants to be removed and applied chemicals that either separates it from wastewater or destruct it or neutralize its harmful effect. Chemical treatment processes can also be applied alone or with physical treatment methods. In textile wastewater, chemical operations either oxidize the pigments in the dyeing and printing wastewater or bleach it. From various chemical oxidation processes, Fenton oxidation and ozone oxidation are often used for the treatment of wastewater.

Oxidizing agents like O_3 and H_2O_2 are used in chemical oxidation methods which forms strong nonselective hydroxyl radicals at high pH. These formed radicals effectively break the conjugated double bonds of chromophore group of dye as well as its functional groups (complex aromatic rings) which ultimately reduces the color of the wastewaters. These oxidizing agents have low degradation rate due to less hydroxyl radicals production as compared to AOP (Asgher et al. 2009). The main advantage of using ozone in ozonation process is its gaseous form which can be used as it is and thus does not raise the volume of wastewater nor produce sludges. Despite of this, disadvantage of ozone is the formation of toxic byproducts from biodegradable dyes in wastewater (Miralles-Cuevas et al. 2016).

5.2.3.1 Ozonation

Ozonation is the most effective and fast treatment method which decolorizes textile wastewater and breaks double bonds of most of the dyes. Ozonation oxidizes the considerable amount of COD and also inhibits or destroys the foaming nature of residual surfactants. It also increases the biodegradability of the wastewaters containing high fraction of non-biodegradable and toxic compounds of the effluents with high fraction of non-biodegradable and toxic compounds also increasing by converting it into effortlessly biodegradable intermediates. This major advantage of this process is neither sludge production nor increase in effluent volume. Sodium hypochlorite has been widely used as an oxidizing agent who initiates and increases azo bond cleavage, but the drawback of this agent is the release of carcinogenic amines and other toxic molecules thus restricting its use.

In this method, ozone is used as a strong and effective oxidizing agent because of its high reactivity which effectively degrades phenols, chlorinated hydrocarbons, aromatic hydrocarbons, and pesticides (Lin and Lin 1993). The main negative aspect of this process is ozone's short half-life as it decomposes in 20 min; thus continuous O_3 supply is required which is very expensive (Gogate and Pandit 2004; Gosavi and Sharma 2014).

5.2.3.2 Fenton Oxidation

This is an oxidation process in which oxidizing agents mainly hydrogen peroxide forming hydroxyl radicals (existing strongest oxidizing agent) are used to potentially decolorize wide range of dyes from textile effluents. In this method, first the Fenton reaction, i.e., formation of hydroxyl radicals from H_2O_2, is activated by adding H_2O_2 in an acidic solution containing Fe^{2+} ions. Fenton reaction is used as pretreatment method for the wastewaters that are anti-biological to treatment or/and toxic to biomass. In large-scale plants, the reactions are carried out at higher temperature by using excess iron and H_2O_2 where ions do not act as a catalyst and large quantity of entirety COD is removed. The most important drawback of this method is the elevated effluent discharge as well as high sludge production.

5.2.3.3 Photocatalytic Oxidation

Nowadays, photochemical or photocatalytic degradation is gaining importance, since this process results in complete oxidation of dye molecules present in textile wastewater at mild temperature and pressure. The chemical reaction is characterized by free radical mechanism which is initiated by the interaction between photons and chemical molecules of wastewater either in the presence or absence of catalyst (Globate and Pandit 2004). This method results in the degradation of dyes into CO_2 and H_2O by treatment with UV in the presence of H_2O_2 group resulting in reduction of foul odor of wastewater and no sludge formation (Forgacs et al. 2004).

5.2.3.4 Sonocatalytic Oxidation

This process is the blend of sonochemistry with catalysts which can be accomplished with quantity of chemical reactions (Gawande et al. 2014). In this process, the heterogeneous reactions are increased by mechanical effects of ionic intermediates, whereas radical reactions are enhanced by sonication. The elementary regulation of sonocatalysis is the diffusion of main compounds on a solid surface by series of heterogenous chemical reactions on active sites (Ince and Ziylan 2015).

5.3 Biological Treatment Methods

Biological treatment approaches are the most apposite methods for the handling of textile wastewaters. These methods are well-organized than clariflocculator treatment methods since it removes the dissolved materials in a way similar to self-purification. The efficient basis for the removal of dissolved matter depends on the ratio linking the organic load and the biomass present in the oxidation ponds and the temperature and oxygen concentration in it. The biomass concentration may increase in the oxidation pond due to the aeration, but it should not reach up to the mixing energy since it can destroy the congregations and inhibit the settlings. The biomass concentration normally ranges from 2500 to 4500 mg/L and oxygen about 2 mg/L but with aeration; the oxygen demand can be abridged up to 99% till 24 h. On the basis of different oxygen demands, the biological treatment processes can be alienated into aerobic and anaerobic treatment methods. Aerobic treatment method has turn into the typical biological treatment process because of its high efficiency and wide applications.

5.3.1 Aerobic Process

The stabilization of textile wastes by decomposing them into harmless inorganic solids is done by involving bacteria during treatment. Bacteria can be divided into aerobic, anaerobic, and facultative bacteria on the source of the oxygen

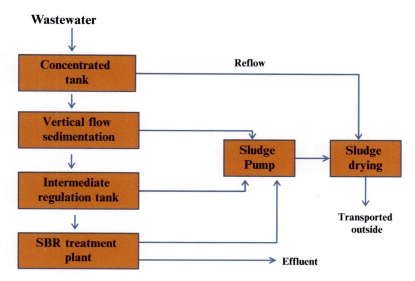

Fig. 11.3 Wastewater treatment process of sequencing batch reactor activated sludge process

requirements by these different bacteria. Aerobic treatment method purifies the water by decomposing the wastes and reducing the unpleasant odors with the help of aerobic and facultative. Aerobic treatment process is performed by activated sludge process and biofilm process.

5.3.1.1 Activated Sludge Process

The most frequently applied biological treatment for textile wastewater at ETP/CTEP in India is activated sludge process (ASP), a kind of colony mainly comprising of microorganisms having strong decomposition and adsorption rate of organic compounds, thus called as "activated sludge." Activated sludge process (ASP) is the most normally applied aerobic wastewater treatment method that not only removes the dissolved organic solids but also the settleable and non-settleable suspended solids. Microorganisms especially bacteria are used in the ASP methods which produces high quality of wastewater by feeding on the organic pollutants present in the wastewaters. The ASP is an effective with higher removal efficiency method which works on the basic principle of microorganisms that grows and clumps together forming a colony which settles down to the bottom of the tank forming an organic material and suspended solids-free clear liquid (Fig. 11.3). Oxidation ditch and sequencing batch reactor (SBR) process are the most commonly used activated sludge methods.

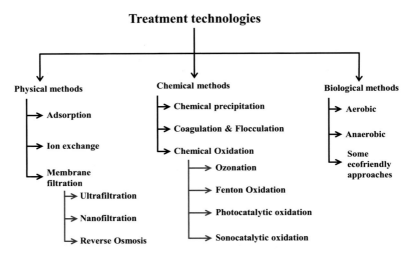

Fig. 11.4 Various treatment technologies for the degradation of dyes from textile wastewater

5.3.1.1.1 Oxidation Ditch Process

It is a special form of wastewater treatment method with extended aeration, which was developed in the last fiftieth century by Health Engineering Research Institute of the Netherlands. It consists of a ditch body, aerated equipment, wastewater influent and effluent sources, diversions, and mixing equipments. The shape of the ditch body is usually circular, but it can also be rectangular, L-shaped, or any other. In the ditch body, the wastewater, activated sludge, and microorganisms are mixed together continuously with a loop for absolute nitrification and denitrification procedure. In comparison with conventional activated sludge process, the equalization tank, primary and secondary sedimentation tank, and sludge digestion tank can be avoided in oxidation ditch process since it has time-consuming hydraulic retention time, low organic load, and long sludge age giving it an advantage over activated sludge process. Oxidation ditch process is money-spinning, provides stumpy energy consumption, is easy maintenance, is reliable, has simple and easy operations and managements, and is most importantly giving high level of water purification.

5.3.1.1.2 Sequencing Batch Reactor Activated Sludge Process

It is a reform and new operation mode of activated sludge process (Fig. 11.4). This process is very efficient in removing the high rate of COD level as well as color from the wastewater, which gives it an advantage over the traditional methods. It is a process with less equipment loads, simple structure, easy operation, and less sludge production.

5.3.1.2 Biofilm

Biofilms are biological effluent treating process involving the microorganisms, which attach to the surface of the fixed object forming a film and purify the flowing wastewater just through the contact. Mainly the biofilm processes are biological contact oxidation, rotating biological contractors, and biological fluidized bed.

5.3.2 Anaerobic Process

In anaerobic treatment process, the anaerobic bacteria are utilized, which decompose the organic matter in anaerobic conditions. Earlier, this process was used in sludge digestion, but now it is used in low as well as high concentration of organic wastewater treatment. The textile industries release effluent from dyeing, wool washing, and textile printing consisting of high concentration of organic matter (1000 mg/L or more) which can be treated anaerobically with good results.

Presently, the hydrolysis acidification process has turn into the first two stages of anaerobic treatment and is mainly used for increasing the biodegradability of the sewage. The anaerobic and facultative bacteria are frequently used for the decomposition of the macromolecules, heterocyclic, and other difficult biodegradable organic matters into small molecular organic matters, thus increasing the biodegradability of the wastewater as well as removing the colored dye molecules from the wastewater. The anaerobic bacteria change the molecular structure of the organic matter over and above the chromophore group of the colored compounds thus making it easy to decolorize as well as decompose under the aerobic conditions which alternatively improves the decolorization effect of the sewage. The pH value of the wastewater decreases up to 1.5 units in the hydrolysis tank, and the organic acid formed during hydrolysis process effectively neutralizes the alkalinity of the effluent up to some extent which drops the pH value of the sewage up to 8 to provide good environment for the treatment.

At present, anaerobic digestion method is effectively used in the biological treatment of the textile effluent. Some other processes such as upflow anaerobic sludge bed (UASB), upflow anaerobic fluidized bed (UABF), anaerobic baffled reactor (ABR), anaerobic biological filter, and so on are used in the treatment of textile dyeing wastewater.

5.3.3 Some Eco-Friendly Approaches

There are numerous treatment approaches for dye-containing textile wastewater as described previously, but choosing the most applicable treatment method or their combinations totally depends on the nature and amount of effluent effluxed from textile processing plants (Karcher et al. 2001; Kurbus et al. 2003; Sen and Demirer 2003; Golob and Ojstrsek 2005; Ojstrsek et al. 2007). In past few years, we became cognizant about some eco-friendly approaches, which have the self-cleaning ability of natural ecosystems through finest physical, chemical, and biological conditions and are considered to have greater importance (Scholz and Xu 2002). Constructed wetland and biocomposting are examples of such systems which are easy to use, are environment friendly, have low construction and operational costs, and are proficient enough to treat diverse wastewaters, though the involvement in treating textile wastewaters is narrow (Bulc and Ojstrsek 2008).

5.3.3.1 Constructed Wetland

A number of researchers have reported different methods for the treatment of textile wastewaters, but these methods are often not practical since these require expertise intensiveness, consistent power demand, and complex components, have an unverified long-term efficiency, and require high asset and maintenance costs (Mbuligwe 2005). But in last little existence, an auspicious result of constructed wetland (CW) has been recorded. Constructed wetlands are considered to be the natural treatment systems of wastewater which are competent of removing both pollutants and nutrients without any additional energy demands through various physical and biochemical mechanisms based on substrate composition, microbial communities and plant ecosystems, and operation strategies (Wang et al. 2008a, b; Zhao et al. 2013; Wu et al. 2014). It can also be shaped for land repossession after mining, refineries, or other ecological turbulences.

Natural wetlands act as biofilters, removing sediments and pollutants, for instance, heavy metals from the water; thus, constructed wetlands can be premeditated to reinvent the features of natural wetlands, and in order to separate the solids from the liquid effluent, it can also be used after a septic tank for primary treatment. During the pollutant degradation processes, various gases such as carbon dioxide, methane, and nitrous oxide are released into the atmosphere which are dangerous for our environment since these are greenhouse gases and mainly responsible for global warming (Mander et al. 2003; Solano et al. 2004; Moir et al. 2005; Bulc 2006; Borin and Tocchetto 2007; VanderZaag et al. 2008; Barbera et al. 2009; Vymazal 2009; Vymazal 2010; O'Geen et al. 2010; Verlicchi and Zambello 2014; Pappalardo et al. 2016).

Constructed wetlands are of two types mainly subsurface flow constructed wetland (SSFCW) and surface flow constructed wetland (SFCW). SSFCW is either of horizontal flow or of vertical flow. Horizontal SSFCWs are made up of gravels or rock beds usually 0.6–0.8 m in depth and planted with wetland vegetation where the wastewater flows from the inlet zone all the way through porous medium until it reaches the collecting and discharging outlet zone (Vymazal et al. 2006). In this type of CWs, the pollutants are removed through physical, chemical, and biological processes by aerobic, anoxic, and anaerobic zones where macrophytes serves as source of substrate for the growth of bacteria, nutrient uptake, radial oxygen loss, and insulation of the bed surface in cold and temperate regions (Brix 1994, 1997; Vymazal and Kropfelova 2008; Salvato and Borin 2010; Nivala et al. 2013; Lee et al. 2013; Vymazal 2016). In this system, organic compounds are removed by microbial degradation under anoxic/anaerobic conditions, suspended solids are recollected by filtration and sedimentation, and nitrogen is removed through denitrification, whereas removal of NH_4 is restricted due to lack of oxygen in the filtration bed because of the permanent drenched situations (Vymazal 2007).

Vertical subsurface flow (VSSF) CWs are mainly a flat bed of graded gravel (30–60 mm) which is topped with 6 mm of sand and planted by way of macrophytes (Vymazal et al. 2006). The effluent is flooded on the shell and steadily percolates down through the bed allowing the air to refill the bed, thus nourishes as a good oxygen transport and appropriate conditions for nitrification but do not provide any denitrification (Vymazal 2010). It is also very effective in removing organic and suspended solids (Maucieri et al. 2016).

Surface flow constructed wetland is as well identified as free water surface constructed wetlands where effluents are used for tertiary treatment or for polishing from wastewater treatment plants. In this system, the pathogens are devastated either by natural decay or through predation from higher organisms and sedimentation and through UV radiations since water are depicted to direct sunlight. The soil layer (bay mud or other silty clays) underneath the water is anaerobic, but the roots of the plants free oxygen around them which allows complex biological and chemical reactions. SFCW may persuade mosquito reproduction and high algal production which ultimately makes it difficult to assimilate them in urban vicinity.

5.3.3.2 Biocomposting

The term composting, in addition recognized as aerobic biological treatment, is used to define biological degradation under controlled aerobic conditions. Biodegradation is a naturally stirring procedure in both man-made and natural environments. The composting practice is used to alleviate the solids from wastewater before their use as a solid modification or mulch in landscaping, horticulture, and agriculture (EPA 2000). Additionally, the process also destroys pathogens, minimizes odors, and reduces vector attraction potential. In order to compost biosolids, three composting skills such as windrow, aerated static pile, and in-vessel composting processes are available. All these techniques are based on the same scientific principals but varies only in procedures and equipment required, and, most essentially, the operational costs of these techniques also differ widely (Fig. 11.5).

5.3.3.2.1 Windrow Composting

In this procedure, rows of elongated piles called windrows are filled with organic wastes and are ventilated by rotating the piles occasionally either by manually or through mechanical methods. The height of pile is kept approximately 4–8 ft and width between 14 and 16 ft which allows sufficient heat generation thus far tiny enough to permit oxygen to flow to the windrow's core. This method is applicable to large capacities of assorted wastes, including yard trimmings, grease, liquids, and animal by-products. However, this process requires frequent revolving of the pile and cautious monitoring.

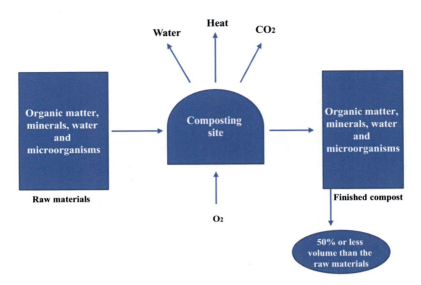

Fig. 11.5 Process of biocomposting

5.3.3.2.2 Aerated Static Pile

In this method, organic wastes are assorted in one large pile instead of many rows. In order to make sure the sufficient flow of oxygen all over the pile, coatings of loosely piled bulking agents, for example, wood chips, are added so that air can pass from bottom to top of the pile. A network of pipes is placed into the piles, and air blowers are used for supply of oxygen mechanically into the pile. This technology works well for outsized capacity generators of yard trimmings and compostable municipal solid waste and is appropriate for a moderately homogenous mixture of organic wastes, thus making it in favor of local governments, farmers, or landscapers. However, this procedure is not apposite for composting animal's by-products or grease from food processing industries.

5.3.3.2.3 In-Vessel Composting

This technique usually takes place within an enclosed vessel, thus allowing the machinist to uphold close control over the process. The main benefit of this procedure over other composting methods is that the consequences of conditions are reduced, the superiority of the resulting artifact is more reliable, fewer manpower is required for operating the system, and public acceptance of the resource may be improved. Due to the minor room necessity, in addition, in-vessel method is extra apposite in suburban and urban technologies in contrast with the other composting technologies, but this technique requires more capital than other methods.

6 Challenges and Future Prospects

The major problem correlated with the treatment of the textile wastewater is the complex nature of the effluent due to the occurrence of the dyes with complex groups and other recalcitrant organic pollutants with poor degradability that inhibits the biological treatment methods. Besides the recalcitrant pollutants, the textile wastewater consists of some toxic heavy metals having a high inhibitory and antimicrobial activity, which hinders the anaerobic digestion process of wastewater. Additionally, the nature and toxicity of the recalcitrant pollutant in textile wastewater need to be explained in details for its complete degradation and detoxification during the common treatment processes applied to the effluents.

7 Conclusions

One of the majority sources of pollution is the textile dyeing effluent which has elevated rate of color, BOD, COD, TDS, and TSS, complex composition of dye compounds, and large emission and is difficult to degrade. If directly discharged into the environment without treatment, it causes serious harm to ecological environment. Therefore, the prevention and treatment of textile dye wastewater is very compulsory. Thus, this chapter has shown the various methods for the prevention and action on textile effluent control and reduction of the pollution load making use of the treated wastewater. Various traditional technologies such as physical and chemical methods are included for the curing of textile effluent, but these require high assets and have high operational expenditure. But, biological treatment processes are considered to be the best alternative that can be approved for eco-friendly treatment processes.

Acknowledgment The authors are highly grateful to the Rajiv Gandhi National fellowship, UGC, New Delhi, India, for providing the financial support to Ms. Sujata and Mr. Pankaj Chowdhary for this work.

References

Abdel-Halim ES, Al-Deyab SS (2013) One-step bleaching process for cotton fabrics using activated hydrogen peroxide. Carbohydr Polym 92:1844–1849

Aguilar MI, Saez J, Llorens M, Soler A, Ortuno JF, Meseguer V, Fuente A (2005) Improvement of coagulationflocculation process using anionic polyacryl amide as coagulant aid. Chemosphere 58:47–56

Ahmad R, Kumar R (2010a) Kinetic and thermodynamic studies of brilliant green adsorption onto activated carbon/iron oxide nanocomposite. J Korean Chem Soc 54(1):125–130

Ahmad R, Kumar R (2010b) Conducting polyaniline/iron oxide composite: a novel adsorbent for the removal of amido black 10 B. J Chem Eng Data 55(9):3489–3493

Ahmad R, Kumar R (2010c) Adsorptive removal of congo red dye from aqueous solution using bale shell carbon. Appl Surf Sci 257(5):1628–1633

Ahmad R, Kumar R (2011) Adsorption of amaranth dyes onto alumina reinforced polystyrene. Clean Soil Air Water 39(1):74–82

Ahmad R, Mondal PK (2009) Application of acid treated almond peel for removal and recovery of brilliant green from industrial wastewater by column operation. Sep Sci Technol 44(7):1638–1655

Ahmad R, Mondal PK (2010) Application of modified water nut carbon as a sorbent in congo red and malachite green dye contaminated wastewater remediation. Sep Sci Technol 45:394–403

Ahmad R, Mondal PK (2012a) Adsorption and photodegradation of methylene blue by using PAni/TiO2 nanocomposite. J Dispers Sci Technol 33(3):380–386

Ahmad R, Mondal PK (2012b) Bioremediation of p-nitrophenol containing wastewater by aerobic granule. J Environ Eng Manag 13(3):493–498

Ali H (2010) Biodegradation of synthetic dyes – a review. Water Air Soil Pollut 213(1–4):251–273

APHA (2012) American Public Health Association, 20th edn. Washington, DC/New York

Arya D, Kohli P (2009) Environmental impact of textile wet processing, India. Dyes and Chemicals

Asgher M, Azim N, Bhatti HN (2009) Decolorization of practical textile industry effluents by white rot fungus Coriolus versicolor IBL-04. Biochem Eng J 47:61–65

Barbera AC, Cirelli GL, Cavallaro V, Di Silvestro I, Pacifici P, Castiglione V, Toscano A, Milani M (2009) Growth and biomass production of different plant species in two different constructed wetland systems in Sicily. Desalination 246:129–136

Bechtold T, Burtscher E, Hung YT (2006) In: Wang LK, Hung YT, Lo HH, Yapijakis C (eds) Treatment of textile wastes, in waste treatment in the process industries. CRC Press, Boca Raton, pp 363–392

Bharagava RN, Mishra S (2017) Hexavalent chromium reduction potential of *Cellulosimicrobium sp.* isolated from common effluent treatment industries. Ecotoxicol Environ Saf 147:102–109

Bizuneh A (2012) Textile effluent treatment & decolorization techniques. Chem Bulg J Sci Educ 21:434–456

Blaise C, Sergey G, Wells P, Bermingham N, van Coillie N (1988) Biological testing-development, application and trends in Canadian environmental protection laboratories. J Toxicol Assess 3:385–406

Borin M, Tocchetto D (2007) Five year water and nitrogen balance for a constructed surface flow wetland treating agricultural drainage waters. Sci Total Environ 380(1):38–47

Brix H (1994) Functions of macrophytes in constructed wetlands. Water Sci Technol 29(4):71–78

Brix H (1997) Do macrophytes play a role in constructed treatment wetlands? Water Sci Technol 35(5):11–17

Bulc TG (2006) Long term performance of a constructed wetland for landfill leachate treatment. Ecol Eng 26:365–374

Chakraborty S, Chowdhury S, Saha PD (2012a) Adsorption of crystal violet from aqueous solution onto sugarcane bagasse: central composite design for optimization of process variables. J Water Reuse Desal 2:55–65

Chakraborty S, Chowdhury S, Saha PD (2012b) Fish (*Labeo rohita*) scales as a new biosorbent for removal of textile dyes from aqueous solutions. J Water Reuse Desal 2:175–184

Chakraborty S, Chowdhury S, Saha PD (2012c) Batch removal of crystal violet from aqueous solution by H_2SO_4 modified sugarcane bagasse: equilibrium, kinetic, and thermodynamic profile. Sep Sci Technol 47:1898–1905

Chakraborty S, Chowdhury S, Saha PD (2012d) Biosorption of hazardous textile dyes from aqueous solutions by hen feathers: batch and column studies. Korean J Chem Eng 29:1567–1576

Chakraborty S, Chowdhury S, Saha PD (2013) Artificial neural network (ANN) modelling of dynamic adsorption of crystal violet from aqueous solution using citric-acid modified rice (Oryza sativa) straw as adsorbent. Clean Techn Environ Policy 15:255–264

Chen KC, Jane YW, Liou DJ, Hwang SCJ (2003) Decolorization of the textile dyes by newly isolated bacterial strains. J Biotechnol 101:57–68

Chowdhary P, Yadav A, Kaithwas G, Bharagava RN (2017a) Distillery wastewater: a major source of environmental pollution and its biological treatment for environmental safety. In: Singh R, Kumar S (eds) Green technologies and environmental sustainability. Springer, Cham, pp 409–435

Chowdhary P, More N, Raj A, Bharagava RN (2017b) Characterization and identification of bacterial pathogens from treated tannery wastewater. Microbiol Res Int 5(3):30–36

Chowdhury S, Chakraborty S, Saha PD (2013a) Removal of crystal violet from aqueous solution by adsorption onto eggshells: equilibrium, kinetics, thermodynamics and artificial neural network modeling. Waste Biomass Valorization 4:655–664

Chowdhury S, Chakraborty S, Saha PD (2013b) Adsorption of crystal violet from aqueous solution by citric acid modified rice straw: equilibrium, kinetics, and thermodynamics. Sep Sci Technol 48:1348–1399

Chowdhury S, Chakraborty S, Saha PD (2013c) Response surface optimization of a dynamic dye adsorption process: a case study of crystal violet adsorption onto NaOH-modified rice husk. Environ Sci Pollut Res 20:1698–1705

Ding S, Li Z, Wangrui (2010) Overview of dyeing wastewater treatment technology. Water Resour Prot 26:73–78

EPA (2000) Bio-solids technology fact sheet: in-vessel composting of biosolids. United States Environmental Protection Agency

Forgacs E, Cserhati T, Oros G (2004) Removal of synthetic dyes from wastewaters: a review. Environ Int 30:953–971

Fu S, Hinks D, Hauser P, Ankeny M (2013) High efficiency ultra-deep dyeing of cotton via mercerization and cationization. Cellulose 20:3101–3110

Gawande MB, Bonifacio VDB, Luque R, Branco PS, Varma RS (2014) Solvent-free and catalysts-free chemistry: a benign pathway to sustainability. ChemSusChem 7:24–44

Ghaly AE, Ananthashankar R, Alhattab M, Ramakrishnan VV (2014) Production, characterization and treatment of textile effluents: a critical review. J Chem Eng Process Technol 5(1):1–19

Gogate PR, Pandit AB (2004) Reviews of imperative technologies for wastewater treatment. I: oxidation technologies at ambient conditions. Adv Environ Res 8:501–551

Golob V, Ojstrsek A (2005) Removal of vat and disperse dyes from residual pad liquors. Dyes Pigments 64:57–61

Gosavi VD, Sharma S (2014) A general review on various treatment methods for textile wastewater. J Environ Sci Comput Sci Eng Technol 3:29–39

Guo J, Zhou J, Wang D, Tian C, Wang P, Salah Uddin M (2007) A novel moderately halophilic bacterium for decolorizing azo dye under high salt condition. Biodegradation 19:15–19

Gupta V, Suhas K (2009) Application of low cost adsorbents for dye removal-a review. J Environ Manag 90(8):2313–2342

Hameed BH, Ahmad AA, Aziz N (2007) Isotherms, kinetics and thermodynamics of acid dye adsorption on activated palm ash. Chem Eng J 133(1–3):195–203

Ince NH, Ziylan A (2015) Single and hybrid applications of ultrasound for decolorization and degradation of textile dye residuals in water. In: Sharma SK (ed) Green chemistry for dyes removal from waste water. Wiley, Hoboken, pp 261–263

Jadhav AJ, Srivastava VC (2013) Adsorbed solution theory based modeling of binary adsorption of nitrobenzene, aniline and phenol onto granulated activated carbon. Chem Eng J 229:450–459

Kant R (2012) Textile dyeing industry an environmental hazard. Nat Sci 4(1):22–26

Karcher S, Kornmuller A, Jekel M (2001) Screening of commercial sorbents for the removal of reactive dyes. Dyes Pigments 5:111–125

Khan R, Bhawana P, Fulekar MH (2012) Microbial decolorization and degradation of synthetic dyes: a review. Rev Environ Sci Biotechnol 12(1):75–97

Koyuncu I (2002) Reactive dye removal in dye/salt mixtures by nanofiltration membranes containing vinyl sulphone dye: effects of feed concentration and cross flow velocity. Desalination 143:243–253

Kumar R, Ahmad R (2010) Adsorption studies of hazardous malachite green onto treated ginger waste. J Environ Manag 91(4):1032–1038

Kumari V, Yadav A, Haq I, Kumar S, Bharagava RN, Singh SK, Raj A (2016) Genotoxicity evaluation of tannery effluent treated with newly isolated hexavalent chromium reducing Bacillus Cereus. J Environ Manag 183:204–211

Kurbus T, Majcen Le Marechal A, BrodnjakVoncina D (2003) Comparison of H_2O_2/UV, H_2O_2/O_3 and H_2O_2/Fe^{2+} processes for the decolorization of vinyl sulphone reactive dyes. Dyes Pigments 58:245–252

Lee SY, Maniquiz MC, Choi JY, Jeong SM, Kim LH (2013) Seasonal nutrient uptake of plant biomass in a constructed wetland treating piggery wastewater effluent. Water Sci Technol 67(6):1317–1323

Lee G, Zhang Y, Shao S (2014) International conference on environment systems science and engineering (ESSE 2014) study on recycling alkali from the wastewater of textile mercerization process by nanofiltration. IERI Procedia 9:71–76

Liang T, Wang L (2015) An environmentally safe and nondestructive process for bleaching birch veneer with peracetic acid. J Clean Prod 92:37–43

Lin SH, Lin CM (1993) Treatment of textile waste effluents by ozonation and chemical coagulation. Water Res 27:1743–1748

Magdum SS, Minde GP, Kalyanraman V (2013) Rapid determination of indirect cod and polyvinyl alcohol from textile desizing wastewater. Pollut Res 32:515–519

Mander U, Kuusemets V, Lohmus K, Mauring T, Teiter S, Augustin J (2003) Nitrous oxide, dinitrogen and methane emission in a subsurface flow constructed wetland. Water Sci Technol 48(5):135–142

Mani S, Bharagava RN (2016) Exposure to crystal violet, its toxic, genotoxic and carcinogenic effects on environmental and its degradation and detoxification for environmental safety. Rev of Environ Conta And Toxicol 237:71–104

Mani S, Bharagava RN (2017) Isolation, screening and biochemical characterization of bacteria capable of crystal violet dye Decolorization. Int J Appl Adv Sci Res 2(2):70–75

Mathur N, Bhatnagar P, Bakre P (2005) Assessing Mutagenicity of Textile Dyes from Pali (Rajasthan) using AMES Bioassay. Environmental Toxicology Unit, Department of Zoology, University of Rajasthan, Jaipur, India

Maucieri C, Mietto A, Barbera AC, Borin M (2016) Treatment performance and greenhouse gas emission of a pilot hybrid constructed wetland system treating digestate liquid fraction. Ecol Eng 94:406–417

Mbuligwe SE (2005) Comparative treatment of dye-rich wastewater in engineered wetland systems (EWSs) vegetated with different plants. Water Res 39:271–280

Miralles-Cuevas S, Oller I, Agüera A (2016) Combination of nanofiltration and ozonation for the remediation of real municipal wastewater effluents: acute and chronic toxicity assessment. J Hazard Mater 323:442–451

Mishra A, Bajpai M (2005) Flocculation behavior of model textile wastewater treated with a food grade polysaccharide. J Hazard Mater 118:213–217

Mishra S, Bharagava RN (2016) Toxic and genotoxic effects of hexavalent chromium in environment and its bioremediation strategies. J Environ Sci Health Part C 34(1):1–34

Moir SE, Svoboda I, Sym G, Clark J, McGechan MB, Castle K (2005) An experimental plant for testing methods of treating dilute farm effluents and dirty water. Biosyst Eng 90(3):349–355

Nivala J, Wallace S, Headley T, Kassa K, Brix H, van Afferden M, Müller R (2013) Oxygen transfer and consumption in subsurface flow treatment wetlands. Ecol Eng 61:544–554

O'Geen AT, Budd R, Gan J, Maynard JJ, Parikh SJ, Dahlgren RA (2010) Chapter one-mitigating nonpoint source pollution in agriculture with constructed and restored wetlands. Adv Agron 108:1–76

Ojstrsek A, Fakin D, Vrhovsek D (2007) Residual dye bath purification using a system of constructed wetland. Dyes Pigments 74:503–507

Pagga U, Brown D (1986) The degradability of dyestuffs: part II behaviour of dyestuffs in aerobic biodegradation tests. Chemosphere 15:479–491

Pappalardo SE, Otto S, Gasparini V, Zanin G, Borin M (2016) Mitigation of herbicide runoff as an ecosystem service from a constructed surface flow wetland. Hydrobiologia 774(1):193–202

Paul SA, Chavan SK, Khambe SD (2012) Studies on characterization of textile industrial waste water in solapur city. Int J Chem Sci 10:635–642

Ramesh Babu B, Parande AK, Raghu S, Prem Kumar T (2007) Textile technology-cotton textile processing: waste generation and effluent treatment. J Cotton Sci 11:141–153

Ratthore JS, Choudhary V, Sharma S (2014) Implications of textile dyeing and printing effluents on groundwater quality for irrigation purpose pali. Rajasthan Eur Chem Bull 3:805–808

Rott U, Minke R (1999) Overview of wastewater treatment and recycling in the textile processing industry. Wat Sci Technol 40(1):137–144

Saha PD, Chakraborty S, Chowdhury S (2012) Batch and continuous (fixed-bed column) biosorption of crystal violet by *Artocarpus heterophyllus* (jackfruit) leaf powder. Colloids Surf B: Biointerfaces 92:262–270

Salvato M, Borin M (2010) Effect of different macrophytes in abating nitrogen from a synthetic wastewater. Ecol Eng 36(10):1222–1231

Sarayu K, Sandhya S (2012) Current technologies for biological treatment of textile wastewater a review. Appl Biochem Biotechnol 167:645–661

Scholz M, Xu J (2002) Performance comparison of experimental constructed wetlands with different filter media and macrophytes treating industrial wastewater contaminated with lead and copper. Bioresour Technol 83:71–79

Schowanck D, Fox K, Holt M (2001) GREATER: a new tool for management and risk assessment of chemicals in river basins. Contribution to GREATER. Wat Sci Technol 43:179–185

Sen S, Demirer GN (2003) Anaerobic treatment of real textile wastewater with a fluidized bed reactor. Water Res 37:1868–1878

Solano ML, Soriano P, Ciria MP (2004) Constructed wetlands as a sustainable solution for wastewater treatment in small villages. Biosyst Eng 87(1):109–118

Sponza DT, Isik M (2006) Anaerobic/aerobic sequential treatment of a cotton textile mill wastewater. J Chem Technol Biotechnol 79(11):1268–1274

Sujata, Bharagava RN (2016) Microbial degradation and decolorization of dyes from textile industry eastewater. Bioremediation Ind Pollutants: 53–90

USEPA (1988) Methods for aquatic toxicity identification evaluations- Phase I. Toxicity characterization procedures. U.S. Environmental Protection Agency, Environmental Research Laboratory, National Effluent Toxicity Assessment Center, Duluth

VanderZaag AC, Gordon RJ, Burton DL, Jamieson RC, Stratton GW (2008) Ammonia emissions from surface flow and subsurface flow constructed wetlands treating dairy wastewater. J Environ Qual 37(6):2028–2036

Verlicchi P, Zambello E (2014) How efficient are constructed wetlands in removing pharmaceuticals from untreated and treated urban wastewaters? A review. Sci Total Environ 470:1281–1306

Vymazal J (2007) Removal of nutrients in various types of constructed wetlands. Sci Total Environ 380(1):48–65

Vymazal J (2009) The use constructed wetlands with horizontal sub-surface flow for various types of wastewater. Ecol Eng 35(1):1–17

Vymazal J (2010) Constructed wetlands for wastewater treatment. Water 2(3):530–549

Vymazal J (2016) Concentration is not enough to evaluate accumulation of heavy metals and nutrients in plants. Sci Total Environ 544:495–498

Vymazal J, Kropfelova L (2008) Is concentration of dissolved oxygen a good indicator of processes in filtration beds of horizontal-flow constructed wetlands? Wastewater treatment, plant dynamics and management in constructed and natural wetlands. Springer, Dordrecht, pp 311–317

Vymazal J, Greenway M, Tonderski K, Brix H, Mander U (2006) Constructed wetlands for wastewater treatment. Wetlands and natural resource management. Springer, Berlin, pp 69–96

Wang YH, Inamori R, Kong HN, Xu KQ, Inamori Y, Kondo T, Zhang JX (2008a) Nitrous oxide emission from polyculture constructed wetlands: effect of plant species. Environ Pollut 152(2):351–360

Wang YH, Inamori R, Kong HN, Xu KQ, Inamori Y, Kondo T, Zhang JX (2008b) Influence of plant species and wastewater strength on constructed wetland methane emissions and associated microbial populations. Ecol Eng 32:22–29

Waring DR, Hallas G (2013) The chemistry and application of dyes. Springer

Wu S, Kuschk P, Brix H, Vymazal J, Dong R (2014) Development of constructed wetlands in performance intensifications for wastewater treatment: a nitrogen and organic matter targeted review. Water Res 57:40–55

Zainith S, Sandhya S, Saxena G, Bharagava RN (2016) Microbes an ecofriendly tools for the treatment of industrial waste waters. Microbes Environ Manag 1:78–103

Zhao YJ, Cheng P, Pei X, Zhang H, Yan C, Wang SB (2013) Performance of hybrid vertical up-and downflow subsurface flow constructed wetlands in treating synthetic high-strength wastewater. Environ Sci Pollut Res 20(7):4886–4894

# Chapter 12
Pesticide Contamination: Environmental Problems and Remediation Strategies

Siddharth Boudh and Jay Shankar Singh

Abstract Pesticides are the chemicals used in the control of weeds and pests. The larger inputs of pesticides and fertilisers contaminate food commodities with trace amounts of chemical pesticides and its invasion in crops causes diseases, which is a growing source of concern for the universal population and environment in today's world. The extensive utilisation of pesticides possibly enhances their accumulation in the agricultural fields and environmental components, such as enlarged farms, field sizes, loss of landscape elements etc. Nevertheless, their low biodegradability has classified these chemical substances as a persistent toxic element. Furthermore, organo-chlorine pesticides have caused multiple problems of health hazards, such as acute and chronic effects including developmental effects and neurological disruptors in humans and animals. The biological stability of pesticides and the higher content of lipophilicity in food products create a significant effect on the physical condition of human beings and animals. As the bio-accumulation and bio-magnification of lethal pesticides are the main cause of the loss of plants, microbes and animal biodiversity, therefore, microbially based bioremediation of toxic pollutants from the polluted sites has been proposed to be a safe and sustainable means of decontaminating the environment. In this communication, we have tried to explain the source of environmental pollution by pesticides, its hazardous effects on living beings and remediation strategies.

Keywords Fertilisers · Pesticide · Bioremediation technologies · Composting

1 Introduction

Environmental exposure to toxic chemicals such as pesticides is a significant health risk to humans and other animals (Azmi et al. 2006; Kiefer and Firestone 2007; Rothlein et al. 2006; Singh et al. 2011). Use of organochlorine pesticides (OCPs) to

S. Boudh · J. S. Singh (✉)
Department of Environmental Microbiology (DEM), Babasaheb Bhimrao Ambedkar University (A Central University), Lucknow, Uttar Pradesh, India
e-mail: jayshankar_1@yahoo.co.in

© Springer Nature Singapore Pte Ltd. 2019
R. N. Bharagava, P. Chowdhary (eds.), *Emerging and Eco-Friendly Approaches for Waste Management*, https://doi.org/10.1007/978-981-10-8669-4_12

control weeds creates resistance to agricultural pests and vector-borne diseases (Abhilash and Singh 2009). Degradation of dichlorodiphenyltrichloroethane (DDT) in soil is estimated to range from 4 to 30 years, whereas some other chlorinated OCPs may remain stable for many years after their use (Afful et al. 2010). Because of their inability to break down in the environment, their degradation is restricted in physical, chemical, biological and microbiological ways (Afful et al. 2010; Darko and Acquaah 2007; NCEH 2005; Swackhamer and Hites 1988; Kumar and Singh 2017). As they are fat-soluble components, they are able to bioaccumulate inside the lipid components of biota including fatty tissues, breast milk and blood within the food chain. As a result, humans and animals are exposed to the harmful effects of these micro-pollutants by eating foods in contact with contaminated soil or water (Belta et al. 2006; Raposo and Re-Poppi 2007; Mishra and Bharagava 2016). These pesticides are also highly toxic to most aquatic life and cause serious diseases in humans and animals. (Aiyesanmi and Idowu 2012) and soil microflora (Megharaj 2002). To tackle these environmental issues, different physico-chemical methods such as land-filling, incineration, composting or burning and chemical amendments have been used to remove pesticide contamination from the environment over the last few decades (Kempa 1997; Wehtje et al. 2000).

Using a broad range of chemicals to destroy pests is an essential aspect of agricultural practice in both developed and developing nations. This has increased crop production and decreased post-harvest losses. However, the extended use of several pesticides expectedly results in residues in foods and caused a worldwide issue over the potential adverse effects of these chemicals on the environment and human health. It is clear that the chance of exposure to pesticides is maximal amongst farm workers. Drinking water and food crops are also contaminated by pesticides, especially fruits and vegetables, which received the largest dosages of pesticides, and are therefore probably serious health hazards to consumers (Pimentel et al. 1992). The pesticides currently used, include a large mixture of chemical compounds, which show great differences in their mode of action, absorbance by the body, metabolism, removal from the body, and toxic effect on humans and other living organisms. Some pesticides show a high acute toxicity, but when they come in contact with the body they are freely metabolised and eliminated, but some others that show lower acute toxicity, and have strong inclination to assemble in the body.

Adverse effects may not only be caused by the critical ingredients and related impurities, but also by solvents, emulsifiers, carriers and other constituents of the formulated products. When lower costs, increased levels of environmental protection, and improved effectiveness are considered, then modern technologies such as bioremediation come into action, which can be more commercial and provide more effective clean-up than recognised treatment technologies (Rigas et al. 2005; Singh and Seneviratne 2017). Presently, the bioremediation tools, particularly microbial-based technologies have been proposed to be safe and sustainable means of decontaminating the toxic pollutants from the polluted sites and environment. Therefore, the present communication explains the various sources of pesticide pollution, its hazardous effects on living beings and how they can be removed from the contaminated sites.

2 Classification of Pesticides

The pesticides are those elements that are used globally as a fungicide, insecticides, herbicide molluscicides, nematicides, rodenticides and plant growth regulators, to control pests, weeds and diseases in crops and for the healthcare of human beings and animals. Millions of tons of pesticides are used every 12 months globally (Pimentel 2009), large amounts of which reach non-agricultural habitats by means of aerial drift, run-off, overspray or endangering organisms living in these regions (Giesy et al. 2000; Lehman and Williams 2010). Pesticides have an effect on behaviour, physiology, development, and ultimately on the survival and reproductive success of non-target organisms through direct toxicity, by means of disrupting endocrine functions and by exerting teratogenic and immune-toxic consequences (Hoffman 2003).

Pesticides are very handy and beneficial agents capable of preventing losses of crops and diseases in humans. Pesticides can be classified as destroying, repelling and mitigating agents. Insects and pests are getting immune to the commercial pesticides owing to over-usage. Recently, pesticides have been developed that target multiple species (Speck-Planche et al. 2012). Nowadays, chemical pesticides and insecticides are becoming a dominant agent for eliminating pests. When these chemical pesticides are used in a combination with an effective natural enemy, then that results in enhanced integrated pest management and acts as a comprehensive prophylactic and remedial treatment (Gentz et al. 2010). At the population level, the effects of pesticides depend on exposure and toxicity, and on different factors such as life history, characteristics, the timing of application, population structure and landscape structure (Schmolke et al. 2010). Nerve targets of insects that are known for the neuron-damaging insecticides include acetylcholinesterase for organophosphates and methyl carbamates, nicotinic acetylcholine receptors for neonicotinoids, gamma-aminobutyric acid receptor channel for polychlorocyclohexanes and fiproles and voltage-gated sodium channels for pyrethroids and DDT (Casida and Durkin 2013). It is an observation that the use of neonicotinoid pesticides is increasing. These pesticides are associated with different types of toxicities (Van Djik 2010).

Worldwide, pesticides are divided into different categories depending upon their target. Some of these categories include herbicides, insecticides, fungicides, rodenticides, molluscicides, nematicides and plant growth regulators. Non-regulated use of pesticides has had disastrous consequences for the environment. Serious concerns about human health and biodiversity are rising owing to the overuse of pesticides (Agrawal et al. 2010). Pesticides are considered to be more water soluble, heat-stable and polar, which makes it very difficult to reduce their lethal nature. Pesticides are not only toxic to people related to agriculture, but they also cause toxicity in industries and areas where pesticides are frequently used. Depending upon the target species, pesticides can cause toxicities in natural flora, natural fauna and aquatic life (Rashid et al. 2010).

3 Good Agricultural Practices and Management of Pesticide Residues

- Keep an inventory of all chemicals. Store all chemicals in their original containers. Never store herbicides with other pesticides.
- Always use only recommended pesticides at the specified doses and frequency and at specific times. Never use banned pesticides.
- Education and training should be provided for pesticide application. The improper use or misuse through lack of understanding creates residue problems.
- Unused pesticide solution and washings generated by cleaning spray pumps contain pesticide residue. Dispose of them properly to avoid pollution.
- An integrated pest management system should be used.
- Use safe pesticides that help in conserving predators/parasites.
- Strictly follow the prescribed waiting period before harvesting.
- Maintain healthy soil with compost and mulch to avoid pest problems.
- Vegetables and fruits should be thoroughly washed with clean water.
- Reduce spray drift in orchards by using lower pressures, larger nozzles and less volatile pesticides.
- Proper precautions must be taken for the control of household insects/stored grain pests.
- Use botanicals/microbial insecticides for the control of various crop pests.
- Get detailed information from the authorised people before mixing various pesticides and purchase pesticides only from authorised dealers.
- Always read the product description carefully.

4 Production and Usage of Pesticides in India

In 1952, pesticide production started in India. The first plant was established near Calcutta to produce benzene hexachloride. After establishment, pesticide production yield has been showing steady increment for example in 1958, a total of 5000 metric tons of technical-grade pesticide were produced which had become 102,240 metric tons by 1998. After China, India is the second largest producer of pesticides in Asia, whereas it is 12th in the global ranking (Mathur 1999). In India, 45% of pesticides are only used for cotton crops followed by paddy crops. Insecticides are mainly used in India rather than fungicides and herbicides. Andhra Pradesh is a major pesticide consumer state in India. India stands at the lowest rank in the world's per hectare pesticide consumption scale with 0.6 kg/ha, whereas Taiwan holds no. 1 position with 17 kg/ha.

5 Impact of Pesticide on Human Health, Environment and Biodiversity

5.1 Human Health

The past few decades have shown an increase in pesticide consumption, which is why their residues are found easily in different environment compartments and several cases have been reported in which pesticide residues cause problems in the environment, human, and all other living creatures. Figures 12.1 and 12.2 show the presence of residues of persistent organic pesticides in different states of India. In 1958, the first report regarding pesticide poisoning in Kerala (India) was reported, in which over 100 people died because of consumption of parathion-contaminated wheat flour (Karunakaran 1958).

Inhalation, ingestion and penetration through the skin are the common routes of pesticide entry into the human body (Spear 1991). Infants, children below the age of 10, pesticide applicators and farm workers are more susceptible to pesticide toxicity than others, Pesticide degradation or elimination is performed by the body, but some residues of are absorbed by blood (Jabbar and Mallick 1994). Hayo and Werf (1996) observed that when pesticide concentration is increased in the body and then its initial concentration in the environment, this causes toxicity. According to the WHO, every year, 220,000 fatalities and 3,000,000 pesticide poisoning cases are

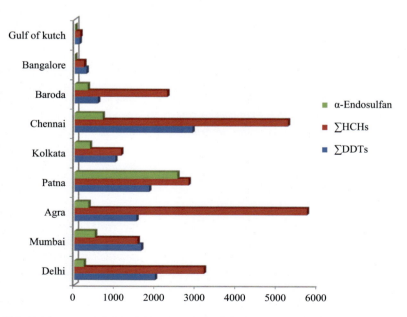

Fig. 12.1 Persistent organic pesticide residues (pg/m^3) in air from different regions of India. (Source: Chakraborty et al. (2010), Zhang et al. (2008), Pozo et al. (2011))

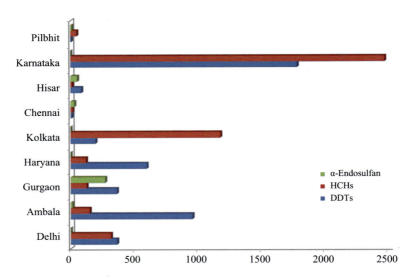

Fig 12.2 Persistent organic pesticide residues (ng/l) in water from different regions of India. (Source: Kaushik et al. (2008, 2010, 2012), Sundar et al. (2010), Ghose et al. (2009), Begum et al. (2009), Malik et al. (2009))

reported and approximately 2.2 million people are at a high risk of pesticide exposure in developing countries (Lah 2011; Hicks 2013).

5.1.1 Acute Effects of Pesticides

Acute effects are skin itching, irritation of the nose and throat, headache, appearance of a rash and blisters on the skin, nausea, vomiting, stinging of the eyes and skin, diarrhoea, dizziness, blindness, blurred vision and very rarely death, which may occur immediately after exposure to a pesticide (usually within 24 h). Most of the time, acute effects of pesticides are not severe enough to require medical attention every time.

5.1.2 Chronic Effects of Pesticides

Chronic effects take several years to appear, sometime years. These effects affect several parts of our body such as the liver, lungs and kidney. It can cause hypersensitivity, allergies, asthma and serious damage to the immune system (Culliney et al. 1992). The affected person can go through several neurological health conditions such as loss of memory and coordination, and their visual ability and motor signalling are reduced. Confusion, nervousness and hypersensitivity to light, sound and touch are symptoms that are also seen in OCP toxicity (Lah 2011). It can also cause oncogenic, mutagenic and carcinogenic effects. It affects the reproductive capabilities of the person by altering the male and female hormone levels. In other words, it causes infertility, spontaneous abortion, birth defects and stillbirth.

Richter (2002) observed that approximately 3 million tonnes of pesticides are used worldwide, which results in 26 million cases of non-fatal pesticide poisoning. Similarly, Hart and Pimentel (2002) reported that from all the cases of pesticide poisonings, 26 million patients are hospitalised and about 750,000 chronic diseases are caused every year. Symptoms of organophosphates and carbamates pesticide exposure are similar to those of another pesticide, which increases acetylcholine levels in the body. Convulsions, coma, improper breathing and death may occur in severe cases. Pyrethroid pesticide also causes reproductive and developmental effects and allergic skin responses.

Pyrethroids can cause an allergic skin response, aggressiveness, hyperexcitation, reproductive or developmental effects, in addition to tremors and seizures (Lah 2011). It is observed that there is a relationship between pesticides and Parkinson's disease/Alzheimer's disease (Casida and Durkin 2013).

5.2 Environment

Pesticide application is harmful in every prospectus. When applied it causes harm to soil; when water drifts from the applied area, it causes water contamination. In other words, pesticide affected not only target organisms, but also others such as birds who eat them, fishes and other aquatic animals in which pesticide residues accumulate, animals and the humans who eat the fishes and aquatic animals, other beneficial insects that died because of pesticide toxicity, and non-target plants. Among all classes of pesticides, insecticides cause most of the toxicity, although herbicides also pose a risk to non-target organisms. Water toxicity caused by pesticide is a major worldwide concern today because water is an essential part of our daily life (Kolpin et al. 1998).

5.2.1 Soil Contamination and Effect on Soil Fertility

Soil pollution has become a worldwide concern. Every day, a large number of contaminants such as pesticides, polycyclic aromatic hydrocarbons (PAHs), chlorophenols, petroleum and related products, and heavy metals, various pollutants enter the soil and pose a serious threat to the environment and human health (Gong et al. 2009; Kavamura and Esposito 2010; Udeigwe et al. 2011; Xu et al. 2012; Hu et al. 2013; Tang et al. 2014; Yadav et al. 2017). Soil contaminated mainly by agricultural and industrial activities has become an area of concern in recent years (Ha et al. 2014). Various transformation products from pesticides have been reported (Barcelo and Hennion 1997; Roberts 1998; Roberts and Hutson 1999). Soil pH also plays an important role in pesticide adsorption. When soil pH decreases, adsorption of ionisable pesticide (e.g. 2,4-D, 2,4,5-T, picloram atrazine) increases (Andreu and Pico 2004).

Overuse of pesticide can kill many beneficial microorganisms in the soil. According to Dr Elaine Ingham "soil will be degraded if we lose both fungi and bacteria". Overuse of pesticide and chemical fertiliser in the case of soil microbiota and overuse of antibiotics in the case of humans, both cases will end on an equal, drastic and damaging effect. Uncontrolled and random use of these chemicals may solve a problem now but because these soils will not able to hold as many beneficial microorganisms in future" (Savonen 1997). Plant needs many soil microorganisms to perform the nitrification process and because of pesticide discrimination, this process is disrupted (Singh 2015a, b, c, d, 2016). Pell et al. (1998) observed that triclopyr inhibits ammonia into nitrite transformation. Similarly, 2,4-D inhibits growth and activity of blue–green algae (Tozum-Çalgan and Sivaci-Guner 1993; Singh and Singh 1989), reduces the nitrogen fixation process (Fabra et al. 1997; Arias and Fabra 1993) and inhibits ammonia into nitrate transformation by soil microorganisms (Martens and Bremner 1993; Frankenberger et al. 1991). Santos and Flores (1995) observed that activity and growth of free-living N_2-fixing bacteria are inhibited by glyphosate.

Mycorrhizal fungi show a symbiotic relationship with plant roots and help them to absorb nutrients. Pesticide overuse also causes damage to these fungi. Trifluralin and oryzalin can inhibit the growth of certain mycorrhizal fungal species (Kelley and South 1978). Similarly, oxadiazon triclopyr and Roundup® show damaging effects on mycorrhizal fungi species (Moorman 1989; Chakravarty and Sidhu 1987; Estok et al. 1989).

5.2.2 Water Contamination

Kole et al. (2001) collected fish and water samples from all streams of Calcutta and found that more than 90% of the sample contained one or more pesticides. Bortleson and Davis (1987–1995) observed all river streams of the USA that flow from urban and agricultural areas and reported that the water of the urban streams contains more pesticide than agricultural river streams. More than 58% of samples of drinking water were found to be contaminated with OCP under a survey conducted around the Bhopal city of Madhya Pradesh (Kole and Bagchi 1995). Clean-up of water is a very complex and costly procedure. Once water becomes polluted with pesticides or other toxic chemicals, its clean-up is difficult and takes years to achieve (US EPA 2001; Waskom 1994; O'Neil and Raucher 1998). Pesticide-contaminated water easily drifts into surface water, which is why the pesticide level is higher in surface water than in groundwater (Anon 1993). Owing to leakages, improper disposal and accidental spills, these pesticides can be transferred to groundwater (Pesticides in Groundwater 2014).

5.3 Biodiversity

5.3.1 Aquatic Biodiversity

The aquatic ecosystem is mainly affected by pesticide, which drifts from the land into rivers, lakes and other bodies of water. Rohr et al. (2008) observe that atrazine shows a toxic effect on some fish and amphibian species. They also found a link between atrazine exposure and variation in the abundance of larval trematodes in northern leopard frogs via an experimental mesocosm study. Relyea (2005) also found that carbaryl and the herbicide glyphosate (Roundup®) are toxic to amphibian species. Asian Amphibian Crisis (2009) state that amphibians species are not majorly affected by overexploitation and habitat loss, but by pesticide-contaminated surface waters. Endosulfan and chlorpyrifos are also toxic for amphibians (Sparling and Feller 2009) The presence of herbicides in aquatic ecosystems also reduces the reproductive abilities of some aquatic animals (Helfrich et al. 2009). Scholz et al. (2012) observed a significant reduction in the fish population when pesticides were overused. Pimentel and Greiner (1997), based on the United States Environmental Protection Agency (US EPA) (1990b), state that large numbers of fishes died every year because of pesticide toxicity in water. The total number of fishes that died of all causes was 141 million fish per year, of which 6–14 million died because of pesticide toxicity.

5.3.2 Terrestrial Biodiversity

Pesticide application affects not only target plants, but also non-target plants. Phenoxy herbicides are toxic for non-targeted trees and shrubs (Dreistadt et al. 1994). Herbicides, sulphonamides, sulfonylureas, and imidazolinones have a profound effect on the productivity of non-targeted crops, plants and associated wildlife (Fletcher et al. 1993). Application of herbicide glyphosate can increase the susceptibility of plants to disease and infection (Brammall and Higgins 1988).

Pesticides such as carbamates, pyrethroids and organophosphates can affect the population of beneficial insects such as beetles and bees. Pilling and Jepson (2006) observed that the synergistic effects of the fungicides pyrethroids and imidazole or triazole are harmful to honey bees. Similarly, neonicotinoid insecticides such as clothianidin and imidacloprid are found to be toxic to bees. A very low dose of imidacloprid can negatively affect the foraging behaviour of bees (Yang et al. 2008) and reduce their learning capability (Decourtye et al. 2003). In the early twenty-first century, neonicotinoids are majorly responsible for the sudden disappearance of honey bees. This has had a profound effect on the food industry, as one third of food production are heavily dependent upon pollination via bees. Several reports show the presence of a significant amount of neonicotinoids in commercial honey and wax. The honeybee population has dropped by 29–36% since 2006.

The bird population is also affected and as experienced a massive decline due to pesticide use. Pesticides enter a bird's body and start accumulating in their tissues, leading to their death. Liroff (2000) reported that DDT and its metabolite exposure are the major reason behind the declining population of the bald eagle in the USA. Fungicides, which are used for killing earthworms, indirectly reduced the birds and mammal population. Granular pesticides look similar to food grains, are swallowed by birds and cause toxicity. Pesticides in sublethal quantities can affect the nervous system and cause behaviour changes.

There are several methods of pesticide application such as spraying on the crop plants or on the soil, mixing in the soil, and can be applied in a granular form. After application, the pesticide can disappear from the target site via dispersion, degradation, leaching into water bodies, rivers or may be consumed by plants and soil microbes (Hayo and Werf 1996). Overuse of pesticide can affect the functioning of soil microbes and indirectly affect soil fertility. Lang and Cai (2009) reported that chlorothalonil and dinitrophenyl fungicides can disrupt nitrification and denitrification processes. Similarly, triclopyr (Pell et al. 1998) and 2,4-D (Frankenberger et al. 1991) affect ammonia-oxidising bacteria that are involved in ammonia into nitrite transformation, whereas glyphosate reduces activity and growth of nitrogen-fixing bacteria in soil (Santos and Flores 1995). In addition to bacteria, herbicides also cause damage to fungal species. The herbicides oryzalin, triclopyr and trifluralin inhibit the growth of mycorrhizal fungi (Kelly and South 1978; Chakravarty and Sidhu 1987,) whereas oxadiazon affects fungal spore production (Moorman 1989).

Earthworms are an inseparable part of the soil ecosystem. They make a major contribution to soil fertility. It is the model organism for testing soil toxicity and also acts as a bio-indicator for soil contamination. Reported that pesticides cause neurotoxic and physiological damage in earthworms. Glyphosate affects the abundance and the feeding activity of earthworms (Casabe et al. 2007). Goulson (2013) studied the harmful effect of neonicotinoids on the ecosystem and reported that it can kill *Eisenia foetida* species of earthworms.

6 Bioremediation Technologies

It is estimated that more than 100 million bacteria (5000–7000 different species) and more than 10,000 fungal colonies are present in only 1 g of soil (Dindal 1990; Melling 1993). Bioremediation approaches are safer and more economical than other commonly used physicochemical strategies (Vidali 2001). There are several compounds that contaminate the soil and require remediation, such as the inorganic compounds nitrates, phosphates and perchlorates (Nozawa-Inoue et al. 2005); explosives such as hexahydro-1,3,5-trinitro-1,3,5-triazine (RDX) and octahydro-1,3,5,7-tetranitro-1,3,5,7-tetrazocine (HMX) (Kitts et al. 1994); monoaromatic hydrocarbons such as benzene, toluene, ethylbenzene and xylene (known as BTEX) (Rooney-Varga et al. 1999); PAHs (Wang et al. 1990); a range of herbicides such as diuron, linuron and chlortoluron (Fantroussi et al. 1999); and heavy metals (Glick 2003). Bioremediation technologies are based on the principles of biostimulation and bioaugmentation for the successful application of bioremediation technology (Singh 2011; Singh and Pandey 2013; Bharagava et al. 2017). Sebate et al. (2004) proposed a protocol for the bio-treatability assays in two phases. Their metabolic activities and inhibitor presence are assessed under the first phase, whereas the second phase deals with the influences of nutrients, surfactants and the amount of inoculum administered at the polluted site. To achieve successful bioremediation, pollutants as substrates must be available and accessible either to microorganisms or to their extracellular enzymes so that metabolism occurs.

There are several microorganisms such as *Streptomyces* sp. strain M7, *Arthrobacter fluorescens* and *Arthrobacter giacomelloi, Chlorella vulgaris* and *Chlamydomonas reinhardtii, Clostridium sphenoides, S. Japonicum* UT26 that have been discovered by researchers that are capable of degrading the organochlorine insecticide lindane (Boudh et al. 2017). *Phanerochaete chrysosporium* and *P. sordida* also have the ability to degrade DDT in contaminated soil using a landfarming approach (Safferman et al. 1995). Different microbial mediated and other approaches such as composting, electro-bioremediation, microbially assisted phytoremediation and bio-augmentation are widely used to achieve successful bioremediation and sustainable environmental development (Singh and Strong 2016; Singh et al. 2016; Vimal et al. 2017).

6.1 Composting

Composting is a natural recycling process in which microorganisms rapidly consume organic matter and use these as an energy source, converting it into CO_2, water, microbial biomass, heat and compost. The feedstock used to fuel the composting process may be obtained from a variety of sources, including crop residues, manure, bio-solids and other agricultural residues. These materials may contain a number of synthetic organic compounds or xenobiotics, including pesticides. Composting provides an optimal environment for pesticide destruction. Compost is well-suited to pesticide degradation because elevated or thermophilic temperatures achieved during composting permit faster biochemical reactions than are possible under ambient temperatures and make pesticides more bioavailable, increasing the chance of microbial degradation. Microorganisms also co-metabolise pesticide during composting. By mixing remediated soil with contaminated soil, the effectiveness of composting can be increased because the remediated soil with acclimated microorganisms significantly influence pollutant degradation in the composting process (Hwang et al. 2001). In the composting matrices, microorganisms degrade pollutants into innocuous compounds, transform more pollutant substances into less toxic substances or help in locking up the chemical pollutants within the organic matrix, thereby reducing pollutant bioavailability. Even in the compost remediation strategy, the bioavailability and biodegradability of pollutants are the two most important factors that determine the degradation efficiency (Semple et al. 2001). The spent mushroom waste from *Pleurotus ostreatus* can degrade and mineralise DDT in soil (Purnomo et al. 2010). On the other hand, Alvey and Crowley (1995) observed that additions of compost can suppress soil mineralisation of atrazine. The critical parameters of composting depend on the type of contaminants and waste materials that may be used for composting. In addition, this composting efficiency essentially depends on the temperature and soil waste amendment ratio (Antizar-Ladislao et al. 2005).

Guerin (2000) recommended for the optimal removal of aged PAH during composting keeping the moisture and amendment ratio constant. Namkoong et al. (2002) studied that the soil amendment with sludge-only or compost-only in a ratio of 1:0.1, 1:0.3, 1:0.5, and 1:1 (soil/amendment, wet weight basis) increased the degradation rates of PAHs, but higher mix ratios did not increase the degradation rates of total PAHs correspondingly. Cai et al. (2007) observed the composting process for bioremediating sewage sludge, which is contaminated with PAHs, and found that intermittently aerated compost treatment showing a higher removal rate of high molecular weight PAHs compared with continuously aerated and manually aerated compost treatments. The nature of waste or soil organic matter, which consists of humic materials, plays an important role in the binding of contaminants such as PAHs and making them bioavailable for degradation. Plaza et al. (2009) reported that during composting, some humic material lost their aliphatic groups, their polarity had been increased and they also entered into aromatic polycondensation, which alters their structural and chemical properties, resulting in a decrease in PAH binding. Humic material acts like a surfactant in compost and plays a crucial role in releasing PAHs sorbed into the soil. It also increased the stability of soil. PAH degradation mostly occurs during the mesophilic stage of composting, whereas the thermophilic stage is inhibitory for biodegradation (Antizar-Ladislao et al. 2004; Haderlein et al. 2006; Sayara et al. 2009). Sayara et al. (2010) reported that stable composts that are present in municipal solid wastes enhanced the biodegradation of PAH, particularly during the initial phase of composting. Similar to any other technology used in bioremediation, composting has its own advantages and limitations. It is a sustainable and most cost-effective remediation method that may also improve the soil structure, nutrient status and microbial activity (Singh 2013a, b, 2014). During composting, the contaminant can degrade through different mechanisms such as mineralisation by microbial activity, transformation to products, volatilisation, and also the formation of non-extractable bound residues with organic matter. One of the critical knowledge gaps of composting is the lack of sufficient knowledge about the microorganisms involved in various stages of composting, especially in the thermophilic stage, which is almost like a black box. In fact, there are conflicting views of researchers about the role of the thermophilic stage of composting in bioremediation of contaminants. For better designing of composting as a bioremediation strategy for contaminated soils, knowledge of the nature and activity of the microorganisms involved in various stages of composting and on the degree of stability of compost and its humic matter content is essential.

6.2 Electro-Bioremediation

Electro-bioremediation is a hybrid technology that combines both technologies of bioremediation and electrokinetics, for the treatment of hydrophobic organic compounds. In this method, we use microbiological phenomena for pollutant degradation and electrokinetic phenomena for the acceleration and orientation of the transport of pollutants or their derivatives and the pollutant degrading microorganisms (Chilingar et al. 1997; Li et al. 2010). In this method, we use weak electric fields of about 0.2–2.0 V cm^{-1} to the soil (Saichek and Reddy 2005) and transport phenomena associated with electrokinetics are electro-osmotic flow, electromigration, and electrophoresis, which can be utilised to effectively deliver nutrients to indigenous bacteria in the soils and to enhance bioavailability. Luo et al. (2005) developed a non-uniform electrokinetic system in which the polarity of an electric field is reversed to accelerate the movement and facilitate higher and more uniform biodegradation of phenol in a sandy loam soil. According to Wick et al. (2007), the impact of the direct current on organism–soil interactions and the organism compound is often neglected. Fan et al. (2007) tested a two-dimensional (2-D) non-uniform electric field on a bench scale with a sandy loam soil and 2,4-dichlorophenol (2,4-DCP) at bidirectional and rotational modes, and observed that about 73.4% of 2,4-DCP was removed at the bidirectional mode and about 34.8% at the rotational mode.

Shi et al. (2008) observed that direct current ($X = 1$ V cm^{-1}; $J = 10.2$ mA cm^{-2}), which is typically applied for electro-bioremediation measures had no negative effect on the activity of a PAH-degrading soil bacterium (*Sphingomonas* sp. LB126), on the other hand, the DC-exposed cells exhibited up to 60% elevated intracellular ATP levels, but remained unaffected by all other levels of cellular integrity and functionality. Niqui-Arroyo and Ortega-Calvo (2007) used an integrated biodegradation and electro-osmosis approach for enhanced removal of PAHs from creosote-polluted soils.

Velasco-Alvarez et al. (2011) applied the low intensity electric current in an electrochemical cell packed with an inert support and observed degradation of hexadecane and higher biomass production by *Aspergillus niger*. Maillacheruvu and Chinchoud (2011) reported the synergistic removal of contaminants by using an electro-kinetically transported aerobic microbial consortium. There are some limitations of electro-bioremediation technologies such as pollutant solubility and its desorption from the soil matrix, availability of suitable microorganisms at the contamination site, the concentration ratio between target and non-target ions, the requirement of a conducting pore fluid to mobilise pollutants, heterogeneity or anomalies found at sites, and toxic electrode effects on microbial metabolism or breakdown of dielectric cell membrane or changes in the physicochemical surface properties of microbial cells (Sogorka et al. 1998; Velizarov 1999; Virkutyte et al. 2002).

6.3 Microbe-Assisted Bioremediation

Physical remediation, chemical remediation and bioaugmentation (the addition of biodegradative bacteria to contaminated soils) techniques are commonly used for the treatment of contaminated soils. These remediation methods are costly and introduced microorganisms often do not survive in the environment; thus, phytoremediation became a good choice for this purpose. It is a cost-effective technique in which plants and their associated microorganisms help to remove, transform, or assimilate toxic chemicals located in soils, sediments, groundwater, surface water, and even the atmosphere (Reichenauer 2008; Glick 2010). There are some beneficial plant–microbe relationships, particularly between plants and plant growth-promoting rhizobacteria, plant endophytic bacteria and mycorrhizal fungi, that exist in nature, which helps in the natural bioremediation process of contaminated soil in which microorganisms increase the availability of contaminants and help plants with the extraction and removal of inorganic and organic compounds by using appropriate degradation pathways and metabolic capabilities (Hare et al. 2017).

Several studies suggest that microbially assisted phytoremediation offers much potential for bioremediation compared with lone phytoremediation (McGuinness and Dowling 2009; Weyens et al. 2009; Glick 2010). Endophytic and rhizobacteria are involved in the degradation of toxic organic compounds in environmental soil. Endophytic bacteria are present naturally in the internal tissues of plants (called endophytes) and rhizobacteria are associated with the rhizosphere of plants. Endophytic bacteria promote plant growth and contribute to enhanced biodegradation of environmental soil pollutants (Weyens et al. 2009). Similarly, rhizobacteria synthesise compounds that protect plants by decreasing plant stress hormone levels, delivering key plant nutrients, protecting against plant pathogens and degrading contaminants (McGuinness and Dowling 2009; Glick 2010). Table 12.1 shows some examples of the successful microbially assisted phytoremediation of pollutants.

Yousaf et al. (2010) isolated hydrocarbon degraders *Pseudomonas*, *Arthrobacter*, *Enterobacter* and *Pantoea* spp. from the root and stem tissues of Italian ryegrass and birds foot trefoil vegetated in hydrocarbon contaminated soil. Similarly, Siciliano et al. (2001) found endophytic hydrocarbon degraders in tall fescue (*Festuca arundinacea*) and rose clover (*Trifolium fragiferum*) at an aged hydrocarbon-contaminated site.

Plants produce many secondary plant metabolites (SPMEs) such as phytohormones, phytoanticipins, allelopathic chemicals, root exudates and phytosiderophores (Hadacek 2002). Gilbert and Crowley (1998) and Kim et al. (2003) reported that SPMEs such as limonene, cymene, carvone and pinene enhanced degradation of polychlorinated biphenyls (PCBs). Kupier et al. (2002) observed that when *Pseudomonas putida* PCL1444 was grown in PAH-polluted soil, it degraded the PAHs. It is isolated from the rhizosphere of *Lolium multiflorum* cv. Narasimhan et al. (2003) applied the rhizosphere metabolomics-driven approach in the rhizosphere of *Arabidopsis* to degrade PCBs.

Table 12.1 Examples of some successful microbial assisted phytoremediation of pollutant

Pollutants	Plant species	Microorganisms	References
Tetrachlorophenol	Wheat (*Triticum* spp.)	*Herbaspirillum* sp *K1*	Mannisto et al. (2001)
Explosives	Popular tissues (*Populus deltoidesnigra*)	*Methylobacterium populi*	van Aken et al. (2004a) and van Aken et al. (2004b)
Hydrocarbons	Pea (*Pisum sativum*)	*Pseudomonas putida*	Germaine et al. 2009
Polycyclic aromatic hydrocarbons	Tall fescue grass (*Festuca arundinacea*)	*Azospirillum lipoferum* sp.	Huang et al. (2004)
		Enterobacter cloacae CAL2	
		Pseudomonas putida UW3	
2,4-dichlorophenoxyacetic acid	Barley (*Hordeum Sativum* L.)	*Burkholderia cepacia*	Jacobsen (1997) and Shaw and Burns (2004)
	Ryegrass (*Lolium perenne* L.)	*Indigenous degraders*	
Pentachlorophenol	Ryegrass (*Lolium perenne* L.)	*Indigenous degraders*	He et al. (2005)
Trichloroethylene	Wheat (*Triticum* spp.)	*Pseudomonas fluorescens*	Yee et al. (1998)

7 Pesticide Application

Irregular and uncontrolled use of pesticide leads to various problems in agriculture such as the development of pesticide resistance to the pest populations that are causing diseases. The timing of pesticide application is mainly linked with extreme and unusual weather events (Johnson et al. 1995; Otieno et al. 2013). For example, in autumn, soil moisture is highly decreased, which limits field work, while an increase in soil moisture in the rainy season also forbids field work (Rosenzweig et al. 2001; Miraglia et al. 2009). Earlier application of the pesticide in autumn can make winter weed control more difficult (Bailey 2003).

The total amount of herbicides and the rate of their application are higher than for insecticides or fungicides in the past (Probst et al. 2005). This may be because of favourable climatic conditions for the pest population (Goel et al. 2005). In general, the increased use of agricultural chemicals appears necessary (Rosenzweig et al. 2001; Hall et al. 2002). For example, infection symptoms that appear frequently after a short time interval lead to frequent pesticide applications so that infection can be prevented (Roos et al. 2011; Noyes et al. 2009). Similarly, the evolution of pesticide-resistance pests requires improvement in the current pest management strategies. Improved biological control tools may be a solution to this problem (Jackson et al. 2011). Poor organic farmers of developing countries needed cheap, easily available, biodegradable and low-risk pesticides (Ntonifor 2011). Thus, some countries increase or re-introduce banned or restricted pesticides in field applications (Macdonald et al. 2005).

8 Conclusion

Nowadays, people are more aware of environmental safety and protection. Organic farming, bio-fertilisers and biopesticides have become the public's favourite research, based on pesticide exposure in the environment. Therefore, more research works are focused on the removal of these pesticides from food chains and different trophic levels. As the pesticides are persistent pollutants and have bioaccumulation and biomagnifications properties at successive trophic levels in an ecosystem, ensuring their long-term presence, this removes organisms at higher trophic levels. Furthermore, as pesticide application affects the biodiversity, human health and the environment, we should learn to avoid their use and try to replace them with natural ones. Use of bio-pesticides and bio-fertilisers to enhance agricultural productivity will be helpful technology to save our biodiversity, agriculture and environmental pollution. Application of efficient microbes for plant growth promotion and bioremediation can add one more step to the development of sustainable agriculture and environment. There is an urgent need to explore and identify more efficient and microbial communities with the potential for the bioremediation of pesticides at contaminated sites, present all over the world.

Acknowledgment We thank our Head for providing facilities and encouragements. Siddharth Boudh is thankful to the University Grants Commission (UGC) for financial support in the form of the Rajiv Gandhi National Fellowship (Award Letter No: F.1-17.1/2013-14/RGNF-2013-14-SC-UTT-37387/(SA-III/Website).

References

Abhilash PC, Singh N (2009) Pesticide use and application: an Indian scenario. J Hazard Mater 165:1–12

Afful S, Anim A, Serfor-Armah Y (2010) Spectrum of organochlorine pesticide residues in fish samples from the Densu Basin. Res J Environ Earth Sci 2(3):133–138

Agrawal A, Pandey RS, Sharma B (2010) Water pollution with special reference to pesticide contamination in India. J Water Res Prot 2(5):432–448

Aiyesanmi AF, Idowu GA (2012) Organochlorine pesticides residues in soil of cocoa farms in Ondo state central district, Nigeria. Environ Nat Resour Res 2(2):65–73

Alvey S, Crowley DE (1995) Influence of organic amendments on biodegradation of atrazine as a nitrogen source. J Environ Qual 24:1156–1162

Andreu V, Picó Y (2004) Determination of pesticides and their degradation products in soil: critical review and comparison of methods. Trends Anal Chem 23(10–11):772–789

Anon (1993) The environmental effects of pesticide drift. English Nature, Peterborough, pp 9–17

Antizar-Ladislao B, Lopez-Real JM, Beck AJ (2004) Bioremediation of polycyclic aromatic hydrocarbon (PAH)-contaminated waste using composting approaches. Crit Rev Environ Sci Technol 34:249–289

Antizar-Ladislao B, Lopez-Real J, Beck AJ (2005) In-vessel composting-bioremediation of aged coal tar soil: effect of temperature and soil/green waste amendment ratio. Environ Int 31:173–178

Arias RN, Fabra PA (1993) Effects of 2, 4-dichlorophenoxyacetic acid on *Rhizobium* sp. growth and characterization of its transport. Toxicol Lett 68:267–273

Azmi MA, Naqvi SN, Azmi MA, Aslam M (2006) Effect of pesticide residues on health and different enzyme levels in the blood of farm workers from Gadap (rural area) Karachi-Pakistan. Chemosphere 64:1739–1744

Bailey SW (2003) Climate change and decreasing herbicide persistence. Pest Manag Sci 60:158–162

Barceló D, Hennion MC (1997) Trace determination of pesticides and their degradation products in water. Elsevier, Amsterdam, p 3

Begum A, HariKrishna S, Khan I (2009) A Survey of persistant organochlorine pesticides residues in some Streams of the Cauvery River, Karnataka, India. Int J Chem Tech Res 1:237–244

Belta GD, Likata P, Bruzzese A, Naccarri C, Trombetta D, Turco VL, Dugo C, Richetti A, Naccari F (2006) Level and congener pattern of PCBs and OCPs residues in blue-fin tuna (*Thunnus thynnus*) from the straits of Messina (Sicily, Italy). Environ Int 32:705–710

Bharagava RN, Chowdhary P, Saxena G (2017) Bioremediation an eco-sustainable green technology, its applications and limitations. In: Bharagava RN (ed) Environmental pollutants and their bioremediation approaches. CRC Press, Taylor & Francis Group, Boca Raton, pp 1–22

Bortleson G, Davis D (1987) U.S. Geological Survey & Washington State Department of Ecology. Pesticides in selected small streams in the Puget Sound Basin, pp 1–4

Boudh S, Tiwar S, Singh JS (2017) Microbial mediated Lindane bioremediation. In: Singh JS, Seneviratne G (eds) Agro-Environmental sustainability: managing environmental pollution, vol II. Springer, pp 213–233

Brammall RA, Higgins VJ (1988) The effect of glyphosate on resistance of tomato to *Fusarium crown* and root rot disease and on the formation of host structural defensive barriers. Can J Bot 66:1547–1555

Cai QY, Mo CH, Wu QT, Zeng QY, Katsoviannis A, Ferard JF (2007) Bioremediation of polycyclic aromatic hydrocarbons (PAHs)-contaminated sewage sludge by different composting processes. J Hazard Mater 142:535–542

Casabé N, Piola L, Fuchs J et al (2007) Ecotoxicological assessment of the effects of glyphosate and chlorpyrifos in an Argentine soya field. J Soils Sediments 7(4):232–239

Casida JE, Durkin KA (2013) Neuroactive insecticides: targets, selectivity, resistance, and secondary effects. Annu Rev Entomol 58:99–117

Chakraborty P, Zhang G, Li J, Xu Y, Liu X, Tanabe S, Jones KC (2010) Selected organochlorine pesticides in the atmosphere of major Indian cities: levels, regional versus local variations, and sources. Environ Sci Technol 44:8038–8043

Chakravarty P, Sidhu SS (1987) Effects of glyphosate, hexazinone and triclopyr on in vitro growth of five species of ectomycorrhizal fungi. Eur J Pathol 17:204–210

Chilingar GV, Loo WW, Khilyuk LF, Katz SA (1997) Electrobioremediation of soils contaminated with hydrocarbons and metals: progress report. Energy Sour 19:129–146

Culliney TW, Pimentel D, Pimentel MH (1992) Pesticides and natural toxicants in foods. Agric Ecosyst Environ 41:297–320

Darko G, Acquaah SO (2007) Levels of organochlorine pesticide residues in meat. Int J Environ Sci Technol 4(4):521–524

Decourtye A, Lacassie E, Pham-Delègue MH (2003) Learning performances of honeybees (*Apis mellifera* L.) are differentially affected by imidacloprid according to the season. Pest Manag Sci 59:269–278

Dindal DL (1990) Soil biology guide. Wiley, New York

Dreistadt SH, Clark JK, Flint ML (1994) Pests of landscape trees and shrubs. An integrated pest management guide. University of California Division of Agriculture and Natural Resources. Publication No. 3359

Estok D, Freedman B, Boyle D (1989) Effects of the herbicides 2,4-D, glyphosate, hexazinone, and triclopyr on the growth of three species of ectomycorrhizal fungi. Bull Environ Contam Toxicol 42:835–839

Fabra A, Duffard R, Evangelista DDA (1997) Toxicity of 2,4-dichlorophenoxyacetic acid in pure culture. Bull Environ Contam Toxicol 59:645–652

Fan X, Wang H, Luo Q, Ma J, Zhang X (2007) The use of 2D non-uniform electric field to enhance in situ bioremediation of 2,4-dichlorophenol-contaminated soil. J Hazard Mater 148:29–37

Fantroussi S, Verschuere L, Verstraete W, Top EM (1999) Effect of phenylurea herbicides on soil microbial communities estimated by analysis of 16S rRNA gene fingerprints and community-level physiological profiles. Appl Environ Microbiol 65:982–988

Fletcher JS, Pfleeger TG, Ratsch HC (1993) Potential environmental risks associated with the new sulfonylurea herbicides. Environ Sci Technol 27:2250–2252

Frankenberger WT, Tabatabai Jr MA, Tabatabai MA (1991) Factors affecting L-asparaginase activity in soils. Biol Fert Soils 11(1):5

Gentz MC, Murdoch G, King GF (2010) Tandem use of selective insecticides and natural enemies for effective, reduced-risk pest management. Biol Control 52(3):208–215

Germaine KJ, Keogh E, Ryan D, Dowling DN (2009) Bacterial endophyte-mediated naphthalene phytoprotection and phytoremediation. FEMS Microbiol Lett 296:226–234

Ghose N, Saha D, Gupta A (2009) Synthetic detergents (surfactants) and organochlorine pesticide signatures in surface water and groundwater of Greater Kolkata, India. J Water Resour Protect 1(4):290–298

Giesy JP, Dobson S, Solomon KR (2000) Ecotoxicological risk assessment for roundup herbicide. Rev Environ Contam Toxicol 167:35–120

Gilbert ES, Crowley DE (1998) Repeated application of carvone-induced bacteria to enhance biodegradation of polychlorinated biphenyl in soil. Appl Environ Biotechnol 50:489–494

Glick BR (2003) Phytoremediation: synergistic use of plants and bacteria to clean up the environment. Biotechnol Adv 21:383–393

Glick BR (2010) Using soil bacteria to facilitate phytoremediation. Biotechnol Adv 28:367–374

Goel A, McConnell LL, Torrents A (2005) Wet deposition of current use pesticides at a rural location on the Delmarva peninsula: impact of rainfall patterns and agricultural activity. J Agri Food Chem 53(20):7915–7924

Gong JL, Wang B, Zeng GM, Yang CP, Niu CG, Niu QY (2009) Removal of cationic dyes from aqueous solution using magnetic multi-wall carbon nanotube nanocomposite as adsorbent. J Hazard Mater 164:1517–1522

Goulson DJ (2013) An overview of the environmental risks posed by neonicotinoid insecticides. J Appl Ecol 50:977

Guerin TF (2000) The differential removal of aged polycyclic aromatic hydrocarbons from soil during bioremediation. Environ Sci Pollut Res 7:19–26

Ha H, Olson J, Bian L, Rogerson PA (2014) Analysis of heavy metal sources in soil using kriging interpolation on principal components. Environ Sci Technol 48:4999–5007

Hadacek F (2002) Secondary metabolites as plant traits: current assessment and future perspectives. Crit Rev Plan Sci 21:273–322

Haderlein A, Legros R, Ramsay BA (2006) Pyrene mineralization capacity increased with compost maturity. Biodegradation 17:293–303

Hall GV, D'Souza RM, Kirk MD (2002) Food borne disease in the new millennium: out of the frying pan and into the fire? Med J Aust 177(11/12):614–619

Hare V, Chowdhary P, Baghel VS (2017) Influence of bacterial strains on *Oryza sativa* grown under arsenic tainted soil: accumulation and detoxification response. Plant Physiol Biochem 119:93–102

Hart K, Pimentel D (2002) Public health and costs of pesticides. In: Pimentel D (ed) Encyclopedia of pest management. Marcel Dekker, New York, pp 677–679

Hayo MG, Werf VD (1996) Assessing the impact of pesticides on the environment. Agric Ecosyst Environ 60:81–96

He Y, Xu J, Tang C, Wu Y (2005) Facilitation of pentachlorophenol degradation in the rhizosphere of ryegrass (*Lolium perenne* L.) Soil Biol Biochem 37:2017–2024

Helfrich LA, Weigmann DL, Hipkins P, Stinson ER (2009) Pesticides and aquatic animals: a guide to reducing impacts on aquatic systems. In: Virginia Polytechnic Institute and State University. Available from https://pubs.ext.vt.edu/420/420-013/420-013.html

Hicks B (2013) Agricultural pesticides and human health. In: National Association of Geoscience Teachers. Available from http://serc.carleton.edu/NAGTWorkshops/health/case_stdies/pesticides.html

Hoffman DJ (2003) Wildlife toxicity testing. In: Hoffman DJ, Rattner BA, Burton GAJ, Cairns JJ (eds) Handbook of ecotoxicology2nd edn. Lewis Publishers, Boca Raton, pp 75–110

Hu G, Li J, Zeng G (2013) Recent development in the treatment of oily sludge from petroleum industry: a review. J Hazard Mater 261:470–490

Huang XD, El-Alawi Y, Gurska J, Glick BR, Greenberg BM (2004) A multi-process phytoremediation system for removal of polycyclic aromatic hydrocarbons from contaminated soils. Environ Pollut 130:465–476

Hwang E, Namkoong W, Park J (2001) Recycling of remediated soil for effective composting of diesel-contaminated soil. Compos Sci Util 9:143–14149

Jabbar A, Mallick S (1994) Pesticides and environment situation in Pakistan (Working Paper Series No. 19). Available from Sustainable Development Policy Institute (SDPI)

Jackson L, Wheeler S, Hollander A, O'Geen A, Orlove B, Si J (2011) Case study on potential agricultural responses to climate change in a California landscape. Clim Chang 109(1):407–427

Jacobsen CS (1997) Plant protection and rhizosphere colonization of barley by seed inoculated herbicide degrading *Burkholderia* (*Pseudomonas*) *cepacia* DBO1(pRO101) in 2,4-D contaminated soil. Plant Soil 189:139–144

Johnson AW, Wauchope RD, Burgoa B (1995) Effect of simulated rainfall on leaching and efficacy of fenamiphos. J Nematol 27(4):555–562

Karunakaran CO (1958) The Kerala food poisoning. J Indian Med Assoc 31:204

Kaushik CP, Sharma HR, Jain S, Dawra J, Kaushik A (2008) Level of pesticide residues in river Yamuna and its canals in Haryana and Delhi, India. Environ Monit Assess 144:329–340

Kaushik A, Sharma HR, Jain S, Dawra J, Kaushik CP (2010) Pesticide pollution of river Ghaggar in Haryana, India. Environ Monit Assess 160:61–69

Kaushik CP, Sharma HR, Kaushik A (2012) Organochlorine pesticide residues in drinking water in the rural areas of Haryana, India. Environ Monit Assess 184:103–112

Kavamura VN, Esposito E (2010) Biotechnological strategies applied to the decontamination of soils polluted with heavy metals. Biotechnol Adv 28:61–69

Kelley WD and South DB (1978) In vitro effects of selected herbicides on growth and mycorrhizal fungi. Weed Science Society America Meeting. Auburn University, Auburn, Alabama, p 38.

Kempa ES (1997) Hazardous wastes and economic risk reduction: case study, Poland. Int J Environ Pollut 7:221–248

Kiefer MC, Firestone J (2007) Neurotoxicity of pesticides. J Agromedicine 12:17–25

Kim BH, Oh ET, So JS, Ahn Y, Koh SC (2003) Plant terpene-induced expression of multiple aromatic ring hydroxylation oxygenase genes in *Rhodococcus* sp. strain T104. J Microbiol 41:349–352

Kitts CL, Cunningham DP, Unkefer PJ (1994) Isolation of three hexahydro-1, 3, 5-trinitro-1, 3, 5-triazine-degrading species of the family Enterobacteriaceae from nitramine explosive-contaminated soil. Appl Environ Microbiol 60:4608–4611

Kole RK, Bagchi MM (1995) Pesticide residues in the aquatic environment and their possible ecological hazards. J Inland Fish Soc Ind 27(2):79–89

Kole RK, Banerjee H, Bhattacharyya A (2001) Monitoring of market fish samples for endosulfan and hexachlorocyclohexane residues in and around Calcutta. Bull Environ Contam Toxicol 67(4):554–559

Kolpin DW, Thurman EM, Linhart SM (1998) The environmental occurrence of herbicides: the importance ofdegradates in ground water. Arch Environ Contam Toxicol 35:385–390

Kuiper I, Kravchenko LV, Bloemberg GV, Lugtenberg BJJ (2002) *Pseudomonas putida* strain PCL1444, selected for efficient root colonization and naphthalene degradation, effectively utilizes root exudates components. Mol Plant-Microbe Interact 15:734–741

Kumar A, Singh JS (2017) Cyanoremediation: a green-clean tool for decontamination of synthetic pesticides from agro- and aquatic ecosystems. In: Singh JS, Seneviratne G (eds), Agro-environmental sustainability: volume 2: managing environmental pollution (pp 59–83). Springer, Cham

Lah K (2011) Effects of pesticides on human health. In: Toxipedia. Available from http://www.toxipedia.org/display/toxipedia/Effects+of+Pesticides+on+Human+Health. Accessed 16 Jan 2017

Lang M, Cai Z (2009) Effects of chlorothalonil and carbendazim on nitrification and denitrification in soils. J Environ Sci 21:458–467

Lehman CM, Williams BK (2010) Effects of current-use pesticides on amphibians. In: Sparling DW, Linder G, Bishop CA, Krest SK (eds) Ecotoxicology of amphibians and reptiles. CRC Press/Taylor & Francis/SETAC, Boca Raton, pp 167–202

Li T, Guo S, Wu B, Li F, Niu Z (2010) Effect of electric intensity on the microbial degradation of petroleum pollutants in soil. J Environ Sci 22:1381–1386

Liroff RA (2000) Balancing risks of DDT and malaria in the global POPs treaty. Pestic Saf News 4:3

Luo Q, Zhang X, Wang H, Qian Y (2005) The use of non-uniform electro kinetics to enhance in situ bioremediation of phenol-contaminated soil. J Hazard Mater 121:187–194

Macdonald RW, Harner T, Fyfe J (2005) Recent climate change in the Arctic and its impact on contaminant pathways and interpretation of temporal trend data. Sci Total Environ 342:5–86

Maillacheruvu K, Chinchoud PR (2011) Electro kinetic transport of aerobic microorganisms under low-strength electric fields. J Environ Sci Health A 46:589–595

Malik A, Ojha P, Singh KP (2009) Levels and distribution of persistent organochlorine pesticide residues in water and sediments of Gomti River (India)a- tributary of the Ganges River. Environ Monit Assess 148:421–435

Mannisto MK, Tiirola MA, Puhakka JA (2001) Degradation of 2,3,4,6-tetrachlorophenol at low temperature and low dioxygen concentrations by phylogenetically different groundwater and bioreactor bacteria. Biodegradation 12:291–301

Martens DA, Bremner JM (1993) Influence of herbicides on transformations of urea nitrogen in soil. J Environ Sci Health B 28:377–395

Mathur SC (1999) Future of Indian pesticides industry in next millennium. Pest Inf 24(4):9–23

McGuinness M, Dowling D (2009) Plant-associated bacterial degradation of toxic organic compounds in soil. Int J Environ Res Pub Health 6:2226–2247

Megharaj M (2002) Heavy pesticide use lowers the soil health. Farming Ahead 121:37–38

Melling Jr FB (1993) Soil microbial ecology: applications in agricultural and environmental management. Marcel Dekker, New York

Miraglia M, Marvin HJP, Kleter GA, Battilani P, Brera C, Coni E (2009) Climate change and food safety: an emerging issue with special focus on Europe. Food Chem Toxico 47(5):1009–1021

Mishra S, Bharagava RN (2016) Toxic and genotoxic effects of hexavalent chromium in environment and its bioremediation strategies. J Environ Sci Health Part C 34(1):1–34

Moorman TB (1989) A review of pesticide effects on microorganisms and microbial processes related to soil fertility. J Prod Agric 2(1):14–23

Namkoong W, Hwang EY, Park JS, Choi JY (2002) Bioremediation of diesel contaminated soil with composting. Environ Pollut 119:23–31

Narasimhan K, Basheer C, Bajic VB, Swarup S (2003) Enhancement of plant–microbe interactions using a rhizosphere metabolomics-driven approach and its application in the removal of polychlorinated biphenyls. Plant Physiol 132:146–153

NCEH (2005) Centers for Disease Control and Prevention. Third national report on human exposure to environmental chemicals. NCEH Pub. No. 05–0570

Niqui-Arroyo JL, Ortego-Calvo JJ (2007) Integrating biodegradation and electroosmosis for the enhanced removal of polycyclic aromatic hydrocarbons from creosote-polluted soils. J Environ Qual 36:1444–1451

Noyes PD, McElwee MK, Miller HD, Clark BW, Van Tiem LA, Walcott KC (2009) The toxicology of climate change: environmental contaminants in a warming world. Environ Int 35(6):971–986

Nozawa-Inoue M, Scow KM, Rolston DE (2005) Reduction of perchlorate and nitrate by microbial communities in vadose soil. Appl Environ Microbiol 71:3928–3934

Ntonifor NN (2011) Potentials of tropical African spices as sources of reduced-risk pesticides. J Entomol 8(1):16–26

O'Neil W, Raucher R (1998, August) Groundwater public policy leaflet series#4: the costs of groundwater contamination. Groundwater Policy Education Project, Wayzata. http://www.dnr.state.wi.us/org/water/dwg/gw/costofgw.htm

Otieno PO, Owuor PO, Lalah JO, Pfister G, Schramm KW (2013) Impacts of climate-induced changes on the distribution of pesticides residues in water and sediment of Lake Naivasha, Kenya. Environ Monit Assess 185(3):2723–2733

Pell M, Stenberg B, Torstensson L (1998) Potential denitrification and nitrification tests for evaluation of pesticide effects in soil. Ambio 27:24–28

Pesticides in Groundwater (2014) In: The USGS water science school. Available from http://water.usgs.gov/edu/pesticidesgw.html. Accessed 17 Jan 2017

Pilling ED, Jepson PC (2006) Synergism between EBI fungicides and a pyrethroid insecticide in the honeybee (*Apis mellifera*). Pestic Sci 39:293–297

Pimentel D (2009) Pesticides and pest control. In: Peshin R, Dhawan AK (eds) Integrated pest management: innovation-development process. Springer, Dordrecht, pp 83–87

Pimentel D, Greine A (1997) Environmental and socioeconomic costs of pesticide use. In: Pimentel D (e) (ed) Techniques for reducing pesticide use: economic and environmental benefits. Wiley, Chichester, pp 51–78

Pimentel D, Acquay H, Biltonen M, Rice P, Silva M, Nelson J, Lipner V, Giordano S, Horowitz A, D'Amore M (1992) Environmental and human costs of pesticide use. Bioscience 42:750–760

Plaza C, Xing B, Fernandez JM, Senesi N, Polo A (2009) Binding of polycyclic aromatic hydrocarbons by humic acids formed during composting. Environ Pollut 157:257–263

Pozo K, Harner T, Lee SC, Sinha RK, Sengupta B, Loewen M, Geethalakshmi V, Kannan K, Volpi V (2011) Assessing seasonal and spatial trends of persistent organic pollutants (POPs) in Indian agricultural regions using PUF disk passive air samplers. Environ Pollut 159:646–653

Probst M, Berenzen N, Lentzen-Godding A, Schulz R (2005) Scenario-based simulation of runoff-related pesticide entries into small streams on a landscape level. Ecotoxicol Environ Saf 62(2):145–159

Purnomo AS, Mori T, Kamei I, Nishii T, Kondo R (2010) Application of mushroom waste medium from *Pleurotus ostreatus* for bioremediation of DDT-contaminated soil. Int Biodeterior Biodegrad 64:397–402

Raposo Jr LJ, Re-Poppi N (2007) Determination of organochlorine pesticides in ground water samples using solid-phase microextraction by gas chromatography electron capture detection. Talanta 72:1833–1841

Rashid B, Husnain T, Riazuddin S (2010) Herbicides and pesticides as potential pollutants: a global problem. In: Plant adaptation phytoremediation, Springer, Dordrecht, pp 427–447

Reichenauer TG, Germida JJ (2008) Phytoremediation of organic pollutants in soil and groundwater. Chem Sustain 1:708–719

Relyea RA (2005) The lethal impact of roundup on aquatic and terrestrial amphibians. Ecol Appl 15:1118–1124

Richter ED (2002) Acute human pesticide poisonings. In: Pimentel D (ed) Encyclopedia of pest management. Dekker, New York, pp 3–6

Rigas F, Dritsa V, Marchant R, Papadopoulou K, Avramides EJ, Hatzianestis I (2005) Biodegradation of lindane by *Pleurotus ostreatus* via central composite design. Environ Int 31:191–196

Roberts TR (1998) Metabolic pathway of agrochemicals. I. In: Herbicides and plant growth regulators. The Royal Society of Chemistry, Cambridge

Roberts TR, Hutson DH (1999) Metabolic pathway of agrochemicals. II. In: Insecticides and fungicides. The Royal Society of Chemistry, Cambridge

Rohr JR, Schotthoefer AM, Raffel TR, Carrick HJ, Halstead N, Hoverman JT, Johnson CM, Johnson LB, Lieske C, Piwoni MD, Schoff PK, Beasley VR (2008) Agrochemicals increase trematode infections in a declining amphibian species. Nature 455:1235–1239

Rooney-Varga JN, Anderson RT, Fraga JL, Ringelberg D, Lovley DR (1999) Microbial communities associated with anaerobic benzene degradation in a petroleum contaminated aquifer. Appl Environ Microbiol 65:3056–3063

Roos J, Hopkins R, Kvarnheden A, Dixelius C (2011) The impact of global warming on plant diseases and insect vectors in Sweden. Eur J Plant Pathol 129(1):9–19

Rosenzweig C, Iglesias A, Yang X, Epstein PR, Chivian E (2001) Climate change and extreme weather events; implications for food production, plant diseases, and pests. Glob Chang Hum Health 2(2):90–104

Rothlein J, Rohlman D, Lasarev M, Phillip J, Muniz J, McCauley L (2006) Organophosphate pesticide exposure and neurobehavioral performance in agricultural and non-agricultural Hispanic workers. Environ Health Perspect 114:691–696

Safferman SI, Lamar RT, Vonderhaar S, Neogy R, Haught RC, Krishnan ER (1995) Treatability study using *Phanerochaete sordida* for the bioremediation of DDT contaminated soil. Toxicol Environ Chem 50:237–251

Saichek RE, Reddy KR (2005) Electrokinetically enhanced remediation of hydrophobic organic compounds in soil: a review. Crit Rev Environ Sci Technol 35:115–192

Santos A, Flores M (1995) Effects of glyphosate on nitrogen fixation of free-living heterotrophic bacteria. Lett Appl Microbiol 20:349–352

Savonen C (1997) Soil microorganisms object of new OSU service. Good Fruit Grower. http://www.goodfruit.com/archive/1995/6other.html.

Sayara T, Sarrà M, Sánchez A (2009) Preliminary screening of co (substrates for bioremediation of pyrene) contaminated soil through composting. J Hazard Mater 172:1695–1698

Sayara T, Pognani M, Sarrà M, Sánchez A (2010) Anaerobic degradation of PAHs in soil: impacts of concentration and amendment stability on the PAHs degradation and biogas production. Int Biodeter Biodegr 64:286–292

Schmolke A, Thorbek P, Chapman P, Grimm V (2010) Ecological models and pesticide risk assessment: current modeling practice. Environ Toxicol Chem 29(4):1006–1012

Scholz NL, Fleishman E, Brown L, Werner I, Johnson ML, Brooks ML, Mitchelmore CL (2012) A perspective on modern pesticides, pelagic fish declines, and unknown ecological resilience in highly managed ecosystems. Bioscience 62(4):428–434

Sebate J, Vinas M, Solanas AM (2004) Laboratory-scale bioremediation experiments on hydrocarbon-contaminated soils. Int Biodeterior Biodegrad 54:19–25

Semple KT, Reid BJ, Fermor TR (2001) Impact of composting strategies on the treatment of soils contaminated with organic pollutants. Environ Pollut 112:269–283

Shaw LJ, Burns RG (2004) Enhanced mineralization of [U-14C]2,4-dichlorophenoxyacetic acid in soil from the rhizosphere of *Trifolium pratense*. Appl Environ Microbiol 70:4766–4774

Shi L, Muller S, Harms H, Wicks LY (2008) Effect of electrokinetic transport on the vulnerability of PAH-degrading bacteria in a model aquifer. Environ Geochem Health 30:177–182

Siciliano SD, Fortin N, Mihoc A, Wisse G, Labelle S, Beaumier D, Ouellette D, Roy R, Whyte LG, Banks MK, Schwab P, Lee K, Greer CW (2001) Selection of specific endophytic bacterial genotypes by plants in response to soil contamination. Appl Environ Microbiol 67:2469–2475

Singh JS (2011) Methanotrophs: the potential biological sink to mitigate the global methane load. Curr Sci 100(1):29–30

Singh JS (2013a) Anticipated effects of climate change on methanotrophic methane oxidation. Clim Chang Environ Sustain 1(1):20–24

Singh JS (2013b) Plant growth promoting rhizobacteria: potential microbes for sustainable agriculture. Resonance 18(3):275–281

Singh JS (2014) Cyanobacteria: a vital bio-agent in eco-restoration of degraded lands and sustainable agriculture. Clim Chang Environ Sustain 2:133–137

Singh JS (2015a) Biodiversity: current perspective. Chang Environ Sustain 3(1):71–72

Singh JS (2015b) Microbes: the chief ecological engineers in reinstating equilibrium in degraded ecosystems. Agric Ecosyst Environ 203:80–82

Singh JS (2015c) Biodiversity: current perspectives. Clim Chang Environ Sustain 2:133–137

Singh JS (2015d) Plant-microbe interactions: a viable tool for agricultural sustainability. Appl Soil Ecol 92:45–46

Singh JS (2016) Microbes play major roles in ecosystem services. Clim Chang Environ Sustain 3:163–167

Singh JS, Pandey VC (2013) Fly ash application in nutrient poor agriculture soils: impact on methanotrophs population dynamics and paddy yields. Ecotoxicol Environ Saf 89:43–51

Singh JS, Seneviratne G (2017) Agro-environmental sustainability: volume 2: managing environmental pollution. Springer, Cham, pp 1–251

Singh JB, Singh S (1989) Effect of 2, 4-dichlorophenoxyacetic acid and maleic hydrazide on growth of blue green algae (cyanobacteria) *Anabaena doliolum* and *Anacystis nidulans*. Sci Cult 55:459–460

Singh JS, Strong PJ (2016) Biologically derived fertilizer: a multifaceted bio-tool in methane mitigation. Ecotoxicol Environ Saf 124:267–276

Singh JS, Singh DP, Dixit S (2011) Cyanobacteria: an agent of heavy metal removal. In: Maheshwari DK, Dubey RC (e) (eds) Bioremediation of pollutants. IK International Publisher Co., New Delhi, pp 223–243

Singh JS, Abhilash PC, Gupta VK (2016) Agriculturally important microbes in sustainable food production. Trends Biotechnol 34:773–775

Sogorka DB, Gabert H, Sogorka BJ (1998) Emerging technologies for soils contaminated with metals-electrokinetic remediation. Hazard Ind Waste 30:673–685

Sparling DW, Feller GM (2009) Toxicity of two insecticides to California, USA, anurans and its relevance to declining amphibian populations. Environ Toxicol Chem 28(8):1696–1703

Spear R (1991) Recognised and possible exposure to pesticides. In: Hayes WJ, Laws ER (eds) Handbook of pesticide toxicology. Academic, San Diego, pp 245–274

Speck-Planche A, Kleandrova VV, Scotti MT (2012) Fragment-based approach for the in silico discovery of multi-target insecticides. Chemom Intell Lab Syst 111:39–45

Sundar G, Selvarani J, Gopalakrishnan S, Ramachandran S (2010) Occurrence of organochlorine pesticide residues in green mussel (*Perna viridis* L.) and water from Ennore creek, Chennai, India. Environ Monit Assess 160:593–604

Swackhamer D, Hites RA (1988) Occurrence and bioaccumulation of organochlorine compounds in fish from Siskiwit Lake, Isle Royale, Lake Superior. Environ Sci Technol 22:543–548

Tang WW, Zeng GM, Gong JL, Liang J, Xu P, Zhang C (2014) Impact of humic/fulvic acid on the removal of heavy metals from aqueous solutions using nanomaterials: a review. Sci Total Environ 468:1014–1027

The Asian Amphibian Crisis (2009) In: IUCN. Available from http://www.iucn.org/about/union/secretariat/offices/asia/regional_activities/asian_amphibian_crisis/. Accessed 19 Feb 2017

Tözüm-Çalgan SRD, Sivaci-Güner S (1993) Effects of 2,4-D and methylparathion on growth and nitrogen fixation in cyanobacterium *Gloeocapsa*. Int J Environ Stud 23:307–311

Udeigwe TK, Eze PN, Teboh JM, Stietiya MH (2011) Application, chemistry, and environmental implications of contaminant-immobilization amendments on agricultural soil and water quality. Environ Int 37:258–267

US EPA (2001) Source water protection practices bulletin: managing small-scale application of pesticides to prevent contamination of drinking water. Office of Water (July), Washington, DC. EPA 816-F-01-031

van Aken B, Peres CM, Doty SL, Yoon JM, Schnoor JL (2004a) *Methylobacterium populi* sp. nov., a novel aerobic, pink-pigmented, facultatively methylotrophic, methane-utilizing bacterium isolated from poplar trees (*Populus deltoides x nigra* DN34). Int J Syst Evolut Microbiol 54:1191–1196

van Aken B, Yoon JM, Schnoor JL (2004b) Biodegradation of nitro-substituted explosives 2,4,6-trinitrotoluene, hexahydro-1,3,5-trinitro-1,3,5-triazine, and octahydro-13,5,7-tetranitro-1,3,5-tetrazocine by a phytosymbiotic *Methylobacterium* sp. associated with poplar tissues (*Populus deltoids nigra* DN34). Appl Environ Microbiol 70:508–517

Van Djik TC (2010) Effects of neonicotinoid pesticide pollution of dutch surface water on non-target species abundance. MSc thesis, Utrecht University, Utrecht. http://www.bijensterfte.nl/sites/default/files/FinalThesisTvD.pdf

Velasco-Alvarez N, Gonzalez I, Matsumura PD, Gutierrez-Rojas M (2011) Enhanced hexadecane degradation and low biomass production by *Aspergillus niger* exposed to an electric current in a model system. Bioresour Technol 102:1509–1515

Velizarov S (1999) Electric and magnetic fields in microbial biotechnology: possibilities, limitations and perspectives. Electro-Magnetobiol 18:185–212

Vidali M (2001) Bioremediation: an overview. Pure Appl Chem 73:1163–1172

Vimal SR, Singh JS, Arora NK, Singh S (2017) Soil-plant-microbe interactions in stressed agriculture management: a review. Pedosphere 27(2):177–192

Virkutyte J, Sillanpaa M, Latostenmaa P (2002) Electrokinetic soil remediation – critical review. Sci Total Environ 289:97–121

Wang X, Xlaobing Y, Bartha R (1990) Effect of bioremediation on polycyclic aromatic hydrocarbon residues in soil. Environ Sci Technol 24:1086–1089

Waskom R (1994) Best management practices for private well protection. Colorado State Univ. Cooperative Extension (August) http://hermes.ecn.purdue.edu:8001/cgi/convertwq?7488

Wehtje G, Walker RH, Shaw JN (2000) Pesticide retention by inorganic soil amendments. Weed Sci 48:248–254

Weyens N, van der Lelie D, Taghavi S, Newman L, Vangronsveld J (2009) Exploiting plant-microbe partnerships to improve biomass production and remediation. Trends Biotechnol 27:591–598

Wick LY, Shi L, Harms H (2007) Electro-bioremediation of hydrophobic organic soil contaminants: a review of fundamental interactions. Electrochim Acta 52: 3441–3443448

Xu P, Zeng GM, Huang DL, Feng CL, Hu S, Zhao MH (2012) Use of iron oxide nanomaterials in wastewater treatment: a review. Sci Total Environ 424:1–10

Yadav A, Chowdhary P, Kaithwas G, Bharagava RN (2017) Toxic metals in environment, threats on ecosystem and bioremediation approaches. In: Das S, Dash HR (eds) Handbook of metal-microbe interactions and bioremediation. CRC Press/Taylor & Francis Group, Boca Raton, p 813

Yang C, Cai N, Dong M, Jiang H, Li J, Qiao C, Mulchandani A, Chen W (2008) Surface display of MPH for organophosphate detoxification surface display of MPH on *Pseudomonas putida* JS444 using ice nucleation protein and its application in detoxification of organophosphates. Biotechnol Bioeng 99(1):30–37

Yee DC, Maynard JA, Wood TK (1998) Rhizoremediation of trichloroethylene by a recombinant, root-colonizing *Pseudomonas fluorescens* strain expressing toluene ortho-monooxygenase constitutively. Appl Environ Microbiol 64:112–118

Yousaf S, Andria V, Reichenauer TG, Smalla K, Sessitsch A (2010) Phylogenetic and functional diversity of alkane degrading bacteria associated with Italian ryegrass (*Lolium multiflorum*) and birds foot trefoil (*Lotus corniculatus*) in a petroleum oil-contaminated environment. J Hazard Mat 184:523–532

Zhang G, Chakraborty P, Li J, Sampathkumar P, Balasubramanian T, Kathiresan K, Takahashi S, Subramanian A, Tanabe S, Jones KC (2008) Passive atmospheric sampling of organochlorine pesticides, polychlorinated biphenyls, and polybrominated diphenyl ethers in urban, rural, and wetland sites along the coastal length on India. Environ Sci Technol 42:8218–8223

Chapter 13
Recent Advances in Physico-chemical and Biological Techniques for the Management of Pulp and Paper Mill Waste

Surabhi Zainith, Pankaj Chowdhary, and Ram Naresh Bharagava

Abstract Pulp and paper industries are one of the major sources of environmental pollution that discharge enormous amount of wastewaters containing recalcitrant pollutants into the environment. Wastewaters have high biological oxygen demand (BOD), chemical oxygen demand (COD), total solids (TS), phenols, lignin and its derivatives. High strength of wastewaters containing dark colour and toxic compounds from pulp paper industries causes serious aquatic and soil pollution. On terrestrial region, pulp and paper mill wastewater at high concentration reduces the soil texture and inhibits seed germination, growth and depletion of vegetation, while in aquatic system, it blocks the photosynthesis and decreases the dissolved oxygen (DO) level which affects both flora and fauna and causes toxicity to aquatic ecosystem. The high pollution load from pulp and paper industrial wastewater gradually increases, and hence, there is a need for adequate treatment to reduce these pollution parameters before final discharge into the environment. Thus, this chapter gives detailed information about sources, characteristics, toxicity and physico-chemical and biological methods for the treatment of pulp and paper mill wastes and wastewaters.

Keywords Pulp and paper mill wastewater · Recalcitrant pollutants · Bioremediation

---

S. Zainith · P. Chowdhary · R. N. Bharagava (✉)
Laboratory for Bioremediation and Metagenomics Research (LBMR), Department of Environmental Microbiology (DEM), Babasaheb Bhimrao Ambedkar University (A Central University), Lucknow, India

© Springer Nature Singapore Pte Ltd. 2019
R. N. Bharagava, P. Chowdhary (eds.), *Emerging and Eco-Friendly Approaches for Waste Management*, https://doi.org/10.1007/978-981-10-8669-4_13

1 Introduction

Industries such as pulp and paper, distillery, tannery and textile are one of the major sources of serious environmental pollution (Bharagava and Mishra 2017). Pulp and paper (P&P) mills are categorized as a core industrial sector and rank third in the world after primary metals and chemical industries in terms of freshwater used (Thompson et al. 2001; Sumathi and Hung 2006; Asghar et al. 2008; Mishra and Bharagava 2016). The pulp and paper industry is one of the most important industries of the North American and US economy (Nemerow and Dasgupta 1991). In terms of production and total earning, pulp and paper industries of Canada play a major role in country's economy, and estimated 50% of pulp and paper wastes are dumped into Canada's waters (Sinclair 1990). At present, more than 759 pulp and paper industries are present in India. The Indian pulp and paper industries are highly water intensive, consuming 100–250 m^3 freshwater/ton paper (Singh. 2004) and generating a subsequent 75–225 m^3 wastewater/ton paper (Ansari 2004). Out of 759 mills, 114 (15%) are large, 303 (40%) are medium and 342 (45%) are small. The pulp and paper industry is divided into the three sectors based on raw materials used: wood and bamboo mills using wood and bamboo produce 3.19 million tonnes, which is 31% of the production, and there are 26 large integrated paper mills. There are 150 agro-based mills using agro-residues like bagasse, wheat and rice straw, etc. that produce 2.2 million tonnes, which is 22% of the total production. Recycled fibre mills using wastepaper contribute almost 47% of the country's current production which is 4.72 million tonnes, and there are 538 recycled paper mills in operation. The environmental effects of agro-residue-based paper mills are of particular concern as these units generate 150–200 m^3 effluent/ton paper with a high pollution load of 90–240 kg suspended solid (SS), 85–370 kg biochemical oxygen demand (BOD) and 500–1100 kg chemical oxygen demand (COD)/ton paper (Mathur et al. 2004). Indian paper mills are highly fragmented on the basis of production, i.e. printing and writing is 38.58%, packaging is 53.61%, while newsprint is 7.81%. According to the Ministry of Environment and Forest, Government of India, the pulp and paper sector is in the "Red Category" list of 17 industries having high polluting potential due to its serious pollution menace, and it is compulsory for pulp and paper mills to follow the appropriate standards set by Central Pollution Control Board (CPCB 2001). Wastewaters of pulp and paper mills impart a dark brown/black colour to the receiving water bodies. Lignin and chlorinated organic compounds are the key environmental pollutants released from pulp and paper industry. The offensive colour of these wastewaters principally of lignin and its degradation products produced during manufacturing process of papermaking reduces the transmission of light in waterways, reduces the aquatic plant photosynthesis and dissolved oxygen (DO) content and ultimately causes the death and putrefaction of aquatic fauna (Sahoo and Gupta 2005; Karrasch et al. 2006; Singh et al. 2016). On terrestrial environment, accumulation of toxic pollutants and metals in soil affects growth and development of plants. Chlorinated compounds released from pulp and paper mills are highly toxic and have carcinogenic, clastogenic, mutagenic and endocrine effects on receiving bodies. Various extractives are also used in pulp and paper manufacturing process, although the concentration of these extractives is present in trace amounts, but some of them are very harmful for environment. Pulp and paper mill wastewaters pollute all aspects of life, i.e. water, soil and air, causing a major threat to the environment. In many developing countries like India, farmers irrigate their fields by wastewaters, which are released by pulp and paper mills, having high levels of several toxic compounds. Hence, the adequate treatment is necessary of these pollutants before the final discharge into the environment. Although several physico-chemical methods (precipitation, sorption, ozonation, ultrafiltration, reverse osmosis and electrochemical treatment) or combination of different methods are available for the treatment of pulp and paper mill wastewater, these methods are more energy intensive, produce secondary sludge and are cost-ineffective (Singhal and Thakur 2009a, b; Raj et al. 2014). The biological treatment is known to be very effective in reducing the organic load and toxic effects of kraft mill wastewaters (Park et al. 2007). Various wood extractives present in wastewaters are removed by biological methods. This treatment also reduces colour, COD, BOD and low-molecular-weight chloro-lignins (Nagarthnamma et al. 1999; Barton et al. 1996). Microorganisms treat the wastewaters generally by the action of enzymes. The various enzymes involved in the treatment of pulp and paper mill wastewaters are lignin peroxidase, manganese peroxidase, laccase, etc. Microorganisms showing good production of these enzymes have the high potential to treat the wastewaters (Hooda et al. 2015; Kumar et al. 2016).

2 Sources and Characteristics of Pulp and Paper Wastewater

Wood is the most abundant source of pulping and papermaking process; this process consists of several toxic compounds which are washed away from the pulp fibres during the washing, dewatering and screening processes. Pulp fibres can be prepared from a vast majority of plants such as woods, straws, grasses, bamboos, canes and reeds. Among various stages of paper manufacturing process, the most considerable sources of pollution from pulp and paper mills are wood preparation, pulping, pulp washing and bleaching. Along with these processes, pulping stage generates a high-strength wastewater called black liquor which is dark brown in colour due to dissolved lignin and its degradation products, hemicelluloses, resin acids, unsaturated fatty acids, etc. (Berryman et al. 2004). About 200 m^3/tonne of pulp is produced during pulping process, and it generates a huge amount of wastewaters (Cecen et al. 1992) which are highly polluted that cannot be easily recovered. During bleaching process, various types of phenolic compounds are produced; lignin and its derivatives are the main source of these compounds (Amat et al. 2005). Pentachlorophenol is generated unintentionally in wastewaters as a byproduct, and other most toxic pollutants such as coloured compounds, chlorinated phenols, chlorinated dibenzo-p-dioxin and dibenzofuran, chlorinated hydrocarbon, etc. are also released. Wastewaters produced from papermaking process contain cellulose (fines)

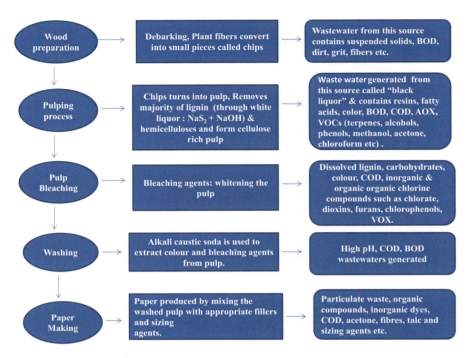

Fig. 13.1 Various stages and produced pollutants of pulp and paper mill wastewater

and other additives; this contaminated wastewater is referred as whitewater, and it can be up to 50% of the total mass. The pollutants released from various stages of the pulping and papermaking process are present in Fig. 13.1. Wastewaters produced from de-inking process contain ink residues, and sludges are also generated from wastewater treatment plant. These wastewaters are the significant concern for the environment because of highly toxic chlorinated compounds (EPA. 2002).

3 Effects of Pulp and Paper Wastewater Pollutants on Environment

Harmful pollutants released from pulp and paper industries affect all aspects of the environment: water, air and land (Makris and Banerjee 2002). Several studies confirmed the toxic and genotoxic effects of pulp and paper mill wastewaters on aquatic as well as on terrestrial environments. On aquatic fauna, the detrimental effects such as liver damage, oxidative and respiratory stress, mutagenic and genotoxic effects and other such lethal effects due to exposure of pulp and paper mill wastewaters were reported by several authors (Owens et al. 1994; Ali and Sreekrishnan. 2001; Pokhrel and Viraraghavan. 2004; Kumari et al. 2016). Mandal and Bandana (1996) reported health impacts such as diarrhoea, vomiting, headaches, nausea and eye irritation on children and workers due to pulp and paper mill wastewater which is discharged into the environment. High level of carbon dioxide in pulp and paper mill wastewaters as a potential source of distress and toxicity to rainbow trout was reported by O'connor et al. (2000). Yen et al. (1996) reported the possibility of sub-lethal effects on aquatic organisms of Dong Nai River in Vietnam due to the discharged wastewaters from pulp and paper industries. Baruah (1997) reported serious concerns on surface plankton population change in Elengabeel's wetland ecosystem in India, because of untreated paper mill wastewater discharged into the receiving system. Howe and Michael (1998) studied the harmful effects of the treated pulp mill wastewater on irrigated soil in northern Arizona, which showed serious change in soil chemistry. Dutta (1999) investigated the toxic effects of paper mill wastewater (treated) applied to paddy field in Assam, India. Gupta (1997) and Singh et al. (1996) reported high levels of organic pollutants derived from paper mill wastewater in Tamil Nadu and Punjab, India, respectively. Skipperud et al. (1998) and Holmbom et al. (1994) reported various trace metals present in pulp and paper mill wastewaters at low concentrations. Pulping process is the most significant potential for environmental pollution. In pulping countries, the effect of the generated wastewater on the environment can be much greater. Most studies found that the bleached and unbleached kraft pulp mill wastewaters cause impaired liver function and also demonstrated a variety of responses in fish population's exposure to bleached wastewaters of kraft mills in the USA and Canada. These include smaller gonads, delayed sexual maturity, depression in secondary sexual characteristics and changes in fish reproduction (Munkittrick et al. 1997). In secondary wastewater treatment plants in the UK, the main problem that occurs is the growth of sewage fungus (Webb 1985), and other problems arise when the treatment of pulp and paper mill wastewater failed due to released suspended solids; it loses the nutrients, such as nitrogen and phosphorus, which can lead to eutrophication in recipient bodies. In agriculture, pulp and paper mill wastewaters are used for irrigation purpose, and it affects not only the crop growth and soil properties but also the mobility of various ions present in soil which is beneficial to plants (Ugurlu et al. 2008; Kumar and Chopra 2012). Chlorinated organic compounds such as dioxins and furans are supposed to cause skin disorders including skin cancer and also show reproductive effects in exposed organisms (Nestmann 1985; Malik et al. 2009), and adsorbable organic halides (AOXs) may bioaccumulate in fish tissues causing a variety of clastogenic, carcinogenic, endocrine and mutagenic effects, which may then also create problems to humans after consumption of the contaminated fish. Chlorinated phenols are also responsible for toxicity to both flora and fauna. During biological treatment process, wood extractives (e.g. resin acids and sterols) can be transformed into other toxic compounds, and severe toxicity may occasionally occur.

4 Treatment Approaches for Pulp and Paper Wastewater

Pulp and paper mill wastewaters hold a number of toxic compounds which are harmful for receiving bodies, are recalcitrant to degradation and can potentially induce toxicity, generally at the reproductive level (Costigan et al. 2012; Hewitt et al. 2008; Waye et al. 2014). Various treatment technologies and its application are shown in Table 13.1 and Fig. 13.2. Pollution from pulp and paper industries can be reduced by various methods as follows:

4.1 Physico-chemical Treatment Approaches

Plenty of literatures are available on various physico-chemical treatment methods that include sedimentation, flotation, adsorption, coagulation, oxidation, ozonation, electrolysis, reverse osmosis, ultrafiltration and nanofiltration technologies to remove suspended solids (SS), colloidal particles, floating matters, colours and toxic compounds from the produced harmful wastewaters.

4.1.1 Sedimentation and Flotation

Primary clarification may be attained by either sedimentation or flotation. Suspended particles present in pulp and paper mill wastewaters primarily consist of bark particles, fibres, fillers and coating materials (Pokhrel and Viraraghavan 2004). Sedimentation was the preferred option in the UK and approximately 80% removal of the suspended solids (Thompson et al. 2001). Dissolved air flotation is a conventional method for the removal of suspended solids and has been widely used in treatment of various types of industrial wastewaters. Gubelt et al. (2000) reported that dissolved air flotation method removes 65–95% total suspended solids (TSS), and it was an unstable unit, while Wenta and Hartmen (2002) stated that dissolved air flotation was able to remove 95% of the TSS. Implementation of such these treatment methods may depend upon the employed pulp and paper production process as well as applied secondary treatment methods (Kamali and Khodaparast 2015).

4.1.2 Coagulation and Precipitation

In case of pulp and paper mill wastewater treatment, coagulation and flocculation methods are basically employed in the tertiary treatment systems but not used in primary treatments. The basic principle of such methods is addition of metal salts to the wastewater stream to generate larger flocs from small particles. Wang et al. (2011) used aluminium chloride as coagulant and a modified natural polymer (starch-g-PAM-g-PDMC) as flocculant for the treatment of wastewaters from

Table 13.1 Various treatment technologies for the remediation of pulp and paper mill wastewaters

Treatment approaches	Properties	Applications	References
Physico-chemical treatments			
Sedimentation and flotation	In sedimentation, the gravity is used to separate solid phase from liquid, while in flotation process, buoyancy is increased of solids by forming gas bubbles	Industrial (paper, food, oil and plastic industries) and domestic wastewater treatment	http://www.gunt.de/images/download/flotation_sedimentation_english.pdf
Coagulation and flocculation	In coagulation, particles aggregate with themselves by the influence of a change in pH, and in flocculation process, the particles aggregate by the use of polymers that bind them together	Removal of organic matter, pathogen removal, removal of inorganics (arsenic and fluoride), all types of wastewater treatment	http://aquarden.com/technology/coagulation-and-flocculation. http://www.iwapublishing.com/news/coagulation-and-flocculation-water-and-wastewater-treatment
Electrochemical methods	They use electron as unique reagent	Colour removal in wastewater treatments and degradation of nonbiodegradable dyes do not produce solid waste residues	Sala and Gutierrez-Bouzan (2012)
Membrane technologies	It generates stable water without the addition of chemicals, low energy use, easy and well-arranged process	Efficient recovery of waste materials, impurities and byproducts from pulp and paper mill effluents prior to discharge. Treatment of industrial wastewaters especially treating wastewater from petrochemical and steel industry and power generation	Ebrahimi et al. (2016), Chen (2008), and Chen et al. (2009)
Adsorption	It is a surface phenomenon. Nature of the bonding between adsorbate and adsorbent: physic sorption (weak van der Waals forces), chemisorption (covalent bonding) and electrostatic attraction	Remove inorganic and organic pollutants, i.e. persistent organic pollutants (POPs)	Rashed (2013)

(continued)

Table 13.1 (continued)

Treatment approaches	Properties	Applications	References
Advanced oxidation process	In this process, the main mechanism is the generation of highly reactive free radicals	Ground remediation, removal of pesticides from drinking water, removal of formaldehyde and phenol and reduction in COD from industrial wastewaters	Ayed et al. (2017)
Biological treatments			
Activated sludge process	Complex mixture of microbiology and biochemistry	Activated sludge process has been the major treatment method for pulp and paper mill effluents in recent years. It removes organic substances (AOX), nutrients, BOD, COD as well as toxic compounds and pathogens from produced wastewaters	Ashrafi et al. (2015) and Wells et al. (2011)
Aerated lagoons		Efficient removal in BOD (over 95%) and chlorinated phenolics (85%) from pulp and paper mill wastewaters and other industrial wastewaters	Kamali and Khodaparast (2015)
Anaerobic digestion	Biochemical reactions occur and convert organic polymers from the feedstock into methane (biogas) and nutrient-rich digestate. It works best at temperatures of 30–60 °C	Used for reducing excess sludge volumes, energy efficient with lower biomass production	Garg and Tripathi (2011) and Metcalf and Eddy et al. (2003)
Bacterial treatment	Bacteria show enhanced biodegradation capability, mainly due to the broad pH range tolerability, biochemical versatility and immense environmental adaptability	Removes all types of pollutants from industrial wastes and wastewaters	Chandra and Singh (2012)
Fungal treatment	Produces extracellular enzymes and can survive at higher effluent load	Degradation of lignin/phenolic compounds from pulp and paper mill wastewaters and also reduction in colour, BOD, COD and AOX from different industrial wastewaters (textile, leather, distillery etc.)	Sankaran et al. (2010)

Algal treatment	Main mechanism for lignin removal by algae is metabolism rather than adsorption, and the main mechanism of colour and organic removal is partially metabolism and partially transformation	Treats several types of industrial wastewaters and also removes metals from wastewaters	Tarlan et al. (2002) and Usha et al. (2016)
Upflow anaerobic sludge blanket reactor (UASB)	Forms agglomerates (0.5–2 mm in diameter), and gas formed causes sufficient agitation in the reactor	Treats various industrial wastewaters like petroleum, distillery, canning industry, heavy metals, paper and pulp, tannery, pharmaceutical, domestic wastewater, etc.	Kaviyarasan (2014)
Sequencing batch reactor (SBR)	Fills and draws activated sludge system for wastewater treatment, uniquely suited for wastewater treatment applications characterized by low or intermittent flow conditions	Treats industrial wastewater containing phenolic compounds, such as p-nitrophenol (PNP) which is a hazardous chemical widely used in agricultural, pharmaceutical and dye industries as a synthetic intermediate in the manufacturing process and also treats both municipal and industrial wastewaters including dairy, pulp and paper, tanneries and textiles	Dutta and Sarkar (2015)
Constructed wetlands	Use natural functions of vegetation, soil and organisms to treat different water streams	They are capable of removing nutrients, biochemical oxygen demand, chemical oxygen demand, total suspended solids, colour, metals and toxic compounds from industrial wastewaters of different origin	Chaudhary et al. (2011)
Enzymatic treatments	Remove pollutants by precipitation or transformation to other value-added products	Application to biorefractory compounds; operation at high and low contaminant concentrations; work over a wide range of pH, temperature and salinity; absence of shock loading effects; reduction in sludge volume (no biomass generated) and the ease of controlling the process. Treat phenolic contaminants and related compounds, pulp and paper wastes, pesticides, cyanide wastes, food-processing wastes, removal of heavy metals, surfactant, oil and grease degradation	Karam and Nicell (1997)

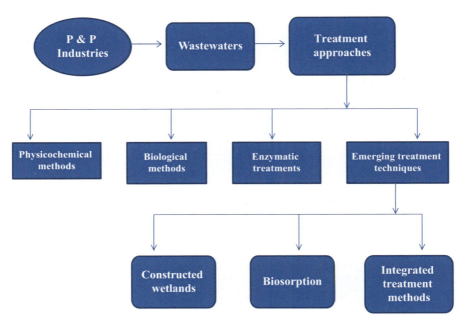

Fig. 13.2 Different methods used in the treatment of pulp and paper mill wastewater

primary sedimentation tank and concluded that at optimum condition, turbidity, lignin removal efficiency and water recovery were 95.7%, 83.4% and 72.7%. Tong et al. (1999) and Ganjidoust et al. (1997) have done a comparative study between horseradish peroxide (chitosan) and other coagulants such as $Al_2(SO_4)_3$), polyethyleneimine (PEI) and hexamethylene diamine epichlorohydrin polycondensate (HE), to remove colour, total organic carbon (TOC) and adsorbable organic halides (AOX), and observed that chitosan was a more better coagulant in removing these pollutants than others. Dilek and Gokcay (1994) reported that by using alum as coagulant, 96% removal of chemical oxygen demand (COD) from papermaking process, 50% from pulping stage and 20% from bleaching wastewaters were attained. Rohella et al. (2001) stated that polyelectrolytes were more effective than the conventional coagulant alum for removal of colour, turbidity and COD. Eskelinen et al. (2010) found that by using chemical precipitation method using 5 g/L of CaO, up to 90% removal of chemical oxygen demand (COD) is attained.

4.1.3 Electrochemical Methods

There has been an increasing attention towards electrochemical techniques (Malkin. 2002), which is an attractive, alternative and eco-friendly process for treating wastewaters in large scale (Soloman et al. 2009). This process includes electrocoagulation and electrooxidation. Different types of technical problems arise in pulp and paper industries, which have been solved by electrochemical method. Electrochemical technologies are of great attention because of their more versatility and environmental compatibility, which makes the treatment of any type of pollutants (liquids, gases and solids) possible. In electrochemical methods, chemicals are not required; the main reagent is the electron, which is a clean reagent (Inan et al. 2004). Chanworrawoot and Hunsom (2012) found that the electrochemical method was much efficient to reduce the pollutants of various types of industries and produce low-density sludge in a very small amount. They also assist the reduction of lignin as well as organic and inorganic compounds, colour, total biological oxygen demand (BOD), chemical oxygen demand (COD), total suspended solids (TSS) and total dissolved solids (TSD). In their batch culture experiment, the removal of colour, BOD and COD were 98%, 98% and 97%, respectively.

4.1.4 Membrane Technologies

Membrane filtration is a separation process that employs semipermeable membrane to divide the supply wastewater stream into two parts: the first part is permeate that contains the material passing through the membranes and the second is retentate in which the species are being left behind (Mallevialle et al. 1996). Membrane filtration achieved the efficient recovery of waste materials, byproducts and several impurities discharged from wastewaters. Recently, various membrane technologies have been used for the treatment of pulp and paper mill wastewaters. Reverse osmosis is an important method having the ability to destroy the pathogens (Asano and Cotruvo 2004). Dube et al. (2000) reported that 88% and 89% removal of BOD and COD was attained by reverse osmosis. The most important applications of membrane filtration techniques are the recovery of lignosulfonate from spent sulphite liquor and lignin from kraft black liquor; it also saves energy and is beneficial for the environment (Olsen 1980). Ciputra et al. (2010) found that nanofiltration technique removes 91% dissolved organic carbon from biologically treated newsprint mill wastewater. Gonder et al. (2011) revealed that at optimized conditions, membrane fouling can be minimized by nanofiltration process. They also investigate by using ultrafiltration membrane method the treatment of pulp and paper mill wastewater which occurs, and they achieved 83%, 97%, 95%, 89% and 50% removal of total hardness, sulphate, spectral absorption coefficient, COD and conductivity, respectively. Krawczyk et al. (2013) recovered high-molecular-mass hemicelluloses, from chemical thermo-mechanical pulping process wastewater, by using a membrane filtration.

4.1.5 Adsorption

Adsorption is one of the most prominent methods for the treatment and removal of inorganic and organic pollutants present in pulp and paper mill wastewaters. Adsorption has many advantages over other conventional methods like simple design, low cost and land requirement. Recently, the low-cost adsorbents such as agricultural wastes, natural wastes and industrial wastes that have pollutant-binding capacities are used and are easily available. Activated carbon can be used as adsorbent for the treatment of water and wastewaters which is produced from these materials (Crini 2005). Various adsorbents such as coal ash, silica, fuller's earth, activated carbon, etc. have revealed adequate results for decolourization and pollutant removals from pulp and paper mill wastewaters, as reviewed by Pokhrel and Viraraghavan (2004). Ciputra et al. (2010) investigated the adsorption mechanism of granular activated carbon and ion exchange resin preferentially acts on the hydrophobic and high-molecular-weight fractions and reported 72% and 76% reductions in dissolved organic carbon by using this method. Das and Patnaik (2000) studied the removal efficiency of lignin from blast furnace dust (BFD) and slag by the adsorption mechanism was 80.4% and 61%. Shawwa et al. (2001) used activated coke as an adsorbent that removes 90% of colour, COD and AOX from bleached wastewater through adsorption process. Xilei et al. (2010) performed the adsorption method (low-cost bentonite as adsorbent) followed by the coagulation tertiary treatment (polyaluminum silicate chloride as coagulant), and they achieved 60.87% and 41.38% removals of COD and colour, respectively, at optimum doses of adsorbent and coagulant, i.e. 450 mg/l and 400 mg/l.

4.1.6 Oxidation

Advanced oxidation process (AOP) is the potential method for the remediation of wastewaters by conversion (oxidation) of complex recalcitrant compounds into inorganic substances (CO_2 and H_2O_2) or partial mineralization and transform them into less toxic forms. Generation of highly reactive free radicals is the main mechanism of advanced oxidation process. Hydroxyl radicals (OH •) are effectively destroying organic pollutants through the action of electrophiles that react rapidly with all electro-rich organic compounds (Covinich et al. 2014). Perez et al. (2002) proved that the combinations of fenton and photo-fenton reaction methods are highly effective for the treatment of bleached kraft mill wastewaters. Sevimli (2005) compared ozonation and combination of ozonation with H_2O_2 oxidation and fenton oxidation for the removal of COD and colour from pulp and paper mill wastewaters and analysed that ozonation and ozonation with hydrogen peroxide successfully remove the colour, while Fenton's oxidation process was more effective in reducing the COD and colour. Ozonation process is also used to oxidize chemicals such as guaiacol, syringaldehyde, vainilline, phenol, trichlorophenol, chlorophenol and cinnamic acid derivatives, which present in pulp and paper mill wastewaters (Fontanier et al. 2005; Hermosilla et al. 2014). Acid orange 7 is a typical dye used in pulp and paper industries, which is degraded by heterogeneous catalytic wet hydrogen peroxide process (Herney-ramirez et al. 2011).

5 Biological Treatment Approaches

Biological methods in which microorganisms such as fungi, bacteria and algae and their enzymes are used, as singly applied or in combination with physical and chemical methods to treat pulp and paper mill wastewaters (Singhal and Thakur 2009a, b). Most of the conventional treatment methods are not very effective for the removal of colour and degradation of recalcitrant compounds such as lignin (Balcioglu et al. 2007), but compared with physico-chemical methods, biological treatment methods are suitable to reduce COD, BOD and lignin from various types of pulp and paper mill wastewaters (Tiku et al. 2010). Detailed list of microorganisms for the treatment of pulp and paper mill wastewater is given in Table 13.2. Biological processes are divided into two categories, namely, aerobic and anaerobic, and it depends on the type of microorganisms.

5.1 Aerobic Process

Aerobic treatment processes such as activated sludge (AS) and aerated lagoons (AL) are the commonly used treatment methods for pulp and paper mill wastewaters. Activated sludge system is the major treatment method capable of removing huge amount of sludge and secondary pollutants generated from pulp and paper industries (Buyukkamaci and Koken 2010). Lots of literature have been published for the treatment of pulp and paper wastewaters through activated sludge system. Chandra (2001) reported the efficient removal of colour, BOD, COD, phenols and sulphide by microorganisms such as *Pseudomonas putida*, *Citrobacter* sp. and *Enterobacter* sp. through activated sludge process. Bengtsson et al. (2008) treated pulp and paper mill wastewaters from recycled fibres by activated sludge system and found 95% removal of COD. Mahmood and Paice (2006) and Ghoreishi and Haghighi (2007) treated various types of pulp and paper wastewaters through aerated stabilization basin and remove 50–70% BOD, 30–40% COD, AOX and chlorinated compounds. Aerated lagoon is the simple and cost-effective biological system which is relevant on both lab scale as well as full scale for pulp and paper mill wastewaters. This system is used to remove BOD, low-molecular-weight AOX and fatty acids at full-scale applications (Bajpai 2001). Schnell et al. (2000) reported that the aerated lagoon system removes BOD, AOX and phenols. Pokhrel and Viraraghavan (2004) reviewed that aerated lagoon system was efficient in removal of chlorinated phenols (85%) and BOD (>95%) from pulp and paper mill wasters.

Table 13.2 Microorganisms involved in the treatment of pulp and paper mill wastewater

Microorganisms	Used for the removal of pollutants	References
Fungal sp.		
Trametes pubescens	Chlorophenols	Gonzalez et al. (2010)
Aspergillus niger	Alkaline peroxide mechanical pulping effluent	Liu et al. (2011)
Emericella nidulans var. nidulans	Colour and lignin	Singhal and Thakur (2009a, b)
Phanerochaete chrysosporium	Colour, lignin and COD	Saritha et al. (2010)
Cryptococcus sp.	Colour, lignin and toxicity of the effluent	Singhal and Thakur (2009a, b)
Trichoderma sp.	Colour	Saravanan and Sreekrishnan (2005)
Aspergillus flavus F10	Colour and lignin	Barapatre and Jha (2016)
Fibrodontia sp. RCK783S	Colour	Kreetachat et al. (2016)
Rhizopus arrhizus	Lignin and chlorophenols	Lokeshwari et al. (2015)
Bacterial sp.		
Pseudomonas fluorescens	Colour, lignin, COD, phenol, chloride content	Chauhan and Thakur (2002)
Paenibacillus sp.	Colour, lignin, BOD, COD, phenol	Raj et al. (2014)
Citrobacter freundii, Serratia marcescens	TOC, COD, lignin	Abhishek et al. (2015)
Alcaligenes faecalis and Bacillus cereus	COD	Mehta et al. (2014)
Bacillus subtilis subsp. inaquosorum	Lignin, colour, COD	Hooda et al. (2016)
Citrobacter freundii and Citrobacter sp.	COD, AOX, colour, lignin	Chandra and Abhishek (2010)
Paenibacillus glucanolyticus	Deconstruct pulping waste	Methews et al. (2014)
Klebsiella sp., Alcaligenes sp. and Cronobacter sp.	Colour, AOX, TDS, TSS	Kumar et al. (2014)
Algal sp.		
Scenedesmus species	Nutrients, organic pollutants, BOD, COD	Usha et al. (2016)
Microalgae	Convert secondary waste into value added products	Kouhia et al. (2015)
Chlorella	COD, colour, AOX, chlorinated compounds	Tarlan et al. (2002)

5.2 Anaerobic Process

Anaerobic process is considered to be more suitable method for the treatment of high-strength organic wastewaters. Lots of literature have been published on anaerobic process along with microbial communities to treat pulp and paper mill wastewaters (Ince et al. 2007). In recent years, a stable biological process, anaerobic digestion (AD), is used for treatment of high loads of pulp and paper mill wastewaters. This approach has several advantages over various conventional treatment methods such as simple design, reduction of produced sludge volume up to 30–70%, destruction of pathogens in the thermophilic region, cost-effective and eco-friendly in nature (Zwain et al. 2013; Ekstrand et al. 2013). The anaerobic treatment of pulp and paper mill wastewater results in the degradation of pollutants such as lignin and their derivatives, fatty acids, resins and organic compounds, which are produced during the various steps of papermaking process (Sumathi and Hung 2006).

5.3 Bacterial Treatment

Several species of bacteria have been used for the remediation of pulp and paper mill wastewaters, and few of them have also been used commercially. Various studies have reported that some bacterial species (anaerobic and aerobic both) could metabolize lignin and their related compounds to low-molecular-weight compounds, and this is due to huge adaptableness and biochemical versatility of bacterial species (Chandra et al. 2007; Abhishek et al. 2015). Chandra et al. (2009) found that B. cereus and Serratia marcescens was capable to reduce colour (45–52%), lignin (30–42%), BOD (40–70%), COD (50–60%), total phenol (32–40%), and PCP (85–90%) in axenic conditions, but in mixed culture conditions, the reduction in colour, lignin, BOD, COD, total phenol and PCP recorded was 62%, 54%, 70%, 90%, 90%, and 100%, respectively. Tyagi et al. (2014) isolated two bacterial strains, *Bacillus subtilis* and *Micrococcus luteus*, and one fungi, *Phanerochaete chrysosporium*, from pulp and paper mill wastewaters and sludge; these microbes were capable in reducing BOD up to 87.2%, COD 94.7% and lignin content 97% after 9 days; pH was down to neutral, and dissolved oxygen increased from 0.8 to 6.8 mg/L. Further in recent studies, Raj et al. (2014) found that bacterial strain, i.e. *Paenibacillus* sp., effectively reduced colour (68%), lignin (54%), phenol (86%), BOD (83%) and COD (78%). Simultaneously, in this study, the toxicity with treated and untreated wastewater on the growth and germination of mung bean (*Vigna radiata* L.) seed was also performed (Kumari et al. 2014). Hooda et al. (2015) studied the degradation of pulp and paper mill wastewater by a rod-shaped Gram-positive bacterium, i.e. *Brevibacillus agri*, in batch culture and in semi-continuous reactor. During batch study, this bacterium reduced COD up to 69%, colour 47%, lignin 37% and AOX 39%, while in semi-continuous reactor study, it reduced COD up to 62%, colour 37%, lignin 30% and AOX 40%.

5.4 Fungal Treatment (Mycoremediation)

Fungi are common in the treatment of pulp and paper mill wastewaters (Yang et al. 2011). In comparison with bacteria, fungi can survive at higher strengths of wastewaters, and like bacteria, they also produce extracellular enzymes (Singhal and Thakur 2009a, b). Malaviya and Rathore (2007) stated the bioremediation of pollutants from pulp and paper mill wastewater by a novel consortium of white-rot and soft-rot fungi which reduced the colour, lignin and COD by 78.6%, 79% and 89.4%. It has been shown that wood-degrading white-rot fungus is very effective for the degradation of lignin and chlorinated compounds, which are mainly responsible for colour and toxicity of pulp and paper mill wastewaters (Saritha et al. 2010). White-rot fungus *T. pubescens* along with TiO_2/UV was used for degradation of chlorophenols, and this combination (biological and advanced oxidation process) allowed up to 100% chlorophenol removal (González et al. 2010).

5.5 Algal Treatment (Phycoremediation)

Algae are important bioremediation agents, and they used natural mechanism for the treatment of wastewaters. Tarlan et al. (2002) were found to remove 58% COD, 84% colour and 80% AOX from pulp and paper wastewaters by algae and also showed that they grew mixotrophically and partially metabolized colour and organic compounds (released from pulping stage) to non-coloured and simple molecules. Several studies have reported wastewater can be used for the cultivation of microalgae (Ramanna et al. 2014). Microalgal cultivation from wastewater has a twin purpose: supply nutrients and minimize the freshwater requirements along with the removal of COD and BOD from wastewaters. Algae can use huge amounts of organic compounds from wastewater for rapid growth in the photoheterotrophic or mixotrophic environment in the presence of light (Li et al. 2011). Usha et al. (2016) stated that microalgal treatment is an efficient tool for the remediation of pulp and paper industry wastewater, and in their lab study, they found maximum removal of BOD (82%) and COD (75%), respectively, through microalgal cultivation in outdoor open pond. Algae have also been used in the removal of heavy metals from wastewaters.

6 Biological Reactors Study Used in the Treatment of Pulp and Paper Mill Wastewaters

Various types of rectors/digesters including SBR, MBR, UASB, etc. are reported for the degradation and treatment of noxious and deleterious pollutant present in pulp and paper mill wastewater. These reactors are principally based on the anaerobic microbial treatment technology that can efficiently reduce high concentration of pollution load and signifies wastewater quality for its reuse in irrigation and other practices. Khan et al. (2011) observed 87% removal of COD, and the turbidity removal was 95% from pulp and paper mill wastewaters through column-type sequencing batch reactor. The alkalinity and pH of the treated wastewaters were in the permissible range and improved the characteristics of produced sludge. Kumar and Subramanian (2014) found through sequential batch reactor (SBR) system the removal efficiencies of COD, BOD, TDS, TSS and organic compounds reached up to 84%, 83%, 85%, 88% and 80 ± 4.5% under the retention time of 24 h. Various studies showed that high loads of organic pollutants present in pulp and paper mill wastewater reduced through SBR (Milet and Duff 1996). Muhamad et al. (2013) reported a significant reduction in the pollutants from recovered fibres of pulp and paper mill wastewater by granular activated carbon-sequencing batch biofilm reactor (GAC-SBBR) and also achieved 97.2% removal of COD, 99.4% of NH_3-N and 100% of DCP. Buyukkamaci and Koken (2010) stated that the activated sludge process is the most important treatment method for the removal of low and medium strength of pulp and paper mill wastewaters but has some drawbacks, which can be improved in combination with membrane bioreactors. Membrane fouling in membrane technologies can increase the maintenance and operational costs, which may also overcome by membrane bioreactors (Le-clech et al. 2006). Removal of COD by moving bed biofilm reactor (MBBR) and the amount of sludge which is produced in secondary treatment from pulp and paper industries can also be reduced by membrane bioreactors (Jahren et al. 2002). Upflow anaerobic sludge blanket (UASB) reactor also known as anaerobic reactor was used in the treatment of various industrial wastewaters like tannery, distillery, pharmaceutical, pulp and paper, etc. Microorganisms living in the sludge blanket of UASB having microbial granules of size 0.5–2 mm break down organic matter by anaerobic digestion into simple compounds and biogas and can be used as energy source. Buyukkamaci and Koken (2010) showed that upflow anaerobic sludge blanket reactor followed by an aeration basin is the most economic and technically feasible treatment for medium and high strength wastewaters. In previous studies, Peerbhoi (2000) investigated that an upflow anaerobic sludge blanket (UASB) reactor was not feasible, as the pollutants were not properly degraded.

7 Other Emerging Treatment Approaches

7.1 Constructed Wetlands (CWs)

Constructed wetlands (CW) are emerging, low-cost and eco-friendly sustainable wastewater treatment systems. These are engineered integrated wastewater treatment systems of plants, water, microorganisms and the environment. Plants, soil, sand and gravels make shallow beds or channels, and a variety of microorganisms grow on these beds or channels to improve the quality of wastewaters (USEPA

2004). Both systems are able to remove high biochemical oxygen demand (BOD), chemical oxygen demand (COD), total suspended solids (TSS), colour, metals and toxic compounds, which are present in various types of industrial wastewaters. Plants directly uptake the nutrients from soil and facilitate indirect aerobic degradation of pollutants. Conventional treatment methods are not very much capable of removing high organic pollutants and colour from pulp and paper mill wastewaters. Constructed wetlands should be a better treatment way for pulp and paper mill wastewaters because their treatment efficiency is higher than conventional treatment methods (Choudhary et al. 2011).

7.2 Biocomposting

Biocomposting is one of the most valuable and green treatment processes for the mitigation of various harmful pollutants of industrial wastewaters. This method is also suitable for the treatment of wastes and sludges especially produced from paper fibres and organic materials. In biocomposting, wastes are dumped with microorganisms; humus like mater is produced which may be used in agriculture, houseplant greenhouse, etc. (Christmas 2002; Gea et al. 2005). Microorganisms converted organic materials into CO_2, humus and heat. The increased temperatures (thermophilic phase) in composts found that the rapid degradation of lignocelluloses occurs and is mainly degraded by thermophilic micro-fungi and actinomycetes (Tuomela et al. 2000).

7.3 Enzymatic Treatments

Several enzymes such as peroxidases, oxidoreductases, cellulolytic enzymes, cyanidase, proteases, amylases, etc. are used to treat industrial wastewaters. Ligninolytic group (laccase, MnP and LiP) of peroxidases from a variety of different sources have been reported to play an important role in waste and wastewater treatment (Chandra and Chowdhary 2015). Recently the demand of these enzymes (laccase, LiP, MnP) has increased gradually due to their prospective applications in the diverse biotechnological areas. Lignin peroxidase (LiP) and Mn peroxidase (MnP) were reported to be very efficient in decolourization of kraft pulp mill wastewaters (Moreira et al. 2003). Predominantly white-rot fungus and their specific enzymes are required for lignin degradation. *Phanerochaete chrysosporium* is one of the most important white-rot fungus and widely studied model for lignin degradation. Bacterial laccases are most considerable in the remediation of pollutants of industrial wastes. Laccases are also involved in the treatment of various industrial wastewaters such as pulp and paper, textile, tannery, distillery, etc. (Sangave and Pandit 2006; Chandra and Chowdhary 2015; Mani and Bharagava 2016; Bharagava et al. 2017; Chowdhary et al. 2017a, b).

7.4 Biosorption

Sorption is a process in which one substance is attached to another and bio means the involvement of living entities, i.e. biosorption is a physiochemical process that can be defined as the involvement of live entity like fungi or bacteria (adsorbent) and chemical or metal (adsorbate) leading to the removal of substances from solution through biological materials (Aksu 2002; Gadd 2009). Biosorption is used to treat wastewaters produced from various industrial sectors, and this treatment is considered clean, efficient, cost-effective and easy to operate (Saiano et al. 2005). Singhal et al. (2015) isolated a fungus, *Emericella nidulans*, was used for biosorption of colour of pulp and paper mill wastewaters, and after treatment the fungus turned dark brown in colour.

8 Challenges

Pulp and paper industrial sectors are facing many critical issues during the wastewater treatment and disposal process. In this study, the following challenges were noted:

- Pulp and paper industry produces large amount of wastewaters causing serious concerns for environment.
- In developing countries including India, lack of advanced treatment techniques and waste disposal are a serious concern.
- High values of physico-chemical parameters such as COD, BOD, TSS, TDS, etc. cause toxic effects in the aquatic system.
- Less involvement of government agencies with the industries.
- Degradation of complex compound lignin, which is a major pollutant in pulp and paper industry.
- Proper dumping of sludge, produced during the papermaking process.

Thus, there is an urgent need to solve these issues for sustainable development.

9 Conclusion

This chapter concluded the following points:

- Pulp and paper industries have high strength of wastewaters, i.e. it contains high COD, BOD, TSS, TDS etc., which causes hazardous effects on living entities.
- Untreated wastewater, when discharged into the green belt, disturbs the ecological balance of environment.
- Effective treatment and proper disposal are thus needed.

- Pulp and paper mill wastewater can also be used for the irrigation purpose with proper dilution.
- Biological approaches for the mitigation/removal of pulp and paper mill wastewater pollutants are gaining its momentum in the arena of wastewater treatment methods.
- Biological treatment methods are commonly used because these are the only treatment technique that produces very useful byproducts, which may be in large demand, and also it is a significant need for sustainable development of eco-friendly environment.
- Physico-chemical treatment strategies are more expensive/costly than biological treatment approaches for the capability of both colour and organic load reduction.
- On the basis of available literature on pulp and paper mill wastewater treatment, it seems that there is a need to address these problems and limitations in conventional treatment approaches and find out an effective solution. Hence, this chapter covers all the issues related to pulp and paper wastes/wastewaters and treatment technology for the sustainable development of environment.

Acknowledgement The University Grant Commission (UGC) to Ms. Surabhi Zainith for his Ph.D. work from UGC, Government of India (GOI), New Delhi, India, is duly acknowledged.

References

Abhishek A, Dwivedi A, Neeraj N et al (2015) Comparative bacterial degradation and detoxification of model and kraft lignin from pulp paper wastewater and its metabolites. Appl Water Sci. https://doi.org/10.1007/s13201-015-0288-9

Aksu Z (2002) Determination of the equilibrium, kinetics and thermodynamic parameters of the batch biosorption of nickel(II) ions onto Chlorella vulgaris. Process Biochem 38:89–99

Ali M, Sreekrishnan TR (2001) Aquatic toxicity from pulp and paper mill effluents: a review. Adv Environ Res 5:175–196

Amat AM, Arques A, Miranda MA et al (2005) Use of ozone and/or UV in the treatment of effluents from board paper industry. Chemosphere 60(8):1111–1117

Ansari PM (2004) Water conservation in pulp and paper, distillery. In: Indo-EU workshop on promoting efficient water use in agro based industries. New Delhi. January 15–16

Asano T, Cotruvo JA (2004) Ground water recharge with recharge with reclaimed municipal wastewater: health and regulatory considerations. Water Res 38:1941–1951

Asghar MN, Khan S, Mushtaq S (2008) Management of treated pulp and paper mill effluent to achieve zero discharge. J Environ Manag 88:1285–1299

Ashrafi O, Yerushalmi L, Haghighat F (2015) Wastewater treatment in the pulp and paper industry: a review of treatment processes and the associated greenhouse gas emission. J Environ Manag 158:146–157

Ayed L, Asses N, Chammem N et al (2017) Advanced oxidation process and biological treatments for table olive processing wastewaters: constraints and a novel approach to integrated recycling process: a review. Biodegradation 28:125–138

Bajpai P (2001) Microbial degradation of pollutants in pulp mill effluents. Adv Appl Microbiol 48:79–134

Balcioglu IA, Tarlan E, Kiyilcimdan et al (2007) Merits of ozonation and catalytic ozonation pretreatment in the algal treatment of pulp and paper mill effluents. J Environ Manag 85:918–926

Barapatre A, Jha H (2016) Decolorization and biological treatment of pulp and paper mill effluent by lignin-degrading fungus Aspergillus flavus strain F10. Int J Curr Microbiol App Sci 5(5):19–32

Barton DA, Lee JW, Buckley DB et al (1996) Biotreatment of kraft mill condensates for reuse. In: Proceedings of Tappi minimum effluent mills symposium, Atlanta, pp 270–288

Baruah BK (1997) Effect of paper mill effluent on plankton population of wetland. Environ Ecol 15(4):770–777

Bengtsson S, Werker A, Christensson M et al (2008) Production of polyhydroxyalkanoates by activated sludge treating a paper mill wastewater. Bioresour Technol 99:509–516

Berryman D, Houde F, Deblois C et al (2004) Non phenolic compounds in drinking and surface waters downstream of their textile and pulp and paper effluents: a survey and preliminary assessment of their potential effects on public health and aquatic life. Chemosphere 56(3):247–255

Bharagava RN, Mishra S (2017) Hexavalent chromium reduction potential of *Cellulosimicrobium* sp. isolated from common effluent treatment industries. Ecotoxicol Environ Saf 147:102–109

Bharagava RN, Chowdhary P, Saxena G (2017) Bioremediation an eco-sustainable green technology, its applications and limitations. In: Bharagava RN (ed) Environmental pollutants and their bioremediation approaches. CRC Press, Taylor & Francis Group, Boca Raton, pp 1–22

Buyukkamaci N, Koken E (2010) Economic evaluation of alternative wastewater treatment plant options for pulp and paper industry. Sci Total Environ 408:6070–6078

Cecan F, Urban W, Haberl R (1992) Biological and advanced treatment of sulfate pulp bleaching effluents. Water Sci Technol 26:435–444

Chandra R (2001) Microbial decolourization of pulp mill effluent in presence of nitrogen and phosphorous by activated sludge process. J Environ Biol 22(1):23–27

Chandra R, Abhishek A (2010) Bacterial decolourization of black liquor in axenic and mixed condition and characterization of metabolites. Biodegradation 22:603–611

Chandra R, Chowdhary P (2015) Properties of bacterial lacasses and their application in bioremediation of industrial wastes. Environ Sci Process Impacts 17:326–342

Chandra R, Singh R (2012) Decolourization and detoxification of rayon grade pulp and paper mill effluent by mixed bacterial culture isolated from pulp and paper mill effluent polluted site. Biochem Eng J 61:49–58

Chandra R, Raj A, Purohit HJ et al (2007) Characterization and optimization of three potential aerobic bacterial strains for kraft lignin degrading from pulp-paper waste. Chemosphere 67:839–846

Chandra R, Raj A, Hj P et al (2009) Reduction of pollutants in pulp and paper mill effluent treated by PCB degrading bacterial strains. Environ Monit Assess 155:1–11

Chanworrawoota K, Hunsom M (2012) Treatment of wastewater from pulp and paper mill industry by electrochemical methods in membrane reactor. J Environ Manag 113:399–406

Chauhan N, Thakur IS (2002) Treatment of pulp and paper mill effluent by pseudomonas fluorescens in fixed film bioreactor. Pollut Res 21:429–434

Chen D (2008) Application of membrane separation technology in the fine chemical industry. Energy Conserv Recycl Fine Spec Chem 11:14–17

Chen P, Zheng J, Zhou Y (2009) The presence and future of membrane industry in china. Environ Prot 8:71–74

Choudhary AK, Kumar S, Sharma C (2011) Constructed wetlands: an option for pulp and paper mill wastewater treatment. EJEAF Che 10(10):3023–3037

Chowdhary P, Yadav A, Kaithwas G, Bharagava RN (2017a) Distillery wastewater: a major source of environmental pollution and its biological treatment for environmental safety. In: Singh R, Kumar S (eds) Green technologies and environmental sustainability. Springer International, Cham, pp 409–435

Chowdhary P, More N, Raj A, Bharagava RN (2017b) Characterization and identification of bacterial pathogens from treated tannery wastewater. Microbiol Res Int 5(3):30–36

Christmas P (2002) Building materials from deinking plant residues- a sustainable solution. In: COST workshop managing pulp and paper residues, Barcelona

Ciputra S, Antony A, Phillips R et al (2010) Comparison of treatment options for removal of recalcitrant dissolved organic matter from paper mill effluent. Chemosphere 81:86–91

Costigan SL, Werner J, Ouellet JD et al (2012) Expression profiling and gene ontology analysis in fathead minnow (Pimephales promelas) liver following exposure to pulp and paper mill effluents. Aquat Toxicol 122–123:44–55

Covinich LG, Bengoechea DI, Fenoglio RJ et al (2014) Advanced oxidation processes for wastewater treatment in the pulp and paper industry: a review. Am J Environ Eng 4(3):56–70

CPCB (2001) Comprehensive industry document for large pulp and paper industry. COINDS/36/2000–2001

Crini G (2005) Recent developments in polysaccharide-based materials used as adsorbents in wastewater treatment. Prog Polym Sci 30:38–70

Das CP, Patnaik LN (2000) Removal of lignin by industrial solid wastes. Pract Period Hazard Toxic Radioact Waste Manag 4(4):156–161

Dilek FB, Gokcay CF (1994) Treatment of effluents from hemp-based pulp and paper industry: waste characterization and physiochemical treatability. Water Sci Technol 29(9):161–163

Dube M, McLean R, MacLatchy D et al (2000) Reverse osmosis treatment: effects on effluent quality. Pulp Pap Can 101(8):42–45

Dutta SK (1999) Study of the physicochemical properties of effluent of the paper mill that affected the paddy plants. J Environ Pollut 6(2 and 3):181–188

Dutta A, Sarkar S (2015) Sequencing batch reactor for wastewater treatment: recent advances. Curr Pollut Rep 1:177–190

Ebrahimi M, Busse N, Kerker S et al (2016) Treatment of the bleaching effluent from sulfite pulp production by membrane filtration. Membranes 6(7):1–15

Ekstrand E, Larsson M, Truong X et al (2013) Methane potentials of the Swedish pulp and paper industry- a screening of wastewater effluents. Appl Energy 112:507–517

EPA (2002) Office of compliance sector notebook project profile of the pulp and paper industry. 2nd ed. Washington, Nov

Eskelinen K, Sarkka H, Kurniawan TA et al (2010) Removal of recalcitrant contaminants from bleaching effluents in pulp and paper mills using ultrasonic irradiation and fenton like oxidation, electrochemical treatment, and/or chemical precipitation: a comparative study. Desalination 255:179–187

Fontanier V, Albet J, Baig S et al (2005) Simulation pulp mill wastewater recycling after tertiary treatment. Environ Technol 26:1335–1344

Gadd GM (2009) Biosorption: critical review of scientific rationale, environmental importance and significance for pollution treatment. J Chem Technol Biotechnol 84:13–28

Ganjidoust H, Tatsumi K, Yamagishi T et al (1997) Effect of synthetic and natural coagulant on lignin removal from pulp and paper wastewater. Water Sci Technol 35(2–3):291–296

Garg SK, Tripathi M (2011) Strategies for decolorization and detoxification of pulp and paper mill effluent. Rev Environ Contam Toxicol. https://doi.org/10.1007/978-1-4419-8453-1-4

Gea T, Artola A, Sanchez A (2005) Composting of deinking sludge from the recycled paper manufacturing industry. Bioresour Technol 96:1161–1167

Ghoreishi SM, Haghighi MR (2007) Chromophores removal in pulp and paper mill effluent via hydrogenation-biological batch reactors. Chem Eng J 127:59–70

Gonder ZB, Arayici S, Barles H (2011) Advanced treatment of pulp and paper mill wastewater by nanofilteration process: optimization of the fouling and rejections. Ind Eng chem Res 51(17):6184–6195

Gonzalez LF, Sarria V, Sanchez OF (2010) Degradation of chlorophenols by sequential biological advanced oxidative process using Trametes pubescens and TiO_2/UV. Bioresour Technol 101:3493–3499

Gubelt G, Lumpe C, Joore L (2000) Towards zero liquid effluents at Niederauer Muhle-the validation of two noval separation technologies. Pap Technol (UK) 41(8):41–48

Gupta A (1997) Pollution load of paper mill effluent and its impact on biological environment. J Ecotoxicol Environ Monit 7(2):101–112

Hermosilla D, Merayo N, Gasco A et al (2014) The application of advanced oxidation technologies to the treatment of effluents from the pulp and paper industry: a review. Environ Sci Pollut Res. https://doi.org/10.1007/s11356-014-3516-1

Herney-Ramirez J, Silva AMT, Vicente MA et al (2011) Degradation of acid orange 7 using as laponite-based catalyst in wet hydrogen peroxide oxidation: kinetic study with the Fermi's equation. Appl Catal B Environ 101:197–205

Hewit LM, Kovacs TG, Dude MG et al (2008) Altered reproduction in fish exposed to pulp and paper mill effluents: roles of individual compounds and mill operating conditions. Environ Toxicol Chem 27:682–697

Holmbom B, Harju L, Lindholm J et al (1994) Effect of a pulp and paper mill on metal concentration in the receiving lake system. Aqua Fenn 24(1):93–110

Hooda R, Nishi K, Bhardwj NK et al (2015) Screening and identification of ligninolytic bacteria for the treatment of pulp and paper mill effluent. Water Air Soil Pollut 226(305):1–11

Hooda R, Bhardwaj NK, Singh P (2016) Decolourization of pulp and paper mill effluent by indigenous bacterium isolated from sludge. Lignocellulose 5(2):106–117

Howe J, Michael RW (1998) Effects of pulp mill effluent irrigation on the distribution of elements in the profile of an arid region soil. Environ Pollut 105:129–135

http://aquarden.com/technology/coagulation-and-flocculation

http://www.gunt.de/images/download/flotation-sedimentation-english.pdf

http://www.iwapublishing.com/news/coagulation-and-flocculation-water-and-wastewater-treatment

Inan H, Dimoglo A, Simsek H et al (2004) Olive oil mill wastewater treatment by means of electrocoagulation. Sep Purif Technol 36(1):23–31

Ince O, Kolukirik M, Cetecioglu Z et al (2007) Methanogenic and sulfate reducing bacterial population levels in a full-scale anaerobic reactor treating pulp and paper industry wastewater using fluorescence in situ hybridization. Water Sci Technol 55(10):183–191

Jahren JS, Rintala JA, Odegaard H (2002) Aerobic moving bed biofilm reactor treating thermomechanical pulping whitewater under thermophilic conditions. Water Res 36:1067–1075

Kamali M, Khodaparast Z (2015) Review on recent developments on pulp and paper mill wastewater treatment. Ecotoxicol Environ Saf 114:326–342

Karam J, Nicell JA (1997) Potential application of enzymes in waste treatment. J Chem Technol Biotechnol 69:141–153

Karrasch B, Parra O, Cid H et al (2006) Effect of pulp and paper mill effluents on the microplankton and microbial self-purification capabilities of the Biobio river, Chile. Sci Total Environ 359(1–3):619–625

Kaviyarasan K (2014) Application of UASB reactor in industrial wastewater treatment- a review. Int J Sci Eng Res 5(1):584–589

Khan NA, Basheer F, Singh D et al (2011) Treatment of paper and pulp mill wastewater by column type sequencing batch reactor. J Ind Res Tech 1(1):12–16

Kouhia M, Holmberg H, Ahtila P (2015) Microalgae- utilizing biorefinery concept for pulp and paper industry:converting secondary streams into value-added products. Algal Res 10:41–47

Krawczyk H, Oinonen P, Jonsson A (2013) Combined membrane filtration and enzymatic treatment for recovery of high molecular mass hemicelluloses from chemithermomechanical pulp process. Water Chem Eng J 225:292–299

Kreetachat T, Chaisan O, Vaithanomsat P (2016) Decolorization of pulp and paper mill effluents using wood rotting fungus Fibrodontia sp. RCK783S. Int J Eng Sci Dev 7(5):321–324

Kumar V, Chopra AK (2012) Fertigation effect of distillery effluent on agronomical practices of Trigonella foenum-graecum L. (Fenugreek). Environ Monit Assess 184:1207–1219

Kumar RR, Subramanian K (2014) Treatment of paper and pulp mill effluent using sequential batch reactor. International Conference on Biological, Civil and Environmental Engineering 39–42

Kumar V, Dhall P, Naithani S et al (2014) Biological approach for the treatment of pulp and paper industry effluent in sequence batch reactor. Bioremed Biodegrad 5(3):1–10

Kumar S, Haq I, Yadav A, Prakash J, Raj A (2016) Immobilization and biochemical properties of purified xylanase from bacillus amyloliquefaciens sk-3 and its application in kraft pulp biobleaching. J Clin Microbiol Biochem Technol 2(1):26–34

Kumari V, Sharad K, Izharul H, Yadav A, Singh VK, Ali Z, Raj A (2014) Effect of tannery effluent toxicity on seed germination á-amylase activity and early seeding growth of mung bean (Vigna Radiata) seeds. Int J Lat Res Sci Technol 3(4):165–170

Kumari V, Yadav A, Haq I, Kumar S, Bharagava RN, Singh SK, Raj A (2016) Genotoxicity evaluation of tannery effluent treated with newly isolated hexavalent chromium reducing Bacillus cereus. J Environ Manag 183:204–211

Leclech P, Chen V, Fane TAG (2006) Fouling in membrane bioreactors used in wastewater treatment. J Membr Sci 284:17–53

Li Y, Chen YF, Chen P et al (2011) Characterization of a microalga chlorella sp. well adapted to highly concentrated municipal wastewater for nutrient removal and biodiesel production. Bioresour Technol 102:5138–5144

Liu T, Hu H, He Z et al (2011) Treatment of poplar alkaline peroxide mechanical pulping (APMP) effluent with aspergillus niger. Bioresour Technol 102:7361–7365

Lokeshwari N, Keshava J, Sangeetha M et al (2015) Optimization and kinetic studies for the degradation of lignin and chlorophenols by using Rhizopus aarhizus. J Phys Chem Sci 3(1):1–6

Mahmood T, Paice M (2006) Aerated stabilization basin design and operating practices in the Canadian pulp and paper industry. J Environ Eng Sci 5:383–395

Makris SP, Banerjee S (2002) Fate of resin acids in pulp mills secondary treatment systems. Water Res 36:2878–2882

Malaviya P, Rathore VS (2007) Bioremediation of pulp and paper mill effluent by a novel fungal consortium isolated from polluted soil. Bioresour Technol 98:4647–4651

Malik MM, Kumar P, Seth R et al (2009) Genotoxic effect of paper mill effluent on chromosome of fish Channa punctatus. Curr World Environ 4:353–367

Malkin VP (2002) Electrochemical methods of treating industrial effluents. Chem Prot Eng 38(9–10):619

Mallevialle J, Odendaal PE, Wiesner MR (1996) Water treatment membrane processes. LyonnaisedesEaux-LdE, New York

Mandal TN, Bandana TN (1996) Studies on physicochemical and biological characteristics of pulp and paper mill effluents and its impact on human beings. J Freshw Biol 8(4):191–196

Mani S, Bharagava RN (2016) Exposure to crystal violet, its toxic, genotoxic and carcinogenic effects on environmental and its degradation and detoxification for environmental safety. Rev Environ Conta Toxicol 237:71–104

Mathur RM, Panwar S, Gupta MK et al (2004) Agro-based pulp and paper mills: environmental status, issues and challenges and the role of central pulp and paper research institute. In: Tewari PK (ed) Liquid asset, proceedings of the indo-EU workshop on promoting efficient water use in agro based industries. TERI Press, New Delhi, pp 99–114

Mehta J, Sharma P, Yadav A (2014) Screening and identification of bacterial strains for removal of COD from pulp and paper mill effluent. Adv Life Sci Health 1(1):34–42

Metcalf, Eddy (revised by Tchobanoglous G, Burton FL, Stensel HD) (2003) Wastewater engineering treatment and reuse, 4th ed. McGraw-Hill, New York

Methews SL, Pawlak JJ, Grunden AM (2014) Isolation of Paenibacillus glucanolyticus from pulp mill sources with potential to deconstruct pulping waste. Bioresour Technol 164:100–106

Milet GM, Duff SJB (1996) Treatment of kraft condensates in a feedback controlled sequencing batch reactor. Water Sci Technol 38(4–5):263–271

Mishra S, Bharagava RN (2016) Toxic and genotoxic effects of hexavalent chromium in environment and its bioremediation strategies. J Environ Sci Health Part C 34(1):1–34

Moreira MT, Feijoo G, Canoval J et al (2003) Semipilot-scale bleaching of kraft pulp with MnP. Wood Sci Technol 37:117–123

Muhamad MH, Abdullah SR, Mohamad AB et al (2013) Application of response surface methodology (RSM) for optimization of COD, NH$_3$-N and 2,4-DCP removal from recycled paper wastewater in a pilot scale granular activated carbon sequencing batch biofilm reactor (GAC-SBBR). J Environ Manag 121:179–190

Munkittrick KR, Servos MR, Carey JH et al (1997) Environmental impacts of pulp and paper wastewater: evidence for a reduction in environmental effects at North American pulp mills since 1992. Water Sci Technol 35:329–338

Nagarthnamma R, Bajpai P, Bajpai PK (1999) Studies on decolourization, degradation and detoxification of chlorinated lignin compounds in kraft bleaching effluents by Ceriporiopsis subvermispora. Process Biochem 34:939–948

Nemerow NL, Dasgupta A (1991) Industrial and hazardous waste management. Van Nostrand Reinhold, New York

Nestmann ER (1985) Detection of genetic activity in effluent from pulp and paper mills: mutagenicity in saccharomyces cerevisiae. In: Zimmerman FK, Taylor-mayor RE (eds) Testing in environmental pollution control. Horwood, London, pp 105–117

O'connor B, Kovacs T, Gibbons S et al (2000) Carbon dioxide in pulp and paper mill effluents from oxygen-activated sludge treatment plants as a potential source of distress and toxicity to fish. Water Qual Res J Can 35(2):189–200

Olsen O (1980) Membrane technology in the pulp and paper industry. Desalination 35:291–302

Owens JW, Swanson SM, Birkholz DA (1994) Environmental monitoring of bleached kraft pulp mill chlorophenolic compounds in a Northern Canadian river system. Chemosphere 29(1):89–109

Park C, Lee M, Lee B et al (2007) Biodegradation and biosorption for decolourization of synthetic dyes by Fulani tragic. Biochem Eng J 36:59–65

Peerbhoi Z (2000) Treatability studies of black liquor by UASBR-Phd thesis. University of Roorkee India

Perez M, Torrades F, Garcia-Hortal JA et al (2002) Removal of organic contaminants in paper pulp treatment effluents under fenton and photo-fenton conditions. Appl Catal B 36:63–74

Pokhrel D, Viraraghavan T (2004) Treatment of pulp and paper mill wastewater: a review. Sci Total Environ 333:37–58

Raj A, Kumar S, Haq I et al (2014) Bioremediation and toxicity reduction in pulp and paper mill effluent by newly isolated ligninolytic Paenibacillus sp. Ecol Eng 71:355–362

Ramanna L, Guldhe A, Rawat L et al (2014) The optimization of biomass and yields of Chlorella sorokiniana when using wastewater supplemented with different nitrogen sources. Bioresour Technol 168:127–135

Rashed MN (2013) Adsorption technique for the removal of Organic from water and wastewater. https://doi.org/10.5772/54048

Rohella RS, Choudhary S, Manthan M et al (2001) Removal of colour and turbidity in pulp ans paper mill effluents using polyelectrolytes. Indian J Environ Health 43(4):159–163

Sahoo DK, Gupta R (2005) Evaluation of ligninolytic microorganisms for efficient decolourization of a small pulp and paper mill effluent. Process Biochem 40:1573–1578

Saiano F, Ciofalo M, Cacciola SO et al (2005) Metal ion adsorption by Phomopsis Sp. biomaterial in laboratory experiments and real wastewater treatments. Water Res 39:2273–2280

Sala M, Gutierrez-Bouzan MC (2012) Electrochemical techniques in textile processes and wastewater treatment. Int J Photoenergy 629103:1–12

Sangave PC, Pandit AB (2006) Enhancement in biodegradability of distillery wastewater using enzymatic pretreatment. J Environ Manag 78:77–85

Sankaran S, Khanal SK, Jastin N et al (2010) Use of filamentous fungi for wastewater and production of high value fungal byproducts: a review. Crit Rev Environ Sci Technol 40:400–449

Saravanan V, Sreekrishnan TR (2005) Bio-physico-chemical treatment for removal of colour from pulp and paper mill effluents. J Sci Ind Res 64:61–64

Saritha V, Maruthi YA, Mukkanti K (2010) Potential fungi for bioremediation of industrial effluents. Bio Resources Com 5(1):8–22

Schnell A, Steel P, Melcer H et al (2000) Enhanced biological treatment of bleached kraft mill effluents: II. Reduction of mixed function oxygenase (MFO) induction in fish. Water Res 34(2):501–509

Sevimli MF (2005) Post-treatment of pulp and paper industry wastewater by advanced oxidation processes. Ozone Sci Eng 27:37–43

Shawwa AR, Smith DW, Sego DC (2001) Colour and chlorinated organics removal from pulp wastewater using activated petroleum coke. Water Res 35(3):745–749

Sinclair WF (1990) Controlling pollution from Canadian pulp and paper manufactures: a federal perspective. Canadian Government Publishing Centre, Ottawa

Singh S (2004) An overview of Indian agro-based paper mills. In: Tewari PK (ed) Liquid asset, proceedings of the indo-EU workshop on promoting efficient water use in agro based industries. TERI Press, New Delhi, pp 31–33

Singh RS, Marwaha SS, Khanna PK (1996) Characteristics of pulp and paper mill effluents. J Ind Pollut Control 12(2):163–172

Singh C, Chowdhary P, Singh JS, Chandra R (2016) Pulp and paper mill wastewater and coliform as health hazards: a review. Microbiol Res Int 4(3):28–39

Singhal A, Thakur IS (2009a) Decolourization and detoxification of pulp and paper mill effluent by Emericella nidulans. Biochem Eng J 171:619–625

Singhal A, Thakur IS (2009b) Decolourization and detoxification of pulp and paper mill effluent by Cryptococcus sp. https://doi.org/10.1016/j.bej.2009.04.007

Singhal A, Kumar PJ, Thakur IS (2015) Biosorption of pulp and paper mill effluent by Emericella nidulans: isotherms, kinetics and mechanism. Dealination Water Treat 1–16

Skipperud L, Salbu B, Hagebo E (1998) Speciation of trace elements in discharges from the pulp industry. Sci Total Environ 217:251–256

Soloman PA, Basha CA, Velan M et al (2009) Augmentation of biodegradability of pulp and paper industry wastewater by electrochemical pre-treatment and optimization by RSM. Sep Purif Technol 69:109–117

Sumathi S, Hung YT (2006) Treatment of pulp and paper mill wastes Waste treatment in the process industries. Wang LK, Hung YT, Lo HH et al 453–497. Taylor & Francis. 0-8493-7233-X. Boca Raton

Tarlan E, Dilek FB, Yetis U (2002) Effectiveness of algae in the treatment of a wood-based pulp and paper industry wastewater. Bioresour Technol 84:1–5

Thompson G, Swain J, Kay M et al (2001) The treatment of pulp and paper mill effluent: a review. Bioresour Technol 77(3):275–286

Tiku DK, Kumar A, Chaturvedi R et al (2010) Holistic bioremediation of pulp mill effluents autochthonous bacteria. Int Biodeterior Biodegrad 64:173–183

Tong Z, Wada S, Takao Y et al (1999) Treatment of bleaching wastewater from pulp-paper plants in China using enzymes and coagulants. J Environ Sci 11(4):480–484

Tuomela M, Vikman M, Hatakka A et al (2000) Biodegradation of lignin in a compost environment: a review. Bioresour Technol 72:169–183

Tyagi SI, Kumar V2, Singh J1 (2014) Bioremediation of pulp and paper mill effluent by dominant aboriginal microbes and their consortium. Int J Environ Res 8(3):561–568

Ugurlu M, Gurses A, Dogar C et al (2008) The removal of lignin and phenol from paper mill effluents by electrocoagulation. J Environ Manag 87:420–428

USEPA (Environmental Protection Agency) (2004) Constructed treatment wetlands. Office of water 843-F-03-013

Usha MT, Chandra TS, Sarada R et al (2016) Removal of nutrients and organic pollution load from pulp and paper mill effluent by microalgae in outdoor open pond. Bioresour Technol 214:856–860

Wang J, Chen Y, Wang Y et al (2011) Optimization of the coagulation-flocculation design and response surface methodology. Water Res 45:5633–5640

Waye A, Annal M, Tang A et al (2014) Canadian boreal pulp and paper feed stocks contain neuroactive substances that interact in vitro with GABA and dopaminergic systems in the brain. Sci Total Environ 468–469:315–325

Webb L (1985) An investigation into the occurrence of sewage fungus in rivers containing paper mill effluents. Removal of sewage fungus nutrients. Water Res 19:961–967

Wells GF, Park H, Eggleston B et al (2011) Fine-scale bacterial community dynamics and the taxa-time relationship with in a full-scale activate sludge bioreactor. Water Res 45:5476–5488

Wenta B, Hartmen B (2002) Dissolved air flotation system improves wastewater treatment at Glatfelter. Pulp Pap 76(3):43–47

Xilei D, Tingzhi L, Weijiang D et al (2010) Adsorption and coagulation tertiary treatment of pulp and paper mills wastewater. In: Proceedings of the 4th International Conference on Bioinformatics and Biomedical Engineering (ICBBE)

Yang Q, Angly FE, Wang Z et al (2011) Wastewater treatment systems harbor specific and diverse yeast communities. Biochem Eng J 58–59:168–176

Yen NT, Oanh NTK, Reutergard LB et al (1996) An integrated waste survey and environmental effects of COGIDO, a bleached pulp and paper mill in Vietnam on the receiving water body. Global Environ Biotechnol 66:349–364

Zwain HM, Hassan SR, Zaman NQ et al (2013) The startup performance of modified anaerobic baffled reactor (MABR) for the treatment of recycled paper mill wastewater. J Environ Chem Eng 1:61–64

Chapter 14
Role of *Rhizobacteria* in Phytoremediation of Metal-Impacted Sites

Reda A. I. Abou-Shanab, Mostafa M. El-Sheekh, and Michael J. Sadowsky

Abstract Phytoremediation is an emerging and eco-friendly technology that has gained wide acceptance and is currently an area of active research in plant biology. A number of metal-hyperaccumulating plants have already been identified as potential candidates to phytoremediate metal-polluted soil. Various strategies have been successfully applied to generate plants able to grow in adverse environmental conditions and accumulate or transfer a number of metals. Recently, biotechnological approaches have opened up new opportunities concerning the application of beneficial rhizospheric and endophytic bacteria for improving plant growth, biological control, and heavy metal remediation from contaminated sites. Further, molecular approaches have been applied to improve the process of phytoremediation efficiently using a transgenic approach. The overexpression of several genes whose protein products are directly or indirectly involved in plant metal uptake, transport, and sequestration, or act as enzymes involved in the biodegradation of hazardous organic wastes, has opened up new possibilities in phytoremediation. This chapter is mainly focused on plant-microbe interactions to phytoremediate metal-contaminated sites and evaluate the progress made thus far in understanding the role of rhizospheric and endophytic bacteria in the phytoremediation of metal-contaminated sites and different phytoremediation technologies. In addition, we also discuss the use of genetic engineering to modify plants for enhanced efficacy

---

R. A. I. Abou-Shanab (✉)
Department of Environmental Biotechnology, Genetic Engineering and Biotechnology Research Institute, City of Scientific Research and Technological Applications, New Borg El Arab City, Alexandria, Egypt

Department of Soil, Water and Climate, Biotechnology Institute, University of Minnesota, St. Paul, MN, USA
e-mail: rabousha@umn.edu

M. M. El-Sheekh
Botany Department, Faculty of Science, Tanta University, Tanta, Egypt

M. J. Sadowsky
Department of Soil, Water and Climate, Biotechnology Institute, University of Minnesota, St. Paul, MN, USA

phytoremediation strategies. These approaches will be helpful to develop phytoremediation technologies for large-scale application to remediate vast areas of metal-polluted sites.

Keywords Organic and inorganic pollutants · Phytoremediation · Bioremediation · Rhizobacteria · Endophytic bacteria · Rhizosphere · Transgenic plants

1 Introduction

Heavy metal (HM) contamination in soils and water is a major environmental concern worldwide, posing significant risks to human and animal health as well as to ecosystem functioning. Consequently, the development of rapid, effective, and inexpensive strategies to remediate metal-polluted soils and water is necessary for environmental protection and to protect human and animal health (Mishra and Bharagava 2016). Phytoremediation using wild-type or transgenic plants and their attendant rhizospheric and endophytic bacteria to remove metal contaminants from polluted sites is being developed as recent tools for the remediation of the polluted environment. This environment-friendly, cost-effective, and plant-based technology is expected to have significant economic, aesthetic, and technical advantages over traditional remediation methods.

Plant-microbe interactions have been explored for their use in HM removal from contaminated environments. Microorganisms can affect metal solubility in soil and their availability to plant. These microbes can also produce chelating compounds and siderophore to assist iron availability, lower soil pH, and/or solubilize metal-phosphate complexes. Improvement in the interactions between plants and microorganisms can promote the plant biomass production and tolerance of the plants to HM and is considered to be an important component of phytoremediation technologies. Recently, overexpression of genes whose protein products are involved in metal uptake, transport, and sequestration, or act as enzymes involved in the degradation of hazardous organic compounds, has opened up new possibilities in phytoremediation technology.

The industrial revolution has led to extensive environmental contamination with metals worldwide. A wide variety of toxic chemicals have been detected in air, soil, and water (Cheng 2003; Turgut 2003; Bharagava et al. 2017; Chowdhary et al. 2017a, 2018). HMs pose a critical concern to living organisms and the environment due to their common occurrence as contaminants, their low solubility, carcinogenicity, and mutagenicity (Diels et al. 2002; Yadav et al. 2017). Vast areas of land around the world are polluted with HMs due to their emissions from waste incinerators, car exhaust, residues from mining activities, and smelting industries, as well as the use of sludge or urban composts, pesticides, and fertilizers (Bharagava and Mishra 2018).

Remediation of HM-contaminated sites is particularly challenging. Unlike organic compounds, metals cannot be degraded and often have limited valence states for biotransformation, and thus, cleanup usually requires physical or chemical removal (Bharagava et al. 2008). Many technologies are available for metal removal from soil, surface, and groundwater. However, current methods are based on either soil removal or replacement, physicochemical extraction of metals, or immobilization in situ. These traditional practices usually produce secondary waste, destroy soil fertility, adversely affect soil physical structure, remove the biological activity from the treated soil, or are very expensive (Pulford and Watson 2003; Bharagava et al. 2017). Therefore, it is recommended to search for alternative safe remediation strategies for decontaminations of metal-contaminated environments. Bioremediation, using microbes to remove metal pollutants from contaminated sites, is being further developed as a practical method for the remediation of contaminated environments (Chaney et al. 2005; Evangelou et al. 2015).

In soil, HMs are bound to organic and inorganic soil constituents as insoluble precipitates. Not all metal contaminants are available for root uptake by field-grown plants. Methods of increasing HM contaminant availability in soil and its uptake by plant roots are vital for the success of phytoremediation (Kukier et al. 2004; Abou-Shanab et al. 2006). The solubility and availability of trace metals to plants are known to be affected by microbial populations through release of chelators, acidification, and redox changes (Abou-Shanab et al. 2003a, b). The rhizobacteria have been reported to increase the concentrations of Cr, Cu, Pb, or Zn in plant tissues (Whiting et al. 2001; Chen et al. 2014; Abou-Shanab et al. 2007a, 2010).

The interaction between beneficial rhizosphere microbes and plants can increase the tolerance of the plants to HMs and is considered to be an important component of phytoremediation technologies (Abou-Shanab et al. 2003a, b; Glick 2003).

2 Heavy Metals and Their Properties

The word heavy metal is a general term that refers to the group of metals and metalloids with atomic density ranging from 3.5 to 7 g/cm^3, atomic weight >22.98 for Na, and atomic numbers from 20 to 92 and has a periodic table position in Groups 3–16 and in periods 4 and greater (Nriagu 1989; Hawkes 1997; Dufus 2002). These metals are also considered as microelements because of their presence in trace concentrations (ppb range to <10 ppm) in different environmental habitats (Kabata-Pendias 2011) and some whose use in biological systems is essential for growth. Their bioavailability is influenced by physical factors such as temperature, phase association, adsorption, sequestration, etc. It is also affected by chemical factors that influence speciation at thermodynamic equilibrium, complexation kinetics, lipid solubility, and octanol/water partition coefficients (Hamelink et al. 1994). Biological factors such as species characteristics, trophic interactions, and biochemical/physiological adaptation also play an important role in their bioavailability (Verkleji 1993).

Excessive deposits of HMs thus may pose harmful effects on biological systems, aquatic and soil macro- and microflora (Yadav et al. 2017). Toxic metals can be distinguished from other contaminants, since they cannot be degraded, persist in the environment for many years, and accumulate in living organisms, causing several diseases and disorders even at lower concentrations (Tangahu et al. 2011; Diels et al. 2002).

3 Sources of Heavy Metal Pollution in Environment

Heavy metal-polluted water and soils, from both anthropogenic and natural origin, are the major sources of metal pollution in environment. Natural sources of metal pollution include weathering of parent rock materials, volcanic eruptions, and forest (Singh and Prasad 2011; Mishra and Bharagava 2016). HM contamination in air, soil, and water is a global problem that is a growing threat to human and animal health. Most of the environmental contamination and human exposure result from anthropogenic activities. This includes mining and smelting operations as well as metal-based industrial operations (i.e., metal processing in refineries, coal combustion in thermal power plants, petroleum combustion, nuclear power stations and high-tension lines, plastics, textiles, tanneries, microelectronics, wood preservation, and paper processing plants) and industrial, agricultural, pharmaceutical, and domestic effluents and atmospheric sources (He et al. 2013; Arruti et al. 2010; Chowdhary et al. 2017b; Bharagava and Mishra 2018). Environmental pollution can also occur through metal corrosion, atmospheric deposition, soil erosion and leaching of HMs, sediment resuspension, and metal evaporation from water resources to soil and groundwater (Nriagu 1989). In addition, fly ash produced during the coal combustion also contains several toxic elements as compared to its parent coal because many of the trace elements present in parent coal (such as Pb, Ni, Zn, and Mn) get vaporized during the combustion process (Baba et al. 2008).

Soil and water contamination with HMs has increased dramatically during the last few decades as a result of mining and smelting activities, metal manufacturing, the use of agricultural fertilizers and pesticides, municipal waste, traffic emissions, as well as industrial effluents (Chibuike and Obiora 2014).

Metals contaminate environment due to atmospheric deposition from industrial activities, disposal of wastes, irrigation, and the utilization of fertilizers or agro-chemicals (Pulford and Watson 2003). Point source (PS) pollution enters waterways from discrete areas and includes any single identifiable pollution source from which contaminants are discharged, such as pipes, ditches, ships, or factory smokestacks (Hill 1997). Oil refineries, pulp and paper mills, timber treatment plant, chemical and electronic factories, and automobile manufacturers discharge many hazardous wastes in their discharged waters.

Nonpoint source (NPS) pollution, unlike pollution from identifiable source of pollution (PS), comes from many diffuse sources. The types of pollutants associated with NPS are diverse, which include herbicides, fertilizers, insecticides, oil, grease, acid drainage, and bacteria and nutrients from agricultural lands, residential area, urban runoff, abandoned mines, livestock, pet wastes, and septic systems. In addition, atmospheric deposition and hydro-modification are also sources of NPS pollution.

4 Toxicological Effects of Metals in Humans and Other Organisms

Heavy metal contamination poses significant risks to living organisms as well as to ecosystems (Nasim and Dhir 2010). Metal pollutants are harmful to the environment and almost all living organisms, i.e., plants, animals, and microorganisms, where their concentration is above certain threshold values (Gao et al. 2014; Zhang et al. 2015). Metals such as Co, Cu, Fe, Mn, Mo, Ni, V, and Zn are required in trace amounts for various biochemical and physiological functions in living organisms, while others, such as Pb, Cd, Al, and Hg, do not have any beneficial effects on organisms and thus generally regarded as toxic pollutants in air, water, and soil (Chang et al. 1996). However, insufficient supply of some essential metals, in very low concentrations, results in a variety of physiological diseases or syndromes (Hall 2002; Todeschini et al. 2011).

Metals accumulate in the ecological food chain through uptake at the primary producer level and then bioaccumulate as HM ions as they move up the food chain. HMs enter the human body via different food chains, inhalation, and ingestion, and once they enter the human body by any means, these may stimulate immune system and may cause nausea, anorexia, vomiting, gastrointestinal abnormalities, and dermatitis (Chui et al. 2013). Harmful metals such as As, Pb, and Hg may be toxic even at low levels of exposure (Clemens and Ma 2016). These metals as soon as entered the body continue to accumulate in vital organs like the brain, liver, bones, and kidneys for years causing serious health hazards (Kamunda et al. 2016). The priority list of HM pollutants which include As, Pb, Hg, and Cd is considered as first, second, and third hazards specified by the US Agency for Toxic Substances and Disease Registry (Park 2010; Wuana and Okieimen 2011). These HMs have serious impacts on terrestrial and aquatic ecosystems, which increase physiological health risks (Pandey 2012).

For example, exposure to arsenic may cause nausea, vomiting, abdominal pain, muscle cramps, and diarrhea and sometimes is associated with peripheral nerve damage and diabetes (NRC 1999; UNEP 2002). In contrast, Pb is considered as a human mutagen and probable carcinogen (Ryan et al. 2000). It induces renal tumors and also disturbs the normal functioning of kidneys, joints, and reproductive and nervous systems (Ogwuegbu and Muhanga 2005). The acute ingestion of Hg potentially causes gastrointestinal disorders, diarrhea, and hemorrhage (Grandjean 2007). While chronic exposure of Hg may seriously affect the brain, kidney, liver, and skin, severe exposure to Cd may result in pulmonary effects such as bronchiolitis, emphysema, and alveolus's (Kabata-Pendias 2011). Cd can also result in bone fracture, kidney dysfunction, and hypertension and even cancer (Khan et al. 2013). Long-term effects from chronic exposure of Cd include arthritis, diabetes, anemia, cardiovascular disease, cirrhosis, reduced fertility, headaches, and strokes. Hexavalent chromium Cr(VI) compounds are known to be mutagenic and carcinogenic. Breathing high levels of Cr (VI) may cause asthma and shortness of breath (Bharagava and Mishra 2018). Long-term exposure may cause damage to the liver and kidney. In contrast, chronic and acute Ni exposure is known to cause both oral and intestinal cancer. It may also cause depression, heart attacks, hemorrhages, and kidney problems (NRC 1999).

In biological systems, HMs have been reported to affect cellular organelles and cellular components, such as cell membrane, mitochondria, the endoplasmic reticulum, nuclei, and some enzymes involved in metabolism, detoxification, and damage repair (Wang and Shi 2001). Metal ions have been found to interact with cell components such as DNA and nuclear proteins, causing DNA damage and conformational changes that may lead to cell cycle modulation, carcinogenesis, or apoptosis (Chang et al. 1996; Kumari et al. 2016; Wang and Shi 2001; Beyersmann and Hartwig 2008). Metals also directly affect microorganisms by reducing their number, biochemical activity, and diversity and altering the community structure (Ellis et al. 2004; Abou-Shanab et al. 2003a, 2010). It has been recognized that high levels of HMs can alter both the qualitative and quantitative structure of microbial communities, resulting in decreased metabolic activity, biomass, and diversity (Chandra et al. 2011; Gremion et al. 2004; Abou-Shanab et al. 2010). Moreover, chronic exposure to HMs has led to the appearance of HM-resistant microorganisms in metal-contaminated sites (Table 14.1).

Microorganisms use several mechanisms to resist and tolerate HMs (Nies 2003; Abou-Shanab et al. 2007b), which may be encoded by chromosomal genes, but usually loci conferring resistance are located on plasmids (Wuertz and Mergeay 1997). A number of gram-negative bacteria belonging to the genus *Ralstonia* showed metal resistance (Schmidt and Schlegel 1994). *Ralstonia eutropha* possesses seven determinants encoding resistance to toxic HMs (Taghavi et al. 1997). These loci were found either within the bacterial chromosome or on one of the two plasmids, pMOL28 or pMOL30 (Mergeay et al. 1985; Siddiqui et al. 1989; Liesegang et al. 1993).

The polymerase chain reaction (PCR) technique used alone or in combination with DNA sequence analysis has been used to investigate the genetic mechanism(s) responsible for metal resistance in some of gram-positive and gram-negative bacteria, which were highly resistant to Hg, Zn, Cr, and Ni. Using PCR, Abou-Shanab et al. (2007b) showed that *czc*, *chr*, *ncc*, and *mer* are the genes, which are responsible for resistance to Zn, Cr, Ni, and Hg, respectively, and are present in some metal-resistant bacteria obtained from *A. murale* rhizosphere and Ni-rich soils.

Plants require certain metals, particularly Zn, Fe, Mn, and Cu, for the proper functioning of metal-dependent enzymes and proteins. However, several metals such as As, Hg, Cd, Al, and Pb are nonessential and potentially toxic (Kramer and Clemens 2005). Plants uptake HMs from soil or water through root system by both

Table 14.1 Minimal inhibitory concentration (MIC) of different heavy metals tested against bacterial strains isolated from rhizosphere of the Ni hyperaccumulator *Alyssum murale* (Abou-Shanab et al. 2007b)

Bacterial strain	MIC (mM)								
	As	Cd	Co	Cr	Cu	Hg	Ni	Pb	Zn
Acidovorax avenae AY512827	2.5	2.5	5	0.5	2.5	0.05	10	10	10
Acidovorax delafieldii AY512826	0.5	0.1	1	0.1	1	0.01	5	5	5
Arthrobacter ramosus AY509238	1	2.5	2.5	5	2.5	0.05	15	10	5
Arthrobacter rhombi AY509239	20	5	5	5	10	0.5	15	10	10
Burkholderia cepacia AY512825	2.5	2.5	10	5	1	0.05	15	15	5
Caulobacter crescentus AY512823	0.5	0.1	0.5	0.1	1	0.005	10	5	0.5
Clavibacter xyli AY509235	5	5	2.5	2.5	5	0.1	15	5	10
Massilia timonae AY512824	20	0.1	5	2.5	1	0.01	15	5	10
Microbacterium arabinogalactanolyticum AY509224	1	2.5	2.5	5	10	0.05	15	10	10
Microbacterium liquefaciens AY509220	5	0.5	0.1	2.5	2.5	0.05	15	10	5
Microbacterium oxydans AY509219	1	1	1	5	5	0.5	15	10	10
Phyllobacterium myrsinacearum AY512821	2.5	0.5	1	0.5	10	0.05	15	10	10
Pseudomonas riboflavina AY512822	1	5	5	0.5	10	0.1	10	10	5
Rhizobium etli AY509210	0.5	0.1	1	0.5	1	0.005	15	15	0.5
Rhizobium galegae AY509213	0.5	0.1	1	0.5	1	0.01	15	2.5	2.5
Rhizobium gallicum AY509211	1	0.1	1	0.1	1	0.01	10	5	10
Rhizobium mongolense AY509212	1	0.1	2.5	0.1	1	0.01	10	10	5
Sphingomonas asaccharolytica AY509241	0.5	0.1	1	0.1	1	0.01	10	5	0.5
Sphingomonas macrogoltabidus AY509243	1	0.5	5	0.5	5	0.5	15	10	5
Variovorax paradoxus AY512828	20	5	5	5	10	0.1	15	10	5

active and passive transport processes (Schützendübel and Polle 2002). The specific ion carriers that are responsible for the influx of essential elements including Zn^{2+}, Fe^{2+}, and calcium (Ca^{2+}) also facilitate the cellular influx of HMs ions (Welch and Norvell 1999; Perfus-Barbeoch et al. 2002). After their uptake, HMs can be translocated from plant roots to the aerial parts through xylem (Clemens 2006). Plants primarily respond to HM stress by accumulating ions in different organs. Based on this responsive strategy, plants are divided into two groups: hyper- and non-accumulating plants (Rascio and Navari-Izzo 2011). The ability of plants to accumulate HMs is one major basis for phytoremediation, which is a plant-based strategy used to clean up the metal-contaminated soils (Chaney et al. 1997). Baker (1987) reported that a few plant species are able to survive and grow on soils heavily polluted with As, Cd, Cr, Cu, Ni, Pb, and Zn. These plant species are divided into pseudometallophytes that can grow on both contaminated and non-contaminated soils whereas absolute metallophytes that grow only on metal-contaminated and naturally metal-rich soil (Baker 1987). Metal tolerance among plants may result from two basic strategies, depending on the plant species: metal exclusion and metal accumulation (Baker and Walker 1990). Hyperaccumulator plants are defined as plant species whose shoots contain >100 mg Cd kg^{-1}; >1000 mg Ni, Pb, and Cu kg^{-1}; or >10,000 mg Zn or Mn kg^{-1} (dry wt.) when grown in metal-rich soils (Baker and Brooks 1989; Baker et al. 2000). The ability of such hyperaccumulating plants to accumulate high amounts of metals in shoots makes them suitable for phytoremediation purposes. Plants that possess high metal uptake and a high biomass production are good candidates for metal removal from polluted sites within a reasonable period (Ebbs and Kochian 1997).

5 Phytoremediation Techniques for Heavy Metal Removal

Phytoremediation is a promising and economically effective technique that uses plant species to decontaminate the aquatic or terrestrial sites polluted with organic or inorganic contaminates (Salt et al. 1998; Abd El Rahman et al. 2008; Abou-Shanab 2010). This technology is simple, cost-effective, and environment-friendly with minimal environmental disruption. It is also agronomically acceptable since it is similar to that used for conventional agricultural practices. Moreover, the technology is likely to be more acceptable to the public than other traditional methods (Baker et al. 1994). Phytoremediation offers the advantage of a solar energy-driven process with a greater potential for public acceptance than many existing technologies. All plant-based techniques are likely to be less costly than those using conventional remediation technologies (Salt et al. 1998).

The metal-enriched plant materials can be removed by concentration of contaminants, and then the metal element recovered and valuable metal recycled. In such cases, some species are very efficient at HM accumulation in shoots and are being exploited for phytoremediation (Abou-Shanab et al. 2003a, 2006; Abou-Shanab 2010). In terms of HM remediation mechanisms, several forms of phytoremediation can be distinguished: phytoextraction, phytostabilization, phytofiltration, and phytovolatilization (Sadowsky 1999; Ghosh and Singh 2005; Chibuike and Obiora 2014).

5.1 *Phytoextraction*

Phytoextraction uses plants with a high capacity for metal accumulation to transport and concentrate metals from soil into the aboveground parts of plants and harvested for recycling or less expensive disposal. The plant materials can be burned for energy/electricity production and the ash processed to recover metals (Dahl et al. 2002). This process may be repeated several times to mitigate contamination up to acceptable levels. The phytoextraction technique may be limited to depths of 1–3 m from soil surface (Cunningham et al. 1997).

There are approximately 400 known plant species, from at least 45 families that are reported to hyperaccumulate metals (Reeves and Baker 2000; Ghosh and Singh

2005). Some of these plants belong to the family of *Brassicaceae, Fabaceae, Euphorbiaceae, Asteraceae, Lamiaceae*, and *Scrophulariaceae* (Salt et al. 1998; Dushenkov 2003). Brooks et al. (1998) suggest that if phytoextraction can be combined with biomass generation and the commercial utilization of plants as energy sources, then this technology can be turned into a profit making operation. The remaining ash can then be used as a "bio-ore" and forms the basis for phytomining.

5.2 Phytostabilization

Phytostabilization is a metal complexation technique that reduces contaminant mobility and availability in soil, sludge, and sediment, but it limits biomagnification via erosion and leaching into groundwater. In this technology, contaminants are absorbed and accumulated by roots, adsorbed onto the roots, or precipitated into the rhizosphere. This reduces or prevents the mobility of contaminants into groundwater or air and reduces their bioavailability and spreading through food chain. This technique can also be used to establish plant communities on sites that have been denuded due to past high levels of metal contamination. Once a community of metal-tolerant plant species has been established on these sites, the potential for wind erosion and thus spread of the pollutant is reduced (Claudia Santibez et al. 2008; Ivano et al. 2008).

5.3 Rhizofiltration

Rhizofiltration is an in situ or ex situ adsorption, absorption, and/or precipitation of inorganic and organic contaminants from impacted environments, especially liquid-based discharges using plant roots. Rhizofiltration is similar in concept to phytoextraction but is concerned with the remediation of contaminated groundwater rather than the remediation of polluted soils. This is ideal for Cd, Cr, Cu, Ni, Pb, U, and Zn contaminants that are retained in the root system (Eapen et al. 2003; Vera Tomé et al. 2008).

Plants used for rhizofiltration are not planted directly in situ but are acclimated to the pollutant first. Plants are hydroponically grown in clean water rather than soil, until a large root system has developed. Once a large root system has developed, the water supply is substituted for one, containing pollutants to remediate and allowing plants to acclimatize to the materials. Plants are then placed into the polluted area where roots uptake the polluted water along with the contaminants. As the roots become saturated, they are harvested and disposed of safely. Repeated treatments of site can reduce contamination to suitable levels as was exemplified in Chernobyl where sunflower (*Helianthus annuus*) was grown in radioactively contaminated pools (Vera Tomé et al. 2008).

5.4 Phytovolatilization

This technique involves the release of pollutants or their metabolites via the leaves to the atmosphere. The contaminants may become modified along the way as water travels along the plant's vascular system from roots to leaves, whereby the contaminants evaporate into the air (Abd El-Rahman et al. 2008; Zhu and Rosen 2009). The major advantage of this method is that the highly toxic contaminant (e.g., mercuric ion) may be transformed into a less toxic substance (elemental Hg). Mercury volatilization by genetically modified *N. tabacum* and *Arabidopsis thaliana* (Meagher et al. 2000) and *Liriodendron tulipifera* (Rugh et al. 1998) and selenium volatilization by *Brassica napus* (Bañuelos et al. 1997) are reported. However, this technique is often viewed as simply a media transfer system and not preferred by many environmentalists.

6 Plant-Microbe Interactions Enhance the Heavy Metal Phytoremediation

Plant-microbe interactions result in higher microbial population density and metabolic activities in the rhizosphere even in stress conditions like metal-contaminated soils. These interactions lead to HM tolerance and their accumulation as well as stimulated plant growth (Khan 2005). The plant roots interact with a large number of different microorganisms, with these interactions being major determinants of the extent of phytoremediation (Glick 2010). Microbial populations are known to affect trace metal mobility and availability to the plants (Abou-Shanab et al. 2003a, b), through acidification, the release of iron chelators and siderophores for ensuring the iron availability and/or mobilizing the unavailable metal phosphates (Fig. 14.1).

A large proportion of metal contaminants is unavailable for the plant root uptake, because HMs in soils are generally bound to soil constituents, or alternatively, present as insoluble precipitates. Improvement of the interactions between plants and beneficial rhizosphere microorganisms can enhance biomass production and tolerance of plants to HMs and is considered to be an important component of phytoremediation technologies (Glick 2010; Kukier et al. 2004; Abou-Shanab et al. 2006, 2010; Abou-Shanab 2010).

6.1 Role of Rhizobacteria in Enhanced Phytoremediation of Heavy Metals

Some rhizosphere bacteria, which are closely associated with roots, have been termed plant growth-promoting *Rhizobacteria* (Glick 2010). Plant growth-promoting *Rhizobacteria* (PGPR) include a diverse group of free-living soil bacteria

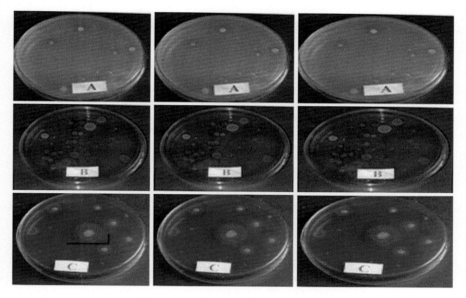

Fig. 14.1 Agar plates showing (A) phosphate solubilizing, (B) acid producing, and (C) siderophore producing bacteria (Abou-Shanab et al. 2003)

that can improve host plant growth and development in HM-polluted soils by alleviating the toxic effects of HMs on plants (Belimov et al. 2005; Abou-Shanab et al. 2010a). The rhizosphere provides a complex and dynamic microenvironment where microorganisms in association with roots form unique communities that have considerable potential for detoxification of hazardous waste compounds (De-Souza et al. 1999; Abou-Shanab et al. 2010a, b).

Microorganisms are ubiquitous in soils in which hyperaccumulators are native, even in those soils containing high metal concentration (Ghaderian et al. 2000). Microorganisms can affect metal solubility and availability to the plant; they can produce iron chelators and siderophores for ensuring iron availability, reduce soil pH, and/or solubilize metal phosphates (Table 14.2). Microbes influence root parameters, such as root morphology and growth. An increase in root exudation of organic solutes may affect the rate of phytosiderophore release. In turn, rhizosphere microorganisms may interact symbiotically with roots to enhance the potential for metal uptake (Burd et al. 2000; Guan et al. 2001). The presence of rhizosphere bacteria has been reported to increase the concentrations of Cr, Cu, Ni, Pb, or Zn in plants (Whiting et al. 2001; Abou-Shanab et al. 2003b, 2007b).

Rhizobacteria have been shown to play important roles in increasing the availability of Ni in soils, thus enhancing their uptake by Ni-hyperaccumulator plant (*Alyssum murale*). Abou-Shanab et al. (2003b) reported that *Sphingomonas macrogoltabidus*, *Microbacterium liquefaciens*, and *M. arabinogalactanolyticum* increased Ni uptake into the shoot of *A. murale* by 17, 24, and 32.4%, respectively, in comparison to uninoculated plants (Fig. 14.2). *M. arabinogalactanolyticum*

Table 14.2 Examples of the phytoremediation of heavy metals assisted by plant growth-promoting rhizobacteria (PGPR)

Bacteria	Plant	Heavy metal	Condition	Role of PGR	References
Azotobacter chroococcum	*Brassica juncea*	Pb and Zn	Pot experiments	Stimulated plant growth	Wu et al. (2006)
Bacillus subtilis	*Brassica juncea*	Ni	Pot experiments	Facilitated Ni accumulation	Zaidi et al. (2006)
Kluyvera ascorbata	Indian mustard	Ni, Zn, and Pb	Pot experiments	Decreased some plant growth inhibition by heavy metals	Burd et al. (2000)
Mesorhizobium huakuii	*Astragalus sinicus*	Cd	Hydroponics	Increased ability of cells to bind Cd^{2+} approximately 9- to 19-fold	Sriprang et al. (2003)
Bacillus subtilis	*Zea mays* and *Sorghum bicolor*	Cr	Pot experiment	Cr accumulation in shoot and biomass	Abou-Shanab et al. (2008)
Bacillus pumilus	*Zea mays*	Zn, Pb, Cr, and Cu	Pot experiment	Heavy metal accumulation	Abou-Shanab et al. (2008)
Brevibacterium halotolerans	*Zea mays* and *S. bicolor*	Cu, Zn, Pb and Cr	Pot experiment	Metal accumulation	Abou-Shanab et al. (2008)
Microbacterium arabinogalactanolyticum	*Alyssum murale*	Ni	Pot experiment	Ni solubilization in soil and uptake by plant	Abou-Shanab et al. (2006, 2003a)
Nitrobacteria irancium	Water hyacinth	Cr and Zn	Hydroponics	Increase chromium and zinc uptake	Abou-Shanb et al. (2007a)

increased foliar Ni contents in *A. murale* from a control concentration of 8500–12,000 mg Ni kg^{-1}. *Microbacterium oxydans* significantly increased Ni uptake of *A. murale* grown in low, medium, and high Ni soils by 36.1, 39.3, and 27.7%, respectively, compared with uninoculated plants (Fig. 14.3). Increased foliar Ni content was found in the same plant grown in low, medium, and high Ni content from 82.9, 261.3, and 2829.3 mg kg^{-1} to 129.7, 430.7, and 3914.3 mg kg^{-1}, respectively, compared with uninoculated plants (Abou-Shanab et al. 2006).

Siderophore production, for Fe uptake, can also be stimulated by the presence of other HMs (van der Lelie et al. 1999) and may affect the bioavailability of many metals. For instance, it was reported that *Azotobacter vinelandii*, siderophore production, is increased in the presence of Zn (II) (Huyer and Page 1988). Iron-chelating hydroxamic acid production in *Bacillus megaterium* is increased by exposure to Cr, Cu, Cd, Zn, and Al was found to increase siderophore production in *Pseudomonas aeruginosa* (Gilis 1993). The same effect was found for Zn and Al in *P. fluorescens*

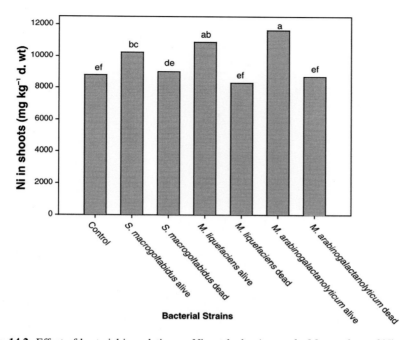

Fig. 14.2 Effect of bacterial inoculation on Ni uptake by *A. murale*. Mean values of Ni uptake with the same letter are not significantly different at $P < 0.05$ (Abou-Shanab et al. 2003a)

ATCC17400 (Gilis 1993). The presence of *Rhizobacteria* increased the concentrations of Zn (Whiting et al. 2001), Ni (Abou-Shanab et al. 2003b), and Se (De-Souza et al. 1999) in *T. caerulescens*, *A. murale*, and *B. juncea*, respectively.

Inoculation of *Brassica campestris* and canola seeds with *Kluyvera ascorbata* SUD165, which produces siderophores and contains the enzyme 1-aminocyclopropane-1-carboxylate (ACC) deaminase, protected the plant against Ni, Pb, and Zn toxicity (Burd et al. 2000). The highest concentrations of Cr (0.4 g kg^{-1}) and Zn (0.18 g kg^{-1}) were accumulated in aerial parts of water hyacinth inoculated with *Nitrobacteria irancium* in hydroponic system (Abou-Shanab et al. 2007a). These results show that these strains are important for decontamination of HM-polluted sites and could potentially be developed as an inoculum for enhancing metal uptake during commercial phytoremediation of metals.

6.2 The Role of Endophytic Bacteria in Enhancing the Phytoremediation of Heavy Metals

Plants and their associated microorganisms are characterized by varied and complex interactions and have been the subject of comprehensive research and various applications (Schulz and Boyle 2006). In comparison with rhizosphere and phyllosphere

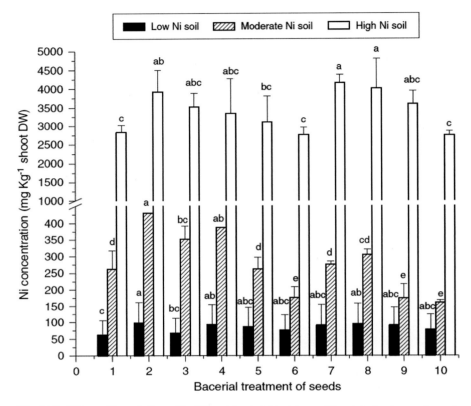

Fig. 14.3 Effects of inoculation with different bacterial isolates on Ni uptake by *A. murale* seedling, expressed as plant shoot Ni concentration. Mean values of Ni uptake with the same letter are not significantly different at $P < 0.05$. (1) Control; (2) *Microbacterium oxydans* AY509223; (3) *Rhizobium galegae* AY509213; (4) *M. oxydans* AY509219; (5) *Clavibacter xyli* AY509236; (6) *Acidovorax avenae* AY512827; (7) *M. arabinogalactanolyticum* AY509225; (8) *M. oxydans* AY509222; (9) *M. arabinogalactanolyticum* AY509226; and (10) *M. oxydans* AY509221. Mean values of Ni uptake with the same letter within the same soil are not significantly different at $P < 0.05$. The bar on the top of column is the SD (Abou-Shanab et al. 2006)

bacteria, endophytes (bacteria living within plant tissues without causing symptoms of infection or negative effects on their host) are likely to interact more closely with their host. Endophytic populations, like rhizospheric populations, are conditioned by biotic and abiotic factors (Hallmann et al. 1997; Seghers et al. 2004), but endophytic bacteria are likely better protected from biotic and abiotic stresses than rhizosphere bacteria (Hallmann et al. 1997). In addition, reinoculation of endophytic bacteria does not affect the indigenous endophyte population in plants (Conn and Franco 2004).

Endophytes enter plant tissue through the root zone, flower, leaf, stem, and cotyledon (Agarwal and Shende 1987; Kobayashi and Palumbo 2000) and may either become localized at entry point or spread throughout the plant (Reinhold-Hurek and Hurek 1998). In accordance with their life strategies, endophytes can form a range of different relationships with their host plant, including symbiotic, mutualistic, commensalistic, and trophobiotic associations. In these very close-knit plant-endophyte interactions, plants provide nutrients and residency for bacteria, in exchange for directly or indirectly improved plant growth and health (Ryan et al. 2008). Direct plant growth-promoting mechanisms may involve production of plant growth regulators such as auxins, cytokinins, and gibberellins, suppression of stress ethylene production by 1-aminocyclopropane-1-carboxylate (ACC) deaminase activity, nitrogen fixation, and the mobilization of unavailable phosphorus and other mineral nutrients. Indirectly, endophytic bacteria can promote plant growth and yield and can act as biocontrol agents by inhibiting the growth of plant pathogens through production of hydrolytic enzymes and antibiosis and induction of plant defense mechanisms (Hamilton and Bauerle 2012; Hamilton et al. 2012). In addition to their beneficial effects on plant growth, endophytes have considerable biotechnological potential to improve the applicability and efficiency of phytoremediation (Newman and Reynolds 2005; Doty et al. 2009).

Endophytic bacteria have been isolated from different plant species, monocotyledonous plants, e.g., Liliaceae, grass, *Zea*, rice, and orchids (Miyamoto et al. 2004; Peng et al. 2006; Gangwar and Kaur 2009; Kelemu et al. 2011; Lin et al. 2012; Rogers et al. 2012), as well as dicotyledonous plants, for example, oak (Basha et al. 2012; Ma et al. 2013), and different tree species (e.g., *Sorbus aucuparia* and *Betula verrucosa*) (Scortichini and Loreti 2007; Krid et al. 2010). The existence of endophytes has also been confirmed in beets, corn, bananas, tomatoes, and rice roots (Brown et al. 1999; Pereira et al. 1999; Cao et al. 2005; Altalhi 2009). These organisms classified as *Bacillus* sp., *Enterobacter* sp., and *Sporosarcina aquimarina* (Rylo sona Janarthine et al. 2011) have been found in roots of *Avicennia marina*.

Hyperaccumulator-associated endophytes may be metal resistant, due to long-term adaptation to the high concentration of metals accumulated in plants (Idris et al. 2004). Thus, previous studies mainly focused on hyperaccumulator-associated endophytes and many metal-resistant endophytes were isolated from hyperaccumulating plants such as *Alyssum bertolonii*, *Alnus firma*, *Brassica napus*, and *Nicotiana*. The reported metal-resistant endophytes belong to a wide range of taxa; in bacteria these include *Arthrobacter*, *Bacillus*, *Clostridium*, *Curtobacterium*, *Enterobacter*, *Leifsonia*, *Microbacterium*, *Paenibacillus*, *Pseudomonas*, *Xanthomonadaceae*, *Staphylococcus*, *Stenotrophomonas*, and *Sanguibacter* (Li et al. 2012).

Some endophytes were found to be able to produce auxins to improve host plant growth in contaminated soils. For instance, the endophytic bacteria *Serratia nematodiphila*, *Enterobacter aerogenes*, *Enterobacter* sp., and *Acinetobacter* sp., isolated from *Solanum nigrum*, *Enterobacter* sp. from *Allium macrostemon*, and *Acinetobacter*, *Agrobacterium tumefaciens*, *Bacillus* sp., *B. subtilis*, and *B. megaterium* from *Commelina communis* can produce IAA to stimulate plant growth and enhance phytoremediation (Chen et al. 2012; Shin et al. 2012; Zhang et al. 2011a, b). In addition, a number of endophytes produce cytokinins and/or gibberellins, which can stimulate the growth of various plants and modify plant morphology under both stress and nonstress conditions (Feng et al. 2006; Oelmüller et al. 2009; Hamayun et al. 2010). Moreover, some endophytes can indirectly improve the plant growth by producing ACC deaminase to modulate the ethylene levels in plants (Ma et al. 2011; Zhang et al. 2011a, b). Many endophytes have been found to be resistant to HMs, and endophyte-assisted phytoremediation has been documented as a promising technology for in situ remediation of HM-contaminated sites (Table 14.3).

Table 14.3 Selected examples of endophytic bacterial strains and their role to enhance the phytoremediation efficiency

Plant species	Endophytic bacterial isolates	Role of endophyte	References
Alyssum bertolonii	*Arthrobacter, Bacillus, Curtobacterium, Paenibacillus, Leifsonia, Pseudomonas*, and *Microbacterium*	Production of siderophores and increase Ni uptake and accumulation	Barzanti et al. (2007)
Alnus firma	*Bacillus thuringiensis* GDB-1	Production of IAA, siderophores, ACCD, and solubilize P. Increased biomass, chlorophyll content, and As, Cu, Pb, Ni, and Zn accumulation	Babu et al. (2013)
Alyssum serpyllifolium	*Pseudomonas* sp. A3R3	Production of IAA, siderophores, ACCD, and solubilize P and increase Ni uptake and accumulation	Ma et al. (2011)
Amaranthus hypochondriacus	*Rahnella* sp. JN27	Production of IAA, siderophores, ACCD, and solubilize P. Enhance plant growth and Cd uptake	Yuan et al. (2014)
Brassica napus	*Pseudomonas fluorescens, Microbacterium*	Production of IAA, siderophores, and ACCD Increased Pb availability in soil solution, and increased biomass production and Pb uptake	Sheng et al. (2008)
B. napus	*Pseudomonas thivervalensis, Pantoea agglomerans, Ralstonia* sp.	Production of IAA, siderophores, ACCD, and solubilize P. Increased the biomass of rape and increase Cu contents	Zhang et al. (2011b)
Lupinus luteus L.	*Burkholderia cepacia*	Ni bioremoval and increased Ni content in roots	Lodewyckx et al. (2001)
Polygonum pubescens	*Rahnella* sp. JN	Production of IAA, siderophores, ACCD, and solubilize P. Promote plant growth and Cd, Pb, Zn uptake by rape	He et al. (2013)
Solanum nigrum	*Serratia nematodiphila, Enterobacter aerogenes*	Production of IAA, siderophores, ACCD, and solubilize P. Increased Cd mobilization in soil, stimulate plant growth, and influence Cd accumulation in plant tissues	Chen et al. (2010)

(continued)

Table 14.3 (continued)

Plant species	Endophytic bacterial isolates	Role of endophyte	References
Solanum nigrum	*Pseudomonas* sp. LK9	Improved metal availability in soil, enhance biomass and Cd, Zn, and Cu uptake.	Chen et al. (2014)
Sorghum bicolor	*Bacillus* sp. SLS18	Production of IAA, siderophores, ACCD. Improved plant biomass and metal uptake	Luo et al. (2012)
Salix caprea	Actinobacterium	Production of siderophores and ACCD. Enhanced plant growth and metal accumulation in leaves	Kuffner et al. (2010)
Thlaspi caerulescens	*Sphingomonas* sp., *Methylobacterium* sp., *Sphingobacterium multivorum*	Enhance Zn and Cd accumulation	Lodewyckx et al. (2002)
Thlaspi goesingense	*Bacillus* sp., *Blastococcus* sp., *Propionibacterium acnes, Flavobacterium* sp., *Desulfitobacterium metallireductans, M. mesophilicum, M. extorquens, Sphingomonas* sp., *Rhodococcus* sp.	Ni resistance	Idris et al. (2004)

7 Genetically Modified Organisms to Enhance Heavy Metal Detoxification

Genetically modified organisms (GMO) are those whose genetic material has been altered to generate more efficient strains for metal detoxification and removal from contaminated sites (Sayler and Ripp 2000). It offers the advantage of GMO, which can withstand under adverse conditions and can be used in bioremediation. Genetic engineering has led to the development of "microbial biosensors" to measure the degree of pollution in contaminated sites quickly and accurately. Various biosensors have been designed to evaluate HM concentrations like As, Cd, Cu, Hg, and Ni (Verma and Singh 2005; Bruschi and Goulhen 2006). Genetic engineering of endophytes and rhizospheric bacteria for plant-associated degradation of pollutants in soil is considered as the most promising recent technologies to decontaminate HM-polluted sites (Divya and Deepak Kumar 2011; Hare et al. 2017). Bacteria like *Escherichia coli* and *Moraxella* sp., expressing phytochelatin 20 on cell surface, have been shown to accumulate 25 times more Cd or Hg than the wild-type strains (Bae et al. 2001, 2003).

Several genes are involved in metal uptake by plants, translocation, and sequestration, and transfer of any of these genes into candidate plants is a possible strategy for plant genetic engineering to improve phytoremediation of metal-contaminated sites (Table 14.4). Metallothionein genes (MT) have been cloned and introduced

Table 14.4 Selected examples of genes introduced into different plant species to improve their efficiency for heavy metal remediation

Gene transferred	Target plant	Response in transgene plants	References
MT2 (metallothionein)	*Nicotiana tabacum*	Enhanced Cd tolerance	Misra and Gedamu (1989)
MT2 (metallothionein)	*Brassica napus*	Enhanced Cd tolerance	Misra and Gedamu (1989)
MT1 (metallothionein)	*Nicotiana tabacum*	Enhanced Cd tolerance	Pan et al. (1994)
MTA (pea metallothionein)	*Arabidopsis*	Enhanced Cu accumulation	Evans et al. (1992)
MTP1, MTP8, MTP11	*Arabidopsis* and/or *Thalaspi*	Metal vacuolar sequestration	Hammond et al. (2006) and Talk et al. (2006)
MsFer (ferritin-iron storage)	*Nicotiana tabacum*	Increased Fe accumulation	Deak et al. (1999)
merApe9 [(Hg (II) reductase)]	*Arabidopsis*	Hg tolerance and volatilization	Rugh et al. (1996)
merApe9 [(Hg (II) reductase)]	Yellow poplar	Hg tolerance and volatilization	Rugh et al. (1998)
merA [(Hg (II) reductase)]	*Nicotiana tabacum*	Hg tolerance and volatilization	Meagher et al. (2000)
merB (organomercurial lyase)	*Arabidopsis*	Hg volatilization	Bizily et al. (2003)
merA and merB	*Arabidopsis*	Hg tolerance and volatilization	Bizily et al. (2000)
NRAMP1, NRAMP, NRAMP5	*Arabidopsis* and/or *Thalaspi*	Metal remobilization from the vacuole	Weber et al. (2004) and Filatov et al. (2006)
NAS, NAS2, NAS3, NAS4 (nicotinamine synthetase)	*Arabidopsis* and/or *Thalaspi*	Synthesis of metal ligands	Becher et al. (2004), Hammond et al. (2006), Talk et al. (2006), and van de Mortel et al. (2008)
AtHMA3 (P-type metal ATPase)	*Arabidopsis* and/or *Thalaspi*	Metal vacuolar sequestration	van de Mortel et al. (2006)
HMA4 (P-type metal ATPase)	*Arabidopsis* and/or *Thalaspi*	Metal remobilization from the vacuole	Hammond et al. (2006)
SAMS1, SAMS2, SAMS3 (S-adenosyl-methionine synthetase)	*Arabidopsis* and/or *Thalaspi*	Synthesis of metal ligands	Talk et al. (2006)
ACC deaminase gene	*Lycopersicon esculentum*	Cd, Co, Cu, Mg, Ni, Pb, or Zn tolerance	Grichko et al. (2000)
CUP-1 gene	Cauliflower	Cd accumulation	Hasegawa et al. (1997)
CUP-1 gene	*Nicotiana tabacum*	Cu accumulation	Thomas et al. (2003)

(continued)

Table 14.4 (continued)

Gene transferred	Target plant	Response in transgene plants	References
Se-cys lyase	Arabidopsis	Se tolerance and accumulation	Pilon et al. (2003)
Ta PCs (phytochelatin synthase)	Nicotiana glauca	Pb accumulation	Gisbert et al. (2003)
CGS gene (cystathionine methyl transferase)	Brassica juncea	Se volatilization	Van Huysen et al. (2004)
Ah MHX (vacuolar metal/proton exchanger)	Nicotiana tabacum	Mg and Zn tolerance	Shaul et al. (1999)
Nt CBP4 (plasma membrane binding transporter protein)	Nicotiana tabacum	Ni tolerance and Pb accumulation	Arazi et al. (1999)
ZAT1 (Zn transporter)	Arabidopsis	Zn accumulation	Van der Zaal et al. (1999)
ZATA (heavy metal transporter)	Arabidopsis	Cd and Pb resistance	Lee et al. (2003)

into *N. tabacum* and *Brassica napus* resulted in plants with enhanced Cd tolerance (Misra and Gedamu 1989) and in *Arabidopsis thaliana* to enhance Cu accumulation (Evans et al. 1992). Transgenic plants with increased phytochelatin (PC) levels through overexpression of cysteine synthase resulted in enhanced Cd tolerance (Harada et al. 2001). In other study, when the CUP-1 gene was transferred to cauliflower (*Brassica oleracea*), it resulted in 16-fold higher Cd tolerance and accumulation (Hasegawa et al. 1997). Similarly, transgenic *B. juncea* overexpressing different enzymes involved in PC synthesis was shown to extract more Cd, Cr, Cu, Pb, and Zn than wild plants (Zhu et al. 1999a, b).

Manipulation of metal transporter genes is known to alter metal tolerance/accumulation in different plant species (Pedas et al. 2009). Proteins (ZIP) from plants are capable of transporting Cd^{2+}, Fe^{3+}/Fe^{2+}, Mn^{2+}, Ni^{2+}, Co^{2+}, Cu^{2+}, and Zn^{2+} (Eckhardt et al. 2001; Grotz and Guerinot 2006; Pedas et al. 2009). More than 100 ZIP family members have been identified and detected in animals, plants, protists, and fungi, but members are also found in archaea and bacteria. Transfer of Zn transporter-ZAT gene from *Thlaspi goesingense* to *A. thaliana* resulted in twofold higher Zn accumulation in roots (Van der Zaal et al. 1999). Introduction of calcium vacuolar transporter (CAX-2) from *A. thaliana* to tobacco resulted in the enhanced accumulation of Ca, Cd, and Mn (Hirschi et al. 2000). Enhanced Ni tolerance was also obtained by transfer of another transporter gene-NtCBP4, which encodes for a calmodulin-binding protein (Arazi et al. 1999).

New metabolic pathways can be introduced into plants for hyperaccumulation or phytovolatilization as in the case of *Mer*A and *Mer*B genes that were introduced into plants like *N. tabacum* (Fig. 14.4), which resulted in plants being severalfold tolerant to Hg and volatilized elemental mercury (Bizily et al. 2000; Abd El-Rahman et al. 2008; Ruiz and Daniell 2009). By transfer of *Escherichia coli*, ars C and g-ECS genes, Dhankher et al. (2002) developed transgenic *Arabidopsis* plants that

Fig. 14.4 Testing the transgenic and wild type (*wt*) *Nicotiana tabacum* on media amended with Phenylmercuric acetate (PMA) and HgCl$_2$

could transport the oxyanion arsenate aboveground, reduced it to arsenite, and sequestered it to thiol peptide complexes. Biomass of known hyperaccumulators can be altered by introduction of genes, which affect phytohormone synthesis resulting in enhanced biomass production. Increased gibberellin biosynthesis in engineered *Populus* and *Eucalyptus* trees promoted growth and biomass production (Erikson et al. 2000).

8 Conclusion and Future Perspectives

Contamination of the biosphere by hazardous metals has accelerated dramatically since the beginning of the industrial revolution, posing worldwide environmental and human health problems. Compared to the conventional methods, the phytoremediation, using hyperaccumulator plants and their associated bacteria to remove metal contaminants from polluted sites, is being developed as a cost-effective technology over traditional engineering techniques. However, there are some limitations for this technology to become efficient on a commercial scale. These limitations needs to be overcome by achieving a better understanding of the mechanisms of metal accumulation in plants; continuous search for more metal hyperaccumulator plant species and engineering of fast growth, higher biomass, and highly branched root system in common plants with suitable functioning genes are required to achieve some of the properties of the hyperaccumulator plants.

Additional microorganisms need to be explored for their application in HM bioremediation through molecular intervention. More knowledge of the population dynamics and activity of metal-resistant endophytes in various hyperaccumulators are also required. Furthermore, in order to improve the applicability of endophyte inoculation in the field level, intensive future research is needed on the role of metal-resistant endophytes on phytoremediation potential of various hyperaccumulators. In order to achieve greater success for metal decontamination, greater efforts and funding are needed to create public awareness regarding metal toxicity and to carefully monitor and regulate the discharge of metal byproducts from different industries. Moreover, there is a growing demand to develop novel environmental monitoring methods to rapidly detect the presence of toxic substances in consumable items.

References

Abd El-Rahman RA, Abou-Shanab RA, Moawad H (2008) Mercury detoxification using genetic engineered *Nicotiana tabacum*. Global NEST J 10:432–438
Abou-Shanab RAI (2003) Ecological and molecular studies on the role of rhizosphere microflora on phytoremediation efficiency. Ph.D. thesis, Faculty of Science, Alexandria University, Egypt C/O Maryland University, College Park USA
Abou-Shanab RAI (2010) Bioremediation: new approaches and trends. In: Khan MS et al (eds) Biomanagement of metal-contaminated soils, environmental pollution. Springer Publications, New York, pp 65–94
Abou-Shanab RI, Angle JS, Delorme TA, Chaney RL, van Berkum P, Moawad H, Ghanem K, Ghozlan HA (2003a) Rhizobacterial effects on nickel extraction from soil and uptake by *Alyssum murale*. New Phytol 158:219–224
Abou-Shanab RA, Delorme TA, Angle JS, Chaney RL, Ghanem K, Moawad H (2003b) Phenotypic characterization of microbes in the rhizosphere of *Alyssum murale*. Int J Phytoremediation 5:367–379
Abou-Shanab RA, Angle JS, Chaney RL (2006) Bacterial inoculants affecting nickel uptake by *Alyssum murale* from low, moderate and high Ni soils. Soil Biol Biochem 38:2882–2889
Abou-Shanab RAI, Angle JS, van Berkum P (2007a) Chromate-tolerant bacteria for enhanced metal uptake by *Eichhornia crassipes* (Mart.) Int J Phytoremediation 9:91–105
Abou-Shanab RAI, van Berkum P, Angle JS (2007b) Heavy metal resistant patterns and further genotypic characterization of metal resistant gene (S) in gram positive and gram negative bacteria isolated from Ni-rich serpentine soil and the rhizosphere of *Alyssum murale*. Chemosphere 68:360–367
Abou-Shanab RAI, Ghanem KM, Ghanem NB, Al-Kolaibe AM (2008) The role of bacteria on heavy-metals extraction and uptake by plants growing on multi-metal contaminated soils. World J Microbiol Biotechnol 24:253–262
Abou-Shanab RAI, Angle JS, Delorme TA, Chaney RL, van Berkum P, Ghozlan HA, Ghanem K, Moawad H (2010) Characterization of Ni-resistant bacteria in the rhizosphere of the hyperaccumulator *Alyssum murale* by 16S rRNA gene sequence analysis. World J Microbiol Biotechnol 26:101–108
Agarwal S, Shende ST (1987) Tetrazolium reducing microorganisms inside the root of Brassica species. Curr Sci India 56:187–188
Altalhi AD (2009) Plasmids profiles, antibiotic and heavy metal resistance incidence of endophytic bacteria isolated from grapevine (*Vitis vinifera* L.) Afr J Biotechnol 8:5873–5882

Arazi T, Sunkar R, Kaplan B, Fromm HA (1999) A tobacco plasma membrane calmodulin binding transporter confers Ni$^+$ tolerance and Pb$^+$ hyper-sensitivity in transgenic plants. Plant J 20:171–182

Arruti A, Fernandez-Olmo I, Irabien A (2010) Evaluation of the contribution of local sources to trace metals levels in urban PM2.5 and PM10 in the Cantabria region (Northern Spain). J Environ Monit 12:1451–1458

Baba A, Gurdal G, Sengunalp F, Ozay O (2008) Effects of leachant temperature and pH on leachability of metals from fly ash. A case study: can thermal power plant, province of Canakkale, Turkey. Environ Monit Assess 139:287–298

Babu AG, Kim JD, Oh BT (2013) Enhancement of heavy metal phytoremediation by *Alnus firma* with endophytic *Bacillus thuringiensis* GDB-1. J Hazard Mater 250–251:477–483

Bae W, Mehra RK, Mulchandani A, Chen W (2001) Genetic engineering of *Escherichia coli* for enhanced uptake and bioaccumulation of mercury. Appl Environ Microbiol 67:5335–5338

Bae W, Wu CH, Kostal J, Mulchandani A, Chen W (2003) Enhanced mercury biosorption by bacterial cells with surface-displayed *MerR*. Appl Environ Microbiol 69:3176–3180

Baker AJM (1987) Metal tolerance. New Phytol 106:93–111

Baker AJM, Brooks RR (1989) Terrestrial higher plants which hyperaccumulate metallic elements. A review of their distribution, ecology and phytochemistry. Biorecovery 1:81–126

Baker AJM, Walker PL (1990) Ecophysiology of metal uptake by tolerant plants. CRC Press, Boca Raton, pp 155–178

Baker AJM, McGrath SP, Sidoli CMD, Reeves RD (1994) The possibility of in situ heavy metal decontamination of polluted soils using crops of metal-accumulating plants. Resour Conserv Recycl 11:41–49

Baker AJM, McGrath SP, Reeves RD, Smith JAC (2000) Metal hyperaccumulator plants: a review of the ecology and physiology of a biochemical resource for phytoremediation of metal-polluted soils. In: Terry N, Banuelos G (eds) Phytoremediation of contaminated soil and water. Lewis Publishers, Boca Raton, pp 85–107

Bañuelos GS, Ajwa HA, Mackey B, Wu LL, Cook C, Akohoue S, Zambrzuski S (1997) Evaluation of different plant species used for phytoremediation of high soil selenium. J Environ Qual 26:639–646

Barzanti R, Ozino F, Bazzicalupo M, Gabbrielli R, Galardi F, Gonnelli C, Mengoni A (2007) Isolation and characterization of endophytic bacteria from the nickel hyperaccumulator plant *Alyssum bertolonii*. Microb Ecol 53:306–316

Basha NS, Ogbaghebriel A, Yemane K, Zenebe M (2012) Isolation and screening of endophytic fungi from Eritrean traditional medicinal plant *Terminalia brownii* leaves for antimicrobial activity. Int J Green Pharm 6:40–44

Becher M, Talke IN, Krall L, Kramer U (2004) Cross-species microarray transcript profiling reveals high constitutive expression of metal homeostasis genes in shoots of the zinc hyperaccumulator *Arabidopsis halleri*. Plant J 37:251–268

Belimov AA, Hontzeas N, Safronova VI, Demchinskaya SV, Piluzza G, Bullitta S, Glick BR (2005) Cadmium-tolerant plant growth-promoting bacteria associated with the roots of Indian mustard (*Brassica juncea* L. Czern.) Soil Biol Biochem 37:241–250

Beyersmann D, Hartwig A (2008) Carcinogenic metal compounds: recent insight into molecular and cellular mechanisms. Arch Toxicol 82:493–512

Bharagava RN, Mishra S (2018) Hexavalent chromium reduction potential of *Cellulosimicrobium sp.* isolated from common effluent treatment industries. Ecotoxicol Environ Saf 147:102–109

Bharagava RN, Chandra R, Rai V (2008) Phytoextraction of trace elements and physiological changes in Indian mustard plants (*Brassica nigra* L.) grow in post methanated distillery effluent (PMDE) irrigated soil. Bioresour Technol 99(17):8316–8324

Bharagava RN, Chowdhary P, Saxena G (2017) Bioremediation: an eco-sustainable green technology, its applications and limitations. In: Bharagava RN (ed) Environmental pollutants and their bioremediation approaches. CRC Press, Taylor & Francis Group, Boca Raton, pp 1–22

Bizily SP, Rugh CL, Meagher RB (2000) Phytodetoxification of hazardous organomercurials by genetically engineered plants. Nat Biotechnol 18:213–217

Bizily SP, Kim T, Kandasamy MK, Meagher RB (2003) Subcellular targeting of methylmercury lyase enhances its specific activity for organic mercury detoxification in plants. Plant Physiol 131:463–471

Brooks RR, Chambers MF, Nicks LJ, Robinson BH (1998) Phytomining. Trends Plant Sci 1:359–362

Brown KB, Hyde KD, Guest DI (1999) Preliminary studies on endophytic fungal communities of *Musa acuminata* species complex in Hong Kong and Australia. Fungal Divers 1:27–51

Bruschi M, Goulhen F (2006) New bioremediation technologies to remove heavy metals and radionuclides using Fe(III)-sulfate- and sulfur reducing bacteria. In: Singh SN, Tripathi RD (eds) Environmental bioremediation technologies. Springer Publication, New York, pp 35–55

Burd GI, Dixon DG, Glick BR (2000) Plant growth-promoting bacteria that decrease heavy metal toxicity in plants. Can J Microbiol 46:237–245

Cao L, Qui Z, You J, Tan H, Zhou S (2005) Isolation and characterization of endophytic Streptomycete antagonists of *Fusarium* wilt pathogen from surface-sterilized banana roots. FEMS Microbiol 247:147–152

Chandra R, Bharagava RN, Kapley A, Purohit JH (2011) Bacterial diversity, organic pollutants and their metabolites in two aeration lagoons of common effluent treatment plant during the degradation and detoxification of tannery wastewater. Bioresour Technol 102:2333–2341

Chaney R, Malik M, Li YM, Brown SL, Brewer EP, Angle JS, Baker AJM (1997) Phytoremediation of soil metals. Curr Opin Biotechnol 8:279–284

Chaney RL, Angle JS, Mcintosh, Reeves RD, Li YM, Brewer EP, Chen KY, Roseberg RJ, Perner H, Synkowski EC, Broadhurst CL, Wang S, Baker AJM (2005) Using hyperaccumulator plants to phytoextract soil Ni and Co. Z Naturforsch 60C:190–198

Chang LW, Magos L, Suzuki T (1996) Toxicology of metals. CRC Press, Boca Raton

Chen WM, Tang YQ, Mori K, Wu XL (2012) Distribution of culturable endophytic bacteria in aquatic plants and their potential for bioremediation in polluted waters. Aquat Biol 15:99–110

Chen L, Luo SL, Li XJ, Wan Y, Chen JL, Liu CB (2014) Interaction of Cd hyperaccumulator *Solanum nigrum* L. and functional endophyte *Pseudomonas* sp. Lk9 on soil heavy metals uptake. Soil Biol Biochem 68:300–308

Chen L, Luo S, Xiao X, Guo H, Chen J, Wan Y, Li B, Xu T, Xi Q, Rao C, Liu C, Zeng G (2010) Application of plant growth-promoting endophytes (PGPE) isolated from *Solanum nigrum* L. for phytoextraction of Cd-polluted soils. Appl Soil Ecol 46:383–389

Cheng S (2003) Heavy metal pollution in China: origin, pattern and control. Environ Sci Pollut Res 10:192–198

Chibuike G, Obiora S (2014) Heavy metal polluted soils: effect on plants and bioremediation methods. Appl Environ Soil Sci 2014:1–12. Article ID 752708, 12 pages

Chowdhary P, Yadav A, Kaithwas G, Bharagava RN (2017a) Distillery wastewater: a major source of environmental pollution and its biological treatment for environmental safety. Green technologies and environmental sustainability. Springer International, Cham, pp 409–435

Chowdhary P, More N, Raj A, Bharagava RN (2017b) Characterization and identification of bacterial pathogens from treated tannery wastewater. Microbiol Res Int 5:30–36

Chowdhary P, Raj A, Bharagava RN (2018) Environmental pollution and health hazards from distillery wastewater and treatment approaches to combat the environmental. Chemosphere 194:229–246

Chui S, Wong YH, Chio HI, Fong MY, Chiu YM et al (2013) Study of heavy metal poisoning in frequent users of Chinese medicines in Hong Kong and Macau. Phototherapy Res 27:859–863

Claudia S, Cesar V, Rosanna G (2008) Phytostabilization of copper mine tailings with biosolids: implications for metal uptake and productivity of *Lolium perenne*. Sci Total Environ 395:1–10

Clemens S (2006) Toxic metal accumulation, responses to exposure and mechanisms of tolerance in plants. Biochimie 88:1707–1719

Clemens S, Ma JF (2016) Toxic heavy metal and metalloid accumulation in crop plants foods. Annu Rev Plant Biol 67:489–512

Conn VM, Franco CMM (2004) Effect of microbial inoculants on the indigenous actinobacterial endophyte population in the roots of wheat as determined by terminal restriction fragment length polymorphism. Appl Environ Microbiol 70:6407–6413

Cunningham SD, Shann JR, Crowley D, Anderson TA (1997) Phytoremediation of soil and water contaminants. American Chemical Society, Washington, DC

Dahl J, Obernberger I, Brummer T, Biedermann F (2002) Results and evaluation of a new heavy metal fractionation technology in grate-fired biomass combustion plants as a basis for an improved ash utilization. In: Proceedings of the 12st European conference on biomass for energy, industry and climate protection, Amsterdam, The Netherlands, pp 690–694

Deak M, Horvath GV, Davletova S, Torok K, Sass L, Vass I, Barna B, Kiraly Z, Dudits D (1999) Plants ectopically expressing the iron-binding protein ferritin, are tolerant to oxidative damages and pathogens. Nat Biotechnol 17:192–196

De-Souza MP, Huang CPA, Chee N, Terry N (1999) Rhizosphere bacteria enhance that accumulation of selenium and mercury in wetland plants. Planata 209:259–263

Dhankher OP, Li Y, Rosen BP, Shi J, Salt D, Senecoff JF (2002) Engineering tolerance and hyperaccumulation of arsenic in plants by combining arsenate reductase and g-glutamylcysteine synthetase expression. Nat Biotechnol 20:1140–1145

Diels L, van der Lelie N, Bastiaens L (2002) New development in treatment of heavy metal contaminated soils. Rev Environ Sci Biotechnol 1:75–82

Divya B, Deepak Kumar M (2011) Plant-microbe interaction with enhanced bioremediation. Res J BioTechnol 6:72–79

Doty SL, Oakley B, Xin G, Kang JW, Singleton G, Khan Z, Vajzovic A, Staley JT (2009) Diazotrophic endophytes of native black cottonwood and willow. Symbiosis 47:23–33

Duffus JH (2002) Heavy metal- a meaningless term? Pure Appl Chem 74:793–807

Dushenkov S (2003) Trends in phytoremediation of radionuclides. Plant Soil 249:167–175

Eapen S, Suseelan K, Tivarekar S, Kotwal S, Mitra R (2003) Potential for rhizofiltration of uranium using hairy root cultures of *Brassica juncea* and *Chenopodium amaranticolor*. Environ Res 91:127–133

Ebbs SD, Kochian LV (1997) Toxicity of zinc and copper to *Brassica* species: implications for phytoremediation. J Environ Qual 5:1424–1430

Eckhardt U, Marques AM, Buckhout TJ (2001) Two iron-regulated cation transporters from tomato complement metal uptake-deficient yeast mutants. Plant Mol Biol 45:437–448

Ellis DR, Sors TG, Brunk DG, Albrecht C, Orser C, Lahner B (2004) Production of S methyl selenocysteine in transgenic plants expressing selenocysteine methyltransferase. BMC Plant Biol 28:4

Erikson ME, Israelsson M, Olsson O, Moritz T (2000) Increased gibberellin biosynthesis in transgenic trees promotes growth, biomass production and xylem fiber length. Nat Biotechnol 18:784–788

Evangelou MWH, Papazoglou EG, Robinson BH, Schulin R (2015) Phytomanagement: phytoremediation and the production of biomass for economic revenue on contaminated land. In: Ansari AA et al (eds) Phytoremediation: management of environmental contaminants, vol 1. Springer International Publishing, Cham

Evans KM, Gatehouse JA, Lindsay WP, Shi J, Tommey AM, Robinson NJ (1992) Expression of the pea metallothionein like gene *Ps MTA* in *Escherichia coli* and *Arabidopsis thaliana* and analysis of trace metal ion accumulation: implications of *Ps MTA* function. Plant Mol Biol 20:1019–1028

Feng Y, Shen D, Song W (2006) Rice endophyte Pantoea agglomerans YS19 promotes host plant growth and affects allocations of host photosynthates. J Appl Microbiol 100:938–945

Filatov V, Dowdle J, Smirnoff N, Ford-Lloyd B, Newbury HJ, Macnair MM (2006) Comparison of gene expression in segregating families identifies genes and genomic regions involved in a novel adaptation, zinc hyperaccumulation. Mol Ecol 15:3045–3059

Gangwar M, Kaur G (2009) Isolation and characterization of endophytic bacteria from endorhizosphere of sugarcane and ryegrass. Int J Microbiol 7:139–144

Gao XL, Zhou FX, Chen CTA (2014) Pollution status of the Bohai Sea, China: an overview of the environmental quality assessment related trace metals. Environ Int 62:12–30

Ghaderian MYS, Anthony JEL, Baker AJM (2000) Seedling mortality of metal hyperaccumulator plants resulting from damping off by *Pythium* spp. New Phytol 146:219–224

Ghosh M, Singh SP (2005) A review on phytoremediation of heavy metals and utilization of its byproducts. Appl Ecol Environ Res 3(1):1–18

Gilis A (1993) Interactie tussen verschillende potentieel toxische metalen (Zn, Cd, Ni en Al) en siderofoor-afhankelijke ijzer-opname in verschillende fluorescerende Pseudomonas stammen. Licentiaatsthesis, Departement Algemene Biologie, Brussels

Gisbert C, Ros R, De Haro A, Walker DJ, Pilar Bernal M, Serrano R (2003) A plant genetically modified that accumulates Pb is especially promising for phytoremediation. Biochem Biophys Res Commun 303:440–445

Glick RB (2003) Phytoremediation: synergistic use of plants and bacteria to clean up the environment. Biotechnol Adv 21:383–393

Glick BR (2010) Using soil bacteria to facilitate phytoremediation. Biotechnol Adv 28:367–374

Grandjean P (2007) Methylmercury toxicity and functional programming. Reprod Toxicol 23:414–420

Gremion F, Chatzinotas A, Kaufmann K, Sigler W, Harms H (2004) Impacts of heavy metal contamination and phytoremediation on a microbial community during a twelve-month microcosm experiment. FEMS Microbiol Ecol 48:273–283

Grichko VP, Filby B, Glick BR (2000) Increased ability of transgenic plants expressing the bacterial enzyme ACC deaminase to accumulate Cd, Co, Cu, Ni, Pb, and Zn. J Biotechnol 81:45–53

Grotz N, Guerinot ML (2006) Molecular aspects of Cu, Fe and Zn homeostasis in plants. Biochim Biophys Acta 1763:595–608

Guan LL, Kanoh K, Kamino K (2001) Effect of exogenous siderophores on iron uptake activity of marine bacteria under iron limited conditions. Appl Environ Microbiol 67:1710–1717

Hall JL (2002) Cellular mechanism of heavy metal detoxification and tolerance. J Exp Bot 53:1–11

Hallmann J, Quadt-Hallmann A, Mahaffee WF, Kloepper JW (1997) Bacterial endophytes in agricultural crops. Can J Microbiol 43:895–914

Hamayun M, Sumera AK, Iqbal I, Ahmad B, Lee I (2010) Isolation of a gibberellin-producing fungus (*Penicillium* sp. MH7) and growth promotion of crown daisy (*Chrysanthemum coronarium*). J Microbiol Biotechnol 20:202–207

Hamelink JL, Landrum PF, Harold BL, William BH (1994) Bioavailability: physical, chemical, and biological interactions. CRC Press, Boca Raton

Hamilton CE, Bauerle TL (2012) A new currency for mutualism? Fungal endophytes alter antioxidant activity in hosts responding to drought. Fungal Divers 54:39–49

Hamilton CE, Gundel PE, Helander M, Saikkonen K (2012) Endophytic mediation of reactive oxygen species and antioxidant activity in plants: a review. Fungal Divers 54:1–10

Hammond JP, Bowen HC, White PJ, Mills V, Pyke KA, Baker AJ, Whiting SN, May ST, Broadley MR (2006) A comparison of the *Thlaspi caerulescens* and *Thlaspi arvense* shoot transcriptomes. New Phytol 170:239–260

Harada E, Choi YE, Tsuchisaka A, Obata H, Sano H (2001) Transgenic tobacco plants expressing a rice cysteine synthase gene are tolerant to toxic levels of cadmium. Plant Physiol 158:655–661

Hare V, Chowdhary P, Baghel VS (2017) Influence of bacterial strains on *Oryza sativa* grown under arsenic tainted soil: accumulation and detoxification response. Plant Physiol Biochem 119:93–102

Hasegawa I, Terada E, Sunairi M, Wakita H, Shinmachi F, Noguchi A (1997) Genetic improvement of heavy metal tolerance in plants by transfer of the yeast metallothionein gene (*CUP1*). Plant Soil 196:277–281

Hawkes SJ (1997) What is a heavy metal? J Chem Educ 74:1374

He H, Ye Z, Yang D, Yan J, Xiao L, Zhong T, Yuan M, Cai X, Fang Z, Jing Y (2013) Characterization of endophytic *Rahnella* sp. JN6 from *Polygonum pubescens* and its potential in promoting growth and Cd, Pb, Zn uptake by *Brassica napus*. Chemosphere 90:1960–1965

Hill MS (1997) Understanding environmental pollution. Cambridge University Press, Cambridge, 316 pp

Hirschi KD, Korenkov VD, Wilganowski NL, Wagner GJ (2000) Expression of *Arabidopsis* CAX2 in tobacco altered metal accumulation and increased manganese tolerance. Plant Physiol 124:125–133

Huyer M, Page W (1988) Zn^{2+} increases siderophore production in *Azotobacter vinelandii*. Appl Environ Microbiol 54:2625–2631

Idris R, Trifonova R, Puschenreiter M, Wenzel WW, Sessitsch A (2004) Bacterial communities associated with flowering plants of the Ni hyperaccumulator *Thlaspi goesingense*. Appl Environ Microbiol 70:2667–2677

Ivano B, Jorg L, Madeleine S, Gunthardt G, Beat F (2008) Heavy metal accumulation and phytostabilisation potential of tree fine roots in a contaminated soil. Environ Pollut 152:559–568

Kabata-Pendias A (2011) Trace elements in soil and plants.4th edn. Taylor & Francis: CRC Press, Boca Raton

Kamunda C, Mathuthu M, Madhuku M (2016) Health risk assessment of heavy metals in soils from Witwatersrand gold mining basin, South Africa. Int J Environ Res Public Health 13:663

Kelemu S, Fory P, Zuleta C, Ricaurte J, Rao I, Lascano C (2011) Detecting bacterial endophytes in tropical grasses of the *Brachiaria* genus and determining their role in improving plant growth. Afr J Biotechnol 10:965–976

Khan AG (2005) Role of soil microbes in the rhizospheres of plants growing on trace metal contaminated soils in phytoremediation. J Trace Elem Med Biol 18:355–364

Khan K, Lu Y, Khan H (2013) Heavy metals in agricultural soils and crops and their health risks in Swat District, northern Pakistan. Food Chem Toxicol 58:449–458

Kobayashi DY, Palumbo JD (2000) Bacterial endophytes and their effects on plants and uses in agriculture. In: Bacon CW, White JF (eds) Microbial endophytes. Marcel Dekker, New York, pp 199–236

Kramer U, Clemens S (2005) Molecular biology of metal homeostasis and detoxification. In: Tamás MJ, Martinoia E (eds) Topics in current genetics. Springer, Berlin, pp 216–271

Krid S, Rhouma A, Mogou I, Quesada JM, Nesme X, Gargouri A (2010) *Pseudomonas savastanoi* endophytic bacteria in olive tree knots and antagonistic potential of strains of *Pseudomonas fluorescens* and *Bacillus subtilis*. J Plant Pathol 92:335–341

Kuffner M, De Maria S, Puschenreiter M, Fallmann K, Wieshammer G, Gorfer M, Strauss J, Rivelli AR, Sessitsch A (2010) Culturable bacteria from Zn- and Cd-accumulating *Salix caprea* with differential effects on plant growth and heavy metal availability. J Appl Microbiol 108:1471–1484

Kukier U, Peters CA, Chaney RL, Angle JS, Roseberg RJ (2004) The effect of pH on metal accumulation in two *Alyssum* species. J Environ Qual 32:2090–2102

Kumari V, Yadav A, Haq I, Kumar S, Bharagava RN, Singh SK, Raj A (2016) Genotoxicity evaluation of tannery effluent treated with newly isolated hexavalent chromium reducing *Bacillus cereus*. J Environ Manag 183:204–211

Lee J, Bae H, Jeong J, Lee JY, Yang YY, Hwang I (2003) Functional expression of heavy metal transporter in Arabidopsis enhances resistance to and decreases uptake of heavy metals. Plant Physiol 133:589–596

Li H-Y, Wei D-Q, Shen M, Zhou Z-P (2012) Endophytes and their role in phytoremediation. Fungal Divers 54:11–18

Liesegang H, Lemke K, Siddiqui RA, Schlegel HG (1993) Characterization of the inducible nickel and cobalt resistance determinant *cnr* from pMOL28 of *Alcaligenes eutrophus* CH34. J Bacteriol 175:767–778

Lin L, Ge HM, Yan T, Qin YH, Tan RX (2012) Thaxtomin A-deficient endophytic *Streptomyces* sp. enhances plant disease resistance to pathogenic Streptomyces scabies. Planta 236:1849–1861

Lodewyckx C, Taghavi S, Mergeay M, Vangronsveld J, Clijsters H, van der Lelie D (2001) The effect of recombinant heavy metal resistant endophytic bacteria in heavy metal uptake by their host plant. Int J Phytoremediation 3:173–187

Lodewyckx C, Vangronsveld J, Porteous F, Moore ERB, Taghavi S, van der Lelie D (2002) Endophytic bacteria and their potential applications. Crit Rev Plant Sci 21:583–606

Luo S, Xu T, Chen L, Chen J, Rao C, Xiao X, Wan Y, Zeng G, Long F, Liu C, Liu Y (2012) Endophyte-assisted promotion of biomass production and metal uptake of energy crop sweet sorghum by plant-growth-promoting endophyte *Bacillus* sp. SLS18. Appl Microbiol Biotechnol 93:1745–1753

Ma Y, Prasad MNV, Rajkumar M, Freitas H (2011) Plant growth promoting rhizobacteria and endophytes accelerate phytoremediation of metalliferous soils. Biotechnol Adv 29:248–258

Ma L, Cao YH, Cheng MH, Huang Y, Mo MH, Wang Y, Yang JZ, Yang FX (2013) Phylogenetic diversity of bacterial endophytes of *Panax notoginseng* with antagonistic characteristics towards pathogens of root-rot disease complex. Antonie van Leeuwen hoek 103:299–312

Meagher RB, Rugh CL, Kandasamy MK, Gragson G, Wang NJ (2000) Engineered phytoremediation of mercury pollution in soil and water using bacterial genes. In: Terry N, Bauelos G (eds) Phytoremediation of contaminated soil and water. Lewis Publishers, Boca Raton, pp 201–221

Mergeay M, Nies D, Schlegel HG, Gerits J, Charles P, van Gijsegem F (1985) *Alcaligenes eutrophus* CH34 is a facultative chemolithotroph with plasmid-bound resistance to heavy metals. J Bacteriol 3:691–698

Mishra S, Bharagava RN (2016) Toxic and genotoxic effects of hexavalent chromium in environment and its bioremediation strategies. J Environ Sci Health Part C 34(1):1–34

Misra S, Gedamu L (1989) Heavy metal tolerant transgenic *Brassica napus* L and *Nicotiana tabacum* L plants. Theor Appl Genet 78:16–18

Miyamoto T, Kawahara M, Minamisawa K (2004) Novel endophytic nitrogen-fixing clostridia from the grass *Miscanthus sinensis* as revealed by terminal restriction fragment length polymorphism analysis. Appl Environ Microbiol 70:6580–6586

Nasim SA, Dhir B (2010) Heavy metals alter the potency of medicinal plants. Rev Environ Contam Toxicol 203:139–149

National Research Council (NRC) (1999) Arsenic in drinking water. National Research Council, Washington, DC, pp 251–257

Newman L, Reynolds C (2005) Bacteria and phytoremediation: new uses for endophytic bacteria in plants. Trends Biotechnol 23:6–8

Nies DH (2003) Efflux-mediated heavy metal resistance in prokaryotes. FEMES Microbiol Rev 27:313–339

Nriagu JO (1989) A global assessment of natural sources of atmospheric trace metals. Nature 338:47–49

Oelmüller R, Sherameti I, Tripathi S, Varma A (2009) *Piriformospora indica*, a cultivable root endophyte with multiple biotechnological applications. Symbiosis 49:1–17

Ogwuegbu MOC, Muhanga W (2005) Investigation of lead concentration in the blood of people in the copper belt province of Zambia. J Environ 1:66–75

Pan A, Yang M, Tie F, Li L, Chen Z, Ru B (1994) Expression of mouse metallothionein-1-gene confers cadmium resistance in transgenic tobacco plants. Plant Mol Biol 24:341–351

Pandey VC (2012) Phytoremediation of heavy metals from fly ash pond by *Azolla caroliniana*. Ecotoxicol Environ Saf 82:8–12

Park JD (2010) Heavy metal poisoning. Hanyang Med Rev 30:319–325

Pedas P, Schjoerring JK, Husted S (2009) Identification and characterization of zinc-starvation induced *ZIP* transporters from barley roots. Plant Physiol Biochem 47:377–383

Peng G, Wang H, Zhang G, Hou W, Liu Y, Wang ET, Tan Z (2006) *Azospirillum melinis* sp. nov., a group of diazotrophs isolated from tropical molasses grass. Int J Syst Microbiol 56:1263–1271

Pereira JO, Carneiro-Vieira ML, Azevedo JL (1999) Endophytic fungi from *Musa acuminata* and their reintroduction into axenic plants. World J Microbiol Biotechnol 15:37–40

Perfus-Barbeoch L, Leonhardt N, Vavasseur A, Forestier C (2002) Heavy metal toxicity: Cd permeates through calcium channels and disturbs the plant water status. Plant J 32:539–548

Pilon M, Owen JD, Garifullina GF, Kurihara T, Mihara H, Esaki N (2003) Enhanced selenium tolerance and accumulation in transgenic Arabidopsis expressing a mouse selenocysteine lyase. Plant Physiol 131:1250–1257

Pulford ID, Watson C (2003) Phytoremediation of heavy metal contaminated land by trees–a review. Environ Int 29:529–540

Rascio N, Navari-Izzo F (2011) Heavy metal hyperaccumulating plants: how and why do they do it? And what makes them so interesting? Plant Sci 180:169–181

Reeves RD, Baker AJH (2000) Metal accumulating plants. In: Raskin I, Ensley BD (eds) Phytoremediation of toxic metals: using plants to clean up the environment. Wiley, New York, pp 193–229

Reinhold-Hurek B, Hurek R (1998) Life in grasses: diazotrophic endophytes. Trends Microbiol 6:39–144

Rogers A, McDonald K, Muehlbauer MF, Hoffman A, Koenig K, Newman L, Taghavi S, van der Lelie D (2012) Inoculation of hybrid poplar with the endophytic bacterium *Enterobacter* sp. increases biomass but does not impact leaf level physiology. GCB Bioenergy 4:364–370

Rugh CL, Wilde D, Stack NM, Thompson DM, Summers AO, Meagher RB (1996) Mercuric ion reduction and resistance in transgenic *Arabidopsis thaliana* plants expressing a modified bacterial *merA* gene. Proc Natl Acad Sci USA 93:3182–3187

Rugh CL, Senecoff JF, Meagher RB, Merkle SA (1998) Development of transgenic yellow poplar for mercury phytoremediation. Nat Biotechnol 16:925–928

Ruiz ON, Daniell H (2009) Genetic engineering to enhance mercury phytoremediation. Curr Opin Biotechnol 20:213–219

Ryan PB, Huet N, Macintoshl DL (2000) Longitudinal investigation of exposure to arsenic, cadmium, and lead in drinking water. Environ Health Perspect 108:731–735

Ryan RP, Germaine K, Franks A, Ryan DJ, Dowling DN (2008) Bacterial endophytes: recent developments and applications. FEMS Microbiol Lett 278:1–9

Rylo Sona Janarthine S, Eganathan P, Balasubramanian T, Vijayalakshmi S (2011) Endophytic bacteria isolated from the pneumatophores of *Avicennia marina*. Afr J Microbiol Res 5:4455–4466

Sadowsky MJ (1999) In phytoremediation: past promises and future practices. Proceedings of the 8th international symposium on microbial ecology. Halifax, Canada, pp 1–7

Salt DE, Smith RD, Raskin I (1998) Phytoremediation. Annu Rev Plant Physiol Plant Mol Biol 49:643–668

Sayler GS, Ripp S (2000) Field applications of genetically engineered microorganisms for bioremediation process. Curr Opin Biotechnol 11:286–289

Schmidt T, Schlegel HG (1994) Combined nickel-cobalt-cadmium resistance encoded by the *ncc* locus of *Alcaligenes xylosoxidans* 31A. J Bacteriol 176:7045–7054

Schulz B, Boyle C (2006) What are endophytes? In: BJE S, CJC B, Sieber TN (eds) Microbial root endophytes. Springer, Berlin, pp 1–13

Schützendübel A, Polle A (2002) Plant responses to abiotic stresses: heavy metal-induced oxidative stress and protection by mycorrhization. J Exp Bot 53:1351–1365

Scortichini M, Loreti S (2007) Occurrence of an endophytic, potentially pathogenic strain of *Pseudomonas syringae* in symptomless wild trees of *Corylus avellana* l. J Plant Pathol 89:431–434

Seghers D, Wittebolle L, Top EM, Verstraete W, Siciliano SD (2004) Impact of agricultural practices on the *Zea mays* L. endophytic community. Appl Environ Microbiol 70:1475–1482

Shaul O, Hilgemann DW, de Almedia-Engler J, Van Montagu M, Inze D, Galili G (1999) Cloning and characterization of a novel Mg^{2+}/H^+ exchanger. EMBO J 18:3973–3980

Sheng XF, Xia JJ, Jiang CY, He LY, Qian M (2008) Characterization of heavy metal resistant endophytic bacteria from rape (*Brassica napus*) roots and their potential in promoting the growth and lead accumulation of rape. Environ Pollut 156:1164–1170

Shin M, Shim J, You Y, Myung H, Bang KS, Cho M, Kamala-Kannan S, Oh BT (2012) Characterization of lead resistant endophytic *Bacillus* sp. MN3-4 and its potential for promoting lead accumulation in metal hyperaccumulator *Alnus firma*. J Hazard Mater 199–200:314–320

Siddiqui RA, Benthin K, Schlegel HG (1989) Cloning of pMOL28- encoded nickel resistance genes and expression of the genes in *Alcaligenes eutrophus* and *Pseudomonas* spp. J Bacteriol 171:5071–5078

Singh A, Prasad SM (2011) Reduction of heavy metal load in food chain: technology assessment. Rev Environ Sci Biotechnol 10:199–214

Sriprang R, Hayashi M, Ono H, Takagi M, Hirata K, Murooka Y (2003) Enhanced accumulation of Cd^{2+} by a *Mesorhizobium* sp. transformed with a gene from *Arabidopsis thaliana* coding for phytochelatin synthase. Appl Environ Microbiol 69:1791–1796

Taghavi S, Mergeay M, van der Lelie D (1997) Genetics and physical map of the *Alcaligenes eutrophus* CH34 mega plasmid pMOL28 and it derivative pMOL50 obtained after temperature induced mutagenesis and mortality. Plasmid 37:22–34

Talk I, Hanikenne M, Kramer U (2006) Zinc dependent global transcriptional control, transcriptional de-regulation and higher gene copy number for genes in metal homeostasis of the hyperaccumulator *Arabidopsis halleri*. Plant Physiol 142:148–167

Tangahu BV, Abdullah SRS, Basri H, Idris M, Anuar N, Mukhlisin M (2011) A review on heavy metals (As, Pb, and Hg) uptake by plants through phytoremediation. Int J Chem Eng. Article ID 939161

Thomas JC, Davies EC, Malick FK, Endreszi C, Williams CR, Abbas M (2003) Yeast metallothionein in transgenic tobacco promotes copper uptake from contaminated soils. Biotechnol Prog 19:273–280

Todeschini V, Lingua G, D'Agostino G, Carniato F, Roc-cotiello E, Berta G (2011) Effects of high zinc concentration on poplar leaves: a morphological and biochemical study. Environ Exp Bot 71:50–56

Turgut C (2003) The contamination with organochlorine pesticides and heavy metals in surface water in Kucuk Menderes River in Turkey, 2000–2002. Environ Int 29:29–32

United Nations Environmental Programme (UNEP) (2002) Global mercury assessment. United Nations, Geneva

Van de Mortel JE, Villanueva LA, Schat H, Kwekkeboom J, Coughlan S, Moerland PD, Ver Loren van Themaat E, Koornneef M, Aarts MGM (2006) Large expression differences in genes for iron and zinc homeostasis, stress response, and lignin biosynthesis distinguish roots of *Arabidopsis thaliana* and the related metal hyperaccumulator *Thlaspi caerulescens*. Plant Physiol 142:1127–1147

Van de Mortel JE, Schat H, Moerland PD, Ver Loren van Themaa E, van der Ent S, Blankestijn H, Ghandilyan A, Tsiatsiani S, Aarts MG (2008) Expression differences for genes involved in lignin, glutathione and sulphate metabolism in response to cadmium in *Arabidopsis thaliana* and the related Zn/Cd-hyperaccumulator *Thlaspi caerulescens*. Plant Cell Environ 31:301–324

Van der Lelie D, Corbisier P, Diels L, Gilis A, Lodewyckx C, Mergeay M, Taghavi S, Spelmans N, Vangronsveld J (1999) The role of bacteria in the phytoremediation of heavy metals. In: Terry N, Bañuelos G (eds) Phytoremediation of contaminated soil and water. Lewis Publishers, Boca Raton, pp 265–281

Van der Zaal BJ, Neuteboom LW, Pinas JE, Chardonnen AN, Schat H, Verkleij JAC (1999) Overexpression of a novel *Arabidopsis* gene related to putative zinc transporter genes from animals can lead to enhanced zinc resistance and accumulation. Plant Physiol 119:1047–1055

Van Huysen T, Terry N, Pilon-Smits EA (2004) Exploring the selenium phytoremediation potential of transgenic Indian mustard over expressing ATP sulfurylase or cystathionine gamma synthase. Int J Phytoremediation 6:111–118

Vera Tomé F, Blanco Rodrguezb P, Lozano JC (2008) Elimination of natural uranium and 226Ra from contaminated waters by rhizofiltration using *Helianthus annuus* L. Sci Total Environ 393:51–357

Verkleji JAS (1993) The effects of heavy metals stress on higher plants and their use as biomonitors. In: Markert B (ed) Plant as bioindicators: indicators of heavy metals in the terrestrial environment. VCH, New York, pp 415–424

Verma N, Singh M (2005) Biosensors for heavy metals. Biometals 18:121–129

Wang S, Shi X (2001) Molecular mechanisms of metal toxicity and carcinogenesis. Mol Cell Biochem 222:3–9

Weber M, Harada E, Vess C, von Roepenack-Lahaye E, Clemens S (2004) Comparative microarray analysis of *Arabidopsis thaliana* and *Arabidopsis halleri* roots identifies nicotianamine synthase, a ZIP transporter and other genes as potential metal hyperaccumulation factors. Plant J 37:269–281

Welch RM, Norvell WA (1999) Mechanisms of Cd uptake, translocation and deposition in plants. In: McLaughlin MJ, Singh BR (eds) Cd in soils and plants. Kluwer Academic Publisher's, Dordretch, pp 125–150

Whiting SN, De Souza M, Terry N (2001) Rhizosphere bacteria mobilize Zn for hyperaccumulator by *Thlaspi caerulescens*. Environ Sci Technol 35:3144–3150

Wu SC, Cheung KC, Luo YM, Wong MH (2006) Effects of inoculation of plant growth-promoting rhizobacteria on metal uptake by *Brassica juncea*. Environ Pollut 140:124–135

Wuana A, Okieimen FE (2011) Heavy metals in contaminated soils: a review of sources, chemistry, risks and best available strategies for remediation. ISRN Ecol. Article ID: 402647

Wuertz S, Mergeay M (1997) The impact of heavy metals on soil microbial communities and their activities. In: van Elsas JD, Wellington EMH, Trevors JT (eds) Modern soil microbiology. Marcel Decker, New York, pp 1–20

Yadav A, Chowdhary P, Kaithwas G, Bharagava RN (2017) Toxic metals in the environment, their threats on ecosystem and bioremediation approaches. In: Das S, Singh HR (eds) Handbook of metal-microbe interaction and bioremediation. CRC Press, Taylor & Francis Group, Boca Raton, pp 128–141

Yuan M, He H, Xiao L, Zhong T, Liu H, Li S, Deng P, Ye Z, Jing Y (2014) Enhancement of Cd phytoextraction by two Amaranthus species with endophytic *Rahnella* sp. JN27. Chemosphere 103:99–104

Zaidi S, Usmani S, Singh BR, Musarrat J (2006) Significance of *Bacillus subtilis* strain SJ-101 as a bioinoculant for concurrent plant growth promotion and nickel accumulation in *Brassica juncea*. Chemosphere 64:991–997

Zhang YF, He L, Chen Z, Zhang W, Wang Q, Qian M, Sheng X (2011a) Characterization of lead-resistant and ACC deaminase-producing endophytic bacteria and their potential in promoting lead accumulation of rape. J Hazard Mater 186:1720–1725

Zhang YF, He LY, Chen ZJ, Wang QY, Qian M, Sheng XF (2011b) Characterization of ACC deaminase-producing endophytic bacteria isolated from copper tolerant plants and their potential in promoting the growth and copper accumulation of *Brassica napus*. Chemosphere 83:57–62

Zhang XY, Chen DM, Zhong TY, Zhang XM, Cheng M, Li XH (2015) Assessment of cadmium (Cd) concentration in arable soil in China. Environ Sci Pollut Res 22:4932–4941

Zhu Y, Rosen BP (2009) Perspectives for genetic engineering for the phytoremediation of arsenic contaminated environments: from imagination to reality? Curr Opin Biotechnol 20:220–224

Zhu Y, Pilon-Smits EAH, Jouanin L, Terry N (1999a) Overexpression of glutathione synthetase in *Brassica juncea* enhances cadmium tolerance and accumulation. Plant Physiol 119:73–79

Zhu Y, Pilon-Smits EA, Tarun AS, Weber SU, Jouanin L, Terry N (1999b) Cadmium tolerance and accumulation in Indian mustard is enhanced by overexpressing g-glutamylcysteine synthetase. Plant Physiol 121:1169–1177

Chapter 15
Remediation of Phenolic Compounds from Polluted Water by Immobilized Peroxidases

Qayyum Husain

Abstract In present chapter, an effort has been done to review the role of immobilized peroxidases in the removal of phenolic pollutants from wastewater or industrial effluents. Immobilized peroxidases were found significantly more stable against several kinds of denaturants such as the broad ranges of pH and temperatures, organic solvents, and other forms of chemical inactivators. These bound peroxidases were found highly resistant to inhibition caused by their specific inhibitors and products compared to their free forms. Peroxidases from several plant sources, horseradish, turnip, soybean seed coat, pointed gourd, and white radish, and some microbial sources have successfully been immobilized on/in various types of organic, inorganic, and nanosupports by using different methods and employed for the treatment of phenolic pollutants in batch processes and continuous reactors. The applications of immobilized peroxidases in targeting such compounds were found remarkably useful in decreasing the cost of the treatment. Herein, the findings have summarized that the immobilized peroxidases have a great potential in the remediation of phenolic pollutants from industrial wastewater.

Keywords Immobilization · Peroxidase · Phenolic compounds · Wastewater · Industrial effluents · Remediation · Treatment

1 Introduction

Since last several decades, the industrial development has resulted into the release of huge amounts of aromatic compounds into environment. Millions of tonnes of such compounds are synthesized every year worldwide for various human consumption and unused compounds moved into environment via industrial wastewater. Water is one of the most essential natural resources, which had badly

Q. Husain (✉)
Department of Biochemistry, Faculty of Life Sciences, Aligarh Muslim University, Aligarh, India

affected by industrial revolution (Chandra et al. 2011; Chowdhary et al. 2018). As concerns over the environmental growth, interest has been directed toward the fate of such compounds, which are commonly found in chemical solvents, pesticides, and petroleum-based products (Das and Chandran 2011). Phenol and its derivatives are aromatic molecules containing hydroxyl groups attached to the benzene ring and are among the most common organic compounds. The origin of these phenolic compounds is both anthropogenic and xenobiotic (Husain and Jan 2000; Bharagava et al. 2017). Xenobiotic sources include industrial wastes derived from fossil fuel extraction; chemical processes in phenol, dye, and pesticide manufacturing plants; paper and pulp production; paint and varnish industries; wood processing; fine chemicals; and pharmaceutical industries (Mishra and Bharagava 2016; Sujata and Bharagava 2016). The contamination of environment by xenobiotic compounds created a serious issue due to their poor aqueous solubility, which posed a big challenge to water sources both locally and far off from the original site of their release (Husain 2010). Aromatic compounds like phenols and their derivatives have been commonly employed for industrial production of a wide range of resins, manufacturing materials for automobiles and appliances, epoxy resins, adhesives, and polyamides for several other purposes (Hollmann and Arenda 2012). Phenolic derivatives included a wide spectrum of pesticides used mainly as algicides, bactericides, herbicides, fungicides, molluscicides, and insecticides (Annadurai et al. 2000). Several other aromatic compounds widely employed at industrial level included nitrophenols like *p*-nitrophenol, used in chemical industry for the manufacturing of analgesics, dyes, and pesticides and in processing of leather; naphthol, the building stones for dyes, plastics, rubbers, and man-made fibers; and chlorophenols, used as biocides (Bharagava and Chandra 2010; Gianfreda et al. 2006; Saxena and Bharagava 2017). Phenol-containing water, when chlorinated during disinfection of water, also resulted in the formation of chlorophenols. The number determined the toxicity of chlorinated phenols and position of chlorine substituents in benzene ring structure, and it was increased with higher chlorine substitution (Czaplicka 2004). Pentachlorophenol (PCP) has chlorine atoms at all of its five substituent positions and has been found most toxic among chlorinated phenols (Torres et al. 2010). It is one of the most recalcitrant pollutants present in the environment, which has been used as a pesticide and wood preservative. Several reports are available on the contamination of soils, lakes, rivers, and groundwaters by PCP (Saxena et al. 2017). As a huge quantity of such compounds has been used at, it is inevitable that a substantial amount will be lost in the environment either due to their excessive usage or accidental spillage (Kim et al. 2006; Hare et al. 2017).

Accumulation of these compounds in nature has resulted in the environmental contamination at a large scale, which has caused too many deleterious effects on living systems. Several phenolic compounds have been considered phytotoxic, antimicrobial, mutagenic, and carcinogenic. For example, the contamination of water bodies with phenol imparted carbolic odor to them and resulted into toxic effects on aquatic flora and fauna (Basha et al. 2010). Phenols are highly toxic to humans and affect several biochemical functions; its acute exposure caused central nervous system disorders (Nuhoglu and Yalcin 2005). These toxic compounds have been strongly regulated in various countries, and therefore, their removal from the wastewater is required prior to their final disposal into environment (Husain and Jan 2000; Regalado et al. 2004; Husain and Qayyum 2013). Some environmental protection agencies have issued a regulation to control phenol level in polluted water from less than 1 mg L^{-1} to several thousand mg L^{-1} (La Rotta et al. 2007).

Recently used techniques for remediation of phenolic pollutants from industrial discharges involved distinct kinds of chemical, physical, and physicochemical procedures, and these methods have their own limitations (Husain 2006). The biological treatment of these compounds has attracted great attention of the biotechnologists due to their potential for partial or complete degradation of phenolic compounds (Pearce et al. 2003; Diez 2010; Yadav et al. 2016). Microbial treatment has shown its potential to degrade phenol and other recalcitrant compounds in order to maintain phenol concentrations below the toxic limit and to reduce operating costs while producing innocuous end products. Most of these compounds have been used as carbon and energy sources by microorganisms (Van Schie and Young 2000; Farhadian et al. 2008; Pradeep et al. 2015). Thus, the microbiological treatment of such compounds is an essential contribution to the global carbon cycle as well as to the detoxification of polluted water and contaminated soils (Beristain-Cardoso et al. 2009). The microbiological remediation of such compounds has its own demerits, i.e., expensive maintenance of microbial culture, problem with the survival of cells in environment and their limited mobility, alternate carbon sources, completion of indigenous population, metabolic inhibition, and synthesis of more toxic by-products even than parent phenols (Husain 2010).

The environmentalists are facing a great challenge to develop efficient, cheap, and eco-friendly techniques to replace existing traditional technology and use a superior technology to clean toxic compounds from the polluted place. Thus, a need arises for the cost-effective and alternative technique to remove such pollutants from the huge volume of wastewater (Kulkarni and Kaware 2013). Recently a lot of efforts have been done to develop an eco-friendly or green technology instead of existing approaches used for the removal of phenolic pollutants from wastewater (Hamid and Rahman 2009; Diao et al. 2010). A main component of the green technology revolution is the application of biocatalysts (enzymes), which exhibited minimal impact on ecosystems. Enzymatic treatment comes between the two commonly used technologies, chemical and biological remediation, since this process engaged chemical reactions based on the action of biological catalysts. Enzymes are biological catalysts which regulate multitude of chemical reactions that occur in the living cells and participate in all vital processes of the cells (Gianfreda 2008; Husain and Ulber 2011; Chandra and Chowdhary 2015). These are preferred over intact organisms because of the isolated enzymes catalyze reactions with high specificity with better optimization, easy to handle and store, and their concentration is not dependent on the growth rate of the microorganisms (Wagner and Nicell 2003; Bodalo et al. 2006).

The enzymatic methods have their own advantages over the traditionally applied techniques for the treatment of persistent and recalcitrant phenolic compounds: operation at very low and high concentrations of contaminants in a broad range of pH, temperature, and salinity, absence of shocking loading effects and delays related to acclimatization of biomass, reduction in sludge volume, the simplicity of controlling process, requirement of bio-acclimatization, and the easy-controlled process among other known procedures (Duran and Eapisito 2000; Gianfreda and Rao 2004; Bhandari et al. 2009). Enzymes catalyze the detoxification of contaminants in a very short time compared to microbial treatment, which needs several days to months to provide similar results (Chowdhary et al. 2016). Enzymes are less likely to be inhibited by compounds and minerals, which may cause toxicity to the living organisms (Gianfreda and Rao 2004; Husain et al. 2009). In recognition of these potential advantages, recent work has emphasized on the development of enzymatic processes. Enzymes have drawn the attention of the environmentalist to investigate new possibilities offered by them for the treatment of polluted water, solid/ hazardous wastes, and wastes containing wide-spectrum organic pollutants (Husain 2006; Yang et al. 2008). In order to apply enzymes for wastewater treatment at large scale to their full potential, various precautions should be taken into consideration. These are the need of inexpensive sources of enzymes, possibility of using enzymes under the conditions of wastewater treatment, feasibility of characterization of reaction products, assessment of their impact on downstream processes or on the environment, search of technology to depollute solid waste, etc. The current research has focused on the advancement of enzyme-based technology to remediate hazardous molecules found in industrial effluents. Most of the recent studies have demonstrated the application of enzymes for the treatment of pollutants present in low amounts in wastewaters (Kulshrestha and Husain 2007; Matto and Husain 2007).

Oxidoreductases, peroxidases and polyphenol oxidases, are the principal biological agents that have demonstrated their significance in the remediation of aromatic compounds such as phenols, aromatic amines, polycyclic aromatic hydrocarbons, polychlorinated compounds, bisphenols, and biphenyls from wastewater without the use of cofactors (Khan and Husain 2007; Husain et al. 2009). These enzymes catalyze conversion of a large number of substrates into free radical, and these free radicals form less toxic insoluble compounds, which can be simply eliminated by centrifugation/filtration (Ahuja et al. 2004; Husain 2006). Many earlier workers have demonstrated that the oxidation of phenolic compounds has been achieved by a large number of enzymes from plants and microorganisms (Duran and Espisito 2000; Chandra and Chowdhary 2015). These included polyphenol oxidase (e.g., tyrosinase) from mushrooms (Ikehata and Nicell 2000), lignin peroxidase (LiP) from white-rot fungus (Ward et al. 2003; Cohen et al. 2009), *Coprinus cinereus* peroxidase (CIP) (Masuda et al. 2001; Ikehata et al. 2005), horseradish peroxidase (HRP) (Wagner and Nicell 2003; Ward et al. 2004), turnip peroxidase (TP) (Duarte-Vazquez et al. 2003), bitter gourd peroxidase (BGP) (Akhtar and Husain 2006; Karim and Husain 2009a; Ashraf and Husain 2010a), soybean peroxidase (SBP) (Bodalo et al. 2006; Watanabe et al. 2011), white radish peroxidase (WRP) (Ashraf and Husain 2009), and chloroperoxidase (La Rotta et al. 2007).

Peroxidase (E.C. 1.11.1.7) catalyzes one electron oxidation of many organic and inorganic substrates employing H_2O_2 as hydrogen acceptor with heme group at its active center (Belcarz et al. 2008). In the last few decades, a lot of work has been done on the treatment of wastewater and soil contaminated with aromatic pollutants by using peroxidases from different plant and microbial sources (Husain and Husain 2008; Barakat et al. 2010). Plant-derived peroxidases are quite cheap, and thus they have been widely employed for bioremediation of phenolics and aromatic amines from polluted water (Karim and Husain 2009a, b; Ashraf and Husain 2010a). Peroxidases have the ability to induce polymerization of phenols via a radical oxidation-reduction mechanism (Nazari et al. 2007). Peroxidases follow a ping-pong mechanism for the oxidation of aromatic molecules (Steevensz et al. 2009). The enzymatic cycle involved during oxidative polymerization of aromatic compounds has been earlier reported (Ghasempur et al. 2007).

Enzyme immobilization, sometimes also referred to as "enzyme insolubilization," is a technique which restricted movement of enzyme in solution. To improve the economic feasibility of an enzyme at industrial level, immobilization of enzymes on a solid support is a common choice. Immobilization as a technique has been increasingly used in industrial processes in order to simplify separation of enzymes from the reaction mixture and purify products without contamination of catalyst (Ozdural et al. 2003; Kumar et al. 2016). It also increases the resistance of enzyme to proteolysis and denaturation against the exposure to extreme conditions of pH, temperature, and substrate concentration swings and reduced susceptibility to various types of contaminants and inhibitors. It provided a longer shelf life and higher productivity per active unit (Matto et al. 2008; Husain and Ulber 2011). These alterations resulted due to structural rigidity provided by the attachment of enzyme to the support at the multiple points. The very early motivation was to replace many industrial processes that were catalyzed with soluble enzymes in solution by immobilized enzymes. Supports used for the immobilization of enzymes may be both organic and inorganic. Some of the used organic and inorganic supports include wood chips; cellulose; agarose; dextran; nylon; acrylamide-, Sephacryl-, styrene-, and maleic anhydride-based polymers; solid glass; silica; alumina; ZnO; CaO; diatomaceous earth; plaster; sand; TiO_2; etc. (Cao and Zhou 2006; Guzik et al. 2014).

Organic supports have been preferred more commonly for the construction of immobilized enzymes due their easy and simple modification. However, inorganic supports have their own advantages such as better stability in comparison to the organic supports like their ability to maintain high mechanical strength, thermal and storage stability, and the ease of regeneration, whereas organic supports are affected by factors like pH and degradation by enzyme and microorganisms. However, the selection of support is not only based on the nature of the enzyme but also depends on particle size, surface area, molar ratio of hydrophilic to hydrophobic groups, and chemical composition (Datta et al. 2013; Mohamad et al. 2015).

In general, an increase in the ratio of hydrophilic groups and the concentration of bound enzymes resulted in a higher activity of immobilized enzymes. Several kinds of methods are in practice to obtain high yield of enzyme immobilization with a minimum loss of activity (Husain and Ulber 2011). Distinct modes of enzyme immobilization are adsorption, chemical bonding, chemical aggregation, entrapment, microencapsulation, and bioaffinity immobilization. It should be kept in mind before selecting any method of immobilization that there should be a minimum loss in enzyme activity upon immobilization. In other words, it is necessary to immobilize an enzyme in such a way that the reactive groups present on its active site should be protected. One of the solutions to protect the groups present on the active site of the enzyme during immobilization of enzyme to the support is the process should be done in the presence of its substrate or competitive inhibitor, which could be removed later on (Guzik et al. 2014).

Peroxidase-based enzymatic approach has shown significant potential and capacity for treating phenol-containing aqueous solutions and wastewaters especially at the lab-scale. For full-scale application, future work should be focused on developing optimal and commercially viable processes for continuous phenol removal. Furthermore, in this chapter author has attempted to review the latest literature based on the treatment of phenol and its derivatives present in polluted water by using immobilized peroxidases in batch processes and different kinds of continuous reactors.

2 Removal of Phenolic Pollutants by Immobilized Peroxidases

2.1 Removal of Phenolic Pollutants by Immobilized HRP

Immobilized peroxidases and their applications for the treatment of phenolic compounds are listed in Table 15.1.

Immobilized peroxidases have been used for the remediation of water contaminated by different kinds of chemical pollutants (Husain and Ulber 2011). Moreover, the immobilized enzymes have successfully been employed for the treatment of aromatic pollutants in batch processes and continuous reactors. Siddique et al. (1993) investigated remediation of *p*-chlorophenol (*p*-CP) from aqueous solution by HRP immobilized onto three different supports: cellulose filter paper, nylon balls, and nylon tubing. They reported that the enzymatic reaction was extremely fast and there was no desorption of enzyme from the reactor matrix. Their results further have demonstrated that above 80% phenol removal was achieved as long as enzyme activity was not limiting in the reactor. Vasudevan and Li (1996) employed HRP immobilized on activated alumina for phenol removal and found that one molecule of HRP has successfully removed about 1100 molecules of phenol when the reaction was done in buffer of pH 8.0 and at room temperature. Magnetite-immobilized HRP catalyzed 100% removal of various chlorophenols, *p*-CP, 2,4-dichlorophenol (2,4-DCP), 2,4,5-trichlorophenol, 2,4,6-trichlorophenol, 2,3,4,6-tetrachlorophenol, and PCP, and also removed complete total organic carbon from the treated wastewater (Tatsumi et al. 1996). Levy et al. (2003) constructed a fused protein consisting of cellulose-binding domain (CBD) and HRP and CBD-HRP was immobilized on microcrystalline cellulose. Immobilized HRP showed increased stability to H_2O_2 and oxidized significantly higher level of *p*-bromophenol (*p*-BP) compared to free CBD-HRP. Lai

Table 15.1 Summarizes immobilized HRP on various supports and type of phenol treated.

Type of support	Mode of immobilization	Type of phenol treated	Reference(s)
Cellulose filter paper, nylon balls, nylon tubing, activated alumina, aluminum-pillared interlayered clay, cinnamic carbohydrate esters, UFM, Fe_3O_4 NPs, porous celite beads, rodlike cellulose, Eupergit@C, gelatin-p(HEMA-GMA) cryogel	Adsorption, covalent attachment	Phenol	Siddique et al. (1993), Vasudevan and Li (1996), Cheng et al. (2006), Rojas-Melgarejo et al. (2006), Cho et al. (2008), Yang et al. (2008), Pramparo et al. (2010), Vasileva et al. (2009), Pradeep et al. (2012), Duan et al. (2014), and Soomro et al. (2016)
Magnetite	Adsorption	p-CP, 4-DCP, 2,4,5-trichlorophenol, 2,4,6-trichlorophenol, 2,3,4,6-tetraphenol, PCP	Tatsumi et al. (1996)
Microcrystalline cellulose, APG, RVC, acrylamide-2-OH-ethyl methacrylate copolymer	Adsorption/covalent	p-BP, p-CP, phenol	Levy et al. (2003), Lai and Lin (2005), Cho et al. (2005), and Shukla and Devi (2005)
Electrode, Con A-Sephadex, Fe_3O_4 sorption-gelatin, polyacrylamide gel	Adsorption, bioaffinity, entrapment	PCP	Kim and Moon (2005), Dalal and Gupta (2007), Zong et al. (2007), and Zhang et al. (2007)
Microporous polypropylene hollow fiber membranes	Impregnated	3,4-Dimethylphenols, 4-ethylphenol, 2-OH-1,2,3,4-tetra-OH-naphthalene, 2-OH-decahydronaphthalene, and 4-OH-biphenyl	Moeder et al. (2004)
Polyethylene-co-acrylic acid film	Adsorption covalent	4-Chloro-1-naphthol	Su et al. (2005)
Cellulose, CS, ethylene vinyl alcohol copolymer, silk fibroin and PAMAM, GF/Ti electrode	Adsorption/covalent attachment	BPA	Maki et al. (2006), Xu et al. (2011), and Zhao et al. (2015)
Magnetic pGMA-co-MMA & GA	Covalent	Phenol, PCP	Bayrumoglu and Arica (2008)

(continued)

Table 15.1 (continued)

Type of support	Mode of immobilization	Type of phenol treated	Reference(s)
Calcium alginate, multifunctional biocapsules	Entrapment	Phenol, BPA	Alemzadeh et al. (2009), Alemzadeh and Nejati (2009a, b), and Ispas et al. (2000)
Glass incorporated TiO$_2$	Covalent	p-BP	Meizler et al. (2011)
GO	Adsorption	o-CP, 2,4-dimethoxyphenol	Zhang et al. (2010)
Montmorillonite coated with humic acid, ZnO nanocrystal	Covalent attachment, adsorption	Phenolic compounds	Kim and Kim (2010) and Zhang et al. (2016)
PVAG-PANI-GA, P(GMA-co-EGDMA), calcium alginate	Covalent/ entrapment	Pyrogallol, resorcinol	Caramori et al. (2012), Prodanovic et al. (2012), and Spasojevic et al. (2014)
Phospholipid-templated titania particles	Encapsulation	Phenolic compounds, p-CP	Jiang et al. (2014)
Silica nanorods, PVA-PAA-SiO$_2$ nanofibrous membrane	Adsorption	2,6-Dimethylphenol, paracetamol	Nanayakkara et al. (2014) and Xu et al. (2015)
Ethyl cellulose polymer, CS-PAN beads modified with ethanediamine, silica-MNPs, PAN-UFM, 3-APTES and GA, MNPs deposited on GO sheet, GO-MNPs and EDC, hydrous titanium, Ti-doped hollow nanofibers	Adsorption/ covalent attachment	2,4-DCP, phenol o-CP, p-CP	Dahili et al. (2015), Wang et al. (2015, 2016a, b), Chang and Tang (2014), Chang et al. (2015, 2016), Ai et al. (2016), and Ji et al. (2016)
Carbon nanosphere and GA	Covalent attachment	Various phenols, 4-methoxyphenol, BPA	Lu et al. (2017)

and Lin (2005) immobilized HRP onto aminopropyl glass and investigated its efficiency for the removal of p-CP from synthetic water. The polymerization of p-CP into insoluble precipitate in the presence of H$_2$O$_2$ was completed within 3 h after the initiation of reaction at pH 7.5 and achieved a maximal removal efficiency of 25%. HRP immobilized onto the surface of reticulated vitreous carbon was employed for the degradation of phenol by in situ generated H$_2$O$_2$-immobilized HRP complex in an electrochemical reactor. The immobilized HRP maintained about 89% of its original activity over 1-month storage. The phenol degradation rate of 86% was found under the optimal experimental conditions (Cho et al. 2005). A fixed bed reactor (17 × 1 cm) filled with HRP covalently coupled to acrylamide-2-hydroxyethyl methacrylate copolymer was employed for the oxidation of phenol at a flow rate of 0.5 cm^3 min^{-1}, 45 °C, and l/d ratio of 6. The phenol was quite efficiently oxidized by HRP in this reactor (Shukla and Devi 2005). HRP impregnated microporous polypropylene hollow fiber membranes used to degrade some selected hydroxylated aromatic compounds: 2-hydroxy-1,2,3,4-tetrahydronaphthalene, 2-hydroxy-decahydronaphthalene, 4-hydroxy-biphenyl, 4-ethylphenol, and 3,4-dimethylphenols. These compounds were effectively degraded 50–100% within 48 h by immobilized HRP, except 2-hydroxy-decahydronaphthalene (Moeder et al. 2004).

Kim and Moon (2005) demonstrated the degradation of PCP by the electroenzymatic method, which involved enzymatic catalysis and electro-generation of H$_2$O$_2$. The experiments were performed in a two-chamber packed-bed reactor using HRP-immobilized electrode. The maximum generation of H$_2$O$_2$ and the current efficiency were noticed at −0.4 V vs. Ag/AgCl and a flow rate of 1 mL min^{-1}. The highest initial degradation rate and efficiency of PCP were recorded at pH 5.0 and 25 °C. The presence of chloride ion demonstrated that PCP was dechlorinated at the initial period of degradation. On the basis of obtained degradation pathway and the intermediates, the electroenzymatic procedure showed higher degradation than the electrochemical method alone. Immobilized HRP preparations obtained via direct adsorption and amine coupling on the surface of poly(ethylene-co-acrylic acid) films were applied for the oxidation of 4-chloro-1-naphthol in the presence of H$_2$O$_2$ (Su et al. 2005). HRP was immobilized on three fiber-forming polymeric materials, cellulose, chitosan (CS), and ethylene vinyl alcohol copolymer, and these immobilized enzyme preparations were used to polymerize native phenol and bisphenol A (BPA) dissolved in aqueous medium (Maki et al. 2006). Aluminum-pillared interlayered clay-immobilized HRP applied for treatment of phenolic compounds was found to be quite successful in removing phenolics significantly over a wide range of pH 4.5–9.3. Immobilized enzyme preparation (20 U) could remove 26 mg of phenol. Hence, this enzymatic mode of treatment has been applied prior to remove a bulk of phenolic pollutants from industrial effluents (Cheng et al. 2006).

Rojas-Melgarejo and co-workers (2006) investigated the immobilization of wild-type and recombinant HRP (rHRP). The immobilization of these enzymes on cinnamic carbohydrate esters involved a process of physical adsorption and intense hydrophobic interactions between cinnamoyl groups of the support and related groups of the enzyme. Immobilized HRP has successfully removed above 70% phenol from polluted water. HRP was immobilized by bioaffinity layering on Con A-Sephadex, and this preparation was applied for the treatment of phenolic compounds. The immobilized HRP has been effectively employed to treat p-CP in a broad spectrum of phenolic concentrations (1.0–15.0 mM) and maintained almost 100% conversion of p-CP after five reuses (Dalal and Gupta 2007). HRP immobilized by Fe$_3$O$_4$ sorption-gelatin embedding cross-linkage method was found to remove PCP repetitively, and even after seven repeated uses, the removal of phenol was more than 39% by 0.05 U mL^{-1} HRP (Zong et al. 2007). Polyacrylamide gel prepared by γ-ray radiation was employed for the immobilization of HRP, and this enzyme preparation was filled in a column to treat PCP-containing wastewater.

Both soluble and immobilized HRP exhibited the same optima at pH 5.15, the immobilized HRP reduced PCP concentration from 13.4 to 4.9 mg L^{-1} within 1 h, and immobilized HRP column was repeatedly and successfully used (Zhang et al. 2007). HRP immobilized on the porous celite beads via aminopropylation with 3-aminopropyltriethoxysilane (3-APTES) and covalent linkage with glutaraldehyde (GA) in a membraneless electrochemical reactor was used to convert phenol. Phenol was oxidized by electrochemical and electroenzymatic methods; however, its oxidation was enhanced by the electroenzymatic method and was converted into *p*-benzoquinone, various organic acids, and CO_2 (Cho et al. 2008). Yang et al. (2008) used peroxidase immobilized on the cyanogen bromide-activated rodlike cellulose nanocrystals for the treatment of chlorinated phenolic compounds. Immobilized peroxidase removed a greater fraction of phenol from the contaminated H_2O compared to soluble counterpart. In another study, Bayramoglu and Arica (2008) covalently immobilized HRP on the magnetic poly(glycidylmethacrylate-co-methylmethacrylate), pGMA-MMA via GA. The immobilized HRP was employed for the polymerization and removal of phenol and *p*-CP in presence of H_2O_2. These phenols were successfully removed from polluted water in a magnetically stabilized fluidized bed reactor. Alemzadeh and Nejati (2009a) used porous calcium alginate bead-entrapped HRP for the treatment of phenol. The maximum conversion of phenol was found at a concentration of 2.0 mM. The encapsulated enzyme showed lower efficiency of phenol removal than its free form; however entrapped enzyme was reused four times without any loss in its initial activity. Moreover, the phenol removal was gradually increased with increasing concentration of enzyme, 0.15–0.8 U g^{-1} alginate. Moreover, these workers studied time course of phenol removal by both calcium alginate-encapsulated and free HRP, and they observed that both free and encapsulated enzyme had similar efficiency of conversion of phenolic compounds. However, the immobilized enzyme retained its full activity after four repeated uses. The ratio of H_2O_2/phenol at which highest phenol removal obtained was found to be dependent on initial phenol concentration, and in the solution of 2 and 8 mM phenol, it was 1.15 and 0.94, respectively. (Alemzadeh and Nejati 2009b; Alemzadeh et al. 2009).

Free and immobilized HRP on modified acrylonitrile copolymer membranes was employed for reducing phenol in water solution. The immobilized HRP oxidized higher level of phenol (100 mg L^{-1}), 95.4% in the presence of H_2O_2 (Vasileva et al. 2009). Ispas et al. (2000) prepared multifunctional biocapsules with immobilized HRP using a layer-by-layer configuration and apply these biocapsules to remove phenol and BPA from polluted water. BPA was removed per capsule 5.6 ppm, whereas phenol was cleaned up to 10 ppm per capsule within 15 h. HRP was covalently immobilized onto epoxy-activated acrylic polymers (Eupergit®C) by applying three different approaches: direct binding to the polymers via their oxirane groups, binding to the polymers via a spacer made from adipic dihydrazide, and binding to hydrazido polymer surfaces via carbohydrate moiety of the enzyme already modified by periodate oxidation. The periodate-mediated covalent attachment of the enzyme on hydrazido Eupergit®C was found to be the best method for the immobilization of biocatalysts. The free and immobilized HRP were independently used to treat phenol in batch reactors. The immobilized HRP removed 50% of the phenol by using only about 100 times lesser enzyme than its free form. HRP covalently immobilized onto epoxy-activated acrylic polymers (Eupergit®C) removed nearly 92% of phenol employing 0.006 U mL^{-1} of the preparation after 4 h of reaction in a stirred reactor (Pramparo et al. 2010). HRP was immobilized on the glass-based support containing TiO$_2$, and this preparation was taken for the treatment of aqueous halogenated phenols. Immobilized HRP was employed under the UVB irradiation without added H$_2$O$_2$, and it worked continuously and maintained 98% 4-BP (0.1 mM) transformation over 16 h (2011). The efficiency of higher removal was noticed when phenolic compounds were treated by HRP immobilized on graphene oxide (GO) compared to soluble enzyme. The highest removal efficiency was found with substrate 2,4-dimethoxyphenol (34.4%), for which the immobilized HRP was found twice more efficient than free HRP and *o*-chlorophenol, *o*-CP (20.4%), which is a major component of industrial wastewater (Zhang et al. 2010). Kim and Kim (2010) described the immobilization of HRP in the nanoscale porous structural material, montmorillonite coated with humic acid (HA), in order to protect the structure and activity of the enzyme. The immobilized HRP catalyzed the oxidative cross-coupling to remove phenolic compounds, and the polymerization of phenolic compounds was almost the same as obtained by free enzyme.

HRP was covalently immobilized on silk fibroin and poly(amidoamine)-bound magnetic Fe$_3$O$_4$ NPs. An electroenzymatic method was used to oxidize BPA in a membraneless electrochemical reactor containing immobilized HRP. The BPA removal efficiency was reached to 80.3% and its degradation was more efficient via electroenzymatic process (Xu et al. 2011). Free and immobilized HRP has been employed for the polymerization of phenol (100–500 mg L^{-1}) at ambient room temperatures between 27 and 32 °C. Free enzyme was used in Erlenmeyer flasks, and immobilized enzyme was fabricated and used for polymerization of phenol in the bed reactor. Native HRP polymerized 84% of phenol when fed with 100 mg L^{-1}, while the immobilized enzyme polymerized the similar phenol concentration to 62%. The greater amount of phenol was polymerized by free enzyme; it might be due to availability of significantly higher number of active sites in free enzyme than the immobilized HRP (Pradeep et al. 2012). Caramori et al. (2012) prepared discs of network polyvinyl alcohol (PVA) coated with polyaniline using GA and employed it for the immobilization of HRP. The immobilized HRP maintained nearly 50% of its pyrogallol oxidation activity, whereas the free enzyme lost almost all its activity when incubated at 70 °C for 15 min. The covalently bound HRP had successfully oxidized resorcinol, m-cresol, catechol, pyrogallol, α-naphthol, β-naphthol, and 4,4′-diaminodiphenyl benzidine from 70% to 90%. The silica-encapsulated HRP was employed for the remediation of phenol from synthetic water, and the maximum phenol removal was 73.1% under optimal experimental conditions. The encapsulated enzyme retained 51.7% phenol removal efficiency even after five repeated uses (Wang et al. 2011). Liu et al. (2012) described immobilization of HRP on honeycomb-patterned microporous polystyrene membranes with carbodiimide (CDI) as the activator. The catalytic performances for oxidation of phenol over immobilized HRP were investigated. It was indicated that the suitable enzyme solution concentration and immobilized time were 1 g L^{-1} and 3 h, respectively, for immobilization of HRP. The obtained phenol removal was around 76% by immobilized HRP, and the immobilized HRP still retained above 95% phenol removal activity after three reuses.

Kim and co-workers (2012) found that the inorganic natural materials, clays, and soil organic matter-immobilized HRP had remarkable potential in removing high concentration of phenol from solution. CS-halloysite hybrid nanotubes were used for the covalent immobilization of HRP by cross-linking with GA. The immobilized HRP exhibited high removal efficiency for phenol from wastewater compared to free form (Zhai et al. 2013). Poly(D,L-lactide-co-glycolide) (PLGA)/PEO-PPO-PEO (F108) electrospun fibrous membranes (EFMs) were employed for the encapsulation of HRP by emulsion electrospinning, and this enzyme preparation was applied for the degradation of PCP. The removal efficiency of PCP reached 83% and 47% for immobilized and free HRP at 25 ± 1 °C, respectively. The presence of humic acid demonstrated inhibition in the activity of HRP and reduced adsorption capacity of PCP due to competitive binding (Niu et al. 2013). Xu et al. (2013) studied covalent immobilization of HRP on a carrier poly(MMA-co-ethyl acrylate) (pMMA-CEA) microfibrous nanomembrane activated by polyethyleneimine (PEI) and GA, and immobilized enzyme was used for the remediation of BPA from polluted water. Immobilized HRP exhibited remarkably high removal efficiency, 93% of BPA in 3 h, compared to free HRP, 61%, and PFM alone, 42%. The high BPA removal was obtained due to improvement in catalytic activity of immobilized HRP with adsorption on modified pMMA-CEA support.

HRP immobilized on a macroporous copolymer of GMA and ethylene glycol dimethacrylate, p(GMA-co-EGDMA) with a mean pore diameter of 120 nm via periodate activation procedure exhibited significantly high specific activity and stability. The immobilized HRP retained 45% pyrogallol oxidation activity after six repeated uses in a batch reactor (Prodanovic et al. 2012). Furthermore the same group has developed an improved method for HRP immobilization into alginate beads by chemical modification of the enzyme and polysaccharide chains; HRP and alginate were oxidized by periodate and subsequently modified with ethylenediamine. The immobilized HRP retained nearly 75% pyrogallol oxidation activity after its repeated uses (Spasojević et al. 2014). Jiang et al. (2014) investigated encapsulation of HRP phospholipid-templated TiO$_2$ particles via biomimetic titanification process and used for the remediation of water polluted with phenolic compounds and dye. The removal efficiency for phenol, p-CP, and direct black-38 by the encapsulated HRP was 92.99%, 87.97%, and 79.72%, respectively. Moreover, the encapsulated HRP exhibited better removal efficiency than free HRP. Silica nanorods were selected for the immobilization of HRP, and modified enzyme was employed for the oxidative polymerization of 2,6-dimethylphenol. The immobilized enzyme successfully catalyzed formation of polymer, poly(2,6-dimethyl-1,4-phenylene oxide), in water-acetone solvent system. The immobilized nanobiocatalyst demonstrated a marked enhancement in enzyme activity toward oxidative polymerization as well as some degree of reusability compared to free HRP

(Nanayakkara et al. 2014). Zhao et al. (2015) described a novel electrochemical approach by combining electro-enzyme and electrocoagulation to precipitate BPA from water containing HA. HRP was immobilized on the graphite felt of Ti electrode as HRP-GF/Ti cathode, with aluminum plate anode containing a pair of working electrodes. BPA was 100% removed, and the reduction of total organic carbon (TOC) reached 95.1% after 20-min sequential treatment with the current density, 2.3 mA/cm^2. From real wastewater (TOC = 28.76 mg L^{-1}, BPA = 4.1 µg L^{-1}), 94% BPA and 52% TOC were minimized after sequential treatment. The electroenzymatic process not only oxidized BPA into dimer and BPA-3,4-quinone but also markedly changed chemical and structural features of HA, where hydrophilic moieties, phenolic and alcohols, transformed into hydrophobic forms: ethers, quinone, and aliphatic. However, these polymerized products were effectively separated from aqueous solution by anodic electrocoagulation. Purified HRP and crude extract from horseradish were covalently bound to the ethyl cellulose (EC) polymer obtained by Nano Spray Dryer B-90 using CDI as a cross-linking agent. The purified HRP and extract from horseradish-immobilized preparations showed better activity in a pH range of 4–10 and retained greater fraction of activity on longer storage than the soluble enzyme. The immobilized HRP appeared highly effective in the elimination of 100% 2,4-DCP which was also due to high adsorbing capacity of the fine particles. The reuse study proved the operational stability of HRP attached to EC after ten reuses (Dahili et al. 2015).

Pan et al. (2015) used carboxyl-functionalized polystyrene (poly(styrene-co-methacrylic acid), PSMAA) nanofibrous membrane (NFM) with average diameters of 250 ± 20 nm for the immobilization of HRP by a chemical method. The obtained removal of *o*-methoxyphenol by immobilized HRP was 80.2% after 2 h. These findings revealed that the present method of HRP immobilization has provided remarkably high utilization in the treatment of phenolic effluents. HRP was adsorbed on PVA/poly(acrylic acid (PAA)-SiO$_2$ electrospinning NFM of diameters of 200–300 nm. The enzyme was also covalently attached on the surface of nanofibers activated with 1,1′-carbonyldiimidazol. The bound HRP maintained 79.4% of its original activity and was applied for the removal of paracetamol from wastewater. Paracetamol removal rate by immobilized HRP was 83.5%, similar to that of free HRP (84.4%), but immobilized HRP showed remarkably very high reusability. The results demonstrated that enzyme immobilized on nanofibers demonstrated its very high potential in the treatment of contaminated water (Xu et al. 2015). Xu and Chen (2016) evaluated the immobilization of HRP on activated carbon/polyvinyl formal composite materials. HRP immobilized on composite materials showed a lot of potential in phenolic wastewater degradation. HRP magnetic nanoparticle system was exploited for the oxidation of phenols into phenoxy radicals in the presence of H$_2$O$_2$. The phenoxy radicals react with each other nonenzymatically and converted into insoluble polymers, which can be easily removed by filtration or centrifugation. The hybrid peroxidase catalyst exhibited three times higher activity than the free HRP and was capable of removing three times more phenol from polluted water compared to free HRP. Moreover, the hybrid biocatalyst decreased substrate inhibition and limited inactivation from reaction products, which were noticed common problems with free or conventionally immobilized enzymes. The performance of the hybrid biocatalyst made them attractive choice for industrial and environmental applications, and their development opened new avenues for treatment of phenolics where the soluble or conventionally immobilized enzymes were not effective (Duan et al. 2014). The SiO_2-coated Fe_3O_4 NPs modified with 3-APTES employed for covalent immobilization of HRP by applying GA as coupling agent. The immobilized HRP was used to activate H_2O_2 for the degradation of 2,4-DCP. The rapid degradation of 2,4-DCP revealed that the immobilized HRP has a lot of potential in the removal of toxic compounds found in polluted water (Chang and Tang 2014). Magnetic Fe_3O_4 NPs deposited on graphene oxide (GO) sheets were used as a support for the immobilization of HRP. Fe_3O_4 NPs-GO sheet-immobilized HRP has been utilized for the remediation of o-CP, p-CP, and 2,4-DCP from contaminated water. The different numbers and positions of electron-withdrawing substituents influenced the chlorophenol removal, and their order of removal was 2-CP<4-CP<2,4-DCP. The oxidation products formed during chlorophenol degradation were analyzed by gas chromatography-mass spectrometry. The NPs were recovered by applying an external magnetic field, and the immobilized HRP exhibited 66% of its activity after four repeated uses. The results demonstrated that the immobilized enzyme has successfully been employed for the treatment of hazardous phenolics found in wastewater (Chang et al. 2015). Furthermore, the same group has carried out covalent binding of HRP onto GO/Fe_3O_4 NPs with 1-ethyl-3-(3-dimethyaminopropyl) carbodiimide (EDC) as a cross-linking agent. The addition of PEG in this reaction mixture enhanced phenol removal efficiency and prevented inactivation of the enzyme. The findings of the work have demonstrated that the immobilized enzyme has efficiently removed over 95% of the phenol from aqueous solution. The nanocomposite bound HRP has been easily separated by employing magnetic field from the reaction mixture and was repeatedly used (Chang et al. 2016).

GA-activated carbon nanosphere-immobilized HRP was used for the remediation of several phenolic compounds from the model wastewater. Immobilized HRP exhibited excellent removal efficiency, especially for chlorophenols, 4-methoxyphenol, and BPA. The biodegradation of phenols with electron donor groups was found to be doubled after immobilization on active carbon nanospheres (Lu et al. 2017). Zhang et al. (2016) evaluated immobilization of HRP on ZnO nanocrystals of varying shapes. HRP immobilized on nanodiscs, nanoflowers, and nanorods retained 42.3, 26.1, and 14.1% activity, respectively, and these immobilized enzyme preparations have efficiently removed higher concentration of phenol, i.e., nanodiscs (86.09%), nanoflowers (79.46%), and nanorods (77.03%) compared to free enzyme (61.52%). The removal of phenolic compounds from aqueous solution using ZnO-immobilized HRP was examined with five additional phenolic pollutants, and ZnO-bound HRP showed greater removal efficiencies for such types of phenolics compared to free HRP. HRP was covalently immobilized onto NaOH- and HCl-treated PAN-based beads modified with ethanediamine and CS. The treatment of 2,4-DCP was performed by taking PAN bead-immobilized HRP in the beakers equipped with magnetic stirrer; about 90% of the 2,4-DCP was eliminated by immobilized HRP. However, the immobilized enzyme showed a lower 2,4-DCP remediation from synthetic water compared to soluble enzyme (Wang et al. 2015). These workers further described the immobilization of HRP on PAN ultrafiltration membranes (UFM) by cross-linking with GA. The potential of HRP-UFM was evaluated in the removal of phenol via oxidation with the addition of H_2O_2. Almost 100% remediation of phenol (1–10 mg L^{-1}) from water was obtained by HRP-UFM. These results showed that the HRP-UFM had a lot of potential for the treatment of polluted water containing phenolics (Wang et al. 2016a). In a recent study, similar researchers have used HRP immobilized on PAN beads to clean phenol from the wastewater. A comparative investigation for in vitro cytotoxicity of phenol/treated solutions was done in HeLa, HepG2, and mcf-7 cells by employing MTT method together with flow cytometry investigation for cell viability and cell cycle distributions. The results demonstrated that the toxicity of phenol solution was remarkably decreased after treatment with immobilized HRP (Wang et al. 2016b). HRP was reversibly immobilized onto gelatin-loaded poly(2-hydroxyethyl methacrylate-GMA) [p(HEMA-GMA)] cryogel discs, and this immobilized enzyme preparation was used for the remediation of phenol from aqueous solutions. The phenol was removed 91% by immobilized HRP in the presence of H_2O_2 within 2 h (Soomro et al. 2016).

Ai et al. (2016) employed hydrous titanium to immobilize HRP and examined catalytic efficiency of immobilized HRP in phenol removal from aqueous solution. The phenol removal potential of immobilized HRP was remarkably higher in the broad ranges of pH and temperatures compared to soluble enzyme. The bound HRP exhibited excellent phenol removal over 90% at 37 ± 3 °C after 15 min. A hybrid catalyst system was then developed by in situ encapsulating HRP inside nanochambers of TiO_2-doped hollow nanofibers via coaxial electrospinning. Such encapsulation effectively avoided UV-induced deactivation of the enzymes; thus the 2,4-DCP degradation efficiency was improved significantly as compared to oxidation obtained by HRP or TiO_2/UV either separately or simultaneously. Furthermore, the higher degradation, 90% of 10 mM 2,4-DCP, was obtained using integrated TiO_2-HRP hybrid biocatalyst system within 3 h. The hybrid catalysts system also showed remarkably high reusability and thermal stability (Ji et al. 2016).

2.2 Removal of Phenolic Pollutants by Immobilized Bitter Gourd Peroxidase

Several workers have already reviewed the work on the remediation of phenolic pollutants by applying HRP from wastewaters (Husain and Husain 2008; Husain et al. 2009). The high cost of purified HRP has limited its application in the treatment of such compounds; in order to minimize the cost of the peroxidase used in the bioremediation of aromatic pollutants, the peroxidases from some other easily available and cheaper sources have been tried for the treatment of aromatic compounds. Peroxidases from an easily available vegetable, bitter gourd, in India have been exploited for the cleaning of phenols, their derivatives, and complex mixtures present in model wastewater (Akhtar and Husain 2006).

Concanavalin A (Con A)-layered Sephadex bioaffinity-bound BGP was extensively used for the removal of a broad range of phenolic compounds and their complex mixtures under optimal experimental conditions. The maximum remediation of phenolics was obtained at pH 5.6 and 40 °C in the presence of 0.75 mM H_2O_2. The results showed that phenols and chlorinated derivatives were markedly removed from the polluted water; however, the removal of other substituted phenols was very low. The loss of total organic carbon was remarkably high when the synthetic wastewater was treated by immobilized BGP (Akhtar and Husain 2006). Furthermore, the similar group has carried out the treatment of *p*-BP by using Con A-layered calcium alginate-cellulose beads adsorbed and cross-linked BGP. The immobilized BGP retained nearly 78% phenolic compound removal efficiency over a period of 1-month storage at 4 °C and maintained about 50% *p*-BP removal efficiency after four repeated uses. Remarkably greater concentration of *p*-BP was cleaned by immobilized enzyme in presence of 0.1 mM $HgCl_2$ and water-miscible organic solvents compared to free enzyme. Two independent reactors filled with immobilized BGP were operated at flow rates of 10 and 20 ml h^{-1} retained 75 and 65% *p*-BP removal efficiency even after 1 month of their continuous operation (Ashraf and Husain (2011). Ahmad et al. (2013) immobilized BGP on APTES-activated TiO_2 NPs (<25 nm). The obtained bioactive nanoconjugates showed higher removal of phenols and dyes from aqueous solutions at 60 °C compared to soluble BGP.

BGP adsorbed on fly ash was successfully applied to clean maximum BPA in the presence of 0.3 mM guaiacol, a redox mediator, 0.75 mM H_2O_2 in sodium phosphate buffer, and pH 7.0 at 40 °C. The obtained degradation of BPA was 61%, 100%, and 100% at 20, 40, and 60 °C in the batch processes, respectively. The adsorbed BGP was more effective in the degradation of BPA than the native enzyme. Immobilized enzyme catalyzed complete degradation of BPA at 40 °C within 3.5 h. The oxidative degradation and polymerization of BPA were also examined in the continuous bed reactors at different flow rates, and the maximum removal of BPA was obtained at a flow rate of 20 mL h^{-1}. HPLC analysis showed two clear peaks, one related to BPA and other related to its degradation product, 4-isopropenylphenol. Plasmid nicking and comet assays have shown that the product, 4-isopropenylphenol, was highly nontoxic (Karim and Husain 2010).

2.3 *Removal of Phenolic Pollutants by Immobilized Soybean Seed Coat Peroxidase*

Gache and co-workers (2003) investigated the immobilization of SBP on the fibrous aromatic polyamide and employed this preparation for the detoxification of phenolics from industrial wastewater. It was noticed that the phenolic concentration was decreased from 0.720 to 0.063 mg mL^{-1} after 5-h treatment. Silica sol-gel/

alginate gel-entrapped SBP was employed for the treatment of phenol. These immobilized SBP preparations have successfully polymerized 85% phenol under optimal experimental conditions (Trivedi et al. 2006). SBP and HRP immobilized on GA-activated APG beads retained about 74% and 78% of their activities, respectively. These immobilized enzyme preparations were used for the removal of phenol from model (Gomez et al. 2006). Immobilization of SBP on its natural support, soybean seed coat, shifted the optimum pH for phenol removal from 4.0 to 6.0. Immobilized enzyme retained its activity over a 4-week period, and reusability assays showed that treated seed coats could be reused for phenol removal. Addition of polyethylene glycol (PEG) was found to increase the stability of phenol degradation. Moreover, adsorption of phenolic polymer on seed coats made their removal easier (Magri et al. 2007). In another study, SBP covalently bound to glass supports with different surface areas was used in a laboratory-scale fluidized bed reactor for phenol removal. The influence of different operational variables on the process was also studied. About 80% phenol was removed by the SBP immobilized on supports with the highest surface area (Gomez et al. 2007).

HRP and SBP were covalently immobilized onto aldehyde glass via amino groups, and these preparations were applied for the treatment of phenols. It was observed that immobilized HRP removed higher concentration of p-CP from aqueous solutions than the native enzyme. Moreover, it has been noticed that at an immobilized enzyme concentration in the reactor of 15 mg L^{-1}, SBP removed 5% more p-CP in a shorter time compared to HRP. The immobilized SBP was noticed less sensitive to inhibition than the free HRP and thus removed higher concentration of p-CP from polluted water (Bodalo et al. 2008). In a further study, similar workers investigated the removal of p-CP from industrial wastewater using a combination of SBP and UV generated by novel excilamps. These workers used free and immobilized SBP and UV to treat p-CP solutions at the concentrations of 50–500 mg L^{-1}. It has revealed that the excilamp has facilitated higher removal efficiencies in all cases with all p-CP removal from 5 to 90 min. About 80% p-CP was removed by both free and immobilized SBP up to the concentrations of 250 mg L^{-1}. At 500 mg L^{-1} the immobilized enzyme showed much higher removal efficiency due to increase in SBP stability by the formation of by-products (Gomez et al. 2009). Gómez et al. (2012) used SBP covalently bound to glass support for continuous removal of two phenolic compounds: phenol and p-CP in stirred tank reactors. The application of two reactors in series, rather than one continuous tank, improved the removal efficiency of these phenols. The distribution of different amounts of enzyme between the two tanks influenced removal efficiency of phenols. The highest removal efficiencies were achieved at the outlet of the second tank for a distribution of 50% of the enzyme in each tank. A continuous tank reactor was used to remove p-CP from aqueous solutions, using immobilized SBP and H_2O_2. The influence of operational variables, enzyme and substrate concentrations and spatial time, on the removal efficiency has been studied (Murcia et al. 2014). SBP was immobilized by GA and periodate method onto series of p(GMA-co-EGDMA) with various surface characteristics and pore size diameters ranging from 44 to 200 nm, and these preparations were employed for the treatment of phenol (Prokopijevic et al. 2014).

In a most recent study, these workers prepared tyramine-modified pectins via periodate oxidation for SBP-induced hydrogel formation and immobilization. SBP immobilized within tyramine-pectin microbeads maintained above 50% of its original pyrogallol oxidation activity after seven repeated uses (Prokopijevic et al. 2017). Chagas and co-workers (2015) investigated covalent immobilization of SBP on CS beads prepared by GA. The free and immobilized SBP preparations were applied for the oxidation of caffeic acid from the synthetic solution and coffee processing wastewater. The CS bead-bound SBP retained 50% caffeic acid oxidation activity after four repeated uses. In a further study, SBP was immobilized onto a magnetic nanosupport, Fe_3O_4-SiO_2 NPs by covalent binding. The enzyme nanocomposite was able to remove 99.67 ± 0.10% of ferulic acid, while the free enzyme could remove only 57.67 ± 0.27% of this acid. The immobilized SBP was easily collected under a magnetic field and reused. On the basis of these results, we concluded that the prepared magnetic NPs can be considered a high-performance nanocatalyst for environmental remediation (Silva et al. 2016).

2.4 Removal of Phenolic Pollutants by Immobilized Turnip Peroxidase

Turnip is one of the most cheapest and easily available vegetable sources all over the globe, and the peroxidases from this source have already demonstrated their potential in remediation of phenol and their derivatives found in polluted water/industrial effluents (Duarte-Vazquez et al. 2003). Silica-bound TP has successfully removed 95% of the phenol, whereas calcium alginate and polyacrylamide gel bead-entrapped enzyme has eliminated only 50% and 60.4% phenol, respectively (Singh and Singh 2002). In a further study, alginate-entrapped TP was able to remove 90% phenol in 3 h. Immobilized TP was quite successful in removing 0.5 mM phenol from synthetic water in a broad range of pH and temperature (Regalado et al. 2004). Calcium alginate spheres entrapped and covalently bound to Affi-Gel 10 TP have been compared for the detoxification of water polluted with phenolic compounds and an industrial effluent from a local paint factory. The oxidative polymerization efficiency of phenolic compounds was evaluated using batch and recycling processes and in the presence and absence of PEG. The presence of PEG enhanced the operational stability of TP. Moreover, the reaction time was reduced to 10 min from 3 h, and remarkably higher phenol removal was observed in the presence of PEG. TP was used to treat industrial effluent containing phenolics for 15 reaction cycles; however, greater than 90% of the phenolic compounds were removed during the first 10 reaction cycles. Modified TP entrapped in calcium alginate beads retained over 65% phenol removal efficiency even after 17 repeated uses (Quintanilla-Guerrero et al. 2008). Azizi et al. (2014) investigated degradation of phenol by soluble and alginate-entrapped TP in a buffer of pH 7 and at 40 °C. The results revealed that the average removal yield under optimal conditions was 93% within 3 h. The highest phenol removal by soluble and entrapped TP was recorded as 80 and 46 mg L^{-1}, respectively.

2.5 Removal of Phenolic Pollutants by Immobilized Other Plant Peroxidases

Peroxidase from chayote (*Sechium edule*) was immobilized onto the divinylbenzene copolymer functionalized by triglycine and activated with 1-1'carbonyldiimidazol, and this peroxidase preparation was applied to convert phenol and 2-methoxyphenol or *m*-chlorophenol into polymers in the presence of H_2O_2. The obtained removal was 75% for native phenol and 100% for 2-methoxyphenol, while these phenols were removed 65% and 80% by simple filtration procedure (Villegas-Rosas et al. 2003). Ashraf and Husain (2010b) treated α-naphthol-contaminated water by diethylaminoethyl cellulose (DEAE cellulose)-immobilized white radish peroxidase (WRP) in a batch process and continuous reactor. An effective removal of α-naphthol at 30 °C, 40 °C, and 50 °C after 5-h treatment was 79%, 87%, and 65% in the stirred batch processes, respectively. The immobilized enzyme retained 58% α-naphthol oxidation activity after six repeated uses. Soluble, encapsulated, and cross-linked forms of peroxidases from *Sapindus mukorossi* leaves were used for the removal of 1.0 mM phenolics in a stirred batch reactor. The maximal removal of *o*-CP was found in the buffers of pH 4–7 and at 30–60 °C in the presence of 1.2 mM H_2O_2 by soluble enzymes, but encapsulated and cross-linked enzymes worked optimally at pH 5 and 50 °C in the presence of 0.8 mM H_2O_2. However, encapsulated and cross-linked enzymes showed a lower efficiency compared to soluble peroxidase, but the immobilized enzyme preparation was successfully reused several times without losing much of its phenol-removing efficiency (Singh et al. 2012). Jamal and Singh (2014) used DEAE cellulose adsorbed pointed gourd peroxidase (PGP) for oxidation of phenol and α-naphthol. Immobilized PGP was more efficient in removing phenol, and α-naphthol was eliminated from synthetic wastewater more efficiently as compared to its soluble counterpart. The reactors containing immobilized peroxidase were continuously operated well for over 1 month for effective removal of phenol and α-naphthol by 54% and 61%, respectively. It revealed that such immobilized PGP systems in reactor have a great future. Basha and Prasada Rao (2017) evaluated removal of phenol and *p*-CP by employing soluble and immobilized green gram root peroxidases (GGP). They found higher removal of phenol by immobilized GGP than *p*-CP. Immobilized peroxidase appeared as a champion tool to clean phenolics from wastewater.

2.6 Removal of Phenolic Pollutants by Immobilized Microbial Peroxidases

The peroxidases from a number of microbial sources have also shown their significance in targeting various kinds of phenolic compounds from wastewater (Bansal and Kanwar 2013). Some of them have successfully been immobilized and employed for the cleaning of hazardous compounds present in wastewater. Grabski et al. (2000) covalently immobilized *Lentinula edodes* MnP via its -COOH groups using an azlactone-functional copolymer derivatized using ethylenediamine and 2-ethoxy-1-ethoxycarbonyl-1,2-dihydroquinoline as cross-linking agents. The tethered enzyme, thus as prepared, was used in a two-stage immobilized MnP bioreactor for catalytic generation of chelated Mn^{III} and subsequent oxidation of chlorophenols. Immobilized MnP present in the enzyme reactor 1 produced Mn^{III}-chelate, which was sent to other chemical reaction vessel reactor 2 containing the pollutant. Reactor 1-generated Mn^{III}-chelates oxidized 2,4-DCP and 2,4,6-TCP in the second reactor, which demonstrated a two-stage enzyme and chemical system. Polyacrylamide matrix immobilized CIP has appeared highly effective in oxidation of 2,6-dichlorophenol (Pezzotti et al. 2004). Patel et al. (2005) prepared a nano-assembly of LiP and MnP from *Phanerochaete chrysosporium* on the flat surfaces and on colloidal particles. LiP and MnP were fabricated with polyelectrolytes—PEI, poly(dimethyldiallylammonium chloride), and poly(allylamine)—employing a layer-by-layer self-assembly technique. Nano-assembled LiP and MnP both enzyme preparations have effectively oxidized veratryl alcohol (VA) to its aldehyde for an extended period of time. Ferapontova et al. (2006) investigated potential of LiP-modified graphite electrodes for electroenzymatic oxidation of phenols, catechols, VA, and some other high-redox-potential lignin model compounds. The bioelectrocatalytic reduction of H_2O_2 mediated by VA and effects of VA as redox mediator on the efficiency of bioelectrocatalytic oxidation of other co-substrates has also been examined. The oxidation of phenol, catechol derivatives, and ABTS by LiP was independent of VA, while the efficiency of LiP bioelectrocatalysis with other lignin model compounds was remarkably enhanced after addition of VA. Table 15.2 depicts immobilized BGP, SBP, TP, and other plant and microbial peroxidases on/in various supports, their mode of binding, and type of phenol treated or removed.

3 Conclusion and Future Outlook

Discharge of effluents containing phenolics and other pollutants is one of the serious ecological/environmental problems these days. There is public demand to receive toxic and hazardous compound-free water; therefore, it has made a top priority to treat industrial effluents prior to their final disposal into environment. Presently available classical and biological methods for the remediation of such contaminants

Table 15.2 Depicts immobilized BGP, SBP, TP, and other plant and microbial peroxidases on/in various supports, their mode of binding, and type of phenol treated or removed

Name of enzyme	Name of support	Treated phenolic compound(s)	Reference(s)
BGP	Con A-Sephadex	Various phenols and their mixtures	Akhtar and Husain (2006)
BGP	Con A-calcium alginate-cellulose beads	p-BP	Asharf and Husain (2011)
BGP	Fly ash	BPA	Karim and Husain (2010)
BGP	TiO$_2$NPs	Phenol	Ahmad et al. (2013)
SBP	Fibrous aromatic polyamide, silica sol-gel/alginate, GA-activated APG beads, natural supports	Phenol	Gache et al. (2003), Trivedi et al. (2006), Gomez et al. (2006), and Mugri et al. (2007)
SBP	Glass support, aldehyde glass	Phenol/o-CP/p-CP	Gomez et al. (2007) and Bodalo et al. (2008)
SBP	CS beads	Phenolic compounds from coffee processing wastewater	Chagas et al. (2015)
SBP	Fe$_3$O$_4$-SiO$_2$ NPs	Ferulic acid	Silva et al. (2016)
		Phenols from refinery wastewater	
TP	Silica/calcium alginate/polyacrylamide,	Phenol and phenolic pollutants from industrial effluents	Singh and Singh (2002), Regalado et al. (2004), Quintanilla-Guerrero et al. (2008), and Azizi et al. (2014)
TP	Affigel		
WRP/PGP	DEAE cellulose	Phenol and α-naphthol	Jamal and Singh (2014)
Sechium edule peroxidase	Polystyrene divinylbenzene-copolymer-triglycine-1,1′ carbonyldiimidazole	Phenol and 2-methoxyphenol	Villegas-Rosas et al. (2003)
Sapindus mukorossi peroxidase	Calcium alginate	Phenolic compounds	Singh et al. (2012)
Lentinula edodes MnP	Azalactone functional copolymer	2,4-DCP, 2,4,6-trichlorophenol	Grabski et al. (2000)
CIP	Polyacrylamide gel	2,6-DCP	Pezzoti et al. (2004)
P. chrysosporium LiP/MnP	Colloidal particles	Veratryl alcohol	Patel et al. (2005)

have their own drawbacks. Therefore, it requires some alternative green and pollution free technology. The treatment of aromatic pollutants by using oxidoreductive enzymes has several advantages over the existing technology. This approach is cost-effective, green and environmentally friendly, and highly specific and does not produce additional wastes. Enzymatic remediation of these compounds has explained as the transformation of phenolic compounds into insoluble products. In the last few decades, researchers have found a lot of interest in peroxidases due to their ability to catalyze phenolic compounds into oxidized products. These oxidized products get easily converted into insoluble complexes, which can be simply removed by centrifugation/decantation or filtration. Peroxidases from various cheap sources have successfully been immobilized and have been employed for the remediation of several kinds of phenolic pollutants present in industrial effluents/wastewater. Immobilized peroxidases have demonstrated their remarkable potential in different types of batch processes and continuous reactors in the remediation such pollutants.

The one of most serious problems in the commercialization of this technology is the shortage of enough amounts of enzymes. A remarkable progress has been made since the last decade to resolve this problem, and it is expected that peroxidase-based technology will supersede the other available effective procedures. There is a need to obtain cheap, active and most stable biocatalyst by means of recombinant DNA technology. Enzymes obtained from cheap sources and immobilized on inexpensive and stable supports have great future for removal of toxic phenolic pollutants from industrial effluents. Moreover, the development of an effective and suitable reactor filled with immobilized peroxidase will be the most cost-effective way to treat huge industrial effluent containing such types of health-threatening compounds.

Acknowledgment The author is highly thankful to the Department of Biochemistry, Faculty of Life Sciences, AMU, Aligarh, India, for all kinds of help during writing of the chapter.

References

Ahmad R, Mishra A, Sardar M (2013) Peroxidase-TiO$_2$ nanobioconjugates for the removal of phenols and dyes from aqueous solutions. Adv Sci Eng Med 5(10):1020–1025

Ahuja SK, Ferreira GM, Moreira AR (2004) Utilization of enzymes for environmental applications. Crit Rev Biotechnol 24:125–154

Ai J, Zhang W, Liao G, Xia H, Wang D (2016) Immobilization of horseradish peroxidase enzymes on hydrous-titanium and application for phenol removal. RSC Adv 6:38117–38123

Akhtar S, Husain Q (2006) Potential applications of immobilized bitter gourd (*Momordica charantia*) peroxidase in the removal of phenols from polluted water. Chemosphere 65:1228–1235

Alemzadeh I, Nejati S (2009a) Removal of phenols with encapsulated horseradish peroxidase in calcium alginate. Iran J Chem Chem Eng 28(2):43

Alemzadeh I, Nejati S (2009b) Phenols removal by immobilized horseradish peroxidase. J Hazard Mat 166:1082–1086

Alemzadeh I, Nejati S, Vossoughi M (2009) Removal of phenols from wastewater with encapsulated horseradish peroxidase in calcium alginate. Eng Lett 17(4):13

Annadurai G, Babu SR, Mahesh KPO, Murugesan T (2000) Adsorption and bio-degradation of phenol by chitosan-immobilized *Pseudomonas putida* (NICM 2174). Bioprocess Eng 22:493–501

Ashraf H, Husain Q (2009) Removal of α-naphthol and other phenolic compounds from polluted water by white radish (*Raphanus sativus*) peroxidase in the presence of an additive, polyethylene glycol. Biotechnol Bioprocess Eng 14:536–542

Ashraf H, Husain Q (2010a) Studies on bitter gourd peroxidase catalyzed removal of *p*-bromophenol from wastewater. Desalination 262:267–272

Ashraf H, Husain Q (2010b) Use of DEAE cellulose adsorbed and crosslinked white radish (*Raphanus sativus*) peroxidase for the removal of α-naphthol in batch and continuous process. Int Biodeter Biodegrad 64:27–31

Ashraf H, Husain Q (2011) Application of immobilized peroxidase for the removal of *p*-bromophenol from polluted water in batch and continuous processes. J Water Reuse Desalin 1(1):52–60

Azizi A, Abouseoud M, Ahmedi A (2014) Phenol removal by soluble and alginate entrapped turnip peroxidise. J Biochem Soc 5(4):795–800

Bansal N, Kanwar SS (2013) Peroxidases in environment protection. Sci World J 2013:714639

Barakat N, Makris DP, Kefalas P, Psillakis E (2010) Removal of olive mill waste water phenolics using a crude peroxidase extract from onion by-products. Environ Chem Lett 8:271–275

Basha SA, Prasada Rao UJ (2017) Purification and characterization of peroxidase from sprouted green gram (*Vigna radiata*) roots and removal of phenol and p-chlorophenol by immobilized peroxidase. J Sci Food Agric 97(10):3249–3260. https://doi.org/10.1002/jsfa.8173

Basha KM, Rajendran A, Thangavelu V (2010) Recent advances in the biodegradation of phenol: a review. Asian J Exp Biol Sci 1:219–234

Bayramoglu G, Arica MY (2008) Enzymatic removal of phenol and *p*-chlorophenol in enzyme reactor: horseradish peroxidase immobilized on magnetic beads. J Hazard Mat 156:148–155

Belcarz A, Ginalska G, Kowalewska B, Kulesza P (2008) Spring cabbage peroxidases-potential tool in biocatalysis and bioelectrocatalysis. Phytochemistry 69:627–636

Beristain-Cardoso R, Texier A-C, Razo-Flores E, Mendez-Pampin R, Gomez J (2009) Biotransformation of aromatic compounds from wastewaters containing N and/or S, by nitrification/denitrification: a review. Rev Environ Sci Biotechnol 8:325–342

Bhandari A, Xu F, Koch DE, Hunter RP (2009) Peroxidase-mediated polymerization of 1-naphthol: impact of solution pH and ionic strength. J Environ Qual 38:2034–2040

Bharagava RN, Chandra R (2010) Biodegradation of the major color containing compounds in distillery wastewater by an aerobic bacterial culture and characterization of their metabolites. Biodegradation J 21:703–711

Bharagava RN, Saxena G, Mulla SI, Patel DK (2017) Characterization and identification of recalcitrant organic pollutants (ROPs) in tannery wastewater and its phytotoxicity evaluation for environmental safety. Arch Environ Contam Toxicol. doi.org/10.1007/s00244-017-0490-x

Bodalo A, Gomez JL, Gomez E, Bastida J, Maximo MF (2006) Comparison of commercial peroxidases for removing phenol from water solutions. Chemsphere 63:626–632

Bodalo A, Bastida J, Maximo MF, Montiel MC, Gomez M, Murcia MD (2008) A comparative study of free and immobilized soybean and horseradish peroxidases for 4-chlorophenol removal: protective effects of immobilization. Bioprocess Biosyst Eng 31:587–593

Cao Q, Zhou W (2006) Immobilization of horseradish peroxidase on a biocompatible titania layer-modified gold electrode for the detection of hydrogen peroxide. Anal Lett 39:2725–2735

Caramori SS, Fernandes KF, Carvalho Jr LBD (2012) Immobilized horseradish peroxidase on discs of polyvinyl alcohol-glutaraldehyde coated with polyaniline. The Scien World J 2012:129706

Chagas PM, Torres JA, Silva MC, Corrêa AD (2015) Immobilized soybean hull peroxidase for the oxidation of phenolic compounds in coffee processing wastewater. Int J Biol Macromol 81:568–575

Chandra R, Chowdhary P (2015) Properties of bacterial laccases and their application in bioremediation of industrial wastes. Environ Sci: Processes Impacts 17:326–342

Chandra R, Bharagava RN, Kapley A, Purohit JH (2011) Bacterial diversity, organic pollutants and their metabolites in two aeration lagoons of common effluent treatment plant during the degradation and detoxification of tannery wastewater. Bioresour Technol 102:2333–2341

Chang Q, Tang H (2014) Immobilization of horseradish peroxidase on NH_2-modified magnetic Fe_3O_4/SiO_2 particles and its application in removal of 2,4-dichlorophenol. Molecules 19(10):15768–15782

Chang Q, Jiang G, Tang H, Li N, Huang J, Wu L (2015) Enzymatic removal of chlorophenols using horseradish peroxidase immobilized on superparamagnetic Fe_3O_4/graphene oxide nanocomposite. Chin J Catal 36(7):961–968

Chang Q, Huang J, Ding Y, Tang H (2016) Catalytic oxidation of phenol and 2,4-dichlorophenol by using horseradish peroxidase immobilized on graphene oxide/Fe_3O_4. Molecules 21(8):1044

Cheng J, Yu SM, Zuo P (2006) Horseradish peroxidase immobilized on aluminum-pillared interlayered clay for the catalytic oxidation of phenolic wastewater. Water Res 40:283–290

Cho S-H, Yeon K-H, Kim G-Y, Shim J-M, Moon S-H (2005) Electrochemical degradation of phenol by using reticulated vitreous carbon immobilized horseradish peroxidase. Kor Soc Environ Eng 27:1263–1263

Cho S-H, Shim J, Yun S-H, Moon S-H (2008) Enzyme-catalyzed conversion of phenol by using immobilized horseradish peroxidase (HRP) in a membraneless electrochemical reactor. Appl Catal A: General 337(1):66–72

Chowdhary P, Saxena G, Bharagava RN (2016) Role of laccase enzyme in bioremediation of industrial wastes and its biotechnological application. In: Bharagava RN, Saxena G (eds) Bioremediation of industrial pollutants1st edn. Write & Print Publications, New Delhi. ISBN: 978-93-84649-60-9

Chowdhary P, Raj A, Bharagava RN (2018) Environmental pollution and health hazards from distillery wastewater and treatment approaches to combat the environmental threats: a review. Chemosphere 194:229–246

Cohen S, Belinky PA, Hadar Y, Dosoretz CG (2009) Characterization of catechol derivative removal by lignin peroxidase in aqueous mixture. Bioresour Technol 100:2247–2253

Czaplicka M (2004) Sources and transformation of chlorophenols in the natural environment. Sci Total Environ 322:21–39

Dahili LA, Kelemen-Horváth I, Feczkó T (2015) 2,4-Dichlorophenol removal by purified horseradish peroxidase enzyme and crude extract from horseradish immobilized to nano spray dried ethyl cellulose particles. Process Biochem 50(11):1835–1842

Dalal S, Gupta MN (2007) Treatment of phenolic wastewater by horseradish peroxidase immobilized by bioaffinity layering. Chemosphere 67:741–747

Das N, Chandran P (2011) Microbial segradation of petroleum hydrocarbon contaminants: an overview. Biotechnol Res Int 2011:941810

Datta S, Christena LR, Rajaram YRS (2013) Enzyme immobilization: an overview on techniques and support materials. 3 Biotech 3(1):1–9

Diao M, Ouedraogo N, Baba-Moussa Lamine Savadogo PW, N'Guesssan AG, Bassole IHN, Dicko MH (2010) Biodepollution of wastewater containing phenolic compounds from leather industry by plant peroxidases. Biodegradation 22(2):389–396

Diez MC (2010) Biological aspects involved in the degradation of organic pollutants. J Soil Sci Plant Nutr 10(3):244–267

Duan X, Corgié SC, Aneshansley DJ, Wang P, Walker LP, Giannelis EP (2014) Hierarchical hybrid peroxidase catalysts for remediation of phenol wastewater. ChemPhysChem 15(5):974–980

Duarte-Vazquez MA, Ortega-Tovar MA, Garcia-Almendarez B, Regalado C (2003) Removal of aqueous phenolic compounds from a model system by oxidative polymerization with turnip (*Brassica napus* L var purple top white globe) peroxidase. J Chem Technol Biotechnol 78:42–47

Duran N, Esposito E (2000) Potential applications of oxidative enzymes and phenoloxidase-like compounds in wastewater and soil treatment: a review. Appl Catal B Environ 28:83–99

Farhadian M, Duchez D, Vachelard C, Larroche C (2008) Monoaromatics removal from polluted water through bioreactors-a review. Water Res 42:1325–1341

Ferapontova EE, Castillo J, Gorton L (2006) Bioelectrocatalytic properties of lignin peroxidase from *Phanerochaete chrysosporium* in reactions with phenols, catechols and lignin-model compounds. Biochim Biophys Acta 1760(9):1343–1354

Gache R, Firdaus Q, Sagar AD (2003) Soybean (Glycine max L.) seed coat peroxidase immobilized on fibrous aromatic polyamide: a strategy for decreasing phenols from industrial wastewater. J Sci Indus Res 62(11):1090–1093

Ghasempur S, Torabi SF, Siadat SOR, Heravi MJ, Ghaemi N, Khajeh K (2007) Optimization of peroxidase-catalyzed oxidative coupling process for phenol removal from wastewater using response surface methodology. Environ Sci Technol 41:7073–7079

Gianfreda L (2008) Enzymes of significance to the restoration of polluted systems: traditional and advanced approaches. J Soil Sci Plant Nutr 8:12–22

Gianfreda L, Rao MA (2004) Potential of extra cellular enzymes in remediation: a review. Enzym Microb Technol 35:339–354

Gianfreda L, Iamarino G, Scelza R, Rao MA (2006) Oxidative catalysts for the transformation of phenolic pollutants: a brief review. Biocatal Biotransformation 24:177–187

Gomez JL, Bodalo A, Gomez E, Bastida J, Hidalgo AM, Gomez M (2006) Immobilization of peroxidases on glass beads: an improved alternative for phenol removal. Enzym Microb Technol 39:1016–1022

Gomez JL, Bodalo A, Gomez E, Hidalgo AM, Gomez M, Murcia MD (2007) Experimental behaviour and design model of a fluidized bed reactor with immobilized peroxidase for phenol removal. Chem Eng J 127:47–57

Gomez M, Matafonovab G, Gomeza JL, Batoevb V, Christofic N (2009) Comparison of alternative treatments for 4-chlorophenol removal from aqueous solutions: use of free and immobilized soybean peroxidase and KrCl excilamp. J Hazard Mat 169:46–51

Gómez E, Máximo MF, Montiel MC, Gómez M, Murcia MD, Ortega S (2012) Continuous tank reactors in series: an improved alternative in the removal of phenolic compounds with immobilized peroxidase. Environ Technol 33(1):103–111

Grabski AC, Grimek HJ, Burgess RR (2000) Immobilization of manganese peroxidase from *Lentinula edodes* and its biocatalytic generation of Mn^{III}-chelate as a chemical oxidant of chlorophenols. Biotechnol Bioeng 60:204–215

Guzik U, Hupert-Kocurek K, Wojcieszyńska D (2014) Immobilization as a strategy for improving enzyme properties-application to oxidoreductases. Molecules 19:8995–9018

Hamid M, Rehman K (2009) Potential applications of peroxidases. Food Chem 115:1177–1186

Hare V, Chowdhary P, Baghel VS (2017) Influence of bacterial strains on Oryza sativa grown under arsenic tainted soil: accumulation and detoxification response. Plant Physiol Biochem 119:93–102

Hollmann F, Arenda IWCE (2012) Enzyme initiated radical polymerizations. Polymer 4:759–793

Husain Q (2006) Potential applications of the oxidoreductive enzymes in the decolorization and detoxification of textile and other synthetic dyes from polluted water: a review. Crit Rev Biotechnol 26:201–221

Husain Q (2010) Peroxidase mediated decolorization and remediation of wastewater containing industrial dyes: a review. Rev Environ Sci Biotechnol 9:117–140

Husain M, Husain Q (2008) Application of redox mediators in the treatment of organic pollutants by using oxidoreductive enzyme: a review. Crit Rev Environ Sci Technol 38:1–42

Husain Q, Jan U (2000) Detoxification of phenols and aromatic amines from polluted wastewater by using phenol oxidases. J Sci Ind Res 59:286–293

Husain Q, Qayyum S (2013) Biological and enzymatic treatment of bisphenol A and other endocrine disrupting compounds: a review. Crit Rev Biotechnol 33(3):260–292

Husain Q, Ulber R (2011) Immobilized peroxidase as a valuable tool in the remediation of aromatic pollutants and xenobiotic compounds: a review. Crit Rev Environ Sci Technol 41(8):770–804

Husain Q, Husain M, Kulshrestha Y (2009) Remediation and treatment of organopollutants mediated by peroxidases: a review. Crit Rev Biotechnol 29:94–119

Ikehata K, Nicell JA (2000) Characterization of tyrosinase for treatment of aqueous phenol. Bioresour Technol 74:191–199

Ikehata K, Buchanan ID, Pickard MA, Smith DW (2005) Purification, characterization and evaluation of extracellular peroxidase from two Coprinus species for aqueous phenol treatment. Bioresour Technol 96:1758–1770

Ispas CR, Ravalli MT, Steere A, Andreescu S (2000) Multifunctional biomagnetic capsules for easy removal of phenol and bisphenol A. Water Res 44(6):1961–1969

Jamal F, Singh S (2014) Application of diethylaminoethyl cellulose immobilized pointed gourd (Trichosanthes dioica) peroxidase in treatment of phenol and α-naphthol. J Bioprocess Biotech 5(1):1000196

Ji X, Su Z, Xu M, Ma G, Zhang S (2016) TiO_2–horseradish peroxidase hybrid catalyst based on hollow nanofibers for simultaneous photochemical-enzymatic degradation of 2,4-dichlorophenol. ACS Sust Chem Eng 4(7):3634–3640

Jiang Y, Tang W, Gao J, Zhou L, He Y (2014) Immobilization of horseradish peroxidase in phospholipid-templated titania and its applications in phenolic compounds and dye removal. Enzym Microb Technol 55:1–6

Karim Z, Husain Q (2009a) Guaiacol-mediated oxidative degradation and polymerization of bisphenol A catalyzed by bitter gourd (*Momordica charantia*) peroxidase. J Mol Catal B Enzym 59:185–189

Karim Z, Husain Q (2009b) Redox-mediated oxidation and removal of aromatic amines from polluted water by partially purified bitter gourd (*Momordica charantia*) peroxidase. Int Biodeter Biodegrad 63:587–593

Karim Z, Husain Q (2010) Application of fly ash adsorbed peroxidase for the removal of bisphenol A in batch process and continuous reactor: assessment of genotoxicity of its product. Food Chem Toxicol 48(12):3385–3390

Khan AA, Husain Q (2007) Decolorization and removal of textile and non-textile dyes from polluted wastewater and dyeing effluent by using potato (*Solanum tuberosum*) soluble and immobilized polyphenol oxidase. Bioresour Technol 98(5):1012–1019

Kim HJ, Kim HS (2010) Removal of phenol using horseradish peroxidase immobilized in the nano-scale porous structural material. Int Conf Biol, Environ Chem 1:105–108

Kim G-Y, Moon S-H (2005) Degradation of pentachlorophenol by an electroenzymatic method using immobilized peroxidase enzyme. Korean J Chem Eng 22:52–60

Kim EY, Choi YJ, Chae HJ, Chu KH (2006) Removal of aqueous pentachlorophenol by horseradish peroxidase in the presence of surfactants. Biotechnol Bioprocess Eng 11:462–465

Kim HJ, Suma Y, Lee SH, Kim JA, Kim HS (2012) Immobilization of horseradish peroxidase onto clay minerals using soil organic matter for phenol removal. J Mole Catal B: Enzym 83:8–15

Kulkarni SJ, Kaware JP (2013) Review on research for removal of phenol from wastewater. Int J Sci Res Publ 3(4):1–5

Kulshrestha Y, Husain Q (2007) Decolorization and degradation of acid dyes medicated by salt fractionated turnip (*Brassica rapa*) peroxidases. Toxicol Environ Chem 89:255–267

Kumar S, Haq I, Yadav A et al (2016) Immobilization and biochemical properties of purified xylanase from bacillus amyloliquefaciens SK-3 and its application in Kraft pulp biobleaching. J Clin Microbiol Biochem Technol 2(1):26–34

La Rotta HCE, D'Elia E, Bon EPS (2007) Chloroperoxidase mediated oxidation of chlorinated phenols using electrogenerated hydrogen peroxide. Electron J Biotechnol 10:24–36

Lai YC, Lin SC (2005) Application of immobilized horseradish peroxidase for the removal of p-chlorophenol from aqueous solution. Process Biochem 40:1167–1174

Levy I, Ward G, Hadar Y, Shoseyov O, Dosoretz CG (2003) Oxidation of 4-bromophenol by the recombinant fused protein cellulose-binding domain-horseradish peroxidase immobilized on cellulose. Biotechnol Bioeng 82:223–231

Liu Y, Shen Y, Fan L, Liu D, Chen X (2012) Degradation of phenol over immobilized horseradish peroxidase on polystyrene microporous membranes. J Shen Inst Chem Technol 1:37–42

Lu Y-M, Yang Q-Y, Wang L-M, Zhang M-Z, Guo W-Q, Cai Z-N, Wang D-D, Yang W-W, Chen Y (2017) Enhanced activity of immobilized horseradish peroxidase by carbon nanospheres for phenols removal. CLEAN–Soil Air Water 45:1600077

Magri ML, Loustau MDLN, Miranda MV, Cascone O (2007) Immobilization of soybean seed coat peroxidase on its natural support for phenol removal from wastewater. Biocatal Biotransformation 5:98–102

Maki F, Yugo U, Yasushi M, Isao I (2006) Preparation of peroxidase-immobilized polymers and their application to the removal of environment-contaminating compounds. Bull Fibre Text Res Found 15:15–19

Masuda M, Sakurai A, Sakakibara M (2001) Effect of reaction conditions on phenol removal by polymerization and precipitation using *Coprinus cinereus* peroxidase. Enzym Microb Technol 28:295–300

Matto M, Husain Q (2007) Decolorization of direct dyes by salt fractionated turnip proteins enhanced in the presence of hydrogen peroxide and redox mediators. Chemosphere 69:338–345

Matto M, Naqash S, Husain Q (2008) An economical and simple bioaffinity support for the immobilization and stabilization of tomato (*Lycopersicon esculentum*) peroxidase. Acta Chim Slov 55:671–676

Meizler A, Roddick F, Porter N (2011) A novel glass support for the immobilization and UV-activation of horseradish peroxidase for treatment of halogenated phenols. Chem Eng J 172(2–3):792–798

Mishra S, Bharagava RN (2016) Toxic and genotoxic effects of hexavalent chromium in environment and its bioremediation strategies. J Environ Sci Health Part C 34(1):1–32

Moeder M, Martin C, Koeller G (2004) Degradation of hydroxylated compounds using laccase and horseradish peroxidase immobilized on microporous polypropylene hollow fiber membranes. J Membr Sci 245(1-2:183–190

Mohamad NR, Marzuki NH, Buang NA, Huyop F, Wahab RA (2015) An overview of technologies for immobilization of enzymes and surface analysis techniques for immobilized enzymes. Biotechnol Biotechnol Equip 29(2):205–220

Murcia MD, Gómez M, Bastida J, Hidalgo AM, Montiel MC, Ortega S (2014) Application of a diffusion-reaction kinetic model for the removal of 4-chlorophenol in continuous tank reactors. Environ Technol 35(13–16):1866–1873

Nanayakkara S, Zhao Z, Patti AF, He L, Kei Saito K (2014) Immobilized horseradish peroxidase (I-HRP) as biocatalyst for oxidative polymerization of 2,6-dimethylphenol. ACS Sustain Chem Eng 2(8):1947–1950

Nazari K, Esmaeili N, Mahmoudi A, Rahimi H, Moosavi-Movahedi AA (2007) Peroxidative phenol removal from aqueous solutions using activated peroxidase biocatalyst. Enzym Microb Technol 41:226–233

Niu J, Xu J, Dai Y, Xu J, Guo H, Sun K, Liu R (2013) Immobilization of horseradish peroxidase by electrospun fibrous membranes for adsorption and degradation of pentachlorophenol in water. J Hazard Mater 246-247:119–125

Nuhoglu A, Yalcin B (2005) Modelling of phenol removal in a batch reactor. Process Biochem 40:1233–1239

Ozdural AR, Tanyolac D, Boyaci IH, Mutlu M, Webb C (2003) Determination of apparent kinetic parameters for competitive product inhibition in packed-bed immobilized enzyme reactors. Biochem Eng J 14:27–36

Pan C, Ding R, Dong L, Wang J, Hu Y (2015) Horseradish peroxidase-carrying electrospun nonwoven fabrics for the treatment of o-methoxyphenol. J Nanomater 2015:616879

Patel DS, Aithal RK, Krishna G, Lvov YM, Tien M, Kuila D (2005) Nano-assembly of manganese peroxidase and lignin peroxidase from *P. chrysosporium* for biocatalysis in aqueous and non-aqueous media. Coll Surf B: Biointerf 43:13–19

Pearce CI, Lloyd JR, Guthrie JT (2003) The removal of colour from textile wastewater using whole bacterial cells: a review. Dye Pigm 58:179–196

Pezzotti F, Okrasam K, Therisodm M (2004) Oxidation of chlorophenols catalyzed by *Coprinus cinereus* peroxidase with in situ production of hydrogen peroxide. Biotechnol Prog 20:1868–1871

Pradeep NV, Anupama US, Hampannavar US (2012) Polymerization of phenol using free and immobilized horseradish peroxidase. J Environ Earth Sci 2(1):31–37

Pradeep NV, Anupama S, Navya K et al (2015) Biological removal of phenol from wastewaters: a mini review. Appl Water Sci 5:105

Pramparo L, Stuber F, Font J, Fortuny A, Fabregat A, Bengoa C (2010) Immobilization of horseradish peroxidase on Eupergit®C for the enzymatic elimination of phenol. J Hazard Mat 177:990–1000

Prodanović O, Prokopijević M, Spasojević D, Stojanović Z, Radotić K, Knežević-Jugović ZD, Prodanović R (2012) Improved covalent immobilization of horseradish peroxidase on macroporous glycidyl methacrylate-based copolymers. Appl Biochem Biotechnol 168:1288–1301

Prokopijevic M, Prodanovic O, Spasojevic D, Stojanovic Z, Radotic K, Prodanovic R (2014) Soybean hull peroxidase immobilization on macroporous glycidyl methacrylates with different surface characteristics. Bioprocess Biosyst Eng 37(5):799–804

Prokopijevic M, Prodanovic O, Spasojevic D, Kovacevic G, Polovic N, Radotic K, Prodanovic R (2017) Tyramine-modified pectins via periodate oxidation for soybean hull peroxidase induced hydrogel formation and immobilization. Appl Microbiol Biotechnol 101:2281–2290

Quintanilla-Guerrero F, Duarte-Vazquez MA, Garcia-Almendarez BE, Tinoco R, Vazquez-Duhalt R, Regalado C (2008) Polyethylene glycol improves phenol removal by immobilized turnip peroxidases. Bioresour Technol 99:8605–8611

Regalado C, Garcia-Almendarez BE, Duarte-Vazquez MA (2004) Biotechnological applications of peroxidases. Phytochem Rev 3:243–256

Rojas-Melgarejo F, Marin-Iniesta F, Rodriguez-Lopez JN, Garcia-Canovas F, García-Ruiz PA (2006) Cinnamic carbohydrate esters show great versatility as supports for the immobilization of different enzymes. Enzym Microb Technol 38:748–755

Saxena G, Bharagava RN (2017) Organic and inorganic pollutants in industrial wastes, their ecotoxicological effects, health hazards and bioremediation approaches. In: Bharagava RN (ed) Environmental pollutants and their bioremediation approaches. CRC Press, Taylor & Francis Group, Boca Raton. ISBN: 9781138628892

Saxena G, Chandra R, Bharagava RN (2017) Environmental pollution, toxicity profile and treatment approaches for tannery wastewater and its chemical pollutants. Rev Environ Contam Toxicol 240:31–69

Shukla SP, Devi S (2005) Covalent coupling of peroxidase to a copolymer of acrylamide (AAm)-2-hydroxyethyl methaacrylate (HEMA) and its use in phenol oxidation. Process Biochem 40:147–154

Siddique MH, St Pierre CC, Biswas N, Bewtra JK, Taylor KE (1993) Immobilized enzyme catalyzed removal of 4-chlorophenol from aqueous solution. Water Res 27:883–890

Silva MC, Torres JA, Nogueira FGE, Tavares TS, Corrêa AD, Oliveira LCA, Ramalho TC (2016) Immobilization of soybean peroxidase on silica-coated magnetic particles: a magnetically recoverable biocatalyst for pollutant removal. RSC Adv 6:83856–83863

Singh N, Singh J (2002) An enzymatic method for removal of phenol from industrial effluent. Prep Biochem Biotechnol 32:127–133

Singh J, Sinha S, Batra N, Joshi A (2012) Applications of soluble, encapsulated and cross-linked peroxidases from *Sapindus mukorossi* for the removal of phenolic compounds. Environ Technol 33(1–3):349–358

Soomro R, Perçin I, Memon N, Iqbal Bhanger M, Denizli A (2016) Gelatin-loaded p(HEMA-GMA) cryogel for high-capacity immobilization of horseradish peroxidase. Artif Cells Nanomed Biotechnol 44(7):1708–1713

Spasojević D, Prokopijević M, Prodanović O, Prodanovic R (2014) Immobilization of chemically modified horseradish peroxidase within activated alginate beads. Chem Ind/Hem Ind 68(1):117–122

Steevensz A, Al-Ansari MM, Taylor KE, Bewtra JK, Biswas N (2009) Comparison of soybean peroxidase with laccase in the removal of phenol from synthetic and refinery wastewater samples. J Chem Technol Biotechnol 84:761–769

Su X, Zong Y, Richter R, Knoll W (2005) Enzyme immobilization on poly(ethylene-co-acrylic acid) films studied by quartz crystal microbalance with dissipation monitoring. J Coll Interf Sci 287:35–42

Sujata, Bharagava RN (2016) Exposure to crystal violet, its toxic, genotoxic and carcinogenic effects on environment and its degradation and detoxification for environmental safety. Rev Environ Contam Toxicol 237:71–104

Tatsumi K, Wada S, Ichikawa H (1996) Removal of chlorophenols from wastewater by immobilized horseradish peroxidase. Biotechnol Bioeng 51:126–130

Torres LG, Hernandez M, Pica Y, Albiter V, Bandala ER (2010) Degradation of di-, tri-, tetra-, and pentachlorophenol mixtures in an aerobic biofilter. Afr J Biotechnol 9:3396–3403

Trivedi UJ, Bassi AS, Zhu J (2006) Investigation of phenol removal using sol-gel/alginate immobilized soybean seed hull peroxidase. Can J Chem Eng 84:239–247

Van Schie PM, Young LY (2000) Biodegradation of phenol: mechanisms and applications. Biorem J 4:1–18

Vasileva N, Godjevargova T, Ivanova D, Gabrovska K (2009) Application of immobilized horseradish peroxidase onto modified acrylonitrile copolymer membrane in removing of phenol from water. Int J Biol Macromol 44:190–194

Vasudevan PT, Li LO (1996) Peroxidase catalyzed polymerization of phenol. Appl Biochem Biotechnol 60:73–82

Villegas-Rosas MLO, Geissler G, Handal-Silva A, Gonzalez-Vergara E (2003) Immobilization of a peroxidase from chayote [*Sechium edule* (Jacq.) SW] and its potential use in the removal of phenolic compounds from contaminated water. Rev Int Contam Amb 19:73–81

Wagner M, Nicell JA (2003) Impact of the presence of solids on peroxidase catalyzed treatment of aqueous phenol. J Chem Technol Biotechnol 78:694–702

Wang C, Jiang Y, Zhou L, Gao J (2011) Horseradish peroxidase encapsulated on nanosilica for phenol removal. J Chem Ind Eng (China) 62(7):2032–2032

Wang S, Fang H, Wen Y, Cai M, Liu W, He S, Xu X (2015) Applications of HRP-immobilized catalytic beads to the removal of 2,4-dichlorophenol from wastewater. RSC Adv 5:57286–57292

Wang S, Liu W, Zheng J, Xu X (2016a) Immobilization of horseradish peroxidase on modified PAN-based membranes for the removal of phenol from buffer solutions. Can J Chem Eng 94(5):865–871

Wang S, Fang H, Yi X, Xu Z, Xie X, Tang Q, Ou M, Xu X (2016b) Oxidative removal of phenol by HRP-immobilized beads and its environmental toxicology assessment. Ecotoxicol Environ Saf 130:234–239

Ward G, Hadar Y, Dosoretz CG (2003) Lignin peroxidase-catalyzed polymerization and detoxification of toxic halogenated phenols. J Chem Technol Biotechnol 78:1239–1245

Ward G, Parales RE, Dosoretz CG (2004) Biocatalytic synthesis of polycatechols from toxic aromatic compounds. Environ Sci Technol 38:4753–4757

Watanabe C, Kashiwada A, Matsuda K, Yamada K (2011) Soybean peroxidase-catalyzed treatment and removal of BPA and bisphenol derivatives from aqueous solutions. Environ Prog Sustain Energy 30(1):81–91

Xu SN, Chen Y (2016) HRP immobilization on activated carbon/polyvinyl formal composite materials for degradation of phenolic wastewater. Mater Sci For 847:256–264

Xu J, Tang T, Zhang K, Ai S, Du H (2011) Electroenzymatic catalyzed oxidation of bisphenol-A using HRP immobilized on magnetic silk fibroin nanoparticles. Process Biochem 46(5):1160–1165

Xu R, Chi C, Li F, Zhang B (2013) Immobilization of horseradish peroxidase on electrospun microfibrous membranes for biodegradation and adsorption of bisphenol A. Bioresour Technol 149:111–116

Xu R, Si Y, Li F, Zhang B (2015) Enzymatic removal of paracetamol from aqueous phase: horseradish peroxidase immobilized on nanofibrous membranes. Environ Sci Pollut Res Int 22(5):3838–3846

Yadav A, Mishra S, Kaithwas G, Raj A, Bharagava RN (2016) Organic pollutants and pathogenic bacteria in tannery wastewater and their removal strategies. In: Singh JS, Singh DP (eds) Microbes and environmental management. Studium Press (India) Pvt. Ltd., New Delhi, pp 101–127

Yang R, Tan H, Wei F, Wang S (2008) Peroxidase conjugates of cellulose nanocrystals for the removal of phenolic compounds in aqueous solutions. Biotechnology 7:233–241

Zhai R, Zhang B, Wan Y, Li C, Wang J, Liu J (2013) Chitosan-halloysite hybrid-nanotubes: horseradish peroxidase immobilization and applications in phenol removal. Chem Eng J 214:304–309

Zhang J, Ye P, Chen S, Wang W (2007) Removal of pentachlorophenol by immobilized horseradish peroxidase. Int Biodeter Biodegrad 59:7–314

Zhang F, Zheng B, Zhang J, Huang X, Liu H, Guo S, Zhang J (2010) Horseradish peroxidase immobilized on graphene oxide: physical properties and applications in phenolic compound removal. J Phys Chem C 114:8469–8473

Zhang F, Zhang W, LiFang Zhao L, Liu H (2016) Degradation of phenol with horseradish peroxidase immobilized on ZnO nanocrystals under combined irradiation of microwaves and ultrasound. Desalin Water Treat 57(51):24406–24416

Zhao H, Zhang D, Du P, Li H, Liu C, Li Y, Cao H, Crittenden JC, Huang Q (2015) A combination of electro-enzymatic catalysis and electrocoagulation for the removal of endocrine disrupting chemicals from water. J Hazard Mater 297:269–277

Zong YR, Zhang JB, Wang SL, Yang YX, Wang WJ (2007) Removal of pentachlorophenol catalyzed by immobilized horseradish peroxidase. Huan Jing Ke Xue 28:2740–2744

Chapter 16
Nanoparticles: An Emerging Weapon for Mitigation/Removal of Various Environmental Pollutants for Environmental Safety

Gaurav Hitkari, Sandhya Singh, and Gajanan Pandey

Abstract Nanotechnology is a recent field of technology and nanoparticle materials are fundamental units that measure within the range 1–100 nm with several types of morphologies. They have exceptional and unique catalytic properties, which are associated with their size and are changed from their bulk materials. These nanoparticle materials are prepared by the various methods such as chemical, physical and biological methods. The prepared nanoparticles are investigated by numerous characterisation techniques such as X-ray diffraction (XRD), scanning electron microscopy (SEM), transmission electron microscopy (TEM), Brunauer–Emmett–Teller (BET) surface area analysis, absorbance spectroscopy, photoluminescence spectroscopy. XRD revealed the crystalline nature of the nanoparticles. SEM and TEM images provides information on the morphology and particle size distribution, and BET revealed the surface properties of the nanoparticles. Optical properties are investigated by absorbance and photoluminescence spectroscopic techniques. The effective photo-catalysis of organic toxic pollutants and heavy metals from the environment have been a challenging subject for human health. Much research has explored the environmental behaviour of nanostructured materials for the effective removal of hazardous organic pollutants and heavy metals, existing both in the surface and underground wastewater. The goal of this chapter is to indicate the outstanding removal capability and environmental remediation of nanostructured materials for various toxic organic pollutants and heavy metal ions.

Keywords Nanoparticles · Environmental contaminants · Nanoparticles synthesis · Waste management · Nanoremediation

---

G. Hitkari · S. Singh · G. Pandey (✉)
Department of Applied Chemistry, Babasaheb Bhimrao Ambedkar University
(A Central University), Lucknow, Uttar Pradesh, India

© Springer Nature Singapore Pte Ltd. 2019
R. N. Bharagava, P. Chowdhary (eds.), *Emerging and Eco-Friendly Approaches for Waste Management*, https://doi.org/10.1007/978-981-10-8669-4_16

1 Introduction

Nanotechnology is an emerging field of technology that covers a wide range of utilisation at nanoscale (1 nm = 10^{-9} m) dimensions, and a size range from 1 to 100 nm (billionths of a meter). Nanoparticles (NPs) are the most fundamental units of nanotechnology and comprise a group of tens of thousands of atoms measuring about 1–100 nm in vastly ordered crystalline way. Such NPs to the atom by atom; thus, the size and frequently the shape of a particle are composed of the preliminary conditions.

The requirement of nanotechnology to synthesise the desired shape and size of the nanomaterials for their convenient applications. Surfactants have been demonstrated to be the best shape-directing agents in the synthesis of nanomaterials, which is basically related to the surface adsorption of active molecules on different crystal planes of nucleating centers and thus, controlling their overall shape. Various types of surfactants have been adopted for the shape-limiting synthesis of nanomaterials, although ionic surfactants express clear shape-directing effects. Owing to the huge surface area to volume ratio, nanostructures display unique properties. The subbranches of nanotechnology in colloidal science, biology, physics, chemistry and other technical fields are associated with the explanation of phenomena and use of materials on the nanoscale (Mansoori 2002). This outcome was found in these materials and systems that often exhibit an arrative and extensively changing physical, chemical and biological properties by cause of their fine size, structure, and morphology. The distinctive properties of these nano-sized materials have resulted in the use of these nanomaterials in various fields (Fig. 16.1) like biomedicine, pharmaceuticals, cosmetics, surface-based sciences (Masciangioli and Zhang 2003).

In circumferential nanoscale science, engineering, technology, and nanotechnology contains investigation of imaging, calculating, modelling, and employing matter at this dimensional scale. Although in the industrial sectors nanostructures such as zinc oxide (ZnO) semiconductors are used in memory storage, display, optical,

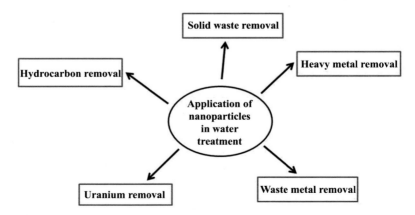

Fig. 16.1 Application of nanoparticles in water treatment in the environment

and photonic technologies; energy along with biotechnology; and in health care produces a considerable number of products containing nanomaterials, there are expanding achievements of nanotechnology to use in the environment as environmental technology to defend the environment from toxic waste, and remove the problems, which are present from the long-term use of hazardous waste materials.

Water is the most important substance for life on the earth and a valuable resource for human development. Finding reliable approaches to uncontaminated and cheap water is considered to be one of the truly fundamental humanitarian goals and remains a considerable worldwide challenge for the twenty-first century. Our present water supply faces many challenges, both old and new. Poisoning of water with lethal heavy metal ions such as Co(II), Cr(III), Cr(VI), (Hg(II), Pb(II), Ni(II), Cu(II), Cd(II), Ag(I), As(V) and As(III)) is attractive a brutal environmental and public health problem (Kumari et al. 2016; Yadav et al. 2017).

In nanotechnology, the profitable production of NPs has the potential to progress the environment, through the direct utilisation of those materials to distinguish, avoid and eliminate contamination, likewise indirectly by applying nanotechnology to intention cleaner manufacturing processes and generate environmentally accountable products. For example, ZnO, iron oxide (Fe_2O_3), cobalt oxide (Co_2O_3) and additional different NPs can eliminate contaminants from topsoil and ground water, and nanostructured sensors hold assurance for improved detection and tracking of contaminants.

These nanomaterials that deliver hygienic water from polluted water sources in both large scale and transportable applications and one that detects and cleans up eco-friendly contaminants (waste and toxic material), that is, remediation (Schrick et al. 2004; Liu et al. 2014). The word "remediate" means to resolve the problem and "bioremediation" means the procedure by which innumerable biological agents, such as bacteria, fungi, protists, or their enzymes are used to degrade the ecological impurities into less toxic forms (Van Dillewijn et al. 2007). The significant advantage of bioremediation over conventional treatments is that it is economical, highly proficient, reduces chemical and biological sludge, discriminates specific metals, there are no accompanying nutrient requirements, regeneration of biosorption and the probability of metal recovery (Kratochvil and Volesky 1998).

Research is required using nanoscale science and technology to identify opportunities and applications to environmental complications, and to estimate the potential ecological impacts of nanotechnology. Among the numerous applications of nanotechnology that have environmental implications, remediation of contaminated groundwater by using NPs containing zero-valent iron is one of the greatest conspicuous examples of a speedily developing technology with considerable potential benefits.

Oil spillage occurring during consideration, transportation, loading and filtering of rudimentary oil is a frequent phenomenon. Accidental oil tankers of crude oil on the surface of an enormous area of ocean currents for spreading and flora and fauna of the coastal shoreline and marine ecosystem have a huge impact. The oil factories and offshore piercing operations also release large amounts of crude oil/sludge, resulted in large-scale contamination of terrestrial and water resources. Unfinished oil/petroleum gunk is a concoction of complexes such as aliphatic, aromatic, asphaltene and resin of hydrocarbons, which are well recognised for their injuriousness, and some of them are reported to be mutagenic and hazardous in nature. Contagion of the soil with lethal polyaromatic hydrocarbon (PAHs) results in extensive destruction of the biodiversity of flora and fauna, leading to reduction of the yield of products on land and contamination of ground water. In view of these problems, valuable remediation of petroleum hydrocarbons to develop a cost-effective and environmentally friendly technology is a very important need.

The dimension of the nanomaterials is involved in particles with at least one dimension measuring between 1.0 and 100 nm. Some definite characteristics such as high surface-to-volume ratio, enhanced magnetic and special catalytic properties etc. (Gupta et al. 2011) make these nanomaterials/ NPs far more advantageous than their majority phase counterparts in the field of remediation technology. The high surface area and surface reactivity of nanomaterials compared with their corresponding bulk material facilitate their remediation of the contamination at a fast rate with reduced amounts of dangerous by-products (Bhattacharya et al. 2013). A miscellaneous arrangement of nanomaterials such as carbon nanotubes (CNTs), nanoscale zeolites, dendrimer enzymes, biometallic particles and metal oxides are now being used for decontamination of polluted sites (Mehndiratta et al. 2013). The application of multi-walled carbon nanotubes has been working to eliminate organic contaminates such as PAHs and polychlorinated biphenyls (PCBs) (Shao et al. 2010). Zhang (2003) reported a comprehensive list of pollutants that can be potentially remediated by using iron (Fe) NPs. By the application of these NPs in enzyme-mediated remediation, technology is steadily gaining ground because NPs provide a biocompatible and inert microenvironment, that interferes least with the inherent properties of the enzymes and helps in recollecting their biological accomplishments (Ansari and Husain 2012). In addition, the magnetic properties of the NPs enable an easy separation of immobilised enzymes or proteins from the reaction mixtures by simply applying a magnetic field. Thus, there is no need for centrifugation or purification which, otherwise, continues the procedural inconsequence and operational complications (Khoshnevisan et al. 2011). In the remediation procedure, there are some selective nanomaterials used for the remediation process, which is a crucial step, as they may be toxic for the microorganisms (Rizwan et al. 2014).

2 Synthesis of Nanomaterials

2.1 Chemical Method for the Synthesis of Nanoparticles

Various chemical methods are applied for the fabrication of nanostructure materials such as controlled precipitation, sol–gel synthesis, hydrothermal reactions, sonochemical reactions, reverse micelles and micro-emulsion technology, hydrolysis and thermolysis of precursors. For the wastewater treatment such as removal of toxic metal applications, a suitable surface alteration of the nanostructures is a critical characteristic concerning both discrimination and aqueous solidity of these materials. To this end, ZnO has been an interesting metal oxide for study by several researchers. Organic and inorganic functionalised Fe_3O_4 NPs have been established and modifications of the synthesis methods mentioned above have been proposed.

2.1.1 Mechanochemical Process

The mechanochemical process is an inexpensive and humble process of finding large amounts of NPs. In this method, high-energy dry milling is used, because it initiates a reaction through ball-powder influences in a ball mill, at a low temperature, and a "thinner" is added to the system in the form of a solid (such as NaCl), which acts as a medium for the reaction and separates the NPs being formed. The most important problem in this process or method is a uniform crushing of the powder and reduction of the particles to the required size, which decreases with increasing time and energy of grinding. Inappropriately, a longer grinding time leads to a greater quantity of impurities. The benefits of this method are the low costs of fabrication, small particle sizes and the restricted tendency of particles to agglomerate, in addition to the high uniformity of the crystalline structure and morphology.

2.1.2 Controlled Precipitation

The controlled precipitation is a commonly used method for the production of ZnO nanostructures as it makes it possible to obtain a product with repeatable properties. This method involves the fast and spontaneous reduction of a zinc precursor solution by introducing a reducing agent, to limit the progression of particles with specified dimensions, followed by the precipitation of a precursor of ZnO from the solution. The next step in this method, the precursor of this salt undergoes thermal treatment, followed by grinding to eliminate impurities. It is very challenging to break down the agglomerates that form; thus, the powder is calcinated at a high temperature to have a high level of agglomeration of particles. In the precipitation process, the precipitation is controlled by parameters such as pH, temperature and time of precipitation.

The aqueous solution of zinc acetate and zinc chloride is applied for the preparation of ZnO (Kołodziejczak-Radzimska et al. 2010). In this technique, the controlled parameters subsumed the concentration of the reagents, the flow of the addition of substrates, and the reaction temperature. A ZnO NP was formed with a uniform particle size distribution and a huge surface area.

Another method of effective precipitation of ZnO was implemented by Wang et al. (2010). Nanometric ZnO NPs were obtained by precipitation from aqueous solutions of NH₄HCO₃ and ZnSO₄·7H₂O by way of the following reactions (16.1) and (16.2):

$$4ZnSO_{4(aq)} + 10NH_4HCO_{3(aq)} \rightarrow Zn_5(CO_3)_2(OH)_{6(s)} 5(NH_4)_2 SO_{4(aq)} + 8CO_{2(g)} + 2H_2O \quad (16.1)$$

$$Zn_5(CO_3)_2(OH)_{6(s)} \rightarrow 5ZnO_{(s)} + 2CO_{2(g)} + 3H_2O_{(g)} \quad (16.2)$$

This study was implemented using a membrane device containing two plates of polytetrafluoroethylene, with stainless steel as a spreading medium. The size of ZnO NPs was 9–20 nm with a narrow range of particle size obtained by this method. On the basis of X-ray diffraction (XRD) investigation, both the precursor and the ZnO itself were revealed to have a wurtzite structure. The size and shape of material was influenced by varying the temperature, calcination time, flow rate and concentration of the supply phase.

The iron oxide or magnetite NPs are synthesised by a stoichiometric combination of the iron precursor of ferrous and ferric salt of iron in an aqueous medium. The residue of Fe_3O_4 is collected at a pH value between 8 and 14. The desired shape and size of the NPs can be made precise by modifying the nature of the salts, ionic strength, temperature and pH (Schwarzer and Peukert 2004; Jolivet et al. 2004). The NPs achieved by this method are rang in size from 5 to 100 nm. The size of the NPs is controlled by the addition of carboxylate ions (e.g. citric, gluconic, or oleic acid) or polymer surface complexing agents (e.g. dextran, carboxy dextran, starch, or polyvinyl alcohol) and surfactant also used for the size control of particles during the formation of magnetite (Laurent et al. 2008).

2.1.3 Sol–Gel Method

The preparation of ZnO nanopowder using the chemical sol–gel method is an interesting subject with regard to the ease, low cost, consistency, repeatability and relatively mild conditions of synthesis, which facilitate the surface variation of ZnO with particular organic compounds. These variations in properties extend its range of applications. The ZnO nanopowder obtained by this method give very advantageous optical properties of NPs and has become a simple and modern topic of research, as reflected in several scientific publications (Mahato et al. 2009). There are two methods of synthesis using the sol–gel method: first, the formation of films from a sol, and second, powder from a sol is transformed into a gel.

The sol–gel method was also used to obtain nanocrystalline ZnO by Ristić et al. (2005). The solution of tetramethylammonium hydroxide (TMAH) was added in to a solution of zinc 2-ethylhexanoate (ZEH) in propan-2-ol. The resulting suspension solution stood for 30 min (otherwise for 24 h) and then washed with water and fine ethanol. TMAH is a powerful organic base, which compared with an inorganic base (e.g. sodium hydroxide [NaOH]) is characterised by a pH of ~14. At such a high pH value, metal oxides are not affected with the cation present in the base, which may have an influence on the ohmic conductance of the oxide material. The consequence of the extent of ZEH used and the growing time of the colloidal solution were determined.

2.1.4 Hydrothermal Reactions

The hydrothermal synthetic technique does not concern the use of organic solvents or additional handling of the product such as crushing and calcination, which makes it a very simple and ecologically friendly technique. The synthesis of the NP by this method takes place in an autoclave, where the combination of substrates is heated gradually to a temperature of 100–300 °C and left for numerous days. As an outcome of heating followed by refrigeration, crystal nuclei are formed, which then propagate. So many advantages of this process, together with the possibility of carrying out the synthesis at very low temperatures, the different shapes and measurements of the resulting crystals depend upon the composition of the starting reaction mixture and the reaction temperature and pressure, the high degree of crystallinity of the product, and the high level of cleanliness of the material gained (Djurisic et al. 2012; Innes et al. 2002).

Chen et al. (1999) reported the synthesis of ZnO NPs by the hydrothermal method by using the reagents $ZnCl_2$ and NaOH in a ratio of 1:2, in an aqueous solution. The process takes place by way of the reaction (16.3):

$$ZnCl_2 + 2NaOH \rightarrow Zn(OH)_2 \downarrow + 2Na^+ + 2Cl^- \qquad (16.3)$$

The white precipitate of $Zn(OH)_2$ was obtained by purification and washing, and then the pH was adjusted to a value of 5–8 using HCl. In the autoclave hydrothermal method, heating takes place at a programmed temperature for a set time, followed by cooling. The end product of the reaction process is ZnO with the following reaction (16.4):

$$Zn(OH)_2 \rightarrow ZnO + H_2O \qquad (16.4)$$

Fe_3O_4 NPs synthesised by the hydrothermal method have been reported in the literature over the last decade (Daou et al. 2006; Wang et al. 2004; Mizutani et al. 2008). Synthesis of these materials using this method is performed in aqueous media in reactors or autoclaves where the pressure can be higher than 2,000 psi and the temperature can be greater than 200 °C. The synthesis of magnetite NPs uses the hydrothermal method in two main ways: hydrolysis and fine oxidation or neutralisation of mixed metal hydroxides. These two reactions are very analogous, except that only ferrous salts are used in the first method. This process, on the basis of reaction conditions, such as solvent, temperature and time, frequently has significant effects on the products or compounds. The particle size of the material is controlled primarily through the rate procedures of nucleation and grain growth in the hydrothermal method.

2.1.5 Solvothermal Method

Zhang et al. (2010) reported the ZnO NPs synthesised using the solvothermal method in the symmetry of globules and non-solid globules with the existence of an ionic liquid such as imidazolium tetrafluoroborate. The authors recommended that the solvothermal process may involve subsequent chemical reactions such as (16.5) and (16.6):

$$2(1-x)OH^- + 2xF^- + Zn^{2+} \rightarrow Zn(OH)_{2-2x} + F_{2x} \quad (16.5)$$

$$Zn(OH)_{2-2x} F_{2x} \rightarrow ZnO_{1-x-y} F_{2x}(OH)_{2y} + (1-x-y)H_2O \quad (16.6)$$

The hollow sphere shape of ZnO NPs is obtained with the diameters of 2–5 μm and channels approximately 10 nm in diameter existing and similarly achieved in the sample or material. The breadth of the wall of such a sphere was approximately 1 μm. The arrangement proposed by Zhang et al. may conglomerate the properties of both a solvothermal hybrid and an isothermal system. It can be estimated that a solvothermal hybrid and an isothermal system may be successfully handed down to prepare novel materials with remarkable properties and morphologies.

2.1.6 Reverse Micelles and Micro-emulsion Technology

The classic definition of an emulsion as a continuous liquid phase in which a second, discontinuous, immiscible liquefied stage is dispersed is far from complete. On the basis of the nature of the external phase, one very appropriate method for categorising emulsions is first to divide them into two broad groups. The two groups are generally called oil-in-water (O/W) and water-in-oil (W/O) emulsions. According to this definition, almost any highly polar, hydrophilic liquid falls in to the "water" category, whereas hydrophobic, non-polar liquids are considered an "oil" (Kołodziejczak-Radzimska et al. 2012; Li et al. 2009).

Vorobyova et al. (2004) used emulsion structures in their research work. ZnO NPs were precipitated in an inter-phase reaction of zinc oleate (liquefied in decane solvent) with NaOH used as a precipitating agent (liquefied in ethanol [C_2H_5OH] or water as a solvent). In the complete process, the following reaction is involved (16.7):

$$Zn(C_{17}H_{33}COO)_2 (decane) + 2NaOH (water\, and\, ethanol) \rightarrow ZnO + H_2O + 2C_{17}H_{33}COONa \quad (16.7)$$

On the basis of scanning electron microscopy (SEM) and XRD, analysis was performed on the ZnO nanopowder obtained by this method, subsequent elimination of the solvents and exposure to air at room temperature. It was found that the reaction may possibly take place in different phases, mutually in water and in the organic phase. The magnitude of the NPs and the situation of their phase are affected by the reaction circumstances such as temperature, substrate and ratio of two-phase components. Vorobyova et al. reported ZnO NPs achieved by this method with dissimilar particle shapes (irregular aggregates of particles, needle shapes, nearly spherical and nearly hexagonal forms, and spherical aggregates) and with the magnitude of: 2–10 µm, 90–600 nm, 100–230 nm and 150 nm respectively, contingent on the reaction process conditions.

There are also numerous additional methods available designed for the synthesis of NPs (ZnO), including development from a gas phase, a pyrolysis spray method, a sonochemical method, synthesis by using microwaves, in addition to several other methods.

2.1.7 Chemical Vapour Deposition

The chemical vapour deposition (CVD) method is mostly used in the semiconductor industry for setting down narrow films of numerous materials. It is essentially a chemical procedure. In this process, the precursors decompose on the substrate or reactant and produce the preferred deposit. In the CVD process, at a prominent temperature, the evaporated precursors are adsorbed onto a substance surface through vaporised precursors hosted in a CVD reactor. In the CVD process, evaporated originators are prepared to adsorb onto a substance surface detained at a high temperature. Crystals are produced by this method through adsorbed molecules, react with other or decompose. Several steps are involved in this process:

- Starting materials are transported on the growth surface by a boundary layer.
- Chemical reactions take place on the growth surface.
- By-products formed by the gas-phase reaction have to be removed from the surface. A homogeneous and assorted nucleation process takes place in the gas phase.

2.2 Physical Method for the Synthesis of Nanoparticles

2.2.1 Physical Vapour Deposition Method

Physical vapour deposition approaches are used for the deposition of narrow films of metals onto different surfaces. This technique includes condensation from the cloud phase. Three main steps are included in this process.

- By the sublimation of the disappearance of a material corresponding to the vapour phase.
- Shipping of the material to the starting material from the source.

- Progress of the narrow film and constituent parts by nucleation and progression. The source disappears because of the consuming electron beams, thermal energy, the sputtering technique and cathode arc plasma.

2.2.2 Mechanical Ball Milling Method

The mechanical ball milling method is used as a solid-state preparation and is frequently performed by the using ball milling equipment that is commonly divided into a "low energy" and a "high energy" category based on the value of the induced mechanical energy to the powder mixture (Boldyrev and Tkáčová 2000). The main purpose of the grinding process in this method is to decrease the size of the particles and unification of the particles in new phases. The different kind of ball milling process can be used for the fabrication of nanomaterials in which the balls leave an impression upon the powder charge (Yadav et al. 2012).

High-energy ball milling is an appropriate technique for fabricating nano-sized powders. Inter-metallic NPs synthesised by this method constitute the most common method reported in the literature. Previously, a motorised milling process was started; the powder of nanomaterial was loaded together with numerous heavy balls (steel or tungsten carbide) into a container. By energetic shaking or high-speed gyration, a high level of mechanical energy was applied to the powder because of collision with heavy balls (Ghorbani 2014).

2.2.3 Laser Ablation Method

Pulsed laser ablation deposition (PLD) is an eye-catching synthetic method owing to its capability to produce NPs with fine size dissemination and near to the ground level of contaminations. In laser ablation, a synthetic method with three main steps is involved and materialisation of nanomaterials from a target absorbed in liquid. In the laser pulse method, the first step is to heat up the target surface to boiling point, and thus, a plasma column containing vapour atoms of the target is generated. Then, the plasma enlarges adiabatically, and as a final point, NPs are produced when the condensation process occurs. So many synthetic parameters such as laser wavelength, laser energy, pulse width, liquid media type and ablation time affect the characteristics of the products formed by this method.

In 2010, aluminium NPs were fabricated using the pulsed laser ablation method of Al targets in ethanol, acetone and ethylene glycol. The comparison between ethanol and acetone decided through the reaction elucidated that acetone medium produced finer NPs (the magnitude of NPs is 30 nm) with thinner size distribution (from 10 to 100 nm) (Baladi and Mamoory 2010). Hur et al. reported the synthesis of Mg-Al and Zn-Al-layered double hydroxide by using laser ablation in the liquid technique. Ordinary thicknesses of these structures were about 500 nm and the breadth of a single layer was approximately 6.0 nm (Hur et al. 2009).

2.2.4 Gas Evaporation Method

This method is most frequently used for the synthesis of NPs by the evaporation of originator materials from the molten state into a chamber occupied by an inert gas, where the vaporous metal condenses. The nanomaterials obtained by this process and its properties are strongly affected by the purity and type of the originator or precursor used as a starting material, and the transparency of the inert gas atmosphere. A modified inert gas evaporation method is known as cryomelting and can also be used for the formation of NPs. In the cryomelting procedure, the disappearing metal is hurriedly condensed in the region and cooled to about 70 K. A magnitude of NPs of 20–200 nm is produced using this method.

In 2010, the materialisation of aluminium NPs by a unique electromagnetic levitational gas condensation (ELGC) system was planned and manufactured. The greatest frequency of flow of argon for the fabrication of aluminium NPs was found to be 10–15 l/min (Kermanpur et al. 2010).

2.3 Biological Method for the Fabrication of Nanoparticles

The recent and the most innovative area of research is synthesis of nanomaterial by the enormous diversity of marine resources. Over a decade, for the fabrication of low-cost, energy-efficient and non-hazardous NPs, so many variations of microorganisms and plants are used. The biological method used for the fabrication of NPs is known as "green synthesis" or "green chemistry" techniques. Selected plant sources such as *Avena sativa*, *Azadirachta indica*, Aloe vera, *Tamarindus indica* leaf extract and *Cinnamomum camphora* are used for the fabrication of NPs. The different microorganisms used on behalf of the fabrication of NPs such as bacteria, fungi, actinomycetes bacteria, yeast and virus have already been reported by many researchers. Few specimens of NPs fabricated by different micro-organisms including *Aspergillus fumigates* (Ag NPs), *Candida glabrata* (CdS NPs), *Fusarium oxysporum* (silver, gold, zirconia, cadmium sulphide, silica and titanium particles, in addition to CdSe quantum dots), *Pseudomonas aeruginosa* (Au NPs), and many others (Nirmala et al. 2013). The broad area of application of biologically synthesised NPs is shown in Fig. 16.2.

2.3.1 Biosynthesis of Gold Nanoparticles

The biological method is used for the fabrication of gold NPs by using an oceanic sponge *Acanthella elongata* at an extracellular level. In this procedure, the sponge extract was added to 10^{-3} M $HAuCl_4$ aqueous solution at 45 °C; then, the colour of this solution changed to pinkish ruby red colour solution. By the continuous stirring of this solution, 95% of the bioreduction of $AuCl_4^-$ ions occurred within 4 h and generated unvarying gold NPs. The morphology and magnitude of these gold NPs

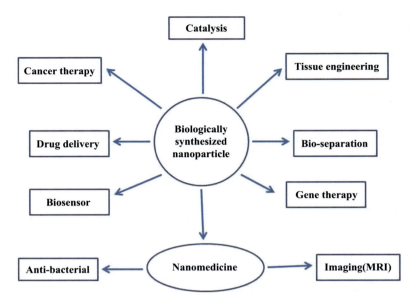

Fig. 16.2 Application of bio-synthesised nanoparticles

obtained by this method is monodisperse, globular, ranging in size from 7 to 20 nm and about 25% are 15 nm in diameter. This was confirmed by high-resolution transmission electron microscope (HR-TEM). From XRD investigation the NPs achieved by this method are crystalline in nature. The extract of aquatic sponge *Acanthella elongata*, possibly acting as the capping agent, is used for the prevention of agglomeration of the NPs and for their stabilisation. It was furthermore identified that the water-soluble organics existing in the extract were included in the reduction of gold ions. Josephine et al. reported that the synthesis of extremely steady gold NPs was achieved by biotransformation using various species of marine sponges (Josephine et al. 2008) (Table 16.1).

3 Nucleation and Growth of Nanoparticles

For many years, the nucleation and growth procedure of NPs have been defined through the LaMer burst nucleation (LaMer and Dinegar 1950) and the variation in the particles size was described by Ostwald ripening (Ostwald 1900). The modern model of this accepted process was developed by the Lifshitz–Slyozov–Wagner (LSW) theory (Lifshitz and Slyozov 1961; Wagner 1961) and was first exhibited by Reiss (1951). This theory was thought to be the only one on nucleation until that of Watzky and Finke (1997), moving toward constant slow nucleation followed by autocatalytic growth. On the basis of empirical formulas, ultraviolet (UV)-visible spectroscopy is a common technique for the determination of the particle size of

Table 16.1 Synthesis of nanoparticles by biological methods

Marine source	Nano-particle synthesised	Size and shape
Sargassum wightii (alga)	Gold	8–12 nm, mostly thin planar structures, some are spherical
Yarrowia lipolytica NCIM 3589 (yeast)	Gold	Particle size varied with varying concentration. Mostly spherical nanoparticles, some hexagonal or triangular nanoplates
Fucus vesiculosus (alga)	Gold	Both size and shape varied according to different initial pH values
Acanthella elongata (sponge)	Gold	7–20 nm, spherical
Penicillium fellutanum (fungi)	Silver	5–25 nm, spherical
Brevibacterium casei MSA19 (sponge associated)	Silver	Uniform and stable
Pichia capsulata (yeast)	Silver	5–25 nm, spherical
Rhodopseudomonas palustris (photo-synthetic bacterium)	Cadmium sulphide	8.01 ± 0.25 nm, cubic crystalline structure

quantum dots. In modern study, by the use of small angle X-ray scattering and a liquid cell within a TEM, it has been possible to further probe NPs in situ and acquire comprehensive information on how NPs grow in solution (Thanh et al. 2014). There are various theories of nucleation and growth, but some of the principles of what happens within the various processes are described.

3.1 LaMer Mechanism

The theoretical separation of the nucleation and growth process into binary stages by using the LaMer mechanism which is the first mechanism. The sulphur sols obtained from the decomposition of sodium thiosulphate was studied by the LaMer, which involved two steps: in the first step, unrestricted sulphur was formed from thiosulphate; in the second step, the sulphur sol is formed in the solution. The procedure of nucleation and development over and done with the LaMer mechanism can be separated into three portions.

1. In the first step the concentration of permitted monomers rapidly increases in the solution.
2. In the second step, these highly concentrated monomers undergoes "burst-nucleation" which is responsible for the reduction in the concentration of free monomer in the solution. The rate of this process is explained as "effectively infinite" and after this point, there is no nucleation occurring because of the low concentration of monomers.
3. In the last step, nucleation growth occurs under the control of the diffusion of the monomers through the solution.

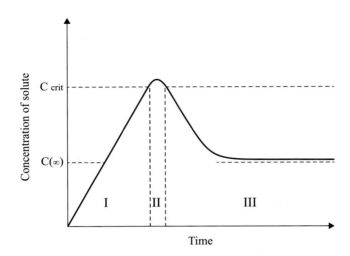

Fig. 16.3 Schematic representation of the LaMer diagram

The three phases are shown in Fig. 16.3 (Sugimoto 2007) where the concentration of the monomers is schematically plotted as a function of time.

3.2 Ostwald Ripening and Digestive Ripening

The Ostwald ripening mechanism (Wagner 1961) was first described in 1900. The growth mechanism is affected by the variation in solubility of NPs dependent on the magnitude of the particles, which is described by the Gibbs–Thomson relation, Eq. (16.8).

$$C_r = C_b \exp\frac{2\gamma v}{rk_b T} \tag{16.8}$$

In this equation, Cr is the solubility of the particle, C_b is the concentration of bulk solution, γ is surface energy, v is the molar volume of the bulk crystal, r is the spherical radius of the particle, k_b is Boltzmann's constant and T is temperature. Owing to the extraordinary solubility and the high surface energy of slighter particles inside the solution, these re-dissolve and allow the larger particles to propagate even more. The calculated theory of Ostwald ripening inside a close system is described by Lifshitz and Slyozov (Reiss 1951) and Wagner (Watzky and Finke 1997).

Digestive ripening is effectively the contradiction of Ostwald ripening. In the former, minor particles propagate at the expense of the larger ones and has been described by Lee et al. (2005), where an applicable form of the Gibbs–Thomson equation, Eq. (16.8), is derived. This manner of formation is controlled once for a second time by the surface energy of the particle inside the solution where the larger particle re-dissolves and in turn less significant particles grow.

4 Characterisation

The characterisation technique of materials is important for understanding their properties and applications. This chapter defines the instruments and experimental set-ups utilised for various measurements toward the characterisation of the synthesised nanomaterials. The techniques adopted to characterise the NPs are: SEM, TEM, XRD, Fourier transform infrared spectroscopy (FTIR), UV-visible spectroscopy, photoluminescence spectroscopy and the Brunauer–Emmett–Teller (BET) technique. TEM and SEM give information about the particle size, shape, topology and their distribution. For the identification of crystal structure and particle size determination, the XRD technique is used. Identification of a functional group existing in the material was characterised by FTIR spectroscopy. Electron microscopy is a powerful and modern technique that allows investigation of the morphology and properties of a solid surface body with a high resolution, that employs beams of fast-tracked electrons and different versions of probe microscopes. Optical properties of material are studied using UV-visible and fluorescence spectra. The BET technique is used for the analysis of the surface area and pore size distribution of material by the N_2 adsorption–desorption isotherm and Barrett–Joyner–Halenda plot.

4.1 X-Ray Diffraction

The physical properties of material, thin film, chemical composition and crystallographic structure is defined by the very convenient characterisation tool of the XRD technique. The XRD characterisation technique has the potential to provide information on an anatomical scale, specifically for crystalline species. The several structural material goods of the crystalline phase such as imperfection structure, strain, particle magnitude and phase composition are measured using this technique. This characterisation determines the thickness of film, but also arrangement in amorphous material. The crystallite size of the nanomaterials is also calculated using this technique. XRD graph is shown in Fig. 16.6.

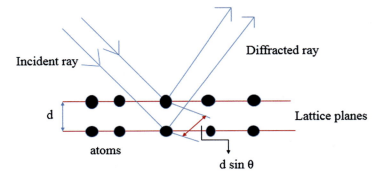

Fig. 16.4 Schematic representation of a Bragg's Low

4.1.1 Basic Principle

The basic principle of XRD is founded on the constructive interference of monochromatic X-rays from a crystalline sample. A cathode ray tube produces X-rays filtered to generated monochromatic radiation, collimated and focused in the direction of the sample. The X-rays first and foremost interact with electrons present in atoms, strike, and some photons from the incident beam are deflected away from the original. The diffraction pattern of the sample is produced by constructively and destructively interfering X-rays formed on the detector. The incident X-ray radiation produces a Bragg peak if their mirror image from the numerous planes interfere constructively. The interference is constructive when the phase shift is a multiple of 2λ; this situation can be expressed by Bragg's law:

$$n\lambda = 2d \sin\theta$$

where n is an integer, λ is the wavelength of the incident wave, d is the spacing between the planes in the atomic lattice and $\sin\theta$ is the angle between the incident ray and the scattering planes. A schematic illustration of XRD is presented in Fig. 16.4.

4.1.2 Instrumentation

The instrumentation of a typical powder X-ray diffractometer includes of a source of radiation, a monochromator to select the wavelength, slits to modify the outline of the beam, a sample and a detector. The position of the detector and the reasonable adjustment of sampling occurs by use of a goniometer. The goniometer mechanism supports the sample and detector, permitting accurate movement. The source of X-rays contains a number of components; the most common being $K\alpha$ and K_β. The particular wavelengths are the main characteristics of the target material (Cu, Fe, Mo, Cr). Monochromators and filters are used to absorb unwanted emission with

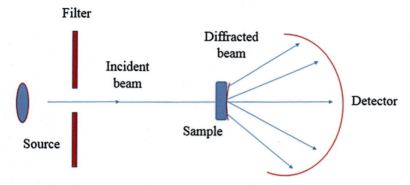

Fig. 16.5 Schematic representations of X-ray diffraction (XRD)

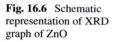

Fig. 16.6 Schematic representation of XRD graph of ZnO

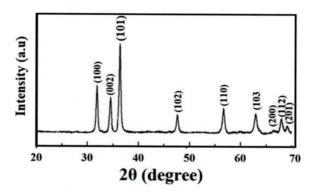

wavelength Kα, while allowing the desired wavelength, K$_β$, to pass through. The X-ray radiation most frequently used in this procedure is that radiated by copper, whose specific wavelength for K$_β$ radiation is equal to 1.5418 Å. The filtered X-rays are collimated and concentrated onto the sample, as presented in Fig. 16.5. When the incident beam attacks a powder sample, diffraction take place in a probable orientation of 2θ. The diffracted beam is detected by using a moveable detector such as a Geiger counter, which is attached to a chart recorder. The counter is fixed for examination over a range of 2θ values at a constant angular velocity. Routinely, a 2θ range of 5°–70° is sufficient to cover the most beneficial part of the powder pattern. The scanning speed of the counter is usually 2θ of 2° min^{-1}. A detector records and progresses this X-ray signal and changes the signal to a count rate, which is then fed into a device such as a printer or computer monitor. The sample must be ground to a fine powder, before loading it into the glass sample holder. The sample should completely occupy the square glass well. In the current work, XRD patterns were noted using Rigaku X-ray diffractometer (RINT-2200) with CuKα radiation at a 0.02°/s step interval.

XRD graphs are shown in Fig. 16.6.

Fig. 16.7 Schematic representation of scanning electron microscopy (SEM)

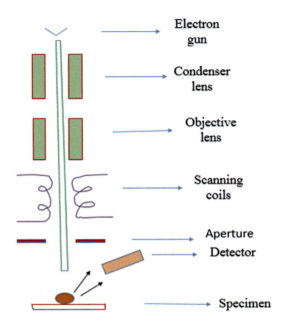

4.2 Scanning Electron Microscope

4.2.1 Basic Principle

Scanning electron microscopy (SEM) is a very valuable imaging technique that applies a ray of electrons to obtain high-intensification images of specimens. The SEM maps the reflected electrons and allows imaging of thick (~mm) samples. SEM images are manufactured by rastering (scanning) a beam over the sample and developing the image point by point.

4.2.2 Instrumentation

The highly magnified image is produced by using electrons instead of light to form an image with the SEM instrument. A schematic illustration of SEM is revealed in Fig. 16.7. The electron gun produces a beam of electrons at the top of microscope. The electron beam monitors a perpendicular path through the microscope, which is held within a vacuum. The stream of electrons travels through electromagnetic fields and lenses, which centre the beam down in the direction of the sample. Once the electron beam forays the sample, electrons and X – rays are turned out from the sample. These X-rays, backscattered electrons and secondary electrons are collected through the detectors and transform them into a signal that is shown on a

Fig. 16.8 Schematic representation of a SEM image

screen such as a television screen. This produces the finished image. In this research work, the powder samples were placed on the carbon tape, which was attached to the sample holder. JEOL JSM 6320F (FESEM), F E I Quanta FEG 200 (HRSEM) were used to study the surface morphology of the sample.

Investigation with SEM is considered to be "non-destructive"; that is, X-rays generated by electron interactions do not lead to volume damage of the sample; thus, it is probable that the same materials are evaluated repeatedly (Egerton 2005). The SEM images the surface structure of bulk samples, from the biological, medical, materials sciences, and earth sciences up to magnifications of ~×100,000. The images have a greater depth of field and resolution than optical micrographs, making them ideal for rough specimens such as fracture surfaces and particulate materials. Some SEM images are shown in Fig. 16.8.

4.3 Transmission Electron Microscopy

4.3.1 Basic Principle

Transmission electron microscopy (TEM is a very useful technique for nanomaterials. It technique provides information about the diameter and a highly magnified transmitted image of the material. In this technique, the electron beam is transmitted with an ultra-thin specimen, then acting together with the specimen as it permits via this interaction the formation of an exceedingly magnified image. The detection of the image by passes electrons through the sample. The shorter wavelength of electrons compared with photons can therefore provide a greater resolution than conventional light microscopes. TEM can provide two distinct types of information about a specimen: a magnified image and a diffraction pattern.

Fig. 16.9 Schematic representation of transmission electron microscopy

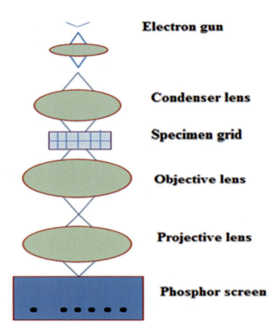

4.3.2 Instrumentation

The instrumentation of TEM includes four parts: electron source, electromagnetic lens system, sample holder and imaging system, as shown in Fig. 16.9. The electron beams generated through the source are forcefully focused by the electromagnetic lenses and the metal apertures. In this system, the electrons allow a slight energy range to pass through, so that the electrons in the electron beam have a definite energy. This electron beam drops on the sample located in the holder. In this system, the electron beam is permitted through the specimen. This transmitted ray reproduces the patterns on the sample. This transmitted beam is projected onto a phosphor screen. In the current work, TEM images were recorded by a JEOL JEM 2100F transmission electron microscope at an accelerating voltage of 200 kV. The images were obtained using this technique, the powdered samples were dispersed in ethanol solvent and it was ultrasonicated for 20 min. The copper grid in this instrument is coated in the dispersed compound and TEM images were found.

4.4 Brunauer–Emmett–Teller Technique

The Brunauer–Emmett–Teller (BET) technique is the most common method for defining the surface area of powders and porous materials. This technique is also used for calculating pore volume, pore radius and pore size dissemination of

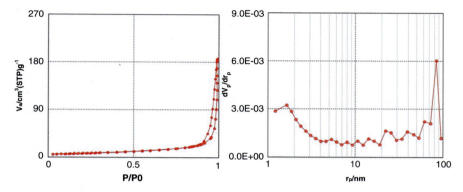

Fig. 16.10 (**a**) N_2 adsorption desorption plot (**b**) Barrett–Joyner–Halenda plot

nanomaterials. Nitrogen is the most commonly employed gaseous adsorbate used for surface probing by BET methods. For this reason, standard BET analysis is most often conducted at the boiling temperature of N_2 (77 K). The surface area of the material is calculated from the measured monolayer capability and knowledge of the cross-sectional area of the molecule being used as a probe. For the instance of nitrogen, the cross-sectional area is taken as 16.2 Å2/ molecule. The adsorption-desorption and BJH plot are shown in Fig 16.10.

4.4.1 Instrumentation

Brunauer–Emmett–Teller experiments are characteristically conducted to a relative pressure, P/P_0, of approximately 0.3 at 77 K, where P_0 is the saturation pressure (Lowell et al. 2004). The relative pressure in terms of relative humidity can be considered, i.e. the experiment is conducted to 30% of the saturation pressure of N_2 at 77 K ($\approx$230 torr). At comparative pressures above the point at which an N2 monolayer has formed on the solid, capillary condensation is found within the pore structure of the material such that the smaller pores are occupied more effortlessly and consecutively larger pores are occupied as pressure is increased. When the saturation point is approached, i.e. P/P_0 is approximately 1.0, the internal pore structure of the material contains condensed (liquid) nitrogen.

4.5 Fourier Transform Infrared Spectroscopy

Fourier transform infrared spectroscopy (FTIR) is one of the most powerful tools for the identification of compounds by the corresponding spectrum of unidentified compounds with a reference spectrum (finger printing) and investigation of functional groups in an unknown compound. The infrared section of the electromagnetic spectrum is considered to cover the range from 50 to approximately 12,500 cm^{-1}.

Fig. 16.11 Schematic representations of a Fourier transfer infrared spectroscopy

4.5.1 Basic Principle

In this instrument, when infrared light is accepted through a sample of an organic compound, particular frequencies are absorbed, whereas other frequencies are transmitted without being absorbed. During this procedure, the transitions involved in the infrared absorption are related to the vibrational variations in the molecule. Different bonds/functional groups occur at different vibrational frequencies and hence the occurrence of these vibrational bands in a molecule can be distinguished by identifying this characteristic frequency as an absorption band in the infrared spectrum. The plot spectrum in infrared spectroscopy between transmittance and frequency is called the infrared spectrum.

4.5.2 Instrumentation

Owing to the remarkable speed and sensitivity of FTIR spectrometers have substituted dispersive instruments for maximum applications. They have significantly prolonged the aptitudes of infrared spectroscopy and have been applied to many areas that are very problematic or impossible to investigate using dispersive instruments. Instead of observing each component frequency sequentially, as in a dispersive IR spectrometer, all frequencies are studied at the same time in FTIR spectroscopy. In this instrument, there are three main basic spectrometer components: radiation source, interferometer and detector. The well-designed block shape of the FTIR spectrometer is shown in Fig. 16.11. The source of IR radiation is a broadband source that is first focused into an interferometer, where it is separated and then recombined after the split beams have travelled dissimilar optical paths to generate constructive and destructive interference. The next step is the resulting beam being permitted through the sample compartment and reaching the detector. The preparation of the sample is very easy. Using this technique, almost any samples such as solids, liquids or gases, can be investigated. The sample to be investigated (minimum of 10 µg) should be crushed into a KBr matrix or dissolved in an appropriate solvent (CCl_4 and CS_2 are desired). The exclusion of water from the sample is possible. In the case of solid samples, these are directly mixed with solid KBr pallets (transparent in the mid-infrared region), then crushed and pressed. FTIR measurements were implemented using a Perkin Elmer FTIR spectrophotometer with a standard KBr pellet technique.

4.6 UV-Visible Spectroscopy

Ultraviolet-visible (UV-Vis) spectroscopy refers to absorption spectroscopy in the ultra-violet and visible spectral region. The electronic transition is found in the molecules as the electromagnetic spectrum was present in this region. When sample molecules are uncovered to light with energy (E = hv where E is energy in joules, h is Planck's constant 6.62 × 10^{-34}Js and v is frequency in Hertz), which matches a probable electronic transition contained by the molecule, a certain amount of the light energy will be absorbed as the electron is promoted to an upper energy orbital. When the absorption occurs by a molecule at a particular wavelength, it is recorded by an optical spectrometer, simultaneously with the amount of absorption at every wavelength. The resultant spectrum exists as a graph of absorbance (A) versus wavelength (λ). Investigation of the optical properties of NPs can be studied with the help of UV-Vis spectra.

4.6.1 Basic Principle

The absorption of light by molecules in the solution is centred on the Beer–Lambert law,

$$A = I / I_0 = \varepsilon bc$$

where, I_0 is the intensity of the reference beam and I is the intensity of the sample beam, ε is the molar absorptivity with units of L mol^{-1} cm^{-1}, b = route length of the sample in centimetres and c = concentration given the solution expressed in mol L^{-1}.

4.6.2 Instrumentation

In this instrument, the main components are a light source, double beams (reference and sample beams), a monochromator, a detector and a recording device. The light source is usually a tungsten filament lamp used for visibility and a deuterium discharge bulb is used for UV measurements. The light coming out from the light source is divided into dual beams, the reference and the sample beams, as shown in the Fig. 16.12. The sample and reference cells are the rectangular quartz/glass containers; they are filled with the solution (to be tested) and pure solvent respectively. This spectrometer records the proportion between the reference and sample beam intensities. The recorder plots the absorbance (*A*) against the wavelength (λ). The sample is prepared by proper grinding into a mortar paste and then liquefied into the solvent to form a dilute sample solution. This sample solution is filled with sample cells to the mark line. In the present work, UV-Vis absorption examinations were carried out by a Varian Cary 100 E UV-Vis-NIR spectrophotometer and a Shimadzu (Japan) 3100 PC spectrophotometer using ethanol as a dispersing medium.

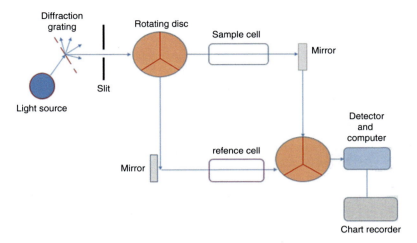

Fig. 16.12 Schematic representation of a ultraviolet-visible spectroscopy

4.7 Photoluminescence Spectroscopy

Photoluminescence spectroscopy is a contactless, non-destructive method for the examination of the electronic structure of materials. This technique is used for the concentration and spectral content of the emitted photoluminescence is an undeviating quantity of immeasurably important material properties, together with band gap determination, contamination levels, imperfection detection and recombination mechanisms.

4.7.1 Basic Principle

Light is concentrated onto a sample, where it is absorbed and imparts additional energy into the material using a method called photo-excitation. Photo-excitation causes electrons contained by a material to transfer into acceptable excited states. These excited electrons come back from their equilibrium states, by a radiative process (the emission of light) or by a non-radiative process, as shown in Fig. 16.13. The quantity of emitted light is associated with the relative contribution of the radiative process.

4.7.2 Instrumentation

The photoluminescence instruments include three elementary items: a cause of light, a sample holder and a detector. A schematic demonstration of a fluorimeter is shown in Fig. 16.14. The light source produces light photons over a broad energy spectrum; in this instrument, the spectra range from 200 to 900 nm. Photons impinge

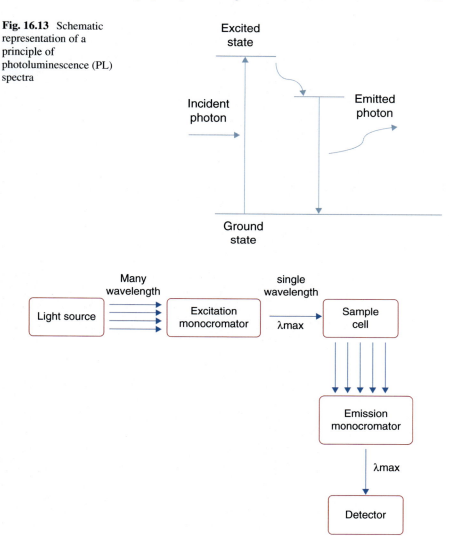

Fig. 16.13 Schematic representation of a principle of photoluminescence (PL) spectra

Fig. 16.14 Schematic representations of PL spectra

on the excitation monochromator, which selectively transfers light within a constricted range centred around the excitation wavelength indicated. The transmitted light is permitted through modifiable slits that govern the magnitude and resolution by further restricting the range of transmitted light. In this instrument, the strained light passes through into the sample cells affecting the fluorescent radiation by fluorophores within the sample. Emitted light enters the emission monochromator, which is situated at a 90° angle from the excitation light path to eradicate background signal and decrease the sound owing to stray light. Yet again, emitted light is transmitted within a contracted range centred around the definite emission wavelength and exiting through adjustable slits, finally entering the photomultiplier tube (PMT). The signal is improved and produces a voltage that is proportional to the measured emitted intensity. The noise arises primarily in the PMT. Therefore, spectral resolutions and signal to noise is directly interconnected to the particular slit widths. The preparation of a sample procedure is the identical to that of UV-Vis spectroscopy. In both cases, it is essential for the sample cell (cuvette) to be free from contaminants. For the present research work, a Fluorolog-3-11 spectrophotometer was employed for photoluminescence measurements.

5 Application of Nanoparticles in Environmental vWaste Management

For environmental waste management, a number of commercial and non-commercial scientific improvements are employed on a daily basis, but nanotechnology has revealed progressive methods of water/wastewater treatment. In the development of nanoscience research, it has to been possible to come up with economically attainable and environmentally stable treatment technologies to dramatically consider water/wastewater contest to improve water quality standards.

Over the past few years, the consumption of NPs as adsorbents, nanosized zero valent ions or nanofiltration membranes has been for the removal/separation of contaminants from water, whereas currently, NPs used as photocatalysts for chemical or photochemical oxidation effect the annihilation of contaminants. The remediation of toxic waste present in air through nanotechnology is carried out in the several ways. One is by the management of nano-photocatalysts, in addition to a simple catalyst with an enhanced surface area for gaseous reactions. Catalysts increase the frequency of chemical reactions that convert harmful gases that have evolved from automobiles and industrial plants into harmless gases. Presently, nanofiber catalysts are applicable and are mostly made of manganese oxide, which discards contaminated organic compounds from industrial smokestacks. The development of other methods is static. An additional way is to approach nanostructure membranes that have smaller pores adequate for separating marsh gas or carbon dioxide from exhausts.

5.1 Role of Nanoparticles in Wastewater Treatment

In terms of wastewater treatment, NPs are applied for the eradication of various forms of waste. Organic contaminants and substantial metal toxic waste pose serious problems to the environment because they are toxic to living beings, including humans, and are not naturally degradable. Several nanomaterials such as TiO_2, ZnO, ceramic membranes, nanowire membranes, polymer membranes, CNTs, submicron nanopowder, metal (oxides), magnetic NPs, nanostructured boron-doped diamond are used for photocatalysis, nanofiltration, adsorption and electrochemical oxidation to resolve or greatly diminish problems involving water excellence in the natural environment.

5.1.1 Role of Nanoparticles in Removal of Organic Pollutants from Wastewater

Nanoparticles are utilised for the elimination of organic pollutants, which have size-dependent properties correlating with the specific surface area, such as fast dissolution, high reactivity, and strong sorption. Most of the applications discussed below are still in the laboratory research phase. The pilot-tested or field-tested exceptions are noted in the text.

5.1.1.1 Photocatalysis

Photocatalytic oxidation is a forward-looking and recently used oxidation technique for the degradation of toxic pollutants and microbial pathogens. It is a relevant pretreatment for harmful and toxic hazardous materials, which are non-biodegradable pollutants, to enhance their biodegradability in the environment. For the degradation of unruly organic pollutants NPs can also be used as photocatalysts. The most important obstacle to its widespread utilisation is limited because of lighting consequences and photocatalytic activity is slow kinetics. Present research concentrates on enhancing the frequency of photocatalytic reaction and the photoactivity area.

5.1.1.2 TiO_2 Metal Oxide Nanoparticles

TiO_2 NPs are mostly used as water/wastewater treatment semiconductor photocatalysts owing to their low toxicity, their stability, low cost, and also the large amount of raw materials present in nature. On the radiation from UV light, its electrons migrated from the valence band to the conduction band. During the movement of electrons from the valence band to the conduction band, holes (h^+) are created in the valance band and electrons (e^-) in the conduction band. This means it produces an electron/hole (e^-/h^+) pair, which either transfers to the surface and reacts with environmental oxygen to form reactive oxygen species (ROS) or go through undesired recombination. The photocatalytic activity of TiO_2 NPs can be enhanced by controlling the diameter and shape of the particle, reducing e^-/h^+ pair recombination by noble/transition metal doping (Han et al. 2009; Murakami et al. 2009; Liu et al. 2011), maximising the reactive characteristic and surface behaviour to improve pollutant adsorption. The shape, size and morphology of TiO_2 plays a momentous role in its solid-phase conversion, sorption and e^-/h^+ recombination. Amid crystalline TiO_2, rutile is the most reliable and larger than 35 nm, while an anatase, more valuable in producing ROS, is the most stable for particles smaller than 11 nm (Fujishima et al. 2008; Banfield et al. 2000). A major cause for the slow rate of photodegradation of pollutants by TiO$_2$ photocatalysis is the rapid recombination of e$^-$ and h$^+$. Decreasing the size of TiO$_2$ NPs diminished its recombination of the e$^-$/h$^+$ pair, and appreciated its interfacial charge carrier transfer (Zhang et al. 2000). However, when the size of the NPs is decreasing at certain nanometres, surface recombination dominated, and increased its photocatalytic activity. Therefore, the photocatalytic degradation efficiency of the TiO$_2$ NPs has been maximised because of the interplay of the above-mentioned mechanisms, which lies in the nanometre rang. In the degradation process of organic compounds, the nanotubes of TiO$_2$ became more capable than TiO$_2$ NPs (Macak et al. 2007). The advanced photocatalytic activity was accredited to the smaller carrier diffusion paths in the tube walls and the rate of mass transfer of reactants toward the surface of the nanotubes was very fast.

5.1.1.3 Nanosorption

Adsorption is usually used as a polishing step in water and wastewater management to eliminate organic and inorganic pollutants. The capability of traditional adsorbents is consistently defined by the surface area or active sites, the absence of distinction and the degree of adsorption. NPs, which are used as nano-adsorbents, attempt powerful enhancement with their large surface area and sorption sites, small intra-particle diffusion distance, pore size and surface chemistry.

5.1.1.4 Carbon-Based Nano-Adsorbents

Recently, CNTs have demonstrated greater adaptability for the exclusion of organic pollutants than activated carbon on adsorption of several organic pollutants (Pan and Xing 2008). Owing to the extraordinarily precise surface area, the adsorption capability of CNTs with pollutant is high. The adsorption phenomena occur on the particular CNTs at finite surface area is the external surface phenomena (Yang and Xing 2010). In the solution phase, owing to the hydrophilic nature of graphitic surface CNTs loosely aggregating and decreasing the effective surface area. In another way, the aggregation of CNTs consists of interstitial spaces that have a tremendous amount of adsorption energy for adsorption of organic molecules (Pan and Xing 2008). In comparison with CNT, activated carbon also acquired an analogous measured specific surface area and accommodated a considerable number of micropores unattainable to heavy organic pollutants such as various antibiotics and pharmaceuticals. Thus, CNTs have considerably more capability for the adsorption of heavy organic pollutants because of their larger pores in bundles and more available sorption sites.

A considerable deficiency of activated carbon is its low adsorption capacity for the small molecular weight of polar organic compounds. Owing to the different types of CNT interaction such as hydrophobic effect, π–π interactions, hydrogen bonding, covalent bonding and electrostatic interactions, CNTs powerfully adsorb polar organic compounds to form varied contaminated CNTs. The easier availability of π electrons on the CNT surface enable it to interact with π–π interactions with organic molecules through C=C bonds or aromatic rings, such as PAHs and polar aromatic compounds. Many organic compounds contain -COOH, -OH and -NH$_2$ as functional groups that could interact through hydrogen bonds with the graphitic CNT surface, which donates electrons. Within an appropriate pH range, the electrostatic attraction encourages the adsorption of positively charged organic chemicals such as some antibiotics (Qu et al. 2013).

5.1.1.5 Zero-Valent Iron

Metal oxide NPs, including semiconductor materials, bimetallic NPs and zero-valence metals, have been applicable for the degradation of environmental pollutants, for instance PCBs, pesticides, and azo dyes, as a result of their greater surface area and shape-reliant properties (Zhao et al. 2011). The NPs, which behave as magnetic nanosorbents, have also been demonstrated to be adequate for the degradation of organic pollutants (Campos et al. 2011). Among the magnetic nanosorbents, iron oxide nanomaterials have demonstrated a superior degradation efficiency for organic contaminants compared with bulk materials (Li et al. 2003a; Wu et al. 2005; Lai and Chen 2001). For the discharge of coloured humic acid from the wastewater, magnetic Fe$_2$O$_3$ NPs have also been applied (Qiao et al. 2003). Usefully, conversion of chlorinated organic contaminants and PCBs has been carried out by applying nano zero valent iron (n-ZVI) (Wang and Zhang 1997; Zhang et al. 1998; Cheng et al. 2007) in addition to inorganic pollutants ions such as nitrate and perchlorate (Choe et al. 2000). NPs of n-ZVI have been revealed as powerful substances for the removal of hazardous chlorinated organic contaminants, for example, 2,2′-dichlorobiphenyl in a minor remediation investigation (Parra et al. 2004). The more reliable n-ZVI NPs could also be a practical technique for in situ remediation of groundwater or industrial toxic wastes. The simple n-ZVI and nanocomposite n-ZVI have appeared as practical redox media for compressing a range of organic contaminants such as PCBs, organic dyes, pesticides, chlorinated hydrocarbon and inorganic toxic anions, e.g. nitrates in water/wastewater because of the large surface areas and high reactivity (Schrick et al. 2002). Many bimetallic NPs such as Ni^0/Fe^0 and Pd^0/Fe^0 were found to be more impressive than general microscale Fe for reducing the halogenation of chlorinated organic pollutants and brominated hydrocarbon, hydro-dechlorination of chlorinated saturated hydrocarbon, chlorinated aromatics and PCBs, as explained by many researchers (Wei et al. 2006; Lim et al. 2007).

5.1.2 Role of Nanoparticles in Elimination of Heavy Metals from Wastewater

Various types of NPs such as CNTs, zeolites, dendrimers and nanosorbents have been identified for the subtraction of heavy metals from water/wastewater because of their exceptional adsorption properties (Amin et al. 2014). The ability of CNTs to adsorb heavy metals has been analysed by many researchers such as Pb^{2+}, Cu^{2+}, Cd^{2+} (Li et al. 2003b), Cr^{3+} (Di et al. 2006) and Zn^{2+} (Lu et al. 2006), and metalloids such as arsenic (As) compounds. The elimination of heavy metal ions by the composites of CNTs with Fe and cerium dioxide (CeO_2) have also been reported in a few studies (Salipira et al. 2007). NPs of cerium dioxide supported on CNTs are effectively applicable to adsorbing arsenic. The fast rate of adsorption on CNTs is primarily because of the high availability of adsorption sites on the surface and the short intra-particle diffusion distance. Metal oxide-based NPs were demonstrated to be superior in discarding toxic heavy metals in comparison with activated carbon, for example, adsorption of arsenic on the surface of nanosized TiO_2 and NPs of magnetite (Deliyanni et al. 2003; Mayo et al. 2007). The application of TiO_2 NPs as photocatalysts has been considered in detail to reduce the harmful metal ions in water (Kabra et al. 2004). In one study, the presentation of NPs of TiO_2 in reducing various forms of arsenic is explained in more detail and they have been shown to be additional impressive photocatalysts compared with generally accessible TiO_2 NPs with a maximum eradication capability of arsenic at about neutral pH value (Pena et al. 2005). The reduction of Cr(VI) to Cr(III) in the presence of sunlight by using a nanocomposite of TiO_2 NPs attached to a graphene sheet and also Cr treatment was implemented by applying palladium NPs in another study (Amin et al. 2014). The efficiency of elimination of poisonous heavy metals such as As (arsenic) is also considered by the use of the low-cost adsorbent iron oxide nanomaterials (Fe_2O_3 and Fe_3O_4) by several researchers (Lai and Chen 2001; Onyango et al. 2003; Oliveira et al. 2004). Further elimination of arsenic was also examined by application of an extraordinary specific surface area of Fe_3O_4 nanocrystals (Yavuz et al. 2006). Implantation of polymer onto Fe_2O_3 was a powerful nanocomposite for the elimination of divalent toxic heavy metal ions for copper, nickel and cobalt in a pH range of up to 3–7 pH (Takafuji et al. 2004). For the elimination of radioactive metal toxins, uranium dioxide (UO_2^{2+}) from water by application of bisphosphonate-modified magnetite NPs was also studied (Inbaraj and Chen 2012). There are more studies that have explained that NPs of n-ZVI or Fe^0 are useful for the conversion of toxic heavy metal ions such as Cd(II), As(V), Pb(II), Cu(II), Ni(II) and Cr(VI) (Zou et al. 2016).

Innovative self-assembled 3D flower-like structures of iron oxide and CeO_2 NPs were also applicable for the good adsorbtion of both As and Cr (Zhong et al. 2006, 2007). The adaptability of NaP1 zeolites was assessed for the elimination of heavy metals (Cr(III), Ni(II), Zn(II), Cu(II) and Cd(II)) from wastewater (Moreno et al. 2001; Alvarez-Ayuso et al. 2003). Polymer-incorporated NPs are also most applicable for the remediation of noxious heavy metal ions (Diallo et al. 2004). The application of mesoporous to support assembled monolayers for the elimination of toxic metal ions was also mentioned by many researchers (Mattigod et al. 1999; Yantasee et al. 2003). Moreover, the application of biopolymers has been reported for heavy metal removal from aqueous waste (Kostal et al. 2001). Chitosan nanomaterials was further reported for the sorption of Pb(II) (Qi and Xu 2004).

5.2 Role of Nanoparticles in the Bioremediation of Petroleum Hydrocarbon

Owing to the hydrophobic nature of the petroleum hydrocarbons that reduces their solubility in water and enhances the sorption properties of soil micelles that inhibit bioremediation of these compounds. Numerous studies have described the fabrication of biosurfactants by the microorganisms that can inhibit the hydrophobicity of these organic pollutants by reducing the surface tension and appreciating the bioavailability of hydrophobic pollutants to microorganisms, which is very significant in the bioremediation process (Cameotra and Makkar 2010). In the adjoining of small nanomaterials, the bioremediation of hydrocarbon can be appreciated, as they reimburse the oxidation of these compounds, decrease their toxic effect and make suitable for the microbial growth. NPs of peroxides (such as calcium peroxide) increase the suspension and kinetics of hydrocarbons. Applications of calcium peroxide (CaO^2) NPs as an attractive oxidant are treated as honest oxygen discharge material (Northup and Cassidy 2008). A mixture of benzene and gasoline (as high as 800 mg L^{-1}) can be completely oxidised within 24 h (Pereira et al. 2005). Application of iron oxide (FeO) and hydrogen peroxide (H_2O_2) in the proportion of 1:33.7 has been applicable to reduce up to 91% of the total petroleum hydrocarbon within 4 h (Kumari and Singh 2016).

The experiential investigation by Jameia et al. (2013) into the degradation of the hydrocarbon chain in the soil registered an enhanced n-ZVI, whereas manganese and cobalt NPs (MnNP and CoNP) promoted the contraction of PAHs (Nador et al. 2010). The regiochemistry and degree of decrease of derivatives of 1-substituted naphthalene were mainly dependent on the nature of metal NPs used. The NPs of Co encourage the contraction of PAHs, most important to the corresponding tetralin products, whereas the Mn NPs permitted the construction of unconjugated 5,8-dihydro derivatives. Co-NP was found to be more conscious than Mn NPs in the conversion of 9,10-dihydrophenanthrene into phenanthrene within 3 h. The conversion of naphthalene into 1,2,3,4-tetrahydronaphthalene by the reaction of Co NPs at room temperature is a qualitative example.

Nano-titanium oxide is applied as a photocatalyst in the existence of UV light for degradation of organic noxious waste from petroleum refinery wastewater (Saien and Shahrezaei 2012). A very low concentration of this catalyst (100 mg L^{-1}) at pH 3.0 and a temperature of 45 °C could remove about 78% of the organic contaminants (after 60–90 min) in the occurrence of UV irradiation. The degradation of petroleum hydrocarbon by the NPs of the TiO_2 was observed by Fard et al. (2013), whereas colloidal NPs of TiO_2 for photocatalysis are applicable for the removal of the seawater-soluble crude oil fraction observed by Ziollia and Jardim (2001). The dissolution of organic carbon in the occurrence of TiO_2 removed 90% crude oil compounds in sea water, which contained about 45 mg of carbon L^{-1} of seawater-soluble after 7 days of non-natural light exposure.

6 Conclusion

This chapter concludes that NPs are the preeminent fundamental unit of nanotechnology. They are a group of tens of thousands of atoms measuring about 1–100 nm of a massively ordered crystalline nature. Such NPs originated atom by atom; thus, the size and regularity of the shape, of a particle, are composed of the preliminary conditions. The prerequisite of nanotechnology is to synthesise the desired shape and size of the nanomaterials for their convenient applications. Surfactants play a significant role during the synthesis of NPs. Owing to the huge surface area to volume ratio, nanostructures display distinctive properties.

In nanotechnology, the cost-effective production of NPs has the potential to advance the environment, both through straight utilisation of those materials to distinguish, avoid and eradicate contamination, and indirectly by using nanotechnology to design cleaner industrial routes and generate environmentally accountable products. For instance, ZnO, iron oxide (Fe_2O_3), cobalt oxide (Co_2O_3) and certain other different NPs can remove contaminants from topsoil and ground water, and nano-sized sensors hold assurance for better-quality detection and tracing of contaminants. These nanomaterials provide clean water from contaminated water sources on both an outsized scale and via a transportable application, in addition to detecting and cleaning up eco-friendly contaminants (waste and toxic material), i.e. remediation.

Acknowledgment The authors wish to thank the Babasaheb Bhimrao Ambedkar University (a central University) of U.P India for providing the opportunity to carry out this work.

References

Alvarez-Ayuso E, Garcıa-Sánchez A, Querol X (2003) Purification of metal electroplating waste waters using zeolites. Water Res 37:4855–4862

Amin MT, Alazba AA, Manzoor U (2014) A review of removal of pollutants from water/wastewater using different types of nanomaterials. Adv Mater Sci Eng 190:208–222

Ansari SA, Husain Q (2012) Potential applications of enzymes immobilized on/in nano materials: a review. Biotechnol Adv 30:512–523

Baladi A, Mamoory RS (2010) Investigation of different liquid media and ablation times on pulsed laser ablation synthesis of aluminum nanoparticles. Appl Surf Sci 256:7559–7564

Banfield JF, Welch SA, Zhang H, Ebert TT, Penn RL (2000) Aggregation-based crystal growth and microstructure development in natural iron oxyhydroxide biomineralization products. Science 289:751–754

Bhattacharya S, Saha I, Mukhopadhyay A, Chattopadhyay D, Chand U, Chatterjee D (2013) Role of nanotechnology in water treatment and purification: potential applications and implications. Int J Chem Sci Technol 3:59–64

Boldyrev VV, Tkáčová K (2000) Mechanochemistry of solids: past, present, and prospects. J Mater Synth Process 8:121–132

Cameotra SS, Makkar RS (2010) Biosurfactant-enhanced bioremediation of hydrophobic pollutants. Pure Appl Chem 82:97–116

Campos AFC, Aquino R, Cotta T, Tourinho FA, Depeyrot J (2011) Using speciation diagrams to improve synthesis of magnetic nanosorbents for environmental applications. Bull Mater Sci 34:1357–1361

Chen D, Jiao X, Cheng G (1999) Hydrothermal synthesis of zinc oxide powders with different morphologies. Solid State Commun 113:363–366

Cheng R, Wang J, Zhang W (2007) Comparison of reductive dechlorination of p-chlorophenol using Fe0 and nanosized Fe0. J Hazard Mater 144:334–339

Choe S, Chang YY, Hwang KY, Khim J (2000) Kinetics of reductive denitrification by nanoscale zero-valent iron. Chemosphere 41:1307–1311

Daou TJ, Pourroy G, Colin SB et al (2006) Hydrothermal synthesis of monodisperse magnetite nanoparticles. Chem Mater 18:4399–4404

Deliyanni EA, Bakoyannakis DN, Zouboulis AI, Matis KA (2003) Sorption of As (V) ions by akaganéite-type nanocrystals. Chemosphere 50:155–163

Di ZC, Ding J, Peng XJ, Li YH, Luan ZK, Liang J (2006) Chromium adsorption by aligned carbon nanotubes supported ceria nanoparticles. Chemosphere 62:861–865

Diallo MS, Christie S, Swaminathan P et al (2004) Dendritic chelating agents. 1. Cu (II) binding to ethylene diamine core poly (amidoamine) dendrimers in aqueous solutions. Langmuir 20:2640–2651

Djurisic AB, Chen XY, Leung YH (2012) Recent progress in hydrothermal synthesis of zinc oxide nanomaterials. Recent Patents Nanotech 6:124–134

Egerton RF (2005) Physical principles of electron microscopy. Springer, New York

Fard MA, Aminzadeh B, Vahidi H (2013) Degradation of petroleum aromatic hydrocarbons using TiO_2 nanopowder film. Environ Technol 34:1183–1190

Fujishima A, Zhang X, Tryk DA (2008) TiO_2 photocatalysis and related surface phenomena. Surf Sci Rep 63:515–582

Ghorbani HR (2014) A review of methods for synthesis of al nanoparticles. Orient J Chem 30:1941–1949

Gupta K, Bhattacharya S, Chattopadhyay D et al (2011) Ceria associated manganese oxide nanoparticles: synthesis, characterization and arsenic (V) sorption behavior. Chem Eng J 172:219–229

Han X, Kuang Q, Jin M, Xie Z, Zheng L (2009) Synthesis of titania nanosheets with a high percentage of exposed (001) facets and related photocatalytic properties. J Am Chem Soc 131:3152–3153

Hur TB, Phuoc TX, Chyu MK (2009) Synthesis of Mg-Al and Zn-Al-layered double hydroxide nanocrystals using laser ablation in water. Opt Lasers Eng 47:695–700

Inbaraj BS, Chen BH (2012) In vitro removal of toxic heavy metals by poly (γ-glutamic acid)-coated superparamagnetic nanoparticles. Int J Nanomedicine 7:4419–4432

Innes B, Tsuzuki T, Dawkins H et al (2002) Nanotechnology and the cosmetic chemist. Cosmetics Aerosols Toiletries Australia 15:10–24

Jameia MR et al (2013) Degradation of oil from soilusing nano zero valent iron. Sci Int 25:863–867

Jolivet JP, Chanéac C, Tronc E (2004) Iron oxide chemistry. From molecular clusters to extended solid networks. Chem Commun 5:481–483

Josephine A, Nithya K, Amudha G, Veena CK, Preetha SP, Varalakshmi P (2008) Role of sulphated polysaccharides from Sargassum Wightii in cyclosporine A-induced oxidative liver injury in rats. BMC Pharmacol 8:4

Kabra K, Chaudhary R, Sawhney RL (2004) Treatment of hazardous organic and inorganic compounds through aqueous-phase photocatalysis: a review. Ind Eng Chem Res 43:7683–7696

Kermanpur A, Dadfar MR, Rizi BN, Eshraghi M (2010) Synthesis of aluminum nanoparticles by electromagnetic levitational gas condensation method. J Nanosci Nanotechnol 10:6251–6255

Khoshnevisan K, Bordbar AK, Zare D et al (2011) Immobilization of cellulase enzyme on superparamagnetic nanoparticles and determination of its activity and stability. Chem Eng J 171:669–673

Kołodziejczak-Radzimska A, Jesionowski T, Krysztafkiewicz A (2010) Obtaining zinc oxide from aqueous solutions of KOH and Zn (CH$_3$COO)$_2$. Physicochem Probl Mineral 44:93–102

Kołodziejczak-Radzimska A, Markiewicz AE, Jesionowski T (2012) Structural characterisation of ZnO particles obtained by the emulsion precipitation method. J Nanomater 2012:15

Kostal J, Mulchandani A, Chen W (2001) Tunable biopolymers for heavy metal removal. Macromolecules 34:2257–2261

Kratochvil D, Volesky B (1998) Advances in the biosorption of heavy metals. Trends Biotechnol 16:291–300

Kumari B, Singh DP (2016) A review on multifaceted application of nanoparticles in the field of bioremediation of petroleum hydrocarbons. Ecol Eng 97:98–105

Kumari V, Yadav A, Haq I, Kumar S, Bharagava RN, Singh SK, Raj A (2016) Genotoxicity evaluation of tannery effluent treated with newly isolated hexavalent chromium reducing Bacillus Cereus. J Environ Manag 183:204–211

Lai CH, Chen CY (2001) Removal of metal ions and humic acid from water by iron-coated filter media. Chemosphere 44:1177–1184

LaMer VK (1952) Nucleation in phase transitions. Ind Eng Chem Res 44:1270–1277

LaMer VK, Dinegar RH (1950) Theory, production and mechanism of formation of monodispersed hydrosols. J Am Chem Soc 72:4847–4854

Laurent S, Forge D, Port M et al (2008) Magnetic iron oxide nanoparticles: synthesis, stabilization, vectorization, physicochemical characterizations, and biological applications. Chem Rev 108:2064–2110

Lee W, Kim MG, Choi J et al (2005) Redox-transmetalation process as a generalized synthetic strategy for core-shell magnetic nanoparticles. J Am Chem Soc 127:16090–16097

Li P, Miser DE, Rabiei S, Yadav RT, Hajaligol MR (2003a) The removal of carbon monoxide by iron oxide nanoparticles. Appl Catal B 43:151–162

Li YH, Ding J, Luan Z et al (2003b) Competitive adsorption of Pb^{2+}, Cu^{2+} and Cd^{2+} ions from aqueous solutions by multiwalled carbon nanotubes. Carbon 41:2787–2792

Li X, He G, Xiao G, Liu H, Wang M (2009) Synthesis and morphology control of ZnO nanostructures in microemulsions. J Colloid Interface Sci 333:465–473

Lifshitz IM, Slyozov VV (1961) The kinetics of precipitation from supersaturated solid solutions. J Phys Chem Solids 19:35–50

Lim TT, Feng J, Zhu BW (2007) Kinetic and mechanistic examinations of reductive transformation pathways of brominated methanes with nano-scale Fe and Ni/Fe particles. Water Res 41:875–883

Liu S, Yu J, Jaroniec M (2011) Anatase TiO$_2$ with dominant high-energy {001} facets: synthesis, properties, and applications. Chem Mater 23:4085–4093

Liu M, Wang Z, Zong S et al (2014) SERS detection and removal of mercury (II)/silver (I) using oligonucleotide-functionalized core/shell magnetic silica sphere@ Au nanoparticles. ACS Appl Mater Interfaces 6:7371–7379

Lowell S, Shields JE, Thomas MA, Thommes M (2004) Surface area analysis from the Langmuir and BET theories. Characterization of porous solids and powders: surface area, pore size and density. Particle Technology Series, vol 16. Springer, Dordrecht

Lu C, Chiu H, Liu C (2006) Removal of zinc (II) from aqueous solution by purified carbon nanotubes: kinetics and equilibrium studies. Ind Eng Chem Res 45:2850–2855

Macak JM, Stein FS, Schmuki P (2007) Efficient oxygen reduction on layers of ordered TiO$_2$ nanotubes loaded with Au nanoparticles. Electrochem Commun 9:1783–1787

Mahato TH, Prasad GK, Singh B, Acharya J, Srivastava AR, Vijayaraghavan R (2009) Nanocrystalline zinc oxide for the decontamination of sarin. J Hazard Mater 165:928–932

Mansoori GA (2002) Advances in atomic & molecular nanotechnology. United Nations Tech Monitor; UN-APCTT Tech Monitor, 2002; Special Issue: 53 59

Masciangioli T, Zhang WX (2003) Peer reviewed: environmental technologies at the nanoscale. Environ Sci Technol 37:102A–108A

Mattigod SV, Feng X, Fryxell GE, Liu J, Gong M (1999) Separation of complexed mercury from aqueous wastes using self-assembled mercaptan on mesoporous silica. Sep Sci Technol 34:2329–2345

Mayo JT, Yavuz C, Yean S et al (2007) The effect of nanocrystalline magnetite size on arsenic removal. Sci Technol Adv Mater 8:71–75

Mehndiratta P, Jain A, Srivastava S, Gupta N (2013) Environmental pollution and nanotechnology. Environ Pollut 2:49

Mizutani N, Iwasaki T, Watano S, Yanagida T, Tanaka H, Kawai T (2008) Effect of ferrous/ferric ions molar ratio on reaction mechanism for hydrothermal synthesis of magnetite nanoparticles. Bull Mater Sci 31:713–717

Moreno N, Querol X, Ayora C, Pereira CF, Janssen-Jurkovicová M (2001) Utilization of zeolites synthesized from coal fly ash for the purification of acid mine waters. Environ Sci Technol 35:3526–3534

Murakami N, Kurihara Y, Tsubota T, Ohno T (2009) Shape-controlled anatase titanium (IV) oxide particles prepared by hydrothermal treatment of peroxo titanic acid in the presence of polyvinyl alcohol. J Phys Chem C 113:3062–3069

Nador F, Moglie Y, Vitale C, Yus M, Alonso F, Radivoy G (2010) Reduction of polycyclic aromatic hydrocarbons promoted by cobalt or manganese nanoparticles. Tetrahedron 66:4318–4325

Nirmala MJ, Shiny PJ, Ernest V et al (2013) A review on safer means of nanoparticle synthesis by exploring the prolific marine ecosystem as a new thrust area in nanopharmaceutics. Int J Pharm Sci 5:23–29

Northup A, Cassidy D (2008) Calcium peroxide (CaO_2) for use in modified Fenton chemistry. J Hazard Mater 152:1164–1170

Oliveira LCA, Petkowicz DI, Smaniotto A, Pergher SBC (2004) Magnetic zeolites: a new adsorbent for removal of metallic contaminants from water. Water Res 38:3699–3704

Onyango MS, Kojima Y, Matsuda H, Ochieng A (2003) Adsorption kinetics of arsenic removal from groundwater by iron-modified zeolite. J Chem Eng Jpn 36:1516–1522

Ostwald W (1900) Über die vermeintliche Isomerie des roten und gelben Quecksilberoxyds und die Oberflächenspannung fester Körper. Z Phys Chem 34:495–503

Pan B, Xing B (2008) Adsorption mechanisms of organic chemicals on carbon nanotubes. Environ Sci Technol 42:9005–9013

Parra S, Stanca SE, Guasaquillo I, Thampi KR (2004) Photocatalytic degradation of atrazine using suspended and supported TiO_2. Appl Catal B 51:107–116

Pena ME, Korfiatis GP, Patel M, Lippincott L, Meng X (2005) Adsorption of As (V) and As (III) by nanocrystalline titanium dioxide. Water Res 39:2327–2337

Pereira KRO et al (2005) Brazilian organoclays as nanostructured sorbents of petroleum-derived hydrocarbons. Mat Res 8:77–80

Qi L, Xu Z (2004) Lead sorption from aqueous solutions on chitosan nanoparticles. Colloids Surf A Physicochem Eng Asp 251:183–190

Qiao S, Sun DD, Tay JH, Easton C (2003) Photocatalytic oxidation technology for humic acid removal using a nano-structured TiO_2/Fe_2O_3 catalyst. Water Sci Tech 47:211–217

Qu X, Alvarez PJJ, Li Q (2013) Applications of nanotechnology in water and wastewater treatment. Water Res 47:3931–3946

Reiss H (1951) The growth of uniform colloidal dispersions. J Chem Phys 19:482–487

Ristić M, Musić S, Ivanda M, Popović S (2005) Sol-gel synthesis and characterization of nanocrystalline ZnO powders. J Alloys Compd 397:L1–L4

Rizwan M, Singh M, Mitra CK, Morve RK (2014) Ecofriendly application of nanomaterials: Nanobioremediation. J Nanopart Res 2014

Saien J, Shahrezaei F (2012) Organic pollutants removal from petroleum refinery wastewater with nanotitania photocatalyst and UV light emission. Int J Photoenergy 2012:1

Salipira KL, Mamba BB, Krause RW, Malefetse TJ, Durbach SH (2007) Carbon nanotubes and cyclodextrin polymers for removing organic pollutants from water. Environ Chem Lett 5:13–17

Schrick B, Blough JL, Jones AD, Mallouk TE (2002) Hydrodechlorination of trichloroethylene to hydrocarbons using bimetallic nickel-iron nanoparticles. Chem Mater 14:5140–5147

Schrick B, Hydutsky BW, Blough JL, Mallouk TE (2004) Delivery vehicles for zerovalent metal nanoparticles in soil and groundwater. Chem Mater 16:2187–2193

Schwarzer HC, Peukert W (2004) Tailoring particle size through nanoparticle precipitation. Chem Eng Commun 191:580–606

Shao D, Sheng G, Chen C, Wang X, Nagatsu M (2010) Removal of polychlorinated biphenyls from aqueous solutions using β-cyclodextrin grafted multiwalled carbon nanotubes. Chemosphere 79:679–685

Sugimoto T (2007) Underlying mechanisms in size control of uniform nanoparticles. J Colloid Interface Sci 309:106–118

Takafuji M, Ide S, Ihara H, Xu Z (2004) Preparation of poly (1-vinylimidazole)-grafted magnetic nanoparticles and their application for removal of metal ions. Chem Mater 16:1977–1983

Thanh NTK, Maclean N, Mahiddine S (2014) Mechanisms of nucleation and growth of nanoparticles in solution. Chem Rev 114:7610–7630

Van Dillewijn P, Caballero A, Paz JA, González-Pérez MM, Oliva JM, Ramos JL (2007) Bioremediation of 2, 4, 6-trinitrotoluene under field conditions. Environ Sci Technol 41:1378–1383

Vorobyova SA, Lesnikovich AI, Mushinskii VV (2004) Interphase synthesis and characterization of zinc oxide. Mater Lett 58:863–866

Wagner C (1961) Theorie der alterung von niederschlägen durch umlösen (Ostwald-reifung). Zeitschrift für Elektrochemie, Berichte der Bunsengesellschaft für physikalische Chemie 65:581–591

Wang CB, Zhang WX (1997) Synthesizing nanoscale iron particles for rapid and complete dechlorination of TCE and PCBs. Environ Sci Technol 31:2154–2156

Wang J, Peng Z, Huang Y, Chen Q (2004) Growth of magnetite nanorods along its easy-magnetization axis of [110]. J Cryst Growth 263:616–619

Wang Y, Zhang C, Bi S, Luo G (2010) Preparation of ZnO nanoparticles using the direct precipitation method in a membrane dispersion micro-structured reactor. Powder Technol 202:130–136

Watzky MA, Finke RG (1997) Transition metal nanocluster formation kinetic and mechanistic studies. A new mechanism when hydrogen is the reductant: slow, continuous nucleation and fast autocatalytic surface growth. J Am Chem Soc 119:10382–10400

Wei J, Xu X, Liu Y, Wang D (2006) Catalytic hydrodechlorination of 2, 4-dichlorophenol over nanoscale Pd/Fe: reaction pathway and some experimental parameters. Water Res 40:348–354

Wu R, Qu J, Chen Y (2005) Magnetic powder MnO-Fe_2O_3 composite-a novel material for the removal of azo-dye from water. Water Res 39:630–638

Yadav TP, Yadav RM, Singh DP (2012) Mechanical milling: a top down approach for the synthesis of nanomaterials and nanocomposites. J Nanosci Nanotechnol 2:22–48

Yadav A, Chowdhary P, Kaithwas G, Bharagava RN (2017) Toxic metals in environment, threats on ecosystem and bioremediation approaches. In: Das S, Dash HR (eds) Handbook of metal-microbe interactions and bioremediation. CRC Press, Taylor & Francis Group, Boca Raton, p 813

Yang K, Xing B (2010) Adsorption of organic compounds by carbon nanomaterials in aqueous phase: polanyi theory and its application. Chem Rev 110:5989–6008

Yantasee W, Lin Y, Fryxell GE, Busche BJ, Birnbaum JC (2003) Removal of heavy metals from aqueous solution using novel nanoengineered sorbents: self-assembled carbamoylphosphonic acids on mesoporous silica. Sep Sci Technol 38:3809–3825

Yavuz CT, Mayo JT, William WY et al (2006) Low-field magnetic separation of monodisperse Fe_3O_4 nanocrystals. Science 314:964–967

Zhang W, Wang CB, Lien HL (1998) Treatment of chlorinated organic contaminants with nanoscale bimetallic particles. Catal Today 40:387–395

Zhang Q, Gao L, Guo J (2000) Effects of calcination on the photocatalytic properties of nanosized TiO_2 powders prepared by $TiCl_4$ hydrolysis. Appl Catal B 26:207–215

Zhang WX (2003) Nanoscale iron particles for environmental remediation: an overview. J Nanopart Res 5:323–332

Zhang J, Wang J, Zhou S et al (2010) Ionic liquid-controlled synthesis of ZnO microspheres. J Mater Chem 20:9798–9804

Zhao X, Lv L, Pan B, Zhang W, Zhang S, Zhang Q (2011) Polymer-supported nanocomposites for environmental application: a review. Chem Eng J 170:381–394

Ziolli RL, Jardim WF (2001) Photocatalytic decomposition of seawater-soluble crude oil fractions using high surface area colloid nanoparticles of TiO_2. J Photochem Photobiol 5887:1–8

Zhong LS, Hu JS, Liang HP, Cao AM, Song WG, Wan LJ (2006) Self-assembled 3D flowerlike iron oxide nanostructures and their application in water treatment. Adv Mater 18:2426–2431

Zhong LS, Hu JS, Cao AM, Liu Q, Song WG, Wan LJ (2007) 3D flowerlike ceria micro/nano-composite structure and its application for water treatment and CO removal. Chem Mater 19:1648–1655

Zou Y, Wang X, Khan A et al (2016) Environmental remediation and application of nanoscale zero-valent iron and its composites for the removal of heavy metal ions: a review. Environ Sci Technol 50:7290–7304

Chapter 17
Biphasic Treatment System for the Removal of Toxic and Hazardous Pollutants from Industrial Wastewaters

Ali Hussain, Sumaira Aslam, Arshad Javid, Muhammad Rashid, Irshad Hussain, and Javed Iqbal Qazi

Abstract Industrial effluents carrying diverse pollutants are discharged freely into the adjacent environments and percolate to the groundwater resources. Currently, the treatment strategies also consider recycling and reuse with the energy recovery. Novel approaches to remove these pollutants include biphasic systems. Different biphasic systems including liquid-liquid two-phase partitioning and solid-liquid partitioning systems have proved successful for the cleaning of effluents containing textile dyes, heavy metals, organic contaminants, pharmaceutical ingredients and many other xenobiotic compounds. The system efficacy is based on the careful selection of the phase-forming substance/polymer as well as control and maintenance of the operational parameters including temperature, pH and hydraulic retention time. Among the biological parameters, selection of the microbes either pure cultures or mixed cultures plays a very important role for the removal of xenobiotics in biphasic systems.

Keywords Aqueous biphasic system · Biotreatment · Hazardous wastes · Remediation · Wastewater treatment · Xenobiotics

---

A. Hussain (✉) · A. Javid
Applied and Environmental Microbiology Laboratory, Department of Wildlife and Ecology, University of Veterinary and Animal Sciences, Lahore, Pakistan
e-mail: ali.hussain@uvas.edu.pk

S. Aslam
Microbiology and Biotechnology Laboratory, Department of Zoology, Government College Women University, Faisalabad, Pakistan

M. Rashid · I. Hussain
General Chemistry Laboratory, Faculty of Fisheries and Wildlife, University of Veterinary and Animal Sciences, Lahore, Pakistan

J. I. Qazi
Microbial Biotechnology Laboratory, Department of Zoology, University of the Punjab, Lahore, Pakistan

1 Introduction

Industrial pollution has been one of the addressable environmental issues causing degradation of the water, air and soil (Govindarajulu 2003; Bharagava et al. 2017; Chowdhary et al. 2017). The growing levels of industrial wastes not only deteriorate the environmental quality but also the human health as well. The nature of the pollutant, its generation source and final destination in the environment are important elements, which determine the level of harmful effects (Ogunfowokan et al. 2005; Jimena et al. 2008; Rajaram and Ashutost 2008).

Nanotechnology is an advanced science, which could play a significant role in environmental cleanup and pollution control (Darnault et al. 2005). But aqueous biphasic system (ABS) is a more attractive process due to its simple, easy and economic scale-up procedures. Recently, ABS composed of polyethylene glycol (PEG) and sodium citrate has successfully been employed for dye removal from the xenobiotics (Ivetic et al. 2013). ABS has been proven to be an efficient and economical process compared to the precipitation, chromatography and other separation systems (Aguilar et al. 2006; Naganagouda and Mulimani 2008; Yazbik and Ansorge-Schumacher 2010).

ABS is a fractionation system with enormous potential for the processing of diverse biomolecules at the industrial sector. The partitioning behaviour of the method is though complicated to predict yet it has been successfully exploited for the detection of drug residues in food, treatment of the wastewaters and metals' extraction. It is an economic and environment-friendly method with high recovery yield and can be up scaled easily (Iqbal et al. 2016). This chapter focuses particularly on the types, thermodynamics, optimization and applications of ABS for the removal of toxicants from industrial effluents.

2 Types of Biphasic Systems

ABS or aqueous two-phase systems (ATPS) consist of two immiscible liquid phases comprising chiefly of water that coexist together in equilibrium under appropriate concentrations of a pair of solutes at ambient temperature and under the limits of various other physical parameters like pH, ionic strength, density, viscosity, dielectric constant, liquid temperature range, etc. (Sen and Chakraborty 2016). Beijernick, in 1896, for the first time observed the formation of ABS when he noticed the incompatibility between the aqueous solutions of two polymers, namely, agar and starch or gelatin. The two solutions of these polymers were separated into two phases upon mixing together. Albertsson (1958) utilized ABS as liquid-liquid extraction system for partition of proteins. Since then researches led to the development of many other ABS comprising of polymer-polymer, polymer-salt, salt-salt and several other combinations (Freire et al. 2012). These systems avoid the use of volatile organic solvents which are costly, toxic and flammable, offer poor miscibility with water and cause disposal problems as in conventional liquid-liquid solvent extraction. On the other hand, ABS are based on green chemistry principle because they consist predominantly of water and their other constituents like polymers and electrolytes and offer low interfacial tension, non-flammability, non-volatility, good biocompatibility with the living cells and bioactive compounds, recyclability, low cost and fast phase separation rates, and all are environment-friendly (Rodrigues et al. 2013; de Souza et al. 2014; Patricio et al. 2016). Therefore, these systems have been extensively investigated and utilized as extraction/separation media for a broad range of compounds including amino acids, phenols, alkaloids, antibiotics, enzymes, proteins, dyes and pigments, cells and organelles, genetic material, bionanoparticles, etc. (Souza et al. 2015; Chandra and Chowdhary 2015). The following are the types of ABS:

2.1 Polymer-Polymer-Based ABS

This type of ABS is formed when aqueous solutions of two hydrophilic and incompatible polymers above their critical concentrations are mixed together. Among the polymers used for the formation of this kind of ABS are PEGs, polyethylene oxides (PEOs), polypropylene glycols (PPGs), polyethylene glycol dimethyl ether (PEGDME) and polyvinyl pyrrolidone (PVP), dextran, starch, gelatin, etc. The driving force for the formation of two phases comes from the loss in entropy during segregation of the components and formation of hydrogen bonds between water and end groups of the polymer units – the monomers (OH in PEG, PPG dextran and starch, OCH_3 in PEGDME, O atoms in PEOs and PVP). Indeed formation of polymer-based ABS is an endothermic process, and phase separation is an entropy-driven phenomenon. However, there is small loss in entropy because of large size of polymers; therefore, the phase separation phenomenon occurs at very low concentrations (Machado et al. 2012; Chakraborty and Sen 2016; Sadeghi and Maali 2016).

2.2 Polymer-Salt-Based ABS

The polymer-polymer systems have been extensively used to study the formation, characterization and extraction systems for variety of substances in many researches. However, they suffer from certain drawbacks, e.g. high viscosities of the two coexisting phases, and some polymers like dextran are also expensive. To overcome these drawbacks, research trends have been shifted to the use of polymer-salt ABS for the same purpose. These systems use cheaper components, offer higher density difference and lower viscosities of the two phases and thus provide faster separation rates. The most widely used polymers in these formulations are PEGs because they are cost-effective and offer attractive features like high water solubility, enhanced biodegradation, negligible volatility, low toxicity and low melting temperature

Table 17.1 Examples of polymer-salt-based ABS

Sr. No.	Phase-forming components	Substance(s) separated	Reference(s)
1.	PEG + K₃PO₄	Proteins	de Belval et al. (1998)
2.	PEG 200, PEG 400, PEG 1000, PEG 2000, PEG 4000, PEG 8000 + Tripotassium citrate +5% I L	Immunoglobulin G	Ferreira et al. (2016)
3.	PEG 600 + potassium citrate +1-butyl-3-methylimidazolium bromide	L-Tyrosine	Hamzehzadeh and Abbasi (2015)
4.	PEG 400 + Na₂SO₄ + H₂O (20 % + 16 % + 64%)	Polysaccharides	Zhou et al. (2014)
5.	PEG 1000 + trisodium citrate	α and β Lactoglobulin	Kalaivani et al. (2015)
6.	PEG 6000 + 1.5 M Na₂SO₄	I⁻, IO₄⁻ and povidone-iodine	Paik and Sen (2016)

(Pereira et al. 2010; Ferreira et al. 2016). Besides PEGs, polyethylene oxides (PEOs) and polypropylene glycols have also been used along with variety of inorganic and organic salts and ionic liquids (ILs) as well. Some common examples of polymer-salt-based ABS are shown in Table 17.1.

2.3 Ionic-Liquid-Based ABS/Salt-Salt-Based ABS

Ionic liquids are basically salts that exist as liquids at temperature below 100 °C and are generally composed of bulky organics like cationic part attached with the anions. A typical example is the 1-ethyl-3-methylimidazolium chloride that melts at −21 °C (Chakraborty and Sen 2017). Rogers and co-workers in 2003 reported the formation of ABS by the addition of various inorganic salts to the aqueous solutions of ILs. Consequently, a wide variety of ionic-liquid-based ABS in combination with inorganic and organic salts, anionic surfactants, amino acids, polymers, polysaccharides and other substrates can be formed (Gao et al. 2014). In the past few years, the use of ILs as extraction systems has been increased because of their unique features such as high thermal and chemical stability; non-flammability; very low volatility; wide liquid temperature range; high solvating capacity for organic, inorganic and organometallic compounds; and their fine-tunable physiochemical properties when different cations and anions are used (Sheikhian et al. 2014). The ILs used in these systems include salts of imidazolium, ammonium, phosphonium, pyridinium, piperidium, pyrrolidinium cations, etc. Some ionic-liquid-based ABS is presented in Table 17.2.

Table 17.2 Examples of ionic-liquid-based ABS

Sr. No.	Phase-forming components	Substance(s) separated	Reference(s)
1.	1-Allyl-3-methylimidazolium chloride + PEG 2000, 1-(2-hydroxyethyl)-3-methylimidazolium + PEG 2000, chloride, 1-methylimidazolium chloride + PEG 2000, 1, 3-dimethylimidazolium chloride + PEG 2000, 1-ethyl-3-methylimidazolium chloride + PEG 2000, 1-ethyl-3-methylimidazolium acetate + PEG 2000, 1-ethyl-3-methylimidazolium methanesulphonate + PEG 2000, 1-ethyl-3-methylimidazolium hydrogenosulphate + PEG 2000, 1-ethyl-3-methylimidazolium dimethylphosphate + PEG 2000	Caffeine, xanthine, nicotine	Pereira et al. (2013)
2.	1-Butyl-3-methylimidazolium tetrafluoroborate + L-lysine, 1-butyl-3-methylimidazolium tetrafluoroborate + D, L-lysine HCl, 1-butyl-3-methylimidazolium tetrafluoroborate + L-proline, 1-butyl-3-methylimidazolium triflate + L-lysine, 1-butyl-3-methylimidazolium triflate + D, L-lysine HCl, 1-butyl-3-methylimidazolium triflate + L-proline, 1-butyl-3-methylimidazolium dicyanamide + L-lysine, 1-butyl-3-methylimidazolium dicyanamide + D, L-lysine HCl, 1-butyl-3-methylimidazolium dicyanamide + L-proline	Caffeine, ciprofloxacin	Domínguez-Pérez et al. (2010)
3.	Sodium polyacrylate NaPA 8000 + PEG 8000	Cytochrome C, chloranilic acid	Santos et al. (2015)
4.	Tetrabutylammonium chloride + buffer ($K_3C_6H_5O_7.H_2O/C_6H_8O_7.H_2O$), tetrabutylphosphonium bromide + buffer ($K_3C_6H_5O_7.H_2O/C_6H_8O_7.H_2O$), tetrabutylphosphonium chloride + buffer ($K_3C_6H_5O_7.H_2O/C_6H_8O_7.H_2O$), tri(isobutyl)methylphosphoniumtosylate + buffer ($K_3C_6H_5O_7.H_2O/C_6H_8O_7.H_2O$), tri(butyl) methylphosphoniummethylsulfate + buffer ($K_3C_6H_5O_7.H_2O/C_6H_8O_7.H_2O$)	Bovine serum albumin	Pereira et al. (2015)

2.4 Surfactant-Based ABS

Chemically, surfactants are sodium salts of alkyl hydrogen sulphates of higher alcohols or alkyl benzene sulphonates. These are of three types, viz. nonionic (lauryl alcohol ethoxylate having formula $CH_3(CH_2)_{10}CH_2(OCH_2CH_2)_8OH$), cationic (cetyltrimethyl ammonium chloride $CH_3(CH_2)_{15}N^+(CH_3)_3Cl$) and anionic detergents, e.g. sodium lauryl sulphate $C_{11}H_{23}CH_2OSO_2Na$ and sodium dodecyl benzene sulphonate (SDS). They are the main constituents of cleansing agents, soaps, detergents and shampoos. They are also used for foam formation in various gels such as emulsifiers, toothpastes, wetting agents and shaving creams, etc. The industrial uses of surfactants range from pharmaceuticals and food to paper and enrichment of minerals and ores of commercially important metals in metallurgy.

Table 17.3 Examples of surfactant-based ABS

Sr. No.	Phase-forming components	Substance(s) separated	Reference(s)
1.	Triton X-45 + Pluronic L-31, Pluronic L-61, Pluronic L-81, Pluronic L-121, Triton X-114 + Pluronic L-31, Pluronic L-61, Pluronic L-81, Pluronic L-121	Extraction of α and β mangostins from *Garcinia mangostana*	Tan et al. (2017)
2.	Tween 20 + trisodium citrate, Tween 20 + magnesium sulphate, Tween 20 + sodium sulphate	Cefalexin	Taghavivand and Pazuki (2014)
3.	Triton X-114 + SDS + NaCl	Vanillin	Safonova et al. (2014)
4.	Decaethylene glycol monotridecyl ether + NaCl, NaSCN, KCl, Na_2SO_4	Bi(III)	Didi et al. (2011)
5.	Triton 114 + Triton 100 + dextran sulphate	Clavulanic acid	Silva et al. (2015)

The molecular structures of surfactants contain two or more groups of atoms that possess different affinities for a solvent within which they establish different interactions with a wide range of molecules ranging from hydrophobic and zwitterionic species to highly charged ions and other biomolecules such as vitamins, amino acids, proteins, lipids, etc. Studies on effects of surfactants with biomolecules have attracted attention of researchers to utilize them as extraction/separation media for bioactive compounds (Table 17.3).

2.5 Alcohol-Based ABS

Alcohols are the organic compounds that contain one or more hydroxyl groups in their structural formula and named as monohydric, dihydric and polyhydric alcohols depending upon the number of hydroxyl groups present. Normally, lower alcohols are miscible with water in any proportion and thus form one phase. However, in the presence of certain additives like salts, alcohols also form two-phase systems. The alcohol-/salt-based ABS were first studied by Greve and Kula (1991). Since then many research groups have opted to use these systems for extraction and purification of biomolecules in downstream processes in biotechnology because these systems offer better and efficient recovery of target compound simply by evaporation method. Alcohol-/salt-based ABS offer other advantages like easy scale-up and low environmental toxicity and cost (Goja et al. 2014). Literature survey reveals a lot of purification studies based on these systems involving glycyrrhizin, betalains, nucleic acids, geniposide, salvianolic acid B, lipase, human interferon, etc. (Lin et al. 2013). Pratiwi et al. (2015) extracted succinic acid, Liu et al. (2006) extracted iodide complex of Cd(II), and Ooi et al. (2009) purified lipase derived from *Burkholderia pseudomallei* using alcohol-based ATPS.

3 Thermodynamics of Biphasic Systems

Thermodynamically when a system is at equilibrium state, each component of the phase has some partial molar free energy, which is always equal to the partial molar free energy of every other phase component of the system. This partial molar free energy of each component of phase system is a function of temperature, pressure and concentration variables (Bhal et al. 2012).

3.1 Criterion of Phase Transitions

The phase transitions in a system at equilibrium can be explained by two synonymous terms, molar Gibbs energy (G_m) of the system and the chemical potential (μ). The chemical potential and molar Gibbs energy have the same meaning for one component system ($\mu = G_m$), but chemical potential has a broader significance in explaining the potential of a substance undergoing physical and chemical changes in a system. The chemical potential of a species in a mixture is defined as the rate of change of its free energy with respect to change in the number of molecules, ions or atoms added to the system initially present at equilibrium. The chemical potential is measured in units of energy per particle or equivalently energy per mole. It is used to take into account chemical reactions and states of bodies that are chemically different from each other in many aspects of equilibrium systems involving phase diagrams in melting, boiling, osmosis, solubility, evaporation, partition coefficient, liquid-liquid extraction, chromatography and electrochemical potential of ions in electrochemical series. The phase transition is based on the fundamental principle that the particles tend to move from higher chemical potential to lower chemical potential (Atkins and de Paula 2010).

3.2 Thermodynamic Criterion of Equilibrium State of System

According to the second law of thermodynamics, *at equilibrium state a substance has the same and uniform chemical potential throughout the system, no matter the system contains how many numbers of phases in it.* In other words, when the liquid and solid phases of a substance are in equilibrium, the chemical potential of the substance is uniform everywhere within the system. To testify this statement, consider a system in which the substance has a chemical potential μ_1 at one point and μ_2 at another point. These points may be present in the same phase or in different phases. When a small amount of matter is taken away from point 1, as a result, the change in Gibbs energy of the system is represented by an amount $-\mu_1 dn$. In a similar way, whenever a little amount of material is put at point 2, then change in Gibbs energy of the system is symbolized by $+\mu_2 dn$. It means that the change in Gibbs free energy of system occurred whenever d*n* amount of substances is shifted from one point to another point. Consequently, the net change in Gibbs energy of the system is dG = (μ₂ − μ₁) dn. If the chemical potential at point 1 is equal to the chemical potential at point 2 (μ₁ = μ₂), then there is no change in Gibbs energy (G) indicating that the system is at equilibrium state. If the chemical potential of a substance is higher at the first point as compared to the second point, the Gibbs energy of system is decreased leading towards spontaneous change in system. The chemical potential of two phases depends upon the transition temperature T_{trs}, and both phases have the same chemical potential at this temperature (Atkins and de Paula 2010).

3.3 The Phase Diagrams and Phase Boundary

For multiphasic systems, the coexistence of various phases at equilibrium as a function of combinations of temperature, pressure, composition or other variables is graphically represented by phase diagrams. For a given temperature and pressure, phase diagrams are used to determine the (1) phases that are present, (2) compositions of the phases and (3) relative fractions of the phases (Atkins and de Paula 2010). The following is an example of a phase diagram for a generic single component system (Fig. 17.1).

Phase diagrams are divided into three single-phase regions that cover the pressure-temperature space over which the matter being evaluated exists in liquid, gaseous and solid states. The lines that separate these single-phase regions are

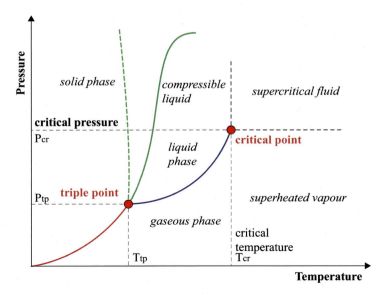

Fig. 17.1 Phase diagram of one component system showing different phases and phase boundaries at ambient temperatures and pressures

known as phase boundaries. Along these boundaries, the matter exists simultaneously in equilibrium between the two states. The two phases can coexist in a state of equilibrium as a function of temperature and pressure. It means that two phases at equilibrium state must have equal chemical potential (μ):

$$\mu_\alpha(p,T) = \mu_\beta(p,T)$$

where α is for one phase and the second phase is denoted by β (Atkins and de Paula 2010).

3.3.1 The Phase Boundary Slope

The ratio in the small change in pressure to the small change in temperature, i.e. dp/dT, is the slope of phase boundary. Since the system is in equilibrium, therefore, it is supposed that the small changes in pressure and temperature do not disturb the phases, α and β, from their equilibrium positions. In the beginning, the chemical potential, μ, of these two phases is equal which is already at equilibrium position. Such conditions continued to behave in a similar way when the state of system is altered to another point on the phase boundary in the system, when two phases keep their state of equilibrium as depicted in the figure given below. Thus, α- and β-phases experience an equal change in chemical potential and are written as dμ_α = dμ_β. Applying the expression for change in Gibbs free energy (dG = Vdp − SdT) to the phases at phase boundary that dμ = − S_mdT + V_mdp, it follows that −$S_{\alpha,m}$dT + $V_{\alpha,m}$dp = − $S_{\beta,m}$dT + $V_{\beta,m}$dp

In the above eq., the $V_{\alpha,m}$ and $V_{\beta,m}$ are the molar volumes, while the $S_{\alpha,m}$ and $S_{\beta,m}$ are the molar entropies of the two phases, respectively.

Hence, ($V_{\beta,m}$ − $V_{\alpha,m}$)dp = ($S_{\beta,m}$ − $S_{\alpha,m}$)dT

Rearrangement of the above equation in the form of Clapeyron equation is as

$$\frac{dP}{dT} = \frac{\Delta_{trs}S}{\Delta_{trs}V}$$

In the above equation, the term $\Delta_{trs}S = S_{\beta,m} - S_{\alpha,m}$ means transition in entropy, while the term $\Delta_{trs}V = V_{\beta,m} - V_{\alpha,m}$ shows the transition in volume, respectively. So, the Clapeyron equation can be applied to any kind of pure substance which is at any equilibrium phase of system. The slope dp/dT of phase boundary can be interpreted well via Clapeyron equation. It means that the shape of phase diagram can be predicted by analysing the thermodynamic data. Furthermore, the boiling and freezing point of the substance can be anticipated when pressure is applied on the system (Atkins and de Paula 2010).

Fig. 17.2 A typical phase diagram showing the relationship between two variables dp and dT. If one of the variables is changed, the system keeps restoring its equilibrium state (Atkins and de Paula 2010)

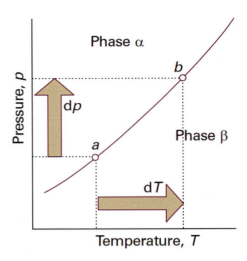

3.3.2 The Solid-Liquid Phase Boundary

The melting or fusion value of the solid phase can be determined by a change of molar enthalpy ($\Delta_{fus}H$) of a substance that takes place at a temperature (T). Therefore, the melting value in terms of molar enthalpy at a temperature (T) is $\Delta_{fus}H/T$, so the Clapeyron equation can be written as

$$\frac{dP}{dT} = \frac{\Delta_{fus}H}{T\Delta_{fus}V}$$

where $\Delta_{fus}V$ is the value of the molar volume change on melting. In this equation, the value of enthalpy of melting temperature of substance is usually positive (excluding the element helium-3), and as a rule the change in volume of substance is also small and positive. Accordingly, the dp/dT shows the sharp steep slope with positive value as shown in Fig. 17.2.

The slope dp/dT for solid-liquid phase boundary can be determined using integration by supposing that there is so small change of $\Delta_{fus}H$ and $\Delta_{fus}V$ as a function of temperature and pressure. For this, both temperature and pressure are considered constant. The integration of the equation between the limits of the pressure p^* at melting temperature T^* and pressure p at the temperature T on the system can be written as

$$\int_{p^*}^{P} dp = \frac{\Delta_{fus}H}{\Delta_{fus}V} \int_{T^*}^{T} \frac{dT}{T}$$

The approximate solution of the above equation is

Fig. 17.3 A phase diagram for solid-liquid phase boundary (Atkins and de Paula 2010)

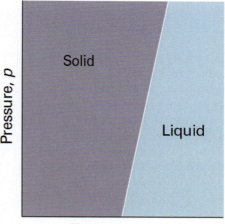

$$p \approx p^* + \frac{\Delta_{fus}H}{\Delta_{fus}V}\ln\frac{T}{T^*}$$

When the value of T is nearer to T^* and the estimated logarithm of above equation is

$$\ln\frac{T}{T^*} = \ln\left(1 + \frac{T-T^*}{T^*}\right) \approx \frac{T-T^*}{T^*}$$

therefore,

$$p \approx p^* + \frac{\Delta_{fus}H}{T^*\Delta_{fus}V}(T-T^*)$$

According to the above equation, when pressure is plotted as a function of T, a sharp and steep straight line in upward direction is obtained as depicted in Fig. 17.3. The slope indicates that when the pressure is increased on a system, the melting temperature of substances rises. Many of the substances follow this trend (Atkins and de Paula 2010).

3.3.3 The Liquid-Vapour Phase Boundary

The term $\Delta_{vap}H/T$ is equal to the vaporization entropy $\Delta_{vap}H$ at a temperature T in the liquid-vapour boundary. The Clapeyron equation for the system comprising of liquid-vapour phase boundary is written as

$$\frac{dP}{dT} = \frac{\Delta_{VAP}H}{T\Delta_{VAP}V}$$

In this equation, the term $\Delta_{vap}H/T$ is equal to the enthalpy of vaporization at temperature T in the liquid-vapour phase boundary. This equation shows that the quantity $\Delta_{vap}V$ is positive and large, so the boiling temperature of the substance responds faster as compared to the freezing temperature at a given pressure. The reason is that molar volume of a gas $V_m(g)$ is many times greater than liquid molar volume; hence,

$\Delta_{vap}V \approx V_m(g)$ where $V_m(g) = RT/p$ if the gas follows the ideal behaviour. Hence the above equation can be rearranged as follows:

$$\frac{dP}{dT} = \frac{\Delta_{VAP}H}{T\left(\frac{RT}{P}\right)}$$

When the vapour pressure p of a system varies with the temperature T, then the above Clapeyron equation can be again rearranged into Clausius-Clapeyron equation as

$$\frac{d\ln p}{dT} = \frac{\Delta_{VAP}H}{RT^2}$$

The variations in vapour pressure with temperature and similarly the variations of boiling temperature with pressure can be predicted by integrating the above relation as follows:

$$\int_{\ln p^*}^{\ln p} d\ln p = \frac{\Delta_{VAP}H}{R}\int_{T^*}^{T}\frac{dT}{T^2} = \frac{\Delta_{VAP}H}{R}\left(\frac{1}{T} - \frac{1}{T^*}\right)$$

In above equation, the vapour pressure is p^* at temperature T^* and in same way the vapour pressure is p at the temperature T. Hence, the integral of the left-hand side of the above equation is $\ln(p/p^*)$, while the relationship between two vapour pressures is shown as

$$p = p^*e^{-\chi} \quad \chi = \frac{\Delta_{VAP}H}{R}\left(\frac{1}{T} - \frac{1}{T^*}\right)$$

A typical phase diagram plot of the liquid-vapour phase boundary is depicted in Fig. 17.4 which is the exact expression of the above equation (Wedler 1997; Atkins and de Paula 2010). The liquid could not exist above the critical temperature T_c. Therefore, the line cannot be extra plotted away from the temperature T_c.

Fig. 17.4 A phase diagram of liquid-vapour phase boundary (Atkins and de Paula 2010)

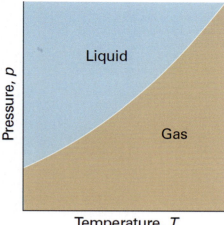

3.3.4 Solid-Vapour Phase Boundary

The main difference between the solid and vapour phase boundary is that the vaporization enthalpy $\Delta_{vap}H$ is substituted by the sublimation enthalpy $\Delta_{sub}H$ of a substance in the Clausius- Clapeyron equation, which for the solid-vapour phase boundary can be written as

$$\frac{dp}{dT} = \frac{\Delta_{sub}H}{T\Delta_{sub}V}$$

Since the value of sublimation enthalpy is much larger than the value of vaporization enthalpy ($\Delta_{sub}H \ggg \Delta_{vap}H$), therefore, this equation reveals that on the same temperature point, the sublimation curve has a steeper slope than the curve of vaporization (Atkins and de Paula 2010). This is closed to a common point where they intersect which is known as the triple point as shown in Fig. 17.5.

4 Factors Influencing Partitioning in Biphasic Systems

The main factors influencing the partitioning of biphasic system include molecular weight, concentration of polymer, hydrophobicity, pH and temperature. By increasing molecular weight, concentration of polymer decreases in biphasic system formation. Hydrophobicity is influenced by tie line length (TLL) and molecular weight of polymer and concentration of salt. Increase in concentration of NaCl influences phase diagram. pH of biphasic system can change charge and surface properties of solute and can affect the partitioning. Change in temperature can also influence biphasic system composition and partitioning by viscosity and density (Iqbal et al.

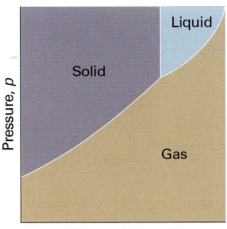

Fig. 17.5 The schematic phase diagram of system closed at the triple point, the temperatures have same values for the slope that was mentioned in the Clausius-Clapeyron equation (Atkins and de Paula 2010)

2016). Silva and Franco (2000) reported that the main factors influencing the partitioning of biphasic system are polymer molecular mass and concentration of polymer and salts. Partitioned proteins are more attracted by smaller polymer size while repelled by larger polymer size when all other factors are constant. When polymer concentration increases, the density of the bottom dextran-rich phase also increases. Salts can affect protein partitioning by changing the physical properties, hydrophobic difference and partitioning of ions in biphasic system. Pei et al. (2007) reported that phase volume ratio, pH, chemical structure of ILs, temperature and salt concentration affect the partitioning of biphasic system. In biphasic systems, the effect of change in pH on partition is more pronounced than phase volume ratio. Similarly, increase in hydrophobicity of the ILs has been reported with increase in length of alkyl substituent on the cation. An improvement in the extraction efficiency of methyl orange has been documented with increase in KCl concentration. Raja et al. (2011) reported that the main factors influencing the partitioning of biphasic system are molecular weight of polymer, pH, neutral salts and surface properties of biomolecules. An increase and decrease in polymer concentration make differences in density, refractive index and viscosity of biphasic system. Expansion in two-phase area is observed with increase in pH and temperature. NaCl concentration alters the partition coefficient, while a linear relationship between hydrophobicity of proteins and partition coefficient has also been observed.

5 Optimization of Biphasic Systems

Various factors are involved in optimum functioning of a biphasic wastewater treatment system, and effects of various physical and chemical characteristics are determined for different contaminants accordingly.

5.1 Hydraulic Retention Time

Hydraulic retention time (HRT) has been found to be a very important parameter for the successful treatment of the textile dye and greatly influences chemical oxygen demand and dye removal (Senthilkumar et al. 2011). The treatment technologies involving sorptive material for bioremediation need complete knowledge of the polarity of the material as well as the target pollutant. The ability of absorption process to recover contaminant does not require higher contact area and is unaffected by reduced surface-to-volume ratios. Prpich et al. (2008) proposed effective solid to liquid ratio between 5% and 10% for this purpose.

5.2 Microbial Processes

An appropriate selection of the biomass source as inocula plays a critical role in determining the degradation of xenobiotics in a two-phase system (TPS). Villemur et al. (2013) used culture as inocula on the basis of microorganisms well adapted to the two-phase portioning bioreactor (TPPB) environment particularly to the immiscible phase for degradation of aromatic molecules. Partitioning rate of toxic material into aqueous phase of a biphasic system is based on metabolism of bacteria. Collins and Daugulis (1997) used TPPB containing *Pseudomonas putida* ATCC 11172 to degrade high concentrations of phenol in batch and fed-batch mode. This partitioning bioreactor employed an aqueous phase and a 500 ml immiscible and biocompatible second organic phase. Marcoux et al. (2000) documented that microbes of interphase are more efficient in degradation of high-molecular-weight polycyclic aromatic hydrocarbons in a two liquid-phase bioreactor.

Both the pure culture and consortia for the treatment of diverse wastewater contaminants in biphasic wastewater treatment systems are well documented (Quijano et al. 2010). Bacterial pure cultures employed in solid-liquid TPPBs successfully degraded the xenobiotics such as PCB, phenolic compounds and benzene (Amsden et al. 2003; Daugulis et al. 2003; Rehmann et al. 2007). Mix cultures of the bacteria, however, appeared as preferred biodegrading agents for complete removal of the toxicants without formation of inhibitory intermediates. Moreover, their ability to adapt variations in the environmental and operational conditions makes them potentially suitable candidates. Diverse contaminants are previously reported to be treated in TPPBs using different microbial consortia (Prpich and Daugulis 2004; Rehmann et al. 2007; Tomei et al. 2011). Microbial metabolic activities generally occur in aqueous phase of the TPPBs (Daugulis et al. 2011). However, later studies showed that microorganisms could grow in the nonaqueous phase with pronounced biodegradation of the xenobiotics (Hernandez et al. 2012).

5.3 Selection of Nonaqueous Phase (NAP)

In biphasic partitioning systems, substrate delivery at optimum levels for the destruction of xenobiotic compounds is a major challenge. This complex situation is narrow range of the substrate levels, while the toxic xenobiotic substrate levels have broad range; thus, appropriate delivery of these materials to the cells is essential. Since, the biphasic partitioning bioreactor concept depends on organic phase of a water-immiscible and biocompatible organic solvent to dissolve high levels of xenobiotic substrates, which later partition into the aqueous phase. This organic phase in a bioreactor is used to feed substrate to the cells as determined by the thermodynamics and metabolism of the cells. The substrate delivery rate is determined by the metabolic activity of the cells. As the biomass increases and cells become adapted to the inhibitory substrate, the demand for more substrate is controlled by equilibrium partitioning. Thus, in this system, the substrate demand and delivery are dependent on the cellular processes (Daugulis 2001).

The selection of NAP in the TPPB is one of the important factors for designing an efficient biphasic treatment system. The optimal performance of NAP depends upon the type of xenobiotic as well as the system configuration. A key feature to be considered while selecting a solid or liquid NAP is its hydrodynamic behaviour (Quijano et al. 2009). While selecting NAP, higher affinity against target contaminant, non-biodegradability, low vapour pressure, densities, thermal resistance and hydrophobicity must be considered (Bruce and Daugulis 1991; Cesario et al. 1992). Octanol/water partition coefficient (K_{ow}) values are important in the selection of liquid NAP. These values determine toxicity against particular groups of bacteria (Ramos et al. 2002; Arriaga et al. 2006). Since wastewaters contain mixtures of contaminants, thus, with those pollutants having different polarities, mixture of NAP is required for optimal treatment (Quijano et al. 2009).

Among different organic solvents to form water-immiscible phase in a two liquid-phase bioreactor by the bacterial consortium, the highest degradation of high-molecular-weight polycyclic aromatic hydrocarbons (HMW PAHs) was observed with silicon oil being hydrophilic, chemically stable and resistant to biodegradation (Marcoux et al. 2000). Silicon oil is considered one of the most promising liquid NAPs commonly used in 10–20% (v/v) in biphasic partitioning systems (Janikowski et al. 2002; Daugulis et al. 2011). Other organic liquid solvents used as NAP include n-hexadecane and perfluorocarbons allowing higher diffusivity of the substrates (Rocha-Rios et al. 2011). Solid NAP is less suitable for degradation of gaseous contaminants as compared to the liquid NAP due to their low interfacial area and lyophobicity (Rehmann et al. 2007; Hernández et al. 2010). However, solid portioning phase formation with commercial granular polymers has successfully been employed for the treatment of wastewaters contaminated with hydrocarbons, phenols and polychlorinated hydrocarbons (PCBs) in solid-liquid biphasic partitioning bioreactors (Prpich et al. 2008; Rehmann et al. 2008; Tomei et al. 2011). Low-cost and successful recycling of solid polymers in the process makes them attractive NAP for treatment of wastewaters (Daugulis et al. 2003; Morrish and Daugulis 2008). Tomei et al. (2016) used specifically selected extruded polymer, Hytrel 8206, and observed 99% removal of 4-chlorophenol.

6 Applications of Biphasic Systems

6.1 Removal of Organic Contaminants

Toxic organic compounds (xenobiotics) pose serious threat to the environment as well as human health worldwide, and biological treatment of these compounds is constrained by their toxic and inhibitory nature. Therefore, great care is required with respect to the rate at which they are provided to cells. The use of a second and distinct organic phase in a bioreactor provides substrate to cells. This technology can be applied to stockpiled xenobiotics as well as contamination of air, water and soil environments (Daugulis 2001). ABS is formed when either two polymers, one polymer and one kosmotropic salt, or two salts are mixed in an appropriate concentration or at a particular temperature. The two phases are mostly composed of water and nonvolatile components, thus eliminating volatile organic compounds. When oil and water are mixed together, they are separated into two phases or layers, because they are immiscible. In general, aqueous- or water-based solutions, being polar, are immiscible with non-polar organic solvents (chloroform, toluene, hexane, etc.) and form a TPS. However, in an ABS, both immiscible components are water-based. The formation of phases depends upon the temperature, pH and ion strength of both components. ATPS has gained an interest because of their great potential for the extraction, separation, purification and enrichment of proteins, membranes, viruses, enzymes, nucleic acids and other biomolecules both in industry and life. This method is also environment-friendly (Iqbal et al. 2016). A fed-batch fermentation was used to degrade phenol in a two-phase portioning system. Using this reactor configuration, operated in batch mode, 10 g phenol was degraded to completion within 84–96 h (Collins and Daugulis 1997).

Yeom et al. (2000) reported significant removal of benzene from benzene-contaminated water in cyclic batch of TPPB technology with repeated use of 1-octadecene, an immiscible solvent. The solvent phase in this biphasic setup worked as a sponge which selectively extracted the contaminant and released it when a decline in the aqueous-phase concentration due to biological benzene mineralization occurred. The study also highlighted the efficacy of the system based on careful selection of NAP (Daugulis and Boudreau 2008). Diesel components also pose a serious threat to the aqueous environments and the biota due to its intrinsically toxic nature. Thus, it is imperative to remove/recover this contaminant from the aqueous environment rapidly to ensure biological safety and water quality. The polymeric sorbents as thermoplastic materials substantially reduce diesel levels in excess. Prpich et al. (2008) proposed automobile tires as recyclable and low-cost materials in a solid-liquid TPPB for reduction of 65% of the initial diesel contamination within a 9-day period, while polymeric sorbents were also biologically regenerated in the process. The polymer sorbent in that setup served as a reservoir to supply diesel to an awaiting microbial consortium and ensured its continuous supply.

6.2 Pharmaceutical Ingredients

Active pharmaceutical ingredient (API) in wastewaters causes serious health problems even at trace levels. Increased use of antibiotics with the corresponding increase in development of antibiotic resistance in bacteria enforces attention towards their removal and monitoring in the environment. Almeida et al. (2016) suggested an IL-based ABS for the removal of pharmaceutical ingredients such as fluoroquinolones, a class of antibiotics. The imidazolium- and phosphonium-based ILs and an aluminium-based salt system of ABS were evaluated for one-step extraction of six fluoroquinolones which induced precipitation and extracted/ cleaned up to 96%. The system also performed excellent with respect to recyclability/ reusability. The ABS were designed previously in such a manner that the phase-forming solute was added during the final stage of the wastewater treatment which removed the APIs from the wastewater (Wilms and Van Haute 1984). Shahriari et al. (2013) also successfully used ABS for the removal of ciprofloxacin, ciprofloxacin HCl and tetracycline from the aqueous media.

The TPPB for the removal of endocrine disrupters from wastewater plant effluents uses three microbial enrichment cultures adapted to a solid-liquid two-phase partitioning. The bacteria associated with such degradation have been identified as *Sphingomonadales* and *Rhodococcus* having great potential of endocrine disrupters' degradation (Villemur et al. 2013). Treatment of the complex wastewaters including two or more steps favours their complete removal (Thompi 2000). High strength wastewaters can be efficiently treated through biomethanation using biphasic systems, which reduce risk of sensitivity to organic shock loadings and maintain balance between organic acid production and consumption (Seth et al. 1995). Jhung and Choi (1995) proved success of up-flow anaerobic sludge blanket (UASB), an anaerobic treatment system for high strength distillery wastewaters.

6.3 Post-methanated Distillery Effluent Treatment

An efficient biphasic treatment system comprising of bacteria and wetland setup is proposed by Yadav (2012). The wetland acts as sink for the large number of pollutants where interactions among the soil, wastewater characteristics, plants and microorganisms determine the function of the setup under specific operational conditions (Benavides et al. 2008). Since it accumulates diversity of the pollutants, the complex industrial wastewaters can be successfully addressed by this biphasic system. The post-methanated distillery effluents (PMDE) contain different pollutants, and a single-step decolourization is not possible. Thus, these PMDE are pretreated with bacteria and are integrated with the constructed wetlands having potential bioremedial plants later where the bacterial biofilm of rhizosphere further supports the process by acting as biofilters (Whiting et al. 2001). The bacteria for the biotreated PMDE provide heavy metals and pollutants as nutrients for the plants. This kind of biphasic system has proved to be a promising approach of treatment for the recalcitrant pollutants in PMDE.

6.4 Biphasic Systems for the Treatment of Textile Effluents

Release of textile industrial effluents loaded with dyes is a matter of economic as well as environmental concern. Novel approaches to remove these dyes include biphasic systems. IL-based ATPS is one such technique, which can efficiently remove pollutants of the textile industry (Ferreira et al. 2014). Ivetic et al. (2013) developed a PEG-salt ATPS model for the removal of Acid Blue. Thus, the ATPS are excellent alternative methods for dye removal. A biphasic mesophilic anaerobic up-flow packed bed reactor was used by Talarposhti et al. (2001) for the treatment of mixed cationic dye in textile effluents and successfully removed up to 90% of the 1000 mg/l of dye. Senthilkumar et al. (2011) successfully treated textile dyeing effluent utilizing Sago as co-substrates in biphasic UASB in acidogenic and methanogenic reactors. Further improvement in the textile effluent treatment systems was brought through the use of anaerobic/aerobic sequential system. The two phases of the treatment system involved an anaerobic UASB reaction system and an aerobic continuously stirred tank reactor system, which efficiently removed up to 97% and 87% of the COD and colour, respectively, in the wool dyeing industrial waste within 3.3 days (Isik and Sponza 2006).

6.5 Biphasic Systems for Removal of Heavy Metals

Heavy-metal-loaded effluents are freely discharged into the adjacent environments from many industries, from where the heavy metals percolate to the groundwater resources. Biological treatment systems of the solid as well as liquid wastes for the heavy metals' removal are long been studied (Bharagava and Mishra 2017). Currently, treatment strategies also involve recycling and reuse with the energy recovery. A biphasic fermentation system was successfully employed for the recovery of cadmium and lead present in the solid waste through biologically mediated hydrolysis and gasification of the heavy-metal-containing solids (Rodriguez et al. 1998). In certain cases, water-soluble complexants are added due to which the metal forms a chelate and is extracted into the upper layer. By this technique, the polymer can be recycled; a variety of ligands have been used to extract selected metal ions for detoxification (Chowdhury and Sidhwani 2009).

ABS of two dissimilar polymers can be used for partitioning of selected metal ions and are well known for separation of biomaterials. The utilization of salt solutions of PEG falls into three categories: (1) separation of the metal ions to the PEG-rich phase by the addition of water-soluble extractant that co-ordinates the metal ions, (2) addition of an inorganic anion that results in the production of metal complex and (3) separation of a metal ion directly from the salt-rich to the PEG-rich phase without addition of component (Rogers 1995).

7 Conclusions

No-doubt biphasic treatment is a novel approach for the removal of toxic and hazardous pollutants from industrial wastewaters. However, various factors need to be sought prior to the establishment of a successful biphasic system. Partitioning, which is the base of the phenomenon, is influenced under varying pH concentrations, polymer molecular weight and salt concentration, nonaqueous-phase material, hydraulic retention time and microbial inocula. Thus, a careful selection of these physical and chemical parameters is necessary for optimal partitioning and treatment in such systems. The biphasic systems have been efficiently employed for the treatment of diverse xenobiotics both organic and inorganic in nature.

References

Aguilar O, Albiter V, Serrano-Carreón L, Rito-Palomares M (2006) Direct comparison between ion-exchange chromatography and aqueous two-phase processes for the partial purification of penicillin acylase produced by *E. coli*. J Chromatogr B 835(1–2):77–83

Albertsson P (1958) Partition of proteins in liquid polymer-polymer two phase systems. Nature 182:709–711

Almeida HFD, Freire MG, Marrucho IM (2016) Improved extraction of fluoroquinolones with recyclable ionic-liquid-based aqueous biphasic systems. Green Chem 18:2717–2725

Amsden BG, Bochanysz J, Daugulis AJ (2003) Degradation of xenobiotics in a partitioning bioreactor in which the partitioning phase is a polymer. Biotechnol Bioeng 84:399–405

Arriaga S, Muñoz R, Hernández S, Guieysse B, Revah S (2006) Gaseous hexane biodegradation by *Fusarium solani* in two liquid phase packed-bed and stirred tank bioreactors. Environ Sci Technol 40:2390–2395

Atkins P, de Paula J (2010) Atkins' physical chemistry. Oxford University Press, New York

Benavides J, Aguilar O, Lapizco-Encinas BH, Rito-Palomares M (2008) Extraction and purification of bioproducts and nanoparticles using aqueous two-phase systems strategies. Chem Eng Technol 31:838–845

Bhal A, Bhal BS, Tuli GD (2012) Essentials of physical chemistry. S. Chand, Chandigarh

Bharagava RN, Chowdhary P, Saxena G (2017) Bioremediation an eco-sustainable green technology, its applications and limitations. In: Bharagava RN (ed) Environmental pollutants and their bioremediation approaches. CRC Press, Taylor & Francis Group, Boca Raton, pp 1–22

Bharagava RN, Mishra S (2017) Hexavalent chromium reduction potential of *Cellulosimicrobium sp.* isolated from common effluent treatment industries. Ecotoxicol Environ Saf 147:102–109

Bruce LJ, Daugulis AJ (1991) Solvent selection strategies for extractive biocatalysis. Biotechnol Prog 7:116–124

Cesario MT, Beeftink HH, Tramper J (1992) Biological treatment of waste gases containing poorly-water soluble compounds. In: Dragt AJ, van Ham J (eds) Biotechniques for air pollution abatements and odour control policies. Elsevier Science Publishers, Amsterdam, pp 135–140

Chakraborty A, Sen K (2016) Impact of temperature and pH on phase diagrams of different aqueous biphasic systems. J Chromatogr A 1433:41–55

Chandra R, Chowdhary P (2015) Properties of bacterial laccases and their application in bioremediation of industrial wastes. Environ Sci Process Impacts 17:326–342

Chakraborty A, Sen K (2017) Ionic liquid vs tri-block copolymer in a new aqueous biphasic system for extraction of Zn-cholesterol complex. J Mol Liq 229:278–284

Chowdhury S, Sidhwani IT (2009) Extraction of toxic metal ions using aqueous biphasic system. 13th Annual green chemistry and engineering conference, College Park, Maryland, June 23–25

Chowdhary P, More N, Raj A, Bharagava RN (2017) Characterization and identification of bacterial pathogens from treated tannery wastewater. Microbiol Res Int 5(3):30–36

Collins LD, Daugulis AJ (1997) Biodegradation of phenol at high initial concentrations in two-phase partitioning batch and fed-batch bioreactors. Biotechnol Bioeng 55:155–162

Darnault C, Rockne K, Stevens A, Mansoori GA, Sturchio N (2005) Fate of environmental pollutants. Water Environ Res 177:2576–2658

Daugulis AJ (2001) Two-phase partitioning bioreactors: a new technology platform for destroying xenobiotics. Trends Biotechnol 19:457–462

Daugulis AJ, Amsden B, Bochanysz J, Kayssi A (2003) Delivery of benzene to *Alcaligenes xylosoxidans* by solid polymers in a two-phase partitioning bioreactor. Biotechnol Lett 25:1203–1207

Daugulis AJ, Boudreau NG (2008) Solid-liquid two phase partitioning bioreactors for the treatment of gas-phase volatile organic carbons by a microbial consortium. Biotechnol Lett 30:1583–1587

Daugulis AJ, Tomei MC, Guieysse B (2011) Overcoming substrate inhibition during biological treatment of mono-aromatics: recent advances in bioprocess design. Appl Microbiol Biotechnol 90:1589–1608

de Belval S, le Breton B, Huddleston J, Lyddiatt A (1998) Influence of temperature upon protein partitioning in poly(ethylene glycol)-salt aqueous two-phase systems close to the critical point with some observations relevant to the partitioning of particles. J Chromatogr B 711:19–29

de Souza RL, Campos VC, Ventura SPM, Soares CMF, Coutinho JAP, Lima AS (2014) Effect of ionic liquids as adjuvants on PEG–based ABS formation and the extraction of two probe dyes. Fluid Phase Equilib 375:30–36

Didi MA, Sekkal AR, Villemin D (2011) Cloud-point extraction of bismuth (III) with nonionic surfactants in aqueous solutions. Colloids Surf 375:169

Domínguez-Pérez M, Tomé LIN, Freire MG, Marrucho IM, Cabeza O, Coutinho JAP (2010) (Extraction of biomolecules using) aqueous biphasic systems formed by ionic liquids and amino acids. Sep Purif Technol 72:85–91

Ferreira AM, Coutinho JA, Fernandes AM, Freire MG (2014) Complete removal of textile dyes from aqueous media using ionic-liquid-based aqueous two-phase systems. Sep Purif Technol 128:58–66

Ferreira AM, Vania FM, Faustino MD, Coutinho JAP, Freire MG (2016) Improving the extraction and purification of immunoglobulin G by the use of ionic liquid as adjuvants in aqueous biphasic systems. J Biotechnol 236:166–175

Freire MG, Cláudio AFM, Araújo JMM, Coutinho JAP, Marrucho IM, Canongia Lopes JN, Rebelo LPN (2012) Aqueous biphasic systems: a boost brought about by using ionic liquids. Chem Soc Rev 41:4966–4995

Gao J, Chen L, Yan ZC (2014) Extraction of dimethyl sulfoxide using ionic-liquid-based aqueous biphasic systems. Sep Purif Technol 124:107–116

Goja MA, Yang H, Cui M, Li C (2014) Aqueous two-phase extraction advances for bioseparation. J Bioproces Biotechniq 4:1–8

Govindarajalu K (2003) Industrial effluents and health status-a case study of Noyyal river basin. In: Proceedings of the 3rd international conference on environment and health. Chennai, India, pp 150–157

Greve A, Kula MR (1991) Phase diagrams of new aqueous phase systems composed of aliphatic alcohols, salts and water. Fluid Phase Equilib 62:53–63

Hamzehzadeh H, Abbasi M (2015) The influence of 1-butyl-3-methyl-imidazolium bromide on the partitioning of L-tyrosine within the {polyethylene glycol 600 + potassium citrate} aqueous biphasic system at T = 298.15 K. J Chem Thermodyn 80:102–111

Hernández M, Quijano G, Muñoz R (2012) Key role of microbial characteristics on the performance of VOC biodegradation in two-liquid phase bioreactors. Environ Sci Technol 46:4059–4066

Hernández M, Quijano G, Thalasso F, Daugulis AJ, Villaverde S, Munoz R (2010) A comparative study of solid and liquid non-aqueous phases for the biodegradation of hexane in two-phase partitioning bioreactors. Biotechnol Bioeng 106:731–740

Iqbal M, Tao Y, Xie S, Zhu Y, Chen D, Wang X, Huang L, Peng D, Sattar A, Shabbir MAB, Hussain HI, Ahmed S, Yuan Z (2016) Aqueous two-phase system (ATPS): an overview and advances in its applications. Biol Proced Online 18:18

Isik M, Sponza D (2006) Biological treatment of acid dyeing wastewater using a sequential anaerobic/aerobic reactor system. Enzym Microb Technol 38:887–892

Ivetic DZ, Sciban MB, Vasic VM, Kukic DV, Prodanovic JM, Antov MG (2013) Evaluation of possibility of textile dye removal from wastewater by aqueous two-phase extraction. Desalin Water Treat 51:1603–1608

Janikowski TB, Velicogna D, Punt M, Daugulis AJ (2002) Use of a two-phase partitioning bioreactor for degrading polycyclic aromatic hydrocarbons by a *Sphingomonas* sp. Appl Microbiol Biotechnol 59:368–376

Jhung JK, Choi E (1995) A comparative study of UASB and anaerobic fixed film reactor in the development of sludge granulation. Water Res 29:271–277

Jimena MG, Roxana O, Catiana Z, Margarita H, Susana M, Ines-Isla M (2008) Industrial effluents and surface waters genotoxicity and mutagenicity evaluation of a river of Tucuman, Argentina. J Hazard Mater 155:403–406

Kalaivani S, Xiuyun Ye IR, Yoshida S, Ng TB (2015) Synergistic extraction of a-Lactalbumin and b-Lactoglobulin from acid whey using aqueous biphasic system: process evaluation and optimization. Sep Purif Technol 146:301–310

Lin YK, Ooi CW, Tan JS, Show PL, Ariff A, Ling TC (2013) Recovery of human interferon α-2b from recombinant *Escherichia coli* using alcohol/salt-based aqueous two-phase systems. Sep Purif Technol 120:362–366

Liu X, Gao Y, Tang R, Wang W (2006) On the extraction and separation of iodide complex of cadmium (II) in propyl-alcohol ammonium sulfate aqueous biphasic system. Sep Purif Technol 50:263–266

Machado FLC, Coimbra JSDR, Zuniga ADG, da Costa AR, Martins JP (2012) Equilibrium data of aqueous two-phase systems composed of poly(ethylene glycol) and maltodextrin. J Chem Eng Data 57:1984–1990

Marcoux J, Deziel E, Villemur R, Lepine F, Bisaillon JG, Beaudet R (2000) Optimization of high-molecular-weight polycyclic aromatic hydrocarbons' degradation in a two-liquid-phase bioreactor. J Appl Microbiol 88:655–662

Morrish JLE, Daugulis AJ (2008) Improved reactor performance and operability in the biotransformation of carveol to carvone using a solid–liquid two-phase partitioning bioreactor. Biotechnol Bioeng 101:946–956

Naganagouda K, Mulimani VH (2008) Aqueous two-phase extraction (ATPE): an attractive and economically viable technology for downstream processing of *Aspergillus oryzae* α-galactosidase. Process Biochem 43(11):1293–1299

Ogunfowokan AO, Okoh EK, Adenuga AA, Asubiojo OI (2005) An assessment of the impact of point source pollution from a university sewage treatment oxidation pond on a receiving stream – a preliminary study. J Appl Sci 5:36–43

Ooi CW, Tey BT, Hii SL, Mazlina S, Kamal M, Lan JCW, Ariff A, Ling TC (2009) Purification of lipase derived from *Burkholderia pseudomallei* with alcohol/salt-based aqueous two-phase systems. Process Biochem 44:1083–1087

Paik SP, Sen K (2016) Species dependent iodine extractions in polymer based aqueous biphasic systems: emerging relations with aggregation number of polymeric micelles. J Mol Liq 223:1062–1066

Patrício PR, Cunha RC, Rodriguez Vargas SJ, Coelho YL, Mendes da Silva LH, Hespanhol da Silva MC (2016) Chromium speciation using aqueous biphasic systems: development and mechanistic aspects. Sep Purif Technol 158:144–154

Pei Y, Wang J, Xuan X, Fan J, Fan M (2007) Factors affecting ionic liquids based removal of anionic dyes from water. Environ Sci Technol 41:5090–5095

Pereira JFB, Lima ÁS, Freire MG, Coutinho JAP (2010) Ionic liquids as adjuvants for the tailored extraction of biomolecules in aqueous biphasic systems. Green Chem 12:1661–1669

Pereira JFB, Ventura SPM, e Silva FA, Shahriari S, Freire MG, JAP C (2013) Aqueous biphasic systems composed of ionic liquids and polymers: a platform for the purification of biomolecules. Sep Purif Technol 113:83–89

Pereira MM, Pedro SN, Quental MV, Lima ÁS, Coutinho JAP, Freire MG (2015) Enhanced extraction of bovine serum albumin with aqueous biphasic systems of phosphonium- and ammonium-based ionic liquids. J Biotechnol 206:17–25

Pratiwi AI, Yokouchi T, Matsumoto M, Kondo K (2015) Extraction of succinic acid by aqueous two-phase system using alcohols/salts and ionic liquids/salts. Sep Purif Technol 155:127–132

Prpich GP, Daugulis AJ (2004) Polymer development for enhanced delivery of phenol in a solid-liquid two-phase partitioning bioreactor. Biotechnol Prog 20:1725–1732

Prpich GP, Rehmann L, Daugulis AJ (2008) On the use, and re-use, of polymers for the treatment of hydrocarbon contaminated water via a solid-liquid partitioning bioreactor. Biotechnol Prog 24:839–844

Quijano G, Hernández M, Thalasso F, Muñoz R, Villaverde S (2009) Two-phase partitioning bioreactors in environmental biotechnology. App Microbiol Biotechnol 84:829–846

Quijano G, Hernández M, Villaverde S, Thalasso F, Muñoz R (2010) A step-forward in the characterization and potential applications of solid and liquid oxygen transfer vectors. Appl Microbiol Biotechnol 85:543–551

Raja S, Murty RM, Thivaharan V, Rajasekar V, Ramesh V (2011) Aqueous two phase systems for the recovery of biomolecules-a review. Sci Technol 1(1):7–16

Rajaram T, Ashutost D (2008) Water pollution by industrial effluents in India: discharge scenario and case for participatory ecosystem specific local regulation. Environment J40:56–69

Ramos JL, Duque E, Gallegos MT, Godoy P, Ramos-Gonzalez MI, Rojas A, Teran W, Segura A (2002) Mechanisms of solvent tolerance in gram-negative bacteria. Annu Rev Microbiol 56:743–768

Rehmann L, Prpich GP, Daugulis AJ (2008) Remediation of PAH contaminated soils: application of a solid–liquid two-phase partitioning bioreactor. Chemosphere 73:798–804

Rehmann L, Sun B, Daugulis AJ (2007) Polymer selection for biphenyl degradation in a solid-liquid two-phase partitioning bioreactor. Biotechnol Prog 23:814–819

Rocha-Rios J, Quijano G, Thalasso F, Revah S, Muñoz R (2011) Methane biodegradation in a two-phase partition internal loop airlift reactor with gas recirculation. J Chem Technol Biotechnol 86:353–360

Rodrigues CD, de Lemos LR, da Silva MCH, da Silva LH (2013) Application of hydrophobic extractant in aqueous two phase systems for selective extraction of cobalt, nickel and cadmium. J Chromatogr A 1279:13–19

Rodriguez ER, Richardson JB, Ghosh S (1998) Removal of heavy metals and pathogens during biphasic fermentation of solid wastes. In: Proceedings of the conference on hazardous waste research. Salt Lake City, UT, pp 363–373

Rogers RD (1995) Metal ion separations in polyethylene glycol-based aqueous biphasic systems. 6th Conference separation of ionic solutes, Slovakia, May 15–19

Sadeghi R, Maali M (2016) Toward an understanding of aqueous biphasic formation in polymer-polymer aqueous systems. Polymer 83:1–11

Safonova EA, Mehling T, Storm S, Ritter E, Smirnova IV (2014) Partitioning equilibria in multicomponent surfactant systems for design of surfactant based extraction processes. Chem Eng Res Des 92:2840–2850

Santos JHPM, Silva FAE, Couinho JAP, Ventura SPM, Pessoa A (2015) Ionic liquids as a novel class of electrolytes in polymeric aqueous biphasic systems. Process Biochem 50:661–668

Sen K, Chakraborty A (2016) A glycine based aqueous biphasic system: application in sequential separation of Ni, Cu and Zn. J Mol Liq 218:106–111

Senthilkumar M, Gnanapragasam G, Arutchelvan V, Nagarajan S (2011) Influence of hydraulic retention time in a two-phase up flow anaerobic sludge blanket reactor treating textile dyeing effluent using sago effluent as the co-substrate. Environ Sci Pollut Res 18:649–654

Seth R, Goyal SK, Handa BK (1995) Fixed film biomethanation of distillery spentwash using low cost porous media. Resour Conserv Recy 14:79–89

Shahriari S, Tome LC, Araújo JMM, Rebelo LPN, Coutinho JAP, Marrucho IM, Freire MG (2013) Aqueous biphasic systems: a benign route using cholinium-based ionic liquids. RSC Adv 3:1835–1843

Sheikhian L, Akhond M, Absalan G (2014) Partitioning of reactive red-120, 4-(2-pyridylazo)-resorcinol, and methyl orange in ionic liquid-based aqueous biphasic systems. J Environ Chem Eng 2:137–142

Silva ME, Franco TT (2000) Liquid-liquid extraction of biomolecules in downstream processing–a review paper. Braz J Chem Eng 17(1):1–17

Silva MSC, Sentos-Ebinuma VC, Lopes AM, Rangel-Yagui CO (2015) Dextran sulfate/triton X two phase micellar systems as an alternative first purification step for clavulanic acid. Fluid Phase Equilib 399:80–86

Souza RL, Lima RA, Coutinho JAP, Soares CMF, Lima AS (2015) Novel aqueous two phase systems based on tetrahydrofuran and potassium phosphate buffer for purification of lipase. Process Biochem 50:1459–1467

Taghavivand M, Pazuki G (2014) A new biocompatible gentle aqueous biphasic system in cefalexin partitioning containing nonionic Tween 20 surfactant and three organic/inorganic different salts. Fluid Phase Equilib 379:62–71

Talarposhti AM, Donnelly T, Andersonm GK (2001) Colour removal from a simulated dye wastewater using a two-phase anaerobic packed bed reactor. Water Res 35(2):425–432

Tan GYT, Zimmermann W, Lee K-H, Lan JC-W, Yim HS, Ng HS (2017) Recovery of mangostins from *Garcinia mangostana* peels with an aqueous micellar biphasic system. Food Bioprod Process 102:233–240

Thompi J (2000) Studies in effluent treatment. PhD thesis, Mumbai University, India

Tomei MC, Mosca Angelucci D, Daugulis AJ (2016) Towards a continuous two-phase portioning bioreactor for xenobiotic removal. J Hazard Mater 317:403–415

Tomei MC, Rita S, Mosca Angelucci D, Annesini MC, Daugulis AJ (2011) Treatment of substituted phenol mixtures in single phase and two-phase solid-liquid partitioning bioreactors. J Hazard Mater 191:190–195

Villemur R, dos Santos SCC, Ouellette J, Juteau P, Lépine F, Déziel E (2013) Biodegradation of endocrine disruptors in solid-liquid two-phase partitioning systems by enrichment cultures. Appl Environ Microbiol 79:4701–4711

Wedler G (1997) Lehrbuch der physikalischen chemie. Wiley-VCH, Weinheim

Whiting SN, de Souza MP, Terry N (2001) Rhizosphere bacteria mobilize Zn for hyperaccumulation by *Thlaspi caerulescens*. Environ Sci Technol 35:3144–3150

Wilms DA, Van Haute AA (1984) Primary flocculation of wastewater with $Al_2(SO_4)_3$ and $NaAlO_2$ salts recuperated from spent aluminium anodising baths. Stud Environ Sci 23:213–220

Yadav S (2012) Degradation and decolourisation of post methanated distillery effluent in biphasic treatment system of bacteria and wetland plant for environmental safety. PhD thesis, Pandit Ravi Shankar Shula University, India

Yazbik V, Ansorge-Schumacher M (2010) Fast and efficient purification of chloroperoxidase from *C. fumago*. Process Biochem 45(2):279–283

Yeom SH, Dalm MCF, Daugulis AJ (2000) Treatment of high concentration gaseous benzene streams using a novel bioreactor system. Biotechnol Lett 22:1747–1751

Zhou XY, Zhang J, Xu RP, Ma X, Zhang ZQ (2014) Aqueous biphasic systems based on low-molecular-weight polyethylene glycol for one-step separation of crude polysaccharides from *Pericarpium granati* using high speed countercurrent chromatography. J Chromatogr A 1362:129–134

Chapter 18
Phycotechnological Approaches Toward Wastewater Management

Atul Kumar Upadhyay, Ranjan Singh, and D. P. Singh

Abstract Phycoremediation technology is a cost-effective and environment-friendly approach, which involves use of microalgae to suck the pollutant present in the soil and water. The accumulation of inorganic and organic pollutants like pesticides, herbicides, insecticide, PCBs, DDT, and metals along with metalloids (Cd, Pb, Se, As) in the aquatic ecosystem can cause deleterious impacts on environment and organisms. Natural and intensive anthropogenic activities are the main factor responsible for the accumulation of these pollutants in plants and animals. These serious problems of pollution can be resolved by the application of phycoremediation technologies using microalgae to remove pollutants from the environment in a sustainable manner. The present chapter focuses on the different techniques used in remediation of pollutants and the mechanism of remediation adopted by micro- and macroalgal species to absorb the organic and inorganic pollutants from wastewater and soil.

Keywords Phycoremediation · Microalgae · Wastewater · Pollutant

1 Introduction

Phycoremediation may be defined as the use of algae for removal and biotransformation of pollutants including xenobiotics from the wastewater. The algae consist of macroalgae, microalgae and marine algae commonly known as the seaweeds. The algae are widely distributed on the earth and are adapted to a variety of habitats. This unique feature of their fast adaptation allows the algae to develop a wide range of tolerance toward different environmental conditions, suited for wastewater treatment and production of biofuel and other valuable products, including food, feed, fertilizer, pharmaceuticals, and lastly biofuel (Menon and Rao 2012; Nigam and Singh 2011; Chowdhary et al. 2017). Algae reduce global warming through photosynthesis

A. K. Upadhyay · R. Singh · D. P. Singh (✉)
Department of Environmental Science, Babasaheb Bhimrao Ambedkar Central University, Lucknow, India

© Springer Nature Singapore Pte Ltd. 2019
R. N. Bharagava, P. Chowdhary (eds.), *Emerging and Eco-Friendly Approaches for Waste Management*, https://doi.org/10.1007/978-981-10-8669-4_18

and could be a carbon-reducing system when applied as integrated biofuel production with waste remediation (McGinn et al. 2011). Microalgae have the capability of removing environmental toxicants such as heavy metals, hydrocarbons, and pesticides through various mechanisms, ranging from biosorption, bioaccumulation, biotransformation, volatilization, degradation, and assimilation (Rath 2012).

With the advent of molecular and functional genomic tools, there has been growing research that aims to improve algal strains for bioremediation by enhancing their photosynthetic efficiency, adaptability, and tolerance to a harsh environment and ability to detoxify pollutants. The algae are endowed with different features that make them potential resource for the cost-effective removal of pollutants, which include ability to grow both autotrophically and heterotrophically, large surface area/volume ratios, high tolerance to heavy metals, phototoxicity, phytochelatin expression, and potential for genetic manipulation (Cai et al. 2013). A range of characteristics should be considered when choosing species for bioremediation. It is essential for species to have:

1. High growth rates which ultimately increased bioremediation capability (Barrington et al. 2009).
2. Species should be able to grow across a wide range of conditions (de Paula Silva et al. 2008).
3. Additionally, species should occur locally and, if possible, have a broad geographic distribution (Barrington et al. 2009).

2 Historical Background

The fascinating idea of remediation begins about 55 years ago in the United States by Oswald and Gotaas (1957). Palmer (1974) surveyed a number of microalgal genera from a wide distribution of waste stabilization ponds. The occurrence of the algae in the order of their abundance and frequency were *Chlorella*, *Ankistrodesmus*, *Scenedesmus*, *Euglena*, *Chlamydomonas*, *Oscillatoria*, *Micractinium*, and *Golenkinia*. Pham et al. (2014) recorded 30 algal species belonging to 5 phyla, Chlorophyta, Chrysophyta, Cryptophyta, Cyanobacteria, and Euglenophyta, from waste stabilization pond and can be used in treatment processes. A list of Some important algae used in remediation process are presented below in the tabular form, (Table 18.1).

3 Strategies of Remediation

The idea of using algae to remove wastewater and metals from contaminated soils and water came from the discovery of different species of algae that accumulate high concentrations of liquid waste and metals (Upadhyay et al. 2016). Various

Table 18.1 Various types of microalgae used in the treatment of different wastes

Algae	Type of waste	References
Spirulina	Anaerobic effluents of pig waste	Lincoln et al. (1996)
Phormidium bohneri	Municipal wastewater	Talbot and De la Noüe (1993)
Chlorella sp.	Municipal wastewater/domestic sewage	Choi and Lee (2015)
Euglena	Domestic wastewater	Mahapatra et al. (2013)
Desmodesmus sp. TAI-1 and *Chlamydomonas*	Industrial wastewater	Wu et al. (2012)
Scenedesmus quadricauda	Campus sewage	Han et al. (2015)

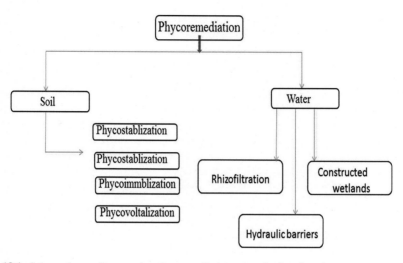

Fig. 18.1 Schematic ray diagram showing remediation strategies in microalgae

techniques have been introduced to exploit the possible role of microalgae for the removal of hazardous compounds such as heavy metal, inorganic and organic waste from contaminated soil and water. A summary of remediation strategies has been mentioned below in Fig. 18.1.

3.1 *Phycoextraction*

The term phycoextraction implies the removal of toxic pollutants of anthropogenic origin from soil and water by means of the uptake, accumulation, and sequestration using microalgae (Raskin et al. 1994). The extracted pollutants either entered inside the cell through active and passive transport along with water or utilized by the algae for its growth and development to avoid toxicity induced due to pollutants. Once the metal crosses the membrane, it sequestered inside vacuole with the aid of different phytochelatins present inside the cell (Mishra et al. 2009; Hare et al. 2017). Resistance to different pollutants like wastewater and heavy metal ions in algae can be achieved through the process of avoidance mechanism, which includes mainly the immobilization of contaminants such as metal and chemicals on the surface and in cell walls (Malik 2004). Tolerance to contaminants in wastewater is based on the process of sequestration from vacuoles, binding them through appropriate ligands such as protein, peptides, organic acids, and enzymes like high levels of metallic ions (Robinson et al. 2009). The phycoextraction of wastewater signifies one of the largest economic and developmental opportunities regarding the scope of environmental problems associated with wastewater and contaminated soils. (Raskin et al. 1994). The ideal algae to be used in phycoextraction must have the following features: tolerant to high levels of metal, fast growth rates, profuse surface system, ability to accumulate high levels of heavy metal in its harvestable parts, and potential to produce a high biomass in the field.

According to Raskin et al. (1994), the level of accumulation of pollutants such as Ni, Cu, and possibly Zn, studies in hyperaccumulators, often reaches 1 ± 5% of their dry weight. Unfortunately, most hyperaccumulators have slow growth rates, and we lack the technology for their huge scale cultivation (Garbisu and Alkorta 2001). According to various authors (Chaney et al. 1997; Rascio and Navari-Izzo 2011; McGrath and Zhao 2003; Chaney et al. 1997), natural pollutant hyperaccumulator phenotype shows to be much more important than higher plant ability, when using plants to remediate contaminated soil and water. Upadhyay et al. (2016), reported that there are, two strategies of phycoextraction: (1) chelate-assisted through surface bonding and (2) continuous phycoextraction. Chelating substances have been worn as soil extractants, a source for macro- and micronutrient fertilizers, to maintain bonding and solubility of micronutrients in hydroponic forms (Salt et al. 1995). Moreover, this process of phycoextraction is based on the application of chemical chelates to the soil and water significantly enhancing the contaminated accumulation by algae (Kayser et al. 2000). Auspiciously, the discovery that the application of certain chelates to the soil increases the translocation of heavy metals from soil and wastewater into the surface and vacuoles of algae has opened a wide new range of possibilities for this field of wastewater and metal phycoextraction (Blaylock et al. 1997). In this view, hyperaccumulators are the most suitable algae for the phycoextraction of chemical and metal-polluted soil and water.

3.2 Phycovolatilization

In phycovolatilization, transformation of toxic substance into their voltalie form are performed with the help of microbes present in the surrounding medium. Phycovolatilization has been extensively used for a number of contaminants, like inorganic, organic, and metal. Volatile forms of several inorganic compounds can be volatilized from algae, including Se, As, and Hg, along with phycovolatilization of radioactive water (T_2O). Chinnasamy et al. (2012) have reported that phycovolatilization can work in two different forms, namely, as direct and indirect phycovolatilization. Direct phycovolatilization is the more spontaneous and better-understanding form, resulting from algae uptake, translocating and transforming the contaminants, finally leading to volatilization of the compound from the surface of cell and vacuoles. Previously, this process is known as "phytovolatilization," or the pollutant is said to be "transpired," as this pathway is similar to the transpiration vascular cycle of water (Chinnasamy et al. 2012). Indirect phycovolatilization is a process in which the increase in volatile compounds fluxes from the subsurface resulting from algae through surface charge activities. Algae move huge amounts of water (globally ~62,000 km3/yr)[14] and simultaneously explore large volumes of water and soil (Agard et al. 2007). These cause the change in subsurface chemical fate and transport. The activities of algae outer surface can increase the flux rate of volatile contaminants from the subsurface through the following mechanisms:

1. Decreasing water table
2. Enlarged soil permeability
3. Chemical transport via surface and vacuole mechanism
4. Advection with water toward the surface
5. Change in rainfall pattern that would otherwise infiltrate to dilute and advect VOCs away from the surface

Direct phycovolatilization has been measured on the basis of both laboratory and field level studies. In laboratory, algae mixed with contaminants in different doses grown hydroponically (Carrasco Gil et al. 2013). After a certain period of time, algal biomass is typically analyzed to perform a mass balance on the contaminant from treatment to control. The amount phycovolatilized may be directly determined by analysis of air passed through a sorbent trap and calculated from the quantity of contaminant mass that remains not present in an open system (Harper 2000). Moreover if mass of contaminant is captured on a sorbent trap, the phycovolatilization rate can be compared on the basis of time and transpiration rate.

3.3 Rhizofiltration

The practice of rhizofiltration is based on the remediation of wastewater and toxic metal-contaminated soil by microalgae and aquatic plants. Inorganic and organic pollutants and metals such as Pb, Zn, Cd, Cu, Ni, and Cr can be extracted with rhizofiltration. A number of unicellular algae, blue-green algae, and plants have been reported to render rhizofiltration such as *Chlamydomonas reinhardtii*, tobacco, spinach, and Indian mustard (Alahuhta et al. 2013). Marine algae and terrestrial plants are widely preferred for rhizofiltration as they have root systems with fast growth (Padmavathiamma and Li 2007). The rhizofiltration system can be built up either as floating rafts on ponds or stagnant water. The main disadvantage of rhizofiltration is growth of plants and algae in a greenhouse and transferring them to the remediation site. Besides, a number of parameters such as maintenance of an optimum pH, temperature, and humidity regularly are also required.

3.4 Phytostabilization

Phytostabilization aims to remove contaminants within the unsaturated zone or land surface and top of phreatic zone through accumulation by roots and surface or precipitation within the rhizosphere (Ul Hassan et al. 2017). Phytostabilization is a phenomenon of establishing algae and plant cover on the surface of the polluted water and soils, which reduces wind, water, and direct contact with human being. Phytostabilization relies on algae and its exudates, to stabilize low levels of contaminants which are present in water and soil (e.g., by absorption or precipitation) to prevent them from leaching (Stevenson et al. 2017). The possible mechanism includes lignification, cell wall sorption, humification and binding to organic matter. Phytostabilization is mostly related to chemicals and metal contamination. Phytostabilization can be increased by soil and water amendments restricting the immobilization of hazardous chemicals and metals (Kumpiene et al. 2009). This technology of phycoremediation does not produce secondary pollutants during treatment. Besides, phytostabilization improves soil fertility leading to ecosystem restoration. However, regular monitoring of the sites is required to ascertain that the optimal stabilizing conditions are maintained (Keller et al. 2005). While some organic and inorganic contaminants undergo microbial or chemical degradation, inorganic contaminants such as metalloids are immutable. Therefore, containment of wastewater, metal, and metalloids by phytostabilization is critical in control contaminated sites (Hua et al. 2017).

3.5 Constructed Wetland

Constructed wetland (CW) has been proven as a cost-effective green technology of waste remediation. In the wetland, green alga and cyanobacteria have greater capacity to fix CO_2 into a useful substrate than other terrestrial flora. This is important because it provides the base of the extensive food web found in wetland communities. Constructed wetland seeded with algae treats different types of pollutants including N, P, organic pollutants, heavy metals, etc. Although nitrogen and phosphorous play a major role in algal growth yet may be a serious pollutant in aquatic system at higher concentration (Pittman et al. 2011). The use of algae to remove excess dissolved nutrients from effluents is widely accepted and a cost-effective wastewater treatment technology (Barrington et al. 2009; Neori et al. 2004). The removal of heavy metal by algae has been known for a long time (Megharaj et al.

2003; Shamsuddoha et al. 2006). Microalgae remove metals from polluted water by two major mechanisms: metabolism-dependent uptake in cells and a non-active adsorption (Matagi et al. 1998). The competence of algae to accumulate and biotransform metals has led to their widespread use as biomonitor in ecosystems (Singh et al. 2017). Wang and Freemark (1995) have reported that the blue-green algae *Phormidium* can significantly accumulate heavy metals including Cd, Zn, Pb, Ni, and Cu. Algal species *Caulerpa racemosa* could be used for the removal of boron (Bursali et al. 2009).

3.6 Hydraulic Barrier

Hydraulic barrier implies reduction of contaminants from large water bodies through means of algal mat. In algal mat, consortium of algae was used to form a thick mat. The mat flows in the water and removes dissolved contaminants. It acts as a trickling filter. Algal mats were prepared by a mixture of filamentous algae, green and blue-green algae. It acts as a barrier of water flow. Water moves across the mat increasing the retention time by slowing down the speed of water. After a certain period of time, algal mats were removed and harvested. Removal of contaminants in algal mat is cost-effective and easy to harvest.

4 Mechanism of Phycoremediation

The basic mechanism involved in removal of contaminants by algae includes flocculation, sedimentation, microbiological activity, and uptake. Pollutants like inorganic and organic and heavy metal reduction mechanisms include physical, biological, and biochemical mechanism methods. Recently, the use of aquatic organism especially micro- and macroalgae has obtained much attention due to its ability to sorb contaminants from the environments. The algae have many characteristic features for the selection and accumulation of metals, including high tolerance to heavy metals, ability to grow both autotrophically and heterotrophically, large surface area/volume ratios, phytochelatin expression, phototaxy, and possibility for genetic manipulation (Kirst et al. 2017).

4.1 Biological Removal Mechanism

In the biological removal, open pond system/waste stabilization pond (WSP) is the best way for the treatment of wastewater of different resources (Bwapwa et al. 2017). In waste stabilization pond, algae grow with the help of the sunlight, water, and nutrients present in waste. Waste stabilization pond systems are particularly favored in region with ample sunlight and large area including Australia, Africa, India, Canada, etc. (Emmett and Nye 2017). In WSP aerobic breakdown of organic contaminants takes place in the presence of good supply of oxygen provided by the algae. Nitrogen present in wastewater is removed by the process of assimilation and nitrogen fixation (Stauch et al. 2017). Phosphorus is present predominantly in unavailable inorganic form, which is degraded by the secretion of different enzymes and secondary metabolites by the algae. This may alter the pH of the medium, which may lead to mineralization and increased availability to the algae in the form of food (Yavuzcan Yildiz et al. 2017).

Algae secrete different types of metal-binding compound like siderophore, which binds with metals and gets precipitated at the bottom of the system (Gupta et al. 2018). The algal cell wall has many functional groups, such as hydroxyl (-OH), phosphoryl (-PO), amino (-NH), carboxyl (-COOH), sulfhydryl (-SH), etc., which confer negative charge to the cell surface. Since, metals are present generally in the cationic form and gets adsorbed onto the cell surface by weak interaction (Grist 1981; Xue et al. 1988), various algae have been genetically modified to remove particular metal by overexpression of a heavy metal-binding protein, such as metallothionein, phytochelatin, and metal transport system. This was first done with an Hg^{++} transport system (Chen and Wilson 1997; Liu et al. 2011).

4.2 Biochemical Mechanism

Biochemical removal by algae includes cations and anion exchange, absorption, precipitation, and oxidation/reduction.

4.2.1 Cation and Anion Exchange

Ion exchange method uses by algae in removal of metals from aquatic system are cost-effective and potential biochemical approach of remediation as well as recovery of strategic metals from wastewater (Malik 2004). In ion exchange, the presence of different functional groups on the cell wall is made accessible for metal adsorption and absorption.

4.2.2 Absorption

Algae present in wastewater accumulate a significant amount of inorganic nutrients such as nitrate, phosphate, and heavy metal. Microalgae convert inorganic nitrogen into organic nitrogen via assimilation (Cai et al. 2013). In the process of assimilation, inorganic nitrogen translocate in the cytoplasm. Inside the cytoplasm, a chain of oxidation and reduction reaction was operated in the presence of nitrate and nitrite reductase and ultimately converted to NH_4 (Staicu and Barton 2017). Thus, the transformed ammonia is being incorporated within the intracellular fluid. Phosphorus is an important constituent of nucleic acids, lipids, and proteins. During algae metabolism, phosphorus is consumed as H_2PO_4 and HPO_4^2. This phosphate compound is further integrated into organic compounds via phosphorylation (Martinez et al. 1999).

4.2.3 Precipitation

This is the insoluble matter left behind after the reaction with soluble reactants. Algae present in wastewater secrete different types of chemicals, organic acids, and secondary metabolites. The release of organic acids resulted into a decrease in the surrounding pH that favors precipitation of different toxic chemicals and phosphorus also. Soluble iron and aluminum or calcium content in wastewater enhances the phosphorus reduction by precipitation (De-Bashan and Bashan 2004).

5 Postharvest Utilization of Microalgae

Microalgae are versatile organisms that convert the contaminants into non-hazardous resources, facilitating the remediated water to be recycled or discharged safely (Rao and Charette 2011). Algae-based treatment is used as secondary treatment with the aid of microbes in symbiosis of algae for an effective treatment. Harvested biomass of algae could be further used as a sustainable alternative of nonrenewable energy resource (Bwapwa et al. 2017). The biomass of algae could be used in the production of different value-added products such as biodiesel, bioethanol, biogas, biohydrogen, fish feed, animal feed, and food supplements. Algae contain high amounts of carbohydrates, fatty acid, minerals, and vitamins, which may be a dietary source for human beings. The dried biomass of algae can be directly used as fertilizer supplements by drying the harvested biomass as green manure in agriculture field. However, proper ecotoxicological analysis must be required prior to application at ground level.

6 Conclusions

Various studies have been done exposing the potential utilization of algae in coming decades as a sustainable means for wastewater treatment and biofuel production. However, using wastewater as a resource and amalgamation of wastewater treatment with the algae-based bioproduct generation can overcome major challenges. Additionally, to reduce the capital cost, implementations of an integrated system of

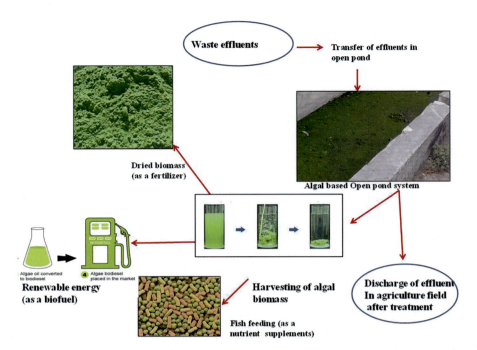

Fig. 18.2 Schematic diagram showing the treatment approach and postharvest utilization of algae

large-scale production of algae and the harvesting of microalgae could be a way to achieve the goal of sustainability. Thus, treatment of wastewater in algal pond system can supply the demand of food, feed, and energy for the future world (Fig. 18.2).

Acknowledgments Authors are thankful to Vice-chancellor, BBA University, Lucknow, for providing facilities. Dr. AKU is also thankful to the Science and Engineering Recruitment Board, New Delhi (Project No. PDF/2016/002432/LS) for providing financial support.

References

Agard J, Angela C, Aquing P, Attzs M, Arias F, Beltran J, Creary M (2007) Caribbean Sea ecosystem assessment. Caribb Mar Stud 8:1–85

Alahuhta J, Joensuu I, Matero J, Vuori K M, Saastamoinen O (2013) Freshwater ecosystem services in Finland

Barrington MJ, Watts SA, Gledhill SR, Thomas RD, Said SA, Snyder GL, Jamrozik K (2009) Preliminary results of the Australasian Regional Anaesthesia Collaboration: a prospective audit of more than 7000 peripheral nerve and plexus blocks for neurologic and other complications. Reg Anesth Pain Med 34:534–541

Blaylock MJ, Salt DE, Dushenkov S, Zakharova O, Gussman C, Kapulnik Y, Ensley BD, Raskin I (1997) Enhanced accumulation of Pb in Indian mustard by soil-applied chelating agents. Environ Sci Technol 31:860–865

Bursali EA, Cavas L, Seki Y, Bozkurt SS, Yurdakoc M (2009) Sorption of boron by invasive marine seaweed: Caulerpa racemosa var. cylindracea. Chem Eng J 150:385–390

Bwapwa JK, Jaiyeola AT, Chetty R (2017) Bioremediation of acid mine drainage using algae strains: a review. South African Journal of Chemical Engineering 24:62–70

Cai T, Park SY, Li Y (2013) Nutrient recovery from wastewater streams by microalgae: status and prospects. Renew Sustain Energ Rev 19:360–369

Carrasco Gil S, Siebner H, Le Duc DL, Webb SM, Millán R, Andrews JC, Hernández LE (2013) Mercury localization and speciation in plants grown hydroponically or in a natural environment. Environ Sci Technol 47:3082–3090

Chaney RL, Malik M, Li YM, Brown SL, Brewer EP, Angle JS, Baker AJ (1997) Phytoremediation of soil metals. Curr Opin Biotechnol 8:279–284

Chen S, Wilson DB (1997) Construction and characterization of Escherichia coli genetically engineered for bioremediation of Hg (2+)-contaminated environments. Appl Environ Microbiol 63:2442–2445

Chinnasamy S, Rao PH, Bhaskar S, Rengasamy R, Singh M (2012) Algae: a novel biomass feedstock for biofuels. Micro biotechnology: energy and environment. CABI, Wallingford, pp 224–239

Choi HJ, Lee SM (2015) Effect of the N/P ratio on biomass productivity and nutrient removal from municipal wastewater. Bioprocess and Biosystems Engineering 38(4):761–766

De Bashan LE, Bashan Y (2004) Recent advances in removing phosphorus from wastewater and its future use as fertilizer (1997–2003). Water Res 38:4222–4246

De Paula Silva PH, McBride S, de Nys R, Paul NA (2008) Integrating filamentous 'green tide' algae into tropical pond-based aquaculture. Aquaculture 284:74–80

Emmett RS, Nye DE (2017) The environmental humanities: a critical introduction. MIT Press, Cambridge

Garbisu C, Alkorta I (2001) Phytoextraction: a cost-effective plant-based technology for the removal of metals from the environment. Bioresour Technol 77:229–236

Grist NR (1981) Hepatitis and other infections in clinical laboratory staff, 1979. J Clin Pathol 34:655–658

Gupta C, Prakash D, Gupta S (2018) Microbes: a tribute to clean environment. Paradigms in pollution prevention. Springer, Cham, pp 17–34

Han L, Pei H, Hu W, Jiang L, Ma G, Zhang S, Han F (2015) Integrated campus sewage treatment and biomass production by *Scenedesmus quadricauda* SDEC-13. Bioresour Technol 175:262–268

Harper M (2000) Sorbent trapping of volatile organic compounds from air. J Chromatogr A 885:129–151

Hua Y, Heal KV, Friesl Hanl W (2017) The use of red mud as an immobiliser for metal/metalloid-contaminated soil: a review. J Hazard Mater 325:17–30

Kayser A, Wenger K, Keller A, Attinger W, Felix HR, Gupta SK, Schulin R (2000) Enhancement of phytoextraction of Zn, Cd, and Cu from calcareous soil: the use of NTA and sulfur amendments. Environ Sci Technol 34:1778–1783

Keller C, Ludwig C, Davoli F, Wochele J (2005) Thermal treatment of metal-enriched biomass produced from heavy metal phytoextraction. Environ Sci Technol 39:3359–3367

Kirst H, Gabilly ST, Niyogi KK, Lemaux PG, Melis A (2017) Photosynthetic antenna engineering to improve crop yields. Planta 245:1009–1020

Kumpiene J, Guerri G, Landi L, Pietramellara G, Nannipieri P, Renella G (2009) Microbial biomass, respiration and enzyme activities after in situ aided phytostabilization of a Pb-and Cu-contaminated soil. Eco Env Saf 72:115–119

Lincoln EP, Wilkie AC, French BT (1996) Cyanobacterial process for renovating dairy wastewater. Biomass Bioenergy 10:63–68

Liu C, Bayer A, Cosgrove SE, Daum RS, Fridkin SK, Gorwitz RJ, Rybak MJ (2011) Clinical practice guidelines by the Infectious Diseases Society of America for the treatment of methicillin-resistant Staphylococcus aureus infections in adults and children. Clin Infect Dis 52:18–55

Mahapatra DM, Chanakya HN, Ramachandra TV (2013) Euglena sp. as a suitable source of lipids for potential use as biofuel and sustainable wastewater treatment. J Appl Phycol 25:855–865

Malik A (2004) Metal bioremediation through growing cells. Env Intern 30:261–278

Martinez ME, Jimenez JM, El Yousfi F (1999) Influence of phosphorus concentration and temperature on growth and phosphorus uptake by the microalga Scenedesmus obliquus. Bioresour Technol 67:233–240

Matagi SV, Swai D, Mugabe R (1998) A review of heavy metal removal mechanisms in wetlands. Afr J Trop Hydrobiol Fish 8:13–25

McGinn PJ, Dickinson KE, Bhatti S, Frigon JC, Guiot SR, O'Leary SJ (2011) Integration of microalgae cultivation with industrial waste remediation for biofuel and bioenergy production: opportunities and limitations. Photosynth Res 109:231–247

McGrath SP, Zhao FJ (2003) Phytoextraction of metals and metalloids from contaminated soils. Current Opinion Biotech 14:277–282

Megharaj M, Avudainayagam S, Naidu R (2003) Toxicity of hexavalent chromium and its reduction by bacteria isolated from soil contaminated with tannery waste. Curr Microbiol 47:51–54

Menon V, Rao M (2012) Trends in bioconversion of lignocellulose: biofuels, platform chemicals & biorefinery concept. Prog Energy Combust Sci 38:522–550

Mishra S, Tripathi RD, Srivastava S, Dwivedi S, Trivedi PK, Dhankher OP, Khare A (2009) Thiol metabolism play significant role during cadmium detoxification by Ceratophyllum demersum L. Bioresour Technol 100:2155–2161

Neori A, Chopin T, Troell M, Buschmann AH, Kraemer GP, Halling C, Shpigel M, Yarish C (2004) Integrated aquaculture: rationale, evolution and state of the art emphasizing seaweed biofiltration in modern mariculture. Aquaculture 231:361–391

Nigam PS, Singh A (2011) Production of liquid biofuels from renewable resources. Prog Energy Combust Sci 37:52–68

Oswald W, Gotass H (1957) Conversion of solar energy to electricity via algae growth. Transactions of the American Society of Civil Engineers (United States)

Padmavathiamma PK, Li LY (2007) Phytoremediation technology: hyper-accumulation metals in plants. Water Air Soil Pollut 184:105–126

Palmer CM (1974) Algae in American sewage stabilization's ponds. Rev Microbiol (S-Paulo) 5:75–80

Pham DT, Everaert G, Janssens N, Alvarado A, Nopens I, Goethals PL (2014) Algal community analysis in a waste stabilisation pond. Ecol Eng 73:302–306

Pittman JK, Dean AP, Osundeko O (2011) The potential of sustainable algal biofuel production using wastewater resources. Bioresour Technol 102:17–25

Rao AM, Charette MA (2011) Benthic nitrogen fixation in an eutrophic estuary affected by groundwater discharge. J Coast Res 28:477–485

Rascio N, Navari-Izzo F (2011) Heavy metal hyperaccumulating plants: how and why do they do it and what makes them so interesting. Plant Sci 180:169–181

Raskin I, Kumar PBAN, Dushenkov S, Salt DE (1994) Bioconcentration of heavy metals by plants. Curr Opin Biotechnol 5:285–290

Rath B (2012) Microalgal bioremediation: current practices and perspectives. J Biochem Technol 3:299–304

Robinson BH, Bañuelos G, Conesa HM, Evangelou MW, Schulin R (2009) The phytomanagement of trace elements in soil. Crit Rev Plant Sci 28:240–266

Salt DE, Blaylock M, Kumar NP, Dushenkov V, Ensley BD, Chet I, Raskin I (1995) Phytoremediation: a novel strategy for the removal of toxic metals from the environment using plants. Nat Biotechnol 13:468–474

Shamsuddoha ASM, Bulbul A, Huq SMI (2006) Accumulation of arsenic in green algae and its subsequent transfer to the soil–plant system. Bangladesh J Microbiol 22:148–151

Singh NK, Upadhyay AK, Rai UN (2017) Algal technologies for wastewater treatment and biofuels production: an integrated approach for environmental management. Algal biofuels. Springer, Cham, pp 97–107

Staicu LC, Barton LL (2017) Bacterial metabolism of selenium – for survival or profit. Bioremediation of selenium contaminated wastewater. Springer, Cham, pp 1–31

Stauch WK, Srinivasan VN, Kuo Dahab WC, Park C, Butler CS (2017) The role of inorganic nitrogen in successful formation of granular biofilms for wastewater treatment that support cyanobacteria and bacteria. AMB Express 7:146

Stevenson LM, Adeleye AS, Su Y, Zhang Y, Keller AA, Nisbet RM (2017) Remediation of cadmium toxicity by sulfidized nano-iron: the importance of organic material. ACS Nano 11:10558

Talbot P, De la Noüe J (1993) Tertiary treatment of wastewater with Phormidium bohneri (Schmidle) under various light and temperature conditions. Water Res 27:153–159

Ul Hassan Z, Ali S, Rizwan M, Ibrahim M, Nafees M, Waseem M (2017) Role of bioremediation agents (bacteria, fungi, and algae) in alleviating heavy metal toxicity. Probiotics in agroecosystem. Springer Nature, Singapore, pp 517–537

Upadhyay AK, Singh NK, Singh R, Rai UN (2016) Amelioration of arsenic toxicity in rice: comparative effect of inoculation of *Chlorella vulgaris* and *Nannochloropsis* sp. on growth, biochemical changes and arsenic uptake. Eco Environ Saf 124:68–73

Wang WC, Freemark K (1995) The use of plants for environmental monitoring and assessment. Eco Environ Saf 30:289–301

Wu LF, Chen PC, Huang AP, Lee CM (2012) The feasibility of biodiesel production by microalgae using industrial wastewater. Bioresour Technol 113:14–18

Xue HB, Stumm W, Sigg L (1988) The binding of heavy metals to algal surfaces. Water Res 22:917–926

Yavuzcan YH, Robaina L, Pirhonen J, Mente E, Domínguez D, Parisi G (2017) Fish welfare in Aquaponic systems: its relation to water quality with an emphasis on feed and Faeces a review. Water 9:13

Printed in the United States
By Bookmasters